Teaching Introductory Chemistry

Teaching Introductory Chemistry

Written by

Scott Milam

Teaching Fast and Slow

Plymouth, MI

2022

1st Edition

Research Consultants:

Katy Dornbos

Ariel Serkin

Teaching Introductory Chemistry

1ˢᵗ Edition

© 2022 by Scott Milam

All rights reserved.

This book or any portion thereof may not be reproduced or used in any manner without the express written consent of the author.

Requests for permission should be sent to:

TeachingFastandSlow@gmail.com

ISBN: 9798842067213

Library of Congress Control Number: 2022911782

It is the great beauty of our science...
that advancement in it, whether in a degree great or small, instead of exhausting the subjects of research, opens the doors to further and more abundant knowledge, overflowing with beauty and utility.

- Michael Faraday

Table of Contents

Introduction 1
Cognitive Science 2
Advanced Cognitive Science 6
Themes in This Book 13

Part I - Chemistry
Ch. 1 – Chemistry Basics 19
Significant Figures 21
Density 25
States of Matter 29
Chemical and Physical Changes 31
Student Struggles 33
Phenomena 35
Flashcards 37
Ch. 2 - Gas Pressure 40
Proportional Relationships, PVnT 41
Measurement 46
Kinetic Molecular Theory 49
Gas Pressure Demos 55
Student Struggles 58
Phenomena 59
Flashcards 62
Ch. 3 - Thermochemistry 65
Temperature and Heat 66
Specific Heat Capacity 69
Heating Curves and Energy Bar Charts (LOL Diagrams) 76
Student Struggles 81
Phenomena 83
Flashcards 86
Ch. 4 - Moles 90
Avogadro's Number 91
Molar Mass 92
% Composition and Empirical Formula 97
Student Struggles 100
Phenomena 101
Flashcards 103
Ch. 5 - Naming and Formula Writing 106
Naming and Formula Writing 107
Ionic and Molecular Compounds 111
Acids 117
Student Struggles 119
Phenomena 121
Flashcards 123

Ch. 6 - Reaction Types 125
Synthesis and Decomposition 126
Single Replacement and Double Replacement 130
Student Struggles 137
Phenomena 138
Flashcards 140
Ch. 7 - Stoichiometry + Extensions 143
BCA Tables 144
Limiting and Excess Reagent 149
Molarity 153
Enthalpy 156
Ideal Gas Law + Partial Pressures 168
Student Struggles 171
Phenomena 173
Flashcards 175
Ch. 8 - Atomic Structure 178
Light and Electrons 179
Atomic Structure 196
Nuclear Chemistry 198
Student Struggles 204
Phenomena 206
Flashcards 208

Part II – Advanced Chemistry
Ch. 9 - Periodic Trends 212
Nuclear Charge, Energy Levels, and Shielding 215
Periodic Trends 218
Student Struggles 226
Phenomena 227
Flashcards 228
Ch. 10 - Bonding 232
Am I Breaking Bonds? 233
Lewis Structures and LEGO 237
Formal Charge 242
Polarization 246
How Things Stick 254
Resonance 259
Hybridization vs. Molecular Orbital Theory 260
Student Struggles 263
Phenomena 265
Flashcards 267

Ch. 11 - Organic Chemistry 270
Basics and Nomenclature 272
Stereochemistry 276
Curly Arrow Mechanisms 283
Spectroscopy 290
Student Struggles 294
Phenomena 296
Flashcards 298

Ch. 12 - Redox Chemistry 302
Oxidation States 303
Batteries 307
Electrolysis 311
Coulometry 314
Student Struggles 317
Phenomena 319
Flashcards 321

Ch. 13 Kinetics 325
Rate 326
Rate Laws 330
Elementary Steps 335
Activation Energy 337
Student Struggles 344
Phenomena 346
Flashcards 348

Ch. 14 - Equilibrium 351
Rate and Concentration 352
Reaction Quotient Q 356
ICE Charts 357
Le Chatelier's Principle 360
Minor Details 367
Student Struggles 368
Phenomena 371
Flashcards 372

Ch. 15 - Acid-Base Chemistry 375
Bonding Model 377
Acid-Base Calculations 380
Buffers 387
Titration Curves 389
Strength Trends 394
Properties and Loose Ends 397
Student Struggles 400
Phenomena 402
Flashcards 404

Ch. 16 - Entropy and Spontaneity 408
Entropy 409
2nd Law of Thermodynamics 413
Gibbs Free Energy and Spontaneity 416
Student Struggles 428
Phenomena 430
Flashcards 432

Part III – Teaching Primer
Ch. 17 - Teaching 437
Grading 437
Standards-Based Grading 439
Modeling Instruction 442
Discussions 444
NGSS 448
POGIL 449
Antiracism 451
Review 452
Written Reflections 453
Start and End of Class 453
Feedback 455
Flashcards 456
Advice for Struggling Students 457
New Teachers 459
Fascinating Stories 461
Phenomena 464
Themes 465
Personal History 470

Acknowledgements 472
Citations and Notes 473
About the Author 485

A Guide to This Book

The intent of this book is to help improve your teaching of chemistry. Every chapter begins with an overview of the key components of chemical **history** that are applicable to that content area. If you follow the citations, you will find plenty of books to elaborate on that history. Every chapter ends with 3 features. **Student struggles** are ideas that students (and sometimes teachers) struggle with. **Phenomena** contains activities, demonstrations, and experiments that you can use to engage students with chemistry content. **Flashcards** comes with a set of flashcards that include a variety of challenging content questions.

At the conclusion of each chapter, I recommend writing down 2-3 summary comments utilizing the following prompts:

- Which student perspectives stood out to me?
- What advice from this chapter can I use in my classroom?
- What did I disagree with?
- Was there anything that did not make sense to me?
- How connections did I see between this chapter and other chemistry content?

Part 1 of this book includes chemistry chapters that are taught in a regular high-school chemistry class. Part 2 gets into more advanced chemistry topics. It is likely that you will teach some of the topics in Part 2 and all of the topics in Part 1. I recommend that you read at least some of the content areas that you do not teach.

Part 3 discusses a variety of teaching topics that are relevant to chemistry classes. If you read a section and desire to learn more about a topic, you should seek out the citations to find more reading. This is only intended to be an introductory sampler, but good teaching includes more than just the chemistry content.

The introduction highlights some basics of cognitive science. Many of these will be used throughout the book. If you are interested in learning more or seeing the experimental evidence that these claims are founded on, please seek out the citations and notes.

A note on safety

Throughout this book there are recommendations for experiments, demonstrations, and various chemistry phenomena. There is not room for procedures for each. It is imperative that the reader locates an appropriate source for safety information before attempting anything from this book. Even though there are some warnings, there is not a dedicated safety component in this book, and it is up to the reader to find a source that will keep themselves and their students safe.

Introduction

Introduction

The late bell rings. I move from my computer to stand in front of a giant periodic table. My head blocks the squares of elements 79 and 80 from the view of the 30 students. I pick up a pair of safety goggles and put them on for no reason. My lungs fill to speak, but instead I pause as one last student scrambles to their seat. I smile for a moment letting the suspense build of where our learning will take us today.

Teaching chemistry isn't a simple task. Be wary of those who tell you that one small thing can make an excellent teacher. Good teaching involves strong content knowledge, quality pedagogy, reflection, an understanding of psychology, and the ability to keep students engaged. None of these in isolation are easy. When combined the task is formidable. When done properly the results are exhausting and exhilarating.

I love teaching chemistry. For many teachers, it is the interaction with the students that causes them joy in teaching science. For me the art of teaching chemistry is my favorite part. I want to be able to productively combine and enhance the initial models in a student's head with explanations, mathematics, particle diagrams, analogies, symbols, and concrete examples. Increasing the students' understanding through so many lenses is the challenge I seek. I have reflected on that challenge for years. The intent of this book is to share the questions, methods, experiences, and explanations that I have found to be critical and/or productive as a chemistry teacher.

What I notice about the best chemistry teachers is they never reduce teaching to just one thing. They see and embrace the complexity. Even when they specialize, they don't avoid other components. This book will focus on content for chemistry teachers, but you will find a mix of strategies throughout the writing.

When I first taught chemistry, my central aim was to explain complicated chemistry as clearly as I could. In my head teaching followed this pattern:

Chemistry + Explanation → Student + Learning

But over time I learned more about each of those components and the role of the brain in all of them. I now understand many of the limitations of this attempted progression.

Much of this book is based on my experiences with teaching and learning. I have been fortunate to learn from experts in chemical education, but the largest influence has been in my own classroom. Some of the claims in this book therefore will not translate to all teachers. I taught for four years at Lincoln Park High School (LPHS) and am in the midst of my 12th season at Plymouth High School (PHS). LPHS had enrollment of about 1200 students while PHS is part of three high schools on one giant plot of land that enrolls about 6500 students. The majority of students that I have taught selected to take chemistry, Advanced Placement (AP) chemistry, or International Baccalaureate (IB) Chemistry instead of an alternative course. I have extensive autonomy over how I choose to instruct and assess. I realize that teachers have a variety of obstacles and restrictions, but my hope is that regardless of structural differences that this book is useful to all of you.

- Part I is what I consider to be the typical high school topics. The sequence in Part I is based on advancing the models of the atom. The first chapters cover

Introduction

topics where a sphere for an atom is sufficient before adding in motion. Later chapters require electrons to explain concepts, and the final chapter requires quantized energy levels for the electrons. The resulting sequence mostly follows in the historical order of chemistry.
- Part II are topics that are more likely to be in an advanced course. Part II starts with the concept "like charges attract" and "opposite charges repel." This fundamental idea is used as a theme throughout the advanced chapters.
- Part III focuses on teaching methods separately from the chemistry material.

The learning that I received from teaching AP Chemistry and IB Chemistry has been impactful on my teaching for regular chemistry. Even if you have no intentions of teaching an advanced course, I encourage you to read some or all the chapters in Part II. Before we begin with chemistry, we should take a moment to clarify some key ideas about how learning works.

Cognitive Science

There would be moments during my first years of teaching where nothing seemed to work. Yet even when I had that perfect explanation, I would still have students (sometimes many of them) that did not understand. This puzzled me. Sometimes they wouldn't understand even if I explained it twice!

Gradually I started to experience the difference between focusing on teaching and focusing on learning. My ability to explain and make sense was limited by what was in their brains, not my words. Some teachers use the phrase, "I can teach it to you, but I can't understand it for you." I was missing the students' side of teaching and learning. I started to read the basics about how the brain works and how learning works.[1,2,3,4,5] Later I expanded into more nuanced components of cognitive science.[6,7,8,9,10] I found a substantial overlap between my successful experiences in teaching and the research behind how learning functions best.

The basics of cognitive science

It can be helpful to consider the goal of teaching to induce permanent changes in the brains of students. Focusing on this goal is aided by understanding the steps required for such changes to occur. New information must be detected by our senses, transferred to short-term memory, encoded in long-term memory, and finally retrieved from long-term memory. Those steps are influenced by prior knowledge, attention, distractions, emotions, repetition, retrieval practice, concrete examples, and much more. Those influences are not binary or discrete. Nor do they function in isolation from one another. Teaching is complicated and we should anticipate that it will be difficult to do well.

Learning begins with sensory.

The teacher speaks. A PowerPoint slide shows a new image. A chemical reaction produces sparks. A student holds a splint next to a bubbling test tube and a small yelp is audible. A student reads a worksheet problem. In order for any of these sensations to transfer into short-term memory the student must be engaged with them. Often chemistry is visually engaging, so this is not difficult to accomplish. When that is not the case, questions can be a key tool for teachers to inspire curiosity.

Introduction

Your brain is inundated with sensory information that gets ignored. If you look up from this book to your right, you will find that there is an object there that your brain had sensed and chosen for you to ignore. If a dangerous bear had been in that exact spot, you would have immediately noticed because your brain would have sent this critical information along the pathway causing a response from you. But since there was something less relevant (for me there's a wall), your brain decided to purge this information so you could continue focusing on this excellent book.

Learning is counterintuitive

When students learn a lot, they often feel confused. When students learn little, they often feel confident.[11] Experienced teachers will recall multiple experiences with a student who walks into a test feeling confident yet gets a poor mark on the exam.

Students frequently study using methods that involve little cognitive effort. They might reread notes, highlight notes or text, reread text, listen to lectures, and watch videos. These are passive for the student and thus result in little change in the students' understanding. Because students naturally gravitate towards study methods that are easier for them, they can develop a false sense of security. These methods build confidence, even though they work less effectively.[12] When students use more effective methods to study, they will feel less confident in their learning. More effective methods of study include self-quizzing, flashcards, and writing down information on a blank sheet of paper.

Many students will argue against using these methods. They might claim they don't know anything to start with. If they had to write what they know they would just have a blank sheet of paper. These are dangerous ideas that breed ineffective mindsets block the learning process and undermine motivation. Students always know things, but often struggle to locate their prior knowledge that is useful. These students that are resistant need practice more than others. The more they seek their prior knowledge, the more they will improve at that practice.

Learning must be simple.

If too many abstract ideas are sensed at once, the short-term memory is unable to process them adequately and the concepts are filtered. If we use multiple vocabulary terms that are new to a student while introducing an abstract concept, it is unlikely the student will be able to encode the information in a meaningful manner. The student will typically maintain an abstract coding where they can identify an association, but they will not understand that association. For example, if a student does not understand the definition of first ionization energy, they still might be able to match the definition with the term. But they do not understand what the definition or the term represent.

Learning must be difficult.

Much like effort is required to strengthen muscles, effort is required to change the structure of the brain. A passive effort does little to change the brain's structure. Effort is involved to take items in short-term memory and encode them with prior knowledge that exists in long-term memory. When I started teaching, I would try hard to make explanations so simple that they would reduce the need for students to think. I was good at reducing the complexity to be more approachable, but instead students would lack understanding and retain little.

Introduction

I was doing too much of the thinking for them. I was also failing to help them identify their own initial ideas or models of how they understood things. This limits their ability to encode into long-term memory. People are naturally curious, but poor learners.[13] Our brains are designed for survival, not for learning. The best ways to invoke curiosity that can lead to learning are asking questions and discrepant events that cause students to seek explanation with their prior knowledge.

Teaching successfully requires navigating this narrow line between too simple and too difficult. If the teaching is too difficult the short-term memory will experience overload and the student shuts down learning. If the teaching is too simple, the information will not be encoded into long-term memory and will quickly be forgotten. Next, we will add in the final step called retrieval.

Retrieval Practice

So far, we've looked at sensory memory, short-term memory, and long-term memory. But at the end of the day, you need to be able to retrieve the information from your brain. So much of education focuses on storage of information. Far too little is focused on retrieval. Successful experiences of retrieving information enhances learning and understanding. Retrieval practice is the most effective tool that students can use to study.[14] Think about when you start a brand-new school year, and you get a list of 150 to 200 new names to learn. How do you learn them?

I print out a seating chart with their pictures and names. Now I could stare at this seating chart for days and not learn much. If all I do is focus on storing the information, I won't remember the names effectively. It will be November and I will still be checking my chart for many of the students. How do we learn names of students? Do we learn them by using them? If so, what exactly does that mean? It means we have to retrieve those names from our brain.

The most effective way for me to learn names is to quiz myself. If I cover up the name and quiz myself, I learn the names much faster. One must practice retrieving the information in order to learn it. Passing back papers is a chance for me to use retrieval practice. Taking attendance is an opportunity for me to use retrieval practice.

What exactly is retrieval practice? Once I was at a graduation ceremony. Two former students sat next to me, but I could only recall one of their names. They had been in my class as juniors one year earlier. And when I learn new names, I can sometimes no longer recall names from previous groups. But I knew her name, I just could not retrieve it from my memory. What I ended up doing was to read the list of graduating seniors and out of the 500+ names it was easy for me to recognize her name. I just could not recall it on my own. Once I saw her name in the ceremony guide, I recognized it immediately. Her name was in my brain, but I needed a prompt to get it.

Retrieval practice is not just about recalling information. Retrieval practice is an active component of learning that often is underutilized. When you retrieve information from your brain you are helping to establish the neural pathway to that information. This strengthens that knowledge. Furthermore, there is a trend in education to dismiss memorization. Students should focus on understanding and skills instead of memorizing information that they can look up on Google. But understanding and skills are not easily separated from memory. An electrician has a set of skills that they can operate from memory. But those skills are intertwined with

Introduction

a substantial amount of professional knowledge. Those skills are memorized in a combination with knowledge.

We should collectively be very wary of normative rhetoric that undermines memory. That does not mean that there have been issues with students memorizing information erratically in classrooms. But the solution to these issues is not going to function well without memory. A student that can do a stoichiometry problem, should be able to do so from memory. Memory is not just knowing the order of the elements on the periodic table.

In order for retrieval practice to be effective, the student must actually retrieve the information. If you ask a student a question and they transfer the answer from a text, they will not be using retrieval practice. Flashcards are an excellent tool for practicing retrieval practice, but the student must actually respond to the flashcard before checking the answer or they will miss the benefits of retrieval practice. It is insufficient for them to read the answer and think, "Oh yeah, I knew that."

Ask an AP chemistry teacher what the best thing to do for review for the AP test is. Most will tell you to have students try doing previous exam questions. Low stakes quizzing is a great means of retrieval practice. So much of learning is about identifying what is known and what is not known yet.

Forgetting and Spaced Practice

Education focuses too much on learning and way too little on forgetting. The instant you stop learning something your brain starts to revert back to how it was.[15] How long does it take you to forget something? Can we interrupt forgetting? Do students forget something even if I explained it perfectly?

Forgetting is predictable. When a student spends several hours the day before an exam cramming, the amount that will be forgotten shortly after the exam is immense. One of the best methods of disrupting forgetting is to forget and relearn the material again. This is called spaced practice and is the opposite of cramming. Cramming involves a long period of study with material, usually right before a major assessment. Spaced practice involves shorter periods of time that are spaced out over intervals where forgetting can take place. Spaced practice has been shown to be more beneficial to long-term retention of information and understanding.[16]

When teachers start class with a warmup or end class with an exit ticket, they are utilizing spaced practice. They are spacing out the material over multiple class periods. A student that learns a lesson one day, does an exit ticket about the lesson on the second day, and does a warmup on the material on a third day will have used spaced practice.

Students do not do this on their own because it is difficult to schedule spaced practice. Once you are behind what option is left besides cramming? But the long-term harm from cramming makes it critical that teachers work to help students schedule their studying to utilize spaced practice. Part of this battle is educating students on the benefits of spaced practice. Would a student rather study for one hour on Thursday evening, or study for twenty minutes on Tuesday, Wednesday, and Thursday? They will get a better benefit from the latter, and those benefits add up over the course of years of education.

If your students struggle with comprehensive exams, they may need more spaced practice. If a student is taking a unit exam and then forgetting the material shortly after they would show a sizable gap on a comprehensive final exam. Teachers

Introduction

can help their students by spiraling through content or just having students do retrieval practice exercises on earlier units throughout the school year.

Cognitive load theory

The short-term memory can hold 5-9 items at once for a period of about 30 seconds. This makes learning new material challenging. It can be especially challenging for teachers to anticipate which items will stress the cognitive load of students.

If a teacher introduces a new algorithm for a new concept with new vocabulary the student is likely to experience cognitive overload. The short-term memory isn't capable of processing large amounts of new information and the student shuts down. This might mean that the student stops attempting to learn. It might also mean that the student copies notes instead of thinking. It could even mean that the student decides to pair the algorithm with the vocab term, but not comprehend any meaning for either.

A powerful tool to avoid cognitive overload is chunking. Our brain can process a chunk of information the same as a single data piece of the chunk is relevant to us in some way. The sequence of numbers 8201983 is a single chunk for me because it represents my birth date. I can store that sequence of digits in my short-term memory equivalently to the digit 5 in a random sequence of numbers I'm trying to recall. There are ways to help students chunk information as they learn complex content that allow them to be successful and avoid cognitive overload.

Advanced cognitive science

The basics of cognitive science are very easily tested for and understood. They are helpful tools, but there are limits to them as we approach more complicated teaching and learning. Unfortunately, it is more difficult to design controlled experiments for nuanced methods of teaching.

System 1 and System 2

The brain is complicated. But the complexity produces predictability. Teachers should aim to understand how the brain works, how memory works, and they should instruct students about the brain. We want students to know what they can do that will work well and what will be inefficient. One of the best approaches to understanding the brain is to start with a model where two systems operate brain function.

System 1 is your automated brain function. We're not talking about your brain controlling your heart rate, we're talking about how your brain responds to stimuli. When someone insults me on the internet, my brain's system 1 kicks into gear. I become angry and my amygdala flares. I don't need to think about this, it just happens as an automatic response. This system 1 takes in an extraordinary amount of information and it filters most of it out. Look around you right now and consider everything your brain noticed but ignored.

System 2 is the voice of reason you hear in your head. Your system 2 makes complex decisions. If you get angry because your mother wrote something mean about you on social media, your system 2 can step in and help you decide to go for a walk until your amygdala has calmed. Your system 2 is usually unaware of what system 1 does.[17]

Introduction

The evidence supporting the existence of these two systems is fascinating. A container with two fruit flies doubles in population every ten minutes. After 2 hours the entire container is full. When was it half full?

Your system 1 responds quickly with the answer of one hour. Your system 2 can correct your system 1 with the actual answer of one hour and fifty minutes. Because your system 1 processes so much and so quickly, it is highly prone to mistakes.

Students were given a test with many tricky questions such as the fruit fly example. Another set of students was given the same test in a hard to read font with poor contrast. 85% of the students with the easy-to-read test made at least one mistake. Yet only 35% of the students with hard to read tests made a mistake.[18] The students whose tests were more difficult to read did substantially better. Why?

The hard to read font with poor contrast overwhelmed system 1. System 1 engaged system 2 to help and system 2 is less likely to make the mistakes.

With experience, system 2 can learn new things that get transferred to system 1. Tying your shoes or playing a video game start off as challenging endeavors. But with experience your system 1 takes over as the primary driver of these activities.

The challenge with learning science is that often the brain's system 1 will take the sensation that has been transferred into short-term memory, and it will classify it according to what the student's current mental model is. This will reinforce their conception. Many of these mental models have errors in them. But it is difficult to undermine the misconception because the brain will quickly sort the new information to match the current student conception.

For example, let's assume a student believes that there is no gravity in space. This incorrect idea will impact other ideas. Perhaps that same student will think that a projectile has no weight while the projectile is in the air. This might make them think that the object is not accelerating at the top of the flight. And even if a student learns to solve mathematical relationships about projectile motion does not mean that these misconceptions will disappear. How would you fix this?

This happens when you introduce a vocabulary term early in the lesson. Let's assume I begin a lesson by defining density as the equation density = mass/volume. Students will lock into the equation for the entire lesson. Because of their confidence of being able to use the equation to solve for the unknown variable, they will connect conceptual questions back to their ability to perform the algorithm properly. This limits them from developing a particulate, conceptual, or mathematical model of density as discussed in chapter 1. This push for overlearning basics is debunked by Ellen Langer's research on mindless and mindful learning.

Many philosophical differences between teachers can be explained by their priorities of these two systems even if they are not aware these systems exist. Some teachers push repetitive practice in order to seek transfer from system 2 to system 1. Some teachers emphasize lessons that seek to maximize the time engaging with system 2 thinking. When we talk about connecting prior knowledge to learn something new, we can translate that as using system 2 to connect a concept in our system 1 to something new. Many educational sales pitches focus too much on one system at the expense of the other. Teachers should seek a balance that maximizes engagement with initial models and transfer to long-term memory by engaging in critical reflection.

Introduction

Style of presentation activates these systems differently. If you show students a new algorithm and explain it to them their thinking consists mostly of, "Does this make sense?" But if you ask students what they notice about a set of data, their thinking will be quite different. Now they will think, "What do I know, what pieces of information are relevant, which pieces are not relevant, is there anything that does not make sense?" This type of thinking is more effective because it is focused on neural pathways that already exist for the student from their experiences and knowledge.

Another important consideration for teachers is that you talk about chemistry for multiple hours every day for years. When I do a calculation, I am using system 1. When a student is seeing that calculation for the first time, they are using system 2. When I come across something new, I have a large base of chemistry knowledge to utilize with my system 2. It is much easier for me to identify a solution than a novice chemistry student. Teachers must intentionally disrupt their expectations of students at times. The more time teachers spend listening to students, the easier this disruption becomes. Teachers that do most of the speaking in their class will struggle to connect with student conceptions.

After my fourth year of teaching AP chemistry, I found an exam from my 400-level physical chemistry course in college. I noticed a multiple-choice question that I had gotten wrong about entropy, enthalpy, and spontaneity. I was dumbfounded. I would have been furious if any single student in my entire AP chemistry class had missed such an easy question. And yet in my 5th year of studying chemistry I had made that error. It was a struggle for me to remember what my chemistry knowledge had been in college prior to so many years of teaching. It is even more difficult to recall what our knowledge had been as a novice chemistry student.

Interleaving

Interleaving requires a tricky balance. The theory here is that the more difficult the task, the more learning will occur. Think of the gas laws (Boyle's, Charles's, Gay Lussac's). Assume we teach students <u>B</u>oyle's Law and then they work on problems with just pressure and volume. Then on day 2 they learn <u>C</u>harles's Law and just work with volume and temperature. On day 3 they learn <u>G</u>ay Lussac's Law and work with just temperature and pressure. These tasks will be much simpler than if we had introduced all three laws on day one and then spent two days having them work on a variety of problems.

Giving students a variety of gas law calculations to work on is an example of interleaving. In the first approach a student might follow an order of BBBBBCCCCGGGG. Using interleaving would be an order like BGCGBBCBGGCBC. We are interleaving the calculation and variables. This forces students to select the appropriate algorithm or equation and execute the algorithm. When a student has been doing just one gas law, they will use patterns to solve the questions. Interleaving forces the student to identify the variables and select the appropriate algorithm.

Research shows that using interleaving results in more effective learning.[19] The difficulty is that there is a line where we approach cognitive overload. If there is too much new learning going on at once then a student will experience cognitive overload and they may shut down. The student will not experience success which decreases motivation. When a student does something correctly and receives affirming feedback, they are more motivated to continue learning.

Introduction
It is up to the teacher to determine when interleaving is appropriate, and when interleaving will overwhelm the students. In parts I and II, I will sometimes recommend one or the other depending on my experience. But these recommendations will not be optimal for everyone. The prerequisite experiences of your students will modify what will function best in your classroom.

Dual Coding and Memory
Try and recall what element 66 is, the number of students you had each of the last five years, and the first sentence of this book. You likely can't remember any of them. You've also likely had a sensory experience with at least two of them. So why don't you remember them? It turns out that our brains are not good at memorizing numbers and words. And yet when I play Minecraft, somehow, I can remember the path to take in the Nether in a world I haven't played for six months. What can our brains remember easily?

Our brains remember images, places, and stories. Picture the classroom that you teach in. What details can you remember? Can you remember specific drawers and the contents of those drawers? How many objects and wall decorations can you easily picture? From an evolutionary perspective, it makes sense that our brains are good at recalling locations.

The author Joshue Fore explores this capability in his book *Moonwalking with Einstein*.[20] He investigates memory competitions where contestants compete for the title of having the best memory. He is surprised to learn that most contestants are not particularly brilliant, they just utilize intensely focused practice that relies on cognitive science about memory. The primary technique used is called a memory palace. The memory palace is built by placing a person doing an action with an object in a location within the memory palace. Marie Curie tripping over a bucket might represent nine digits where Marie Curie represents the first 3, the tripping action is the next 3, and the bucket represents the final 3 digits. But to me I only have to recall the one image of Marie Curie tripping over a bucket by the light switch to my classroom instead of those nine digits. After the light switch comes the bookshelf with another unusual character, action, and object to represent the next nine digits. The record is over 7000 digits memorized in 30 minutes.

In education, we use a similar concept called dual coding. Dual coding involves creating a visual image to represent a chunk of information. This chunking reduces strain on system 2 because system 2 can only manage so many chunks at once. The visual nature also increases the ease of retrieval of the information by creating a single retrieval cue. The concern with dual coding is that it can be distracting. If you have students sew a mole, the visual image should function as a retrieval cue. But the student must put in effort to link the visual image to the information. If the student has no idea what a mole is in chemistry, all they will remember is the image of Moley Cyrus Wrecking Mole or Cray"Mole"a.

Can you easily picture Albert Einstein or Marie Curie? Those visual images can be used for dual coding. If students are learning about separations, they can use an image of Marie Curie working diligently in a shed to recrystallize a sample of radium that she is separating from pitchblende. If students are learning about atomic structure, they might recall Hans Geiger complaining to Ernest Rutherford about how boring it is to count scintillations in a dark basement for hours on end. Students will remember Red Cat and An Ox for years after their redox unit. Giant ants and small

Introduction

cats is a simple visual for anions being larger and cations being smaller. As teachers we want to encourage dual coding while being clear that the image is not the end goal.

Mental Models and Elaborative Interrogation

Mental models are constructions we use in our brain to represent something in a useful manner. Scientific models are mental models that a group (of scientists) use to represent something. In order for a mental model to be a scientific model, consensus is needed and must be based on evidence. A model should be useful in making testable predictions or explaining phenomena. Models can be revised and improved in a cycle where the model is developed, enhanced, deployed, and then evaluated. Upon evaluation we cycle back to the beginning where further improvements and uses can be found. The Science and Engineering Practices (SEP) for the Next Generation Science Standards (NGSS) include developing and using models as one of the eight practices.

Modeling is difficult to describe as an abstract concept. But we use models frequently. When I multiply numbers by 9, I drop the number by one for the tens place and the ones place must combine with the tens place digit to add to 9. 8 times 9 means that I would put a 7 in the tens place and I would need to add a 2 in the ones place. But this model has limitations. When I multiply 9 by 11, I run into difficulties. I can still add the digits to make 18, which add again to make 9. But the rest has stopped functioning. I would either need to expand my model to include a new set of rules for digits above 10, or I would abandon my model at that point.

In chemistry we frequently use particle models. These have tremendous explaining power for phenomena that students observe. When a solid is heated to a higher temperature how would the particle representation change? When a liquid changes to a gas how would the particle representation change? When iron reacts with oxygen how would the particle representation change? But within the particle model of matter, there are many concepts teachers use that students have not developed yet.

Johnstone's Triangle (Fig. 0-1) points out that there are three common representations of chemistry phenomena. The particle-level representations show what the atoms or molecules are doing if we could zoom in like we were on The Magic School Bus. The macroscopic view is what we actually see happening in the lab. The symbolic view is how we represent the macroscopic and submicroscopic using symbols. Sn (s) → Sn (l) is a symbolic representation of tin melting. The macroscopic view would be us watching a pile of solid tin melt in real life. The particle level representation would show what the tin atoms do as the material changes from the solid to liquid state. Chemistry classes tend to use symbols too frequently while underutilizing the particle-level representations.

Introduction

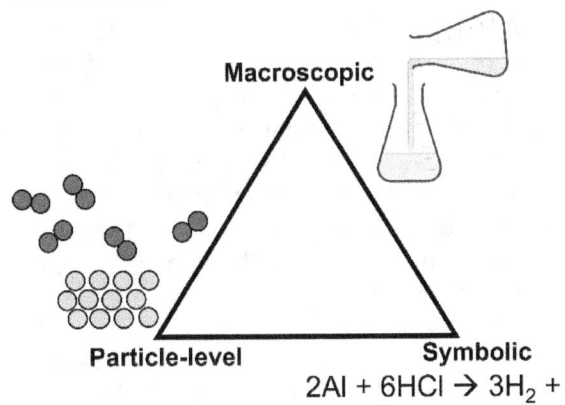

Figure 0-1: Johnstone's Triangle includes three perspectives towards chemistry concepts[21]

Teachers should consider equations and algorithms as components of models. We use these to make predictions, we can evaluate their shortcomings, and we can revise them. Using dimensional analysis allows me to predict the amount of product formed from the amount of reactant. I can revise the model to incorporate coulometry by using Faraday's constant. I can use the mole ratio to explain the proportionality of the coefficients. A teacher familiar with BCA tables (see stoichiometry chapter) knows they function as an alternative representation of the same model.

One method of challenging students' models is through elaborative interrogation. This method involves asking students a series of questions that challenge and clarify their models. Ideally this will lead students to use multiple models together. This technique is highly effective at engaging system 2 thinking. This is the opposite of Ben Stein's character in Ferris Bueller's Day Off. That character was trying to test system 1 being able to identify connections between a term and a phrase. We want to push students to articulate their mental models and finding their limitations of understanding.

Teacher: What happened when we put the hot metal into the cool water?
Student 1: The metal cooled down and the water warmed up.
Teacher: Tell me more about that.
Student 1: The metal cools down, so that means that the particles slowed down. The water got hotter, so that means the particles sped up.
Teacher: Who can tell me what happens that causes the particles to change speed?
Student 2: The energy transfers from the hot particles to the cold ones.
Teacher: What do you mean by energy?
Student 2: How fast they are going. Kinetic energy.
Teacher: I don't understand (this is a lie, secretly the teacher is a genius)
Student 2: The fast particles have a lot of kinetic energy. They are moving fast. Then they transfer that energy to the slower water particles.
Teacher: How does a particle transfer energy?
Student 3: Like this (student smashes one hand into the other)
Student 2: Right, so the fast hand is the hot metal particle, and the slow hand is the cold-water particle. The hot metal particle hits the cold-water particle and the energy transfers.

Introduction

Notice how the teacher avoids vocabulary terms and guides the students into explaining their thinking by using simple questions. Ideally the teacher would utilize long pauses, or "wait time," to allow students to think. If a teacher responds quickly after a response is given the students will provide shorter explanations. If a teacher pauses after students respond, more students will participate in the conversation and the responses will increase in quality and length.[22] This conversation could go much deeper by next exploring why the water particles did not change speed by as much as the metal particles. The teacher could also ask if all the water particles move the same speed, or they could explore the collisions in greater detail.

Abstract vs. concrete

Concrete details are things that are visible or can be sensed. Abstract details are not physical objects. Brains work better with concrete details. Abstract ideas are best learned by connecting the abstract idea to concrete details. Most of what your brain has as prior knowledge is concrete. This is particularly true for younger students. When people complain about "memorizing facts" they are often referring to association between two abstract details where the learner has minimal understanding of either. Imagine someone who has never seen a cello.

"It's like a big violin!" you explain to them.

"What's a violin?" they ask.

"It's like a small cello!" you enthusiastically respond.

They don't know what either object is, but they know they are related. This abstract association is common. A student might not understand a term nor a definition, but they know they go together.

Rate is an example of an abstract idea. In chemistry, "rate" refers to how quickly a reaction takes place. What concrete examples can strengthen our concept of rate? Trial 1 has three particle diagrams at time 0 s, 10 s and 20s and shows particles changing from A to B. Trial 2 has three particle diagrams with the same times, but there are more B particles and fewer A particles at 10 s and 20 s (Fig. 0-2).

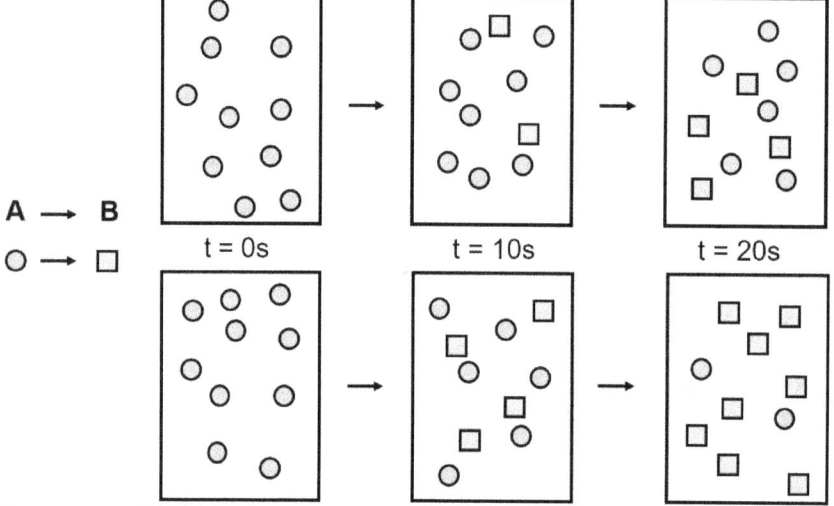

Figure 0-2: The reaction A → B at times 0, 10 s, and 20 s for two sets

Introduction

We could show an example of a fast reaction such as elephant toothpaste (Ch. 13 phenomenon #5). Here the rapid bubble formation is evidence of a rapidly changing amount of chemical. We could connect the concepts of temperature to rate. We could look at other examples of rates that are not based in chemistry. What does driving at a fast rate mean?

When using concrete examples to highlight an abstract concept, we want to use more than one example. We also want our examples to differ whenever possible. If I use elephant toothpaste as an example, and then follow up with a baking soda and vinegar mixture, students will confuse the overlap as the rapid generation of bubbles or gas. Instead, we could use a reaction between bleach and different colors of food coloring where the color change is indicative of the rate of reaction. These two examples overlap with changing amounts of chemical instead of how much bubbling.

Because students spend so much time solving problems in chemistry class, we must ensure that these efforts are not entirely at the abstract level. When a student is given an assessment question, they start by identifying what they know. But teachers often present what we know as identifying the variables and the units. What else do students know? Can the student construct a macroscopic visual image of the question? Can they construct a particle-level representation of the question? Can they identify symbols to use? What do they know about the chemicals involved? What units do they identify? What experiences in the classroom do they connect this to?

When students are stuck on an abstract concept, the fix is to give them multiple concrete examples. Then enhance their understanding by having them describe what these examples share and what knowledge or experience the student has with them. What do they see?

Themes in this book

This book shares student struggles with chemistry content while providing ideas and explanations for helping those students. There are some consistent themes that emerge.

Sequencing

The order with which people learn things matters. Density can be described with an equation, the slope of a mass vs. volume plot, particle representations, a for-every statement, with units, or a proportion. If you learn the equation form first, this will influence everything else because new learning connects to your prior knowledge. The more concrete the initial learning is the better because the brain will be more adaptable.

The sequence of content is an expansion of this idea. As you read this book you may find that some of the sequences do not mesh well with how you currently teach. This is fine. The intent here is not to claim a single sequence, but to be intentional about the sequence. If you have a rationale that fits well in your classroom, please continue to use what works. But you should have a rationale for the order you go in, and part of teaching is evaluating and adjusting based on evidence.

Prior knowledge

Knowledge is stored in long-term memory. New information (if sensed) will go to short-term memory and learning happens when the short-term memory encodes information into long-term memory. There are two methods of transfer. In

Introduction

the first, information is repeated until the response becomes more automatic. If you're familiar with the phrase drill and kill that would be one example of how the transfer occurs. If a student does a large number of questions, they will eventually acquire algorithms that they know. The second method is through connecting to knowledge in the long-term memory. This would happen when a student observes a discrepant event that they must search for knowledge that might explain the discrepancy.

Both can be valuable in education, and both carry risks. Repetition can dull motivation and also leave knowledge as abstract and meaningless. This increases forgetting as soon as the material stops being practiced.

Johnstone's triangle: particles, symbols, and macroscopic representation (Fig.0-2).

Teachers easily move between representations of particles, symbols, and macroscopic level observations. If a student dissolves salt into a beaker of water (macroscopic), the teacher can picture the particles changing from a crystal lattice to solvated ions (particle level). The teacher has both of those images in mind when they write the symbolic representation:

$NaCl\ (s) \rightarrow NaCl\ (aq)$ or $NaCl\ (s) \rightarrow Na^+\ (aq) + Cl^-\ (aq)$

Students require extensive practice flipping between those three modes. These three levels also provide teachers an opportunity to make abstract ideas (symbolic) into concrete representations (macroscopic and particle level).

Concept first, vocabulary/algorithm second.

A vocabulary term is abstract. Most algorithms are abstract. "Good students" are able to take new terms and connect them to concrete ideas and experiences. But as the teacher you must show all students how to do this and model it. The best way to do this is to flip the sequence of instruction. First give the students a phenomenon of some sort (lab experiment, demonstration, data, an equation, etc.) and then have the students search for connections.

The purpose of a vocabulary term is to consolidate a concept into a single piece of recall. Instead, we focus on getting students to associate terms and definitions. But our goal should be that when we bring up a term, they recall a series of connected representations of a model.

Cognitive overload

Cognitive overload is when the short-term memory has too much to be able to make sense of. This is the danger of learning that involves too many abstract ideas, or too many new ideas. Younger learners are particularly vulnerable to cognitive overload because they lack experiences and concrete examples to build abstract ideas. They also lack understanding for many abstract ideas. When a student experiences frequent cognitive overload in a chemistry class, it might be due to the student having a base set of knowledge acquisition that was abstract.

Cognitive overload is a tremendous challenge for teachers. The more difficult the learning is, the more it is retained. But if we push too hard the brain can't keep pace and the student is likely to shut down.

The first step in learning is sensory. The student has a sensation, and that sensation needs to move into short-term memory where it can be meaningfully engaged with. The feedback I receive from struggling students suggests that they do

Introduction

not have organized systems from the very beginning. When they sense new information, they struggle to make meaningful observations, they struggle to characterize what they do know and what they do not know, they struggle to store information effectively, they struggle to connect new concepts with previous knowledge.

Help students organize a plan on how they should interact with new ideas, terms, symbols, and macroscopic observations in the class. A struggling student might need explicit statements about what steps they should take and in what order to help them learn how to learn.

Slope

How quickly two variables change relative to another is a powerful way to relate two variables. Many common units in science are slopes. Molar mass is the slope of mass vs. moles for a substance. Molarity is the slope of moles of solute vs. liters of solution. Density is the slope of mass vs. volume for a substance. Pi (π) is the slope of how far around a circle vs. how far across a circle.

The units for these slopes are units that students often struggle with. Some know how to use them to "cancel units" and find answers, but students need help articulating what the units represent.

A "for every" statement is a key tool to incorporating slope successfully into the chemistry classroom.[23] For every 1 cm across, a circle will be 3.14 cm around. For every 1 hour, the car will travel 70 miles. For every 1 °C the temperature of the gas increases, the pressure increases by 0.34 kPa.

Using slope in the chemistry class helps students by creating a central model they can use for multiple different content pieces. Students may also benefit their mathematical understanding by using slope in chemistry class.

Proportional Reasoning

Dimensional analysis is a more abstract algorithm than proportional reasoning. We use dimensional analysis in chemistry education as a tool. But proportional reasoning allows students to better connect with the content. Proportional reasoning tends to be slower and more difficult. Because so much chemistry is learned through solving problems, I find proportional reasoning allows for more students to connect the content with those problems.

51.4 g Cu	1 mol Cu
	63.5 g Cu

$$\frac{x \; mol \; Cu}{51.4 \; g \; Cu} = \frac{1 \; mol \; Cu}{63.5 \; g \; Cu}$$

The mathematics looks equivalent, but the key with proportional reasoning is that we can draw students' attention to the proportion. For every 1 mol of Cu, there is 63.5 g of Cu. These two amounts are equivalent. Since we have 51.4 g, we know that we have less than 1 mol of Cu since we have less than 63.5 g. When we calculate an answer of 0.809 mol Cu we get a result that matches our prediction of less than 1 mol.

Proportional reasoning matches well with the macroscopic level. If we have a sheet of Cu metal that is 63.5 g, 51.4 g must be a smaller sheet. It is easy for students

Introduction

to compare the relative amounts from the proportion set up. These comparisons can be made from dimensional analysis but tend to be more obscured.

A great introduction to proportional reasoning involves the teacher drawing Mr. Short and Mr. Tall. Mr. Tall is 6 paper clips tall while Mr. Short is only 4 paper clips tall.[24] If Mr. Tall is 8 buttons tall, how many buttons is Mr. Short? What would be some incorrect responses and how might a student arrive at an answer such as 6?

History

What is your evidence for your claim? Being knowledgeable about the history of chemistry allows the teacher to introduce stories into their classroom. Cognitive science shows that our brains connect better with stories than any other medium.

Each chapter will begin with a short description of key historical events that altered the trajectory of chemical theory. But know that these descriptions are merely a starting point. The intention behind these is to introduce why we believe what we believe as well as try to help the reader understand the lack of clarity at the time of these shifts. Some ideas seem obvious to us now, but at the time required boldness and confidence.

Unfortunately, the history of chemistry is a chaotic mess. The development is not sequential and linear. Rather scientists would progress forward in one arena while multiple incorrect assumptions remained in others. Inconsistent beliefs transcend geography and time.

I find the puzzling together of chemical knowledge without our present-day models to be formidable and intimidating, but also something that forces me to engage deeply with the content. If I were to travel back in time 200 years, how would I be able to convince the top chemists of the time period of current understandings? Could I do so without technology and terminology?

These challenges are helpful because they limit the curse of knowledge. When you have taught a subject for multiple class periods for years upon years it becomes very simple. Your brain is now using system 1 for the methods, and this makes it difficult to perceive information through the lens of a novice student.

Eric Scerri proposed a new philosophy about chemistry history in his book "A Tale of 7 Scientists." He pushes for a shift in narrative from assigning a single person to represent a critical shift in chemical theory towards a narrative that recognizes the cooperative efforts of many.[25] When looking back in time, we find that major discoveries usually involved a wide array of experiments and ideas from a collection of scientists working towards the advancement of our understanding.

As a chemistry instructor, you want to be familiar with the key stories, experiments, and people involved in the development of new ideas in chemistry. Without them, we are reduced to relying on our authority to transmit ideas in the absence of evidence for our claims.

We also want to grow over time to realize that many of these stories that we will have been reduced a great deal from how they originally occurred. To state that Boyle discovered the relationship between pressure and volume ignores the fact that he was working on advancing the work of Van Helmont. To state that Lavoisier redirected the narrative away from the phlogiston theory that was hindering chemical progress ignores the fact that Lavoisier was utilizing the experimental evidence and methods performed by Joseph Priestley. To say that Dmitri Mendeleev constructed the first periodic table ignores the contributions of Stanislao Cannizzaro and the

Introduction

many scientists who constructed periodic tables prior to Mendeleev that were not as good.

Right now, we are at a moment in time where we have a strong understanding of how most things function. There are a few lingering mysteries about highly technical phenomena. Our knowledge has advanced very far from when Dalton proposed his atomic theory or when Berzelius started to run electrical current through chemicals. But we can do better by understanding how we progressed from those initial experiments to where we are today.

Part I

General Chemistry

"I have often noticed with regret the great amount of labour which an earnest student expends in noting down the reactions and the names and formulae of substances which are presented to his notice in the lecture-theatre. He is thus greatly interrupted in following the arguments and explanations of the speaker, and he often loses more important generalizations in securing a record of details. One of my chief objects in the preparation of this book has been to relieve him from such distractions."

- Edward Frankland **Lecture notes for Chemical students** (1866)

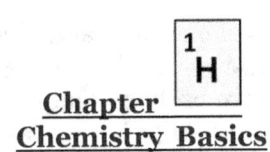

Chapter
Chemistry Basics

History

The history of chemistry is sloppy. There was so much we didn't yet know that when a major discovery was made, the advancement was often limited by unsettled components. It is challenging to communicate the progression of chemical knowledge.

Antoine Lavoisier starts his book (Elements of Chemistry) by stating how critical it is that chemists base their ideas off evidence from nature and not storytelling. Lavoisier describes this by saying that "the series of facts which are the objects of science, the ideas which represent these facts, and the words by which these ideas are expressed."[1] Today in education this triangle is similar to Johnstone's triangle where facts are macroscopic observations, ideas are particulate representations, and words are symbolic representations.[2]

Reading older chemistry texts reveals claims that are justified using experimental evidence. Newer chemical education texts often focus on simplifying explanations and frequently omit evidence. Teachers should prepare themselves by knowing the evidence and history of why we believe what we currently teach. It is in your best interest to know how we know what we know and the progression that took us from the beginning to our current state.

Famous chemistry teacher Edward Frankland authored "How to Teach Chemistry" (1875) and it stands in stark contrast with today's lesson plans.[3] There is a heavy focus on convincing students of the composition of compounds using elements. Why is such a focus left out of our current education progression and should some of this evidence be re-introduced? By knowing how we progressed to where we are today, we can often find struggles that students have conceptually in discarded models of the past. The phlogiston model, the four elements (earth, air, water, and fire), the caloric model, the Bohr model, and others are all found in student intuition about how the world works.

Teachers who know the history of chemistry can justify content they teach with evidence rather than authority. Teachers can use stories from the past as a means to help their students form stronger cognitive memories of chemistry content. The conflicts, challenges, and disputes all add drama that makes memories stronger and holds attention better through frequent transitions.

The history of chemistry often involved intentional exclusion of groups of people. The effects of this linger today. Had this been addressed in a systematic fashion, all of us would have benefited. Knowing and communicating the racist and sexist components of our history sets up to be able to challenge the residual impacts that persist today. How a teacher describes Henry Moseley matters. How they tell the story of Otto Hahn's Nobel prize matters. And whether they choose to concern themselves with current racial discrepancies or to dismiss that evidence under the pretense that chemistry is race neutral matters.[4]

*****

The beginning of chemistry is debatable. The four elements theory of the Greeks (earth, air, fire, and water) is more extensive than you would expect. Vedas

Chapter 1 – Chemistry Basics

(about 1000 BC) had a system of 5 elements that included light in addition to the other four. Well before Aristotle there was chemical analysis being done in Egypt. But the first organized chemists were probably Jābir ibn Hayyān (720-813 AD), al Razī (866-925 AD), and ibn Sina (980-1036) who was also known as Avicenna. Jabir produced writings about chemical experiments, al-Razi worked primarily on medical advances using chemistry, and ibn Sina worked on both medicine and alchemy even though he did not pursue transmutation. Roger Bacon (1214-1292) was an early alchemist who used ibn Sina's work.[5]

From there the chain of knowledge continues to Paracelsus (1493-1541) who worked on chemistry via medicine. His famous quote still relevant today is "All things are poison and nothing is without poison. It is the dose that makes a thing not a poison." Van Helmont (1579-1644) was the first to study gases while also experimenting with other chemistry. He held a greater focus on quantitative measurement than his predecessors. His work was advanced by Robert Boyle (1627-1691).

This chain could have countless others added in. During these progressions it's important to note that communication between scientists was limited. The journals we use now didn't begin to exist until the 1600s. Even the nomenclature of the chemicals and experimental methods were unique to the individuals. These scientists spoke a variety of languages, but the chemical nomenclature was distinct and would not share translation.

A noteworthy tale for those about to teach measurement, statistics, and significant figures is the tale of Lord Rayleigh analyzing nitrogen gas. Rayleigh dedicated his career to measuring precise densities of gases. He produced nitrogen gas in two manners. For one, he took air and removed the oxygen by reacting hot copper metal with the oxygen gas. The other, he produced nitrogen using ammonia and water.

Keep in mind that he spent immense effort in making these measurements as precisely as possible. When he would measure the mass of the empty flask, he could detect the mass difference due to buoyancy as the glass contracted slightly when there was no internal pressure. Interestingly he found that the nitrogen gases had slightly (1/1000) different densities when he removed oxygen from air and when he produced nitrogen gas chemically.[6]

He asked for ideas, and eventually he began removing the nitrogen gas as well. As the nitrogen reacted and was taken out, the density of the remaining nitrogen would increase steadily. No matter how much he tried to remove all the nitrogen, there remained a bubble about 1% of the original volume of gas. That bubble was argon. When a voltage was applied to the bubble, the emission spectra confirmed that he had discovered a new element.[7]

Most students who had a 0.1% disagreement in two methods of determining the density of nitrogen would have celebrated such a consensus. Instead, Raleigh worked to decipher the exact cause of that difference. With precise measurement, this led to a monumental discovery. Not only did he find a new element, but this was the first discovery of a noble gas. The placement of argon proved difficult as the mass was between potassium and calcium.

Chapter 1 – Chemistry Basics

Significant Figures

I would be lying if I said that my students easily transition into the world of measurement. My students abandon significant figures shortly after we get to unit 2. If I tell them to round using significant figures, their anxiety heightens. Teachers have a choice to make. Will their class be about significant figures or about chemistry? If you choose to always require significant figures, students will improve, but it will occupy part of their cognitive load. But why is it so difficult for students to make this transition in the first place?

For students to truly understand significant figures, they must know that measurements differ from numbers. Students have spent their entire lives in a world with numbers. Not only have they been taught mathematics for ten years prior to chemistry, the math they have been taught has been abstract. We begin on an unstable foundation and must put in effort to distinguish the two ideologies. In mathematics, numbers are numbers. They are points. In math, 6.70 and 6.7 are identical. But significant figures are about measurements. And measurements are not numbers. They are ranges. 6.7 mL means a range between 6.65 and 6.75 (or 6.6-6.8). 6.70 mL means a range between 6.695 and 6.705 (or 6.69-6.71). 6.7 mL and 6.70 mL are completely different even though they look mathematically equivalent.[8]

When I was in high school, I remember interrupting (I'd like to offer a sincere apology for being an obnoxious student at times) my teacher when she first introduced us to the rules for significant figures. I pointed out, "According to these rules, six times four would be twenty."

She stared at the rules for a moment, doing some thinking in her head, and then finally stated, "Yes, six times four would be twenty." The class was audibly confused and distressed. Some students put their heads down in frustration while others dropped their pencils as if their efforts for the rest of the hour would be futile.

And this frustration all makes sense. We were living in a world where six times four was twenty-four. We had been taught this all our lives and that made sense to us. Why would you make the answer worse if you had this knowledge? Why would six times four be the same result as five times four? Why would you pretend to be stupid like you didn't know it was twenty-four?

Herein lies the key problem with teaching significant figures. Too many teachers start the lesson by showing students the rules, giving an abstract definition, and having students do practice problems. Instead, you need to disconnect the students from their traditional world of mathematics. We need to make clear that this is not math, this is measurement. These are not numbers, they are ranges. They look just like numbers, but as soon as the unit is attached, they are no longer numbers.

Because people are curious, one method that has some success in accomplishing this disconnect is an activity based on the show "The Price is Right." Choose a few random students, give them each something to write on, and have them write down what they think various pieces of glassware cost. Starting with a 250 mL beaker and then progress through a 100 mL beaker, a 1000 mL beaker, a 50 mL beaker, and a graduated cylinder. The prices of the mid-range beakers are the cheapest.

Chapter 1 – Chemistry Basics

Beaker capacity	Cost
10 mL	$3.95
100 mL	$3.70
250 mL	$3.55
600 mL	$4.60
1000 mL	$8.65

As beakers get larger, more glass is required. Thus, it makes sense (and cents!) that they are more expensive. And a 1-Liter beaker is more expensive than a 250 mL beaker. But why is a 100 mL beaker more expensive than a 250 mL beaker? This phenomenon invokes curiosity. There is something else at play here.

It turns out that the gradations on a small beaker are every 10 mL. And even though beakers are notoriously poor measuring tools, the precision required to produce these 10 mL increments instead of 50 or 100 mL increments are more expensive (Fig. 1-1). To heighten the landing point of this lesson, it can be helpful to then ask students what they think the cost of a 50 mL buret is. The buret will cost five to twenty times as much ($128.00) as other less precise glassware. Why do they think the buret is so expensive to produce? What do they notice about the number of lines and line spacing on a buret? A 50 mL graduated cylinder costs $7.10.

The next challenge with significant figures is what to do when we take two measurements and manipulate them. If we add 50 mL from a beaker to 50.00 mL from a buret, what do we have? Do we have 100.00 mL, or 100 mL, or something else? There are two keys to help students transition into understanding. The first is that we must emphasize that these are not numbers, but measurements. We must make this emphasis again and again. But we also need to offer something for students to take on that displacement. And ranges are highly effective.

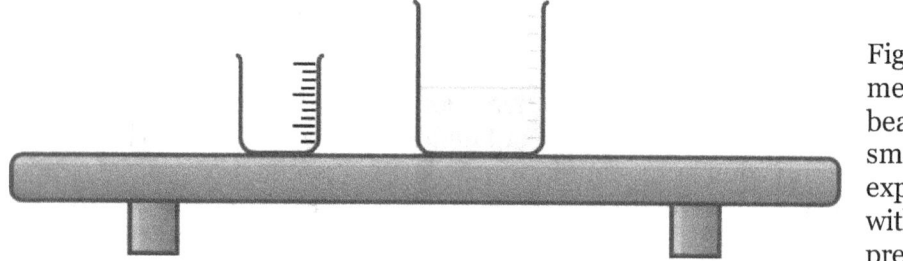

Figure 1-1: A medium sized beaker and a smaller more expensive beaker with greater precision

The question we start with is, "If I have about five thousand dollars, and someone gives me a quarter, how much money do I have now?" The response we're hoping for is that you still have about five thousand dollars. Our uncertainty of the initial amount makes the quarter irrelevant. But I definitely have more money now, and yet my reporting of money has not changed. That makes sense when we have a range. If I have between $4500 and $5500 and you add $0.25, the range of possibilities is now $4500.25 to $5500.25. You still have $5000 ish.

Chapter 1 – Chemistry Basics

Let's return to the example from when I was in high school. When I say six times four, the number six as a measurement represents a very large range. It is between 5.5 and 6.5. The second measurement is between 3.5 and 4.5. If I multiply the bottom ends of those ranges (5.5x3.5 = 19.25) or the top ends of those ranges (6.5x4.5 = 29.25) I get a wildly different result. For me to claim that 6x4 = 24 does not work in ranges because my answer is too precise given my poor measurements. I cannot stipulate that the range of 19.25-29.25 is 24 because 24 implies that it is not 27 or 22. But if I make a more precise measurement things change. If instead I multiply 6.0 x 4.0 to get 24, now my answer properly reflects the measurements that I made because I am working with a narrower set of ranges (3.9x5.9=23.01; 4.1x6.1=25.01).

Students are initially concerned about rounding their answers too much because it feels dishonest to them. It feels like they are hiding part of their knowledge. But instead, we are appropriately communicating the range of our results. Poor measurements cannot result in a highly precise final answer. We must communicate our solution, but also how precise that solution is.

A good lesson to help with this point is to construct three identical boards, each board with different line spacing. The first board has lines at 0 and 100 cm. The second board has lines every 10 cm. The third board has lines every 1 cm. Each board has a dot at the same location. Have the students report what the measurement is for the first board, then the second, and finally the third. They should be different. Emphasize that the first board has a large range, maybe between 60 and 70. But the second board has a much smaller range, maybe 63 to 64. And the third board has a much narrower range, maybe 63.2 to 63.3.

If you were to manipulate those measurements, the final answer should reflect the quality of ruler used. A square that is 60 cm x 60 cm is 4000 cm². This square has a side length that is somewhere between 50 cm and 70 cm. If we multiply those extremes, we get 50cm*50cm = 2500 cm² or 70cm*70cm = 4900 cm². This makes it improper for us to claim the area is 3600 cm². That doesn't reflect our range of possibilities with fidelity. 62 cm x 62 cm on the other hand gives us an area of 3800 cm². This does adhere to the range of minimum and maximum values we could have for those measurements (3720-3970 cm²).

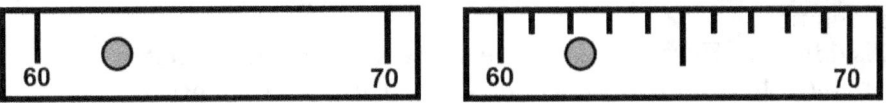

Figure 1-2: Two rulers with the same dot, but different line spacing precision

The dots are in the same location (Fig. 1-2). But the line spacing is different on the two rulers. The first ruler we can see clearly that the dot is between 60 and 70. It is less than halfway. We might say the center of the dot measures to 63 or 64 units. But the second ruler gives us a much narrower range. Now the dot is clearly between 63 and 64. The center of the dot is closer to 63 than 64. So now our measurement is 63.2 or 63.3 (assuming we are measuring the center of the dot). The dot is in the same location, but we end up with two different measurements.

Our measurements must communicate about the quality of the instrument used to obtain them. Let's say that I use a really bad ruler to measure something as four feet long. If I break that object into three pieces, it would be improper for me to

Chapter 1 – Chemistry Basics

say that one of those pieces is 1.33333 feet long. That measurement is deceptive because a chemist reading that measurement would think that I used a highly precise ruler. But the one I used was obviously a piece of junk. That ruler would have had markings at 10 ft and 0. Such a ruler is unlikely to sell consistently.

The primary conclusion we want students to make about measurement is that the number of lines or gradations on the measuring apparatus is the determining factor for the precision. The more lines there are, the more precise the ranges of our measurements will be. This is obscured with digital measurements because that precision has been determined by the manufacturer.

With that said, I have never been highly successful in getting these ideas to stick. Students seem to walk away from the lesson that they should be stressed about rounding in this class, but they often don't materialize that stress into learning. They have a lot of neural connections from math class, and they quickly default back to those when they see numbers. This is also a time of year when students have not adjusted to the large amount of interplay between abstract ideas and concrete examples that chemistry provides. I end up telling students to try their best to round correctly, but that it won't impact their grade as long as they are close. In the end I prefer not to occupy too much cognitive load with significant figures when students are learning other topics. Perhaps as the demand for students to learn more statistics grows, this will become less of an issue.

If you are going to commit to keeping significant figures as an evaluation metric all year, I would advise that you to spend some of that effort on the concept. Make sure that students are continuing to understand the difference between a measurement and a number. Push them to explain how the number of decimal places relates to the precision of the measuring apparatus.

From the numerical perspective, any digit 1-9 is always from a measurement and hence a significant figure. A zero on the other hand depends. Some zeroes are meant to hold places to show mathematical value, and others are intended to show the measurement's precision. The zeroes in the measurement 40,000 cm are needed to make the measurement reflect its mathematical value. If I got rid of the zeroes to make 4 cm, I would have the wrong value. These zeroes are not significant and do not reflect the precision of the measurement. This is a bad measurement. It has 1 significant figure, and the precision is in the ten thousands place. A highly unusual ruler is again the culprit.

4.00 cm is different. Here the zeroes are not serving any mathematical function. They are there to communicate about the measurement. Hence, they are significant. 4.00 cm has three significant figures and is precise to the hundredths decimal place.

You can use shortcuts with the students. One method is to tell the students that if there is a decimal count from the right until you get to the last digit 1-9. That is how many significant figures there are. If there is not a decimal, count from the first digit 1-9 from the left to the last digit 1-9. For 0.002030 mL we would start on the far right and count until the 2 (0.00**2030** mL) for four significant figures. For 20,350 kg we would start with the 2 and count until the 5 (**20,35**0 kg) and again we would have 4 significant figures.

Some teachers use the Atlantic-Pacific rule for this. If a decimal place is Absent, then you ignore zeroes on the Atlantic (the right side). If a decimal is Present, then you ignore zeroes on the Pacific (the left side). I find the Atlantic-Pacific rule to

Chapter 1 – Chemistry Basics

be too abstract and recommend trying to focus on the why for significant figures. Either way the concept should be addressed first. Shortcuts such as Atlantic/Pacific must be last during the instruction for students to understand why this method works.

One issue to anticipate is that students will focus on the right and left of the decimal. They will need to be told that the decimal's existence is the criteria, and from there it is to the right or left of the digits. For example, for 40.0 cm they will identify the 4 and 0 as significant without trouble (**40.0** cm). But some students will be uncertain if the middle 0 should be significant since it is to the left of the decimal (it is significant).

For 0.04 students may think the zero to the right of the decimal place is significant because it is to the right of the decimal. Instead, they must see that the decimal is present, and separately see that the 0 is to the left of all of the digits 1-9. Students tend to group those two separate assertions.

Density

Initially I was skeptical that density was too simple for chemistry. Now I find density to be a key lesson. The algorithms to use density are simple, but density has a wide number of representations that serve as a sturdy foundation for models in your classroom. You can represent density using an equation, proportions, particle diagrams, slope, units, and students can experience the phenomena of density in a variety of discrepant events. One fun demonstration is to have students predict whether a polystyrene sphere or marble is heavier based on feel. The sensation is based on the pressure on your skin, not the weight. The higher density of the marble makes it feel heavier even though the weight is much lower.

We want students to connect a variety of representations in chemistry, and density is one of the simplest ways to introduce that lesson. Try showing a wooden block floating in water. Then ask students what would happen to a wooden block that has large holes drilled in it. Will it float or sink? If it floats, will the percentage of the block that's above the water level change?

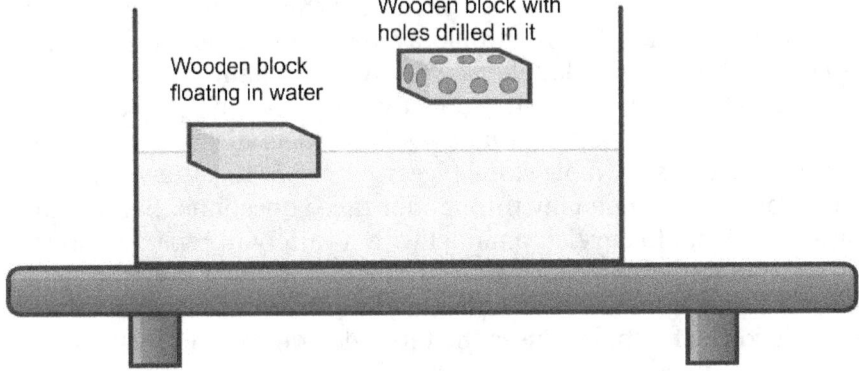

Figure 1-3: A wood block floats in water. If the block has holes drilled in it, will it sink or float?[9]

Chapter 1 – Chemistry Basics

The first density lesson should begin with students measuring mass and volume. You can even mislead them into thinking this is going to be a lesson on significant figures. Have them measure the same substance with different amounts. I have sets of blocks where each set has about 5 blocks of each material. Sand and water can be used although water is not ideal because its density is so close to 1 g/mL. Hardware made of the same alloy work using water displacement to find volume. They should measure a variety of volumes and masses and use those data points to construct a mass vs. volume plot. From the plot they should find a line of best fit in the form of y = mx + b. Optionally they could also draw particle diagram representations to accompany their plots.

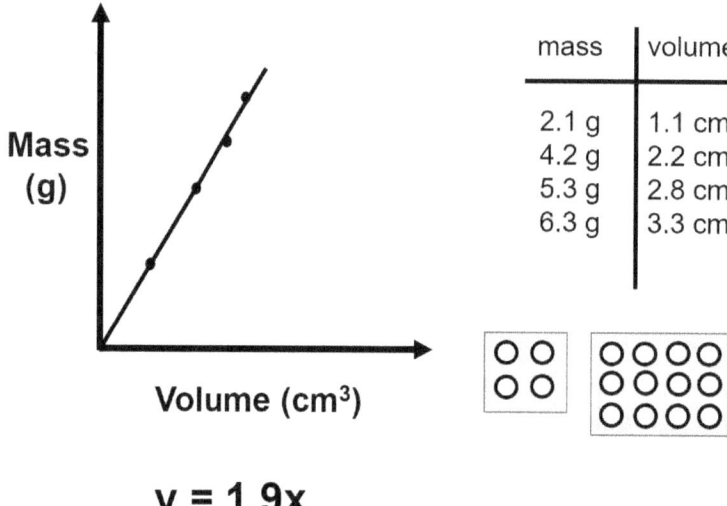

Figure 1-4: Sample whiteboard of experimental mass and volume data

From this plot and line of best fit a wealth of questions are available to ask students. What is the y-intercept on your plot? What does that y-intercept represent? Should the y-intercept be zero? Why is the slope constant? What are the units for the slope? What would a large slope mean? What would a small slope mean? If I double the volume, should I expect the mass to double as well? Are the changes in mass proportional to the changes in volume? Are they directly proportional? What would a particle diagram look like for a large slope? What would a particle diagram look like for a small slope? Are the particles for a big slope bigger, or closer together, or both?

You'll want to avoid the term density throughout these questions. Before you use that term, you want students to develop a model with a variety of representations for density and the term can undermine that. Many students who hear the term density will immediately think of mass over volume, and this will limit their ability to think through these questions. If a student uses the term, do your best to have them explain again without using the term.

Prior to introducing the term density, introduce "For every" statements to your class. A "For every" statement is a means to explain what slope represents. They must begin with the words "For every" and they must include the value of the slope in them.[10]

If we construct a plot of mass vs. volume for sand, the line of best fit will be something like y = 1.6 x. The "For every" statement would then read, "For every

Chapter 1 – Chemistry Basics

change in one of x, the y changes by 1.6." This can be improved to say, "For every one x, there are 1.6 y." Those statements are similar but have a key difference based on the inclusion of the phrase "change." The inclusion of change is needed when there is a non-zero y-intercept. But when the variables are directly proportional that is not needed since 0,0 is a point.

But we can improve more than this. We could also add in what x and y were. Then we would have, "For every 1 mL of sand, there are 1.6 g of sand." Think about the versatility of a well-crafted "For every" statement. You're setting up the students to do proportional reasoning, connect the slope to the equation for density, and you could easily make a visual representation for dual coding (Fig. 1-3). Perhaps the most popular perk is that you are setting students up to use their mathematical knowledge and number sense. If I have 1.9 g of sand, do I have more or less than 1 mL? Obviously, we have more than 1 mL, so if a student divides incorrectly their error is much more readily corrected.

Perhaps you can anticipate that the "For every" statement covers many topics in chemistry. Molar mass jumps out as a critical topic that we will discuss in chapter 4. One of the key lessons from "For every" statements is that students connect the term "per" in units to mean "per one." 70 miles per hour means you travel 70 miles for every 1 hour. 12 grams per mole means you have 12 grams for every 1 mole.

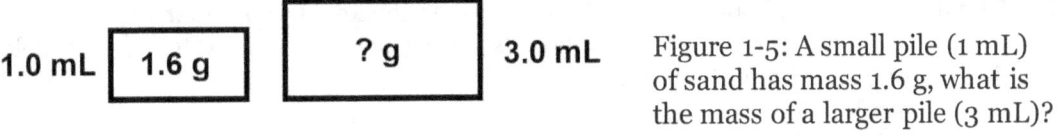

Figure 1-5: A small pile (1 mL) of sand has mass 1.6 g, what is the mass of a larger pile (3 mL)?

Once students have established the framework for density through a "For Every" statement, you can now introduce the term density. Teachers often stipulate that the purpose of vocabulary is for communication between experts. But what needs to be more explicit is that vocabulary is used to chunk ideas into a single retrieval term. This minimizes the strain on short-term memory that can only work with 7 +/- 2 terms at one time. But the most effective chunking occurs when students have made several connections prior to the introduction of the terminology. This is referred to as "concept first, vocabulary last."

Ask the students, "What part of their y = mx + b equation is the density?" The slope is the density. The slope has units of g/mL and represents the amount of mass in a given volume. If you have them plug in the variables for y and x you will get the equation **mass = density * volume**. For sand, the equation is
mass = 1.6 g/mL * volume. The density would apply equally to large and small quantities of the sand.

Providing students with the equation and definition at the beginning of the lesson will undercut the students' motivation to think critically about their data, their observations, and the conceptual components of density. Try to frontload thinking and confusion. Then as students ponder, gradually narrow the scope of focus.

At this point students are now ready to do problems. But let them do them. I assure you that they have sufficient mathematical abilities to figure out whether they should divide or multiply. And if they don't, they will need experience in figuring that out. Instead of working through a sample problem, let them do the math. Be prepared

Chapter 1 – Chemistry Basics

for some students to be disgruntled, but you want them to use their mathematical tools and not rely on copying algorithms provided by teacher examples.

Trying a new problem without a sample is a highly effective way to learn.[11] It provides students an opportunity to retrieve what they know, identify what they do not know, and work towards a solution. But teachers frequently undercut this learning opportunity for the sake of time or for fear of students struggling. Struggle can build competence and confidence if we teach students how to respond to it, but avoidance leads to anxiety when struggle happens.

For students to solve a problem on their own; they must read the problem, analyze what they know, analyze what they do not know yet, and develop a plan to find the unknown information. After completing a problem, they should evaluate and reflect on their solution and process. Don't wait for the chemical symbols to become more complex before students figure out this process. It will take them practice, but that makes density the perfect starting point for them to learn how to solve problems. After students find solutions on their own, have them provide feedback to other students based on how they constructed their solutions.

There are several good extensions for density. The density of aluminum metal is 2.70 g/mL. This can be used to determine the thickness of aluminum foil. Students can compare the thickness of regular and heavy-duty foil. The results can be used to set an upper limit on the size of an aluminum particle (we know the foil has to be at least 1 particle thick). Heavy duty is about 1.5 times thicker than regular duty.

The relative densities of solids, liquids, and gases can be used to infer particle spacing differences. A simple means to measure the density of a gas is to displace water from a graduated cylinder using a butane lighter. The graduated cylinder should be submerged upside down into a bucket of water. The atmospheric pressure will hold the water in the cylinder. Then the butane from the lighter can be bubbled into the cylinder to find the volume. The loss of mass from the lighter provides the mass of the gas. While the density of a gas is highly variable, this can still provide evidence to show that gases have much larger spacing between particles than solids and liquids. The misconception that liquids have spacing in the middle of gases and liquids will take multiple attempts to deconstruct.

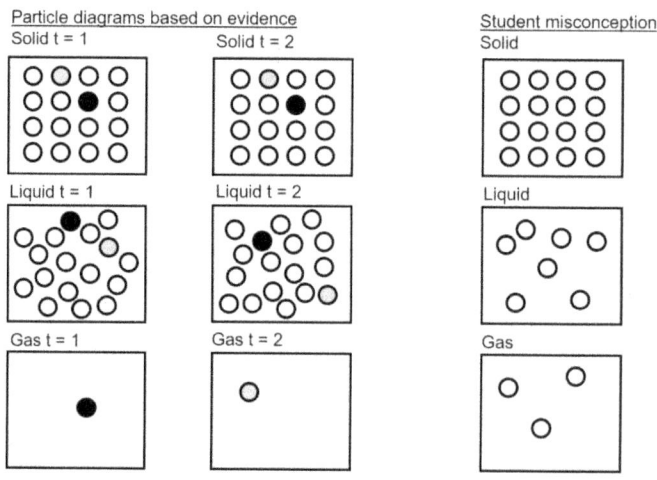

Figure 1-6: Particle spacing of solids, liquids, and gases based on density evidence. Students hold misconceptions shown on the right.

Chapter 1 – Chemistry Basics

States of Matter

Students enter chemistry with poor conceptions about solids, liquids, and gases. They often picture particle spacing that has solids touching, gases spaced apart and liquids as being in between the two. They also will not know that most objects are not a solid, liquid, nor a gas. Only pure substances (elements and compounds) should have those states of matter assigned. Mixtures are either solutions, colloids, or suspensions. Most objects are suspensions that have properties of at least two of the states of matter.

A common internet meme claims that cats are a liquid. The justification provided is that a cat takes the shape of its container. Teachers should be wary of using containment as a means to assign states of matter. What happens at the particle level is the best means to assign phases. If the particles are locked into a position, the substance is a solid. If the particles are in contact but can change positions, the substance is a liquid. If the particles are not in contact except during collisions, the substance is a gas. This sets up for a better understanding of suspensions where particles are doing all those things. A cat has solid, liquid, and gaseous parts. A cat is not a single phase. Please stop tagging me with this meme on social media.

Many teachers use slime to evaluate student understanding of states of matter. Is it a solid or a liquid? There was even a Wendy's commercial that asked if a Frosty was a solid or a liquid before claiming it was a "soquid."[12] Many of these discussions give little resolution to students because they never reach a conclusion that is useful. Teachers present these to get the students thinking, but they do not reach a landing point. A landing point is when students mostly find a consensus that they agree upon and feel secure in their understanding for the time being. Instead, poorly developed conceptions are reinforced. Students might have attended a science show where the presenter freezes a banana in liquid nitrogen. You can't freeze a banana though. If we had to assign a single state of matter to a regular banana, it would probably be a solid. But we can't because a banana is a complicated suspension. When the banana is cooled down, the banana becomes more brittle, and the structure of the particles become more rigid.

Ice cream is a little bit trickier because a substantial portion of the ice cream is water. The water would be capable of freezing and melting if it were pure. The ice cream goes through changes that are similar to melting and freezing. But the structural changes are more complex. There isn't a single temperature where freezing and melting occur. As the ice cream becomes more rigid there can be changes in the mixture where separations occur.

All examples that involve states of matter in your class should be pure substances. You want to give your students the potential for closure by acknowledging that they have likely seen examples that are inappropriate. Talk to them about some examples of suspensions that are not a single phase and let them work out why it might be inappropriate to say that a tree is a solid.

There are some mixtures that can be assigned a single phase. Brass in a tuba is a solid (at typical temperatures at least). Air is a gas (ignoring things like smoke or clouds). Juice is a liquid (ignoring pulp). But carbonated beverages are not as simple. They have gases temporarily suspended in them. Keep in mind that many students have not experienced a liquid that is not water or a solution with water. Some will know gasoline or oils, but if you have a sample of bromine show it to them. They can

Chapter 1 – Chemistry Basics

use the exposure to more pure liquids. Gallium will melt in your hand since its melting point is just below body temperature.

The teacher should also be aware of one of the most common misconceptions that exists in chemistry. Steam is invisible always. You cannot see steam. It does not interact with visible light. When you see a *"cloud"* arising from hot soup, or you see your *"breath"* in the cold of winter, that is not steam that you are seeing. There is steam there, but what you are seeing is actually tiny drops of liquid water.[13] Some people will hear that and calmly nod along saying, "Ahhh, yes. It is condensed water vapor." Many of those people do not think critically about that and realize that condensed means liquid. Hence why you only see your breath or exhaust from cars when it is cold out and the steam can condense into tiny droplets. A cloud is also tiny droplets of liquid and not a gas.

Many examples in worksheets show gases by showing a cloud-like substance or wavy lines coming off of hot soup. There is a demonstration that can be effectively used to counteract this notion. A coil of copper wire that runs through a rubber stopper called "superheated steam" can be purchased (or constructed if you are talented). The rubber stopper fits into a flask that is filled with water that is heated to a boil. The steam goes through the copper piping to the coil, around the coil, and out the other end. When the steam emerges, it cools as it expands and forms the liquid droplets, and a cloud is visible. But the coil is separated in a way that it can be heated using a Bunsen burner. When the steam gets much hotter it emerges from the copper so hot that it cannot quickly form liquid droplets. The cloud disappears (Fig. 1-7). But the hot steam can be used to ignite a match (or flash cotton), or to char paper.

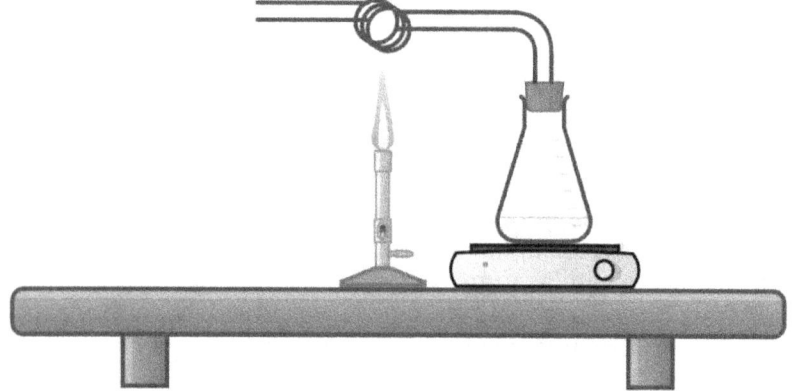

Figure 1-7: Superheated steam demonstration. Steam rises from boiling water in a flask. The steam travels through copper tubing where it is heated, and the cloud disappears as the steam remains in the gaseous state.

The first unit taught should address the misconceptions about particle spacing for liquids, and the misconception that impure mixtures have a single state of matter assigned to them. The introduction into states of matter can also delve into more complex ideas. In this book, these will be addressed in future chapters such as thermochemistry and bonding.

Chapter 1 – Chemistry Basics

Chemical and Physical Changes

Chemical and Physical changes is a controversial topic in chemical education. In my personal opinion (most disagree), there is little to gain from teaching this topic and it should be removed from curriculums. There is no long-term benefit from the topic as it does not warrant consideration at higher levels of chemistry. But there is also a short-term conflict in that the information presented is inconsistent and does not work toward understanding.

Many teachers disagree on how to characterize dissolving. According to most texts dissolving is considered a physical change. But dissolving of ionic substances results in a coordinate covalent bond forming, can result in color and magnetic property changes, and fits the criterion for a Lewis acid-base reaction. Many point to the easy reversibility of dissolution, but many chemical reactions are reversible with the same ease. In fact, there are many acid-base reactions that are easily reversible, and many dissolutions are acid-base reactions.

I question the understanding gained from categorizing dissolving as either chemical or physical. It would be better to delve into the particle level characterizations of what is happening and/or the electronic changes that guide those changes. And these are better developed in a unit on chemical reactions or acid-base theory. If the phrases chemical and physical changes did not exist, would we lose any understanding?

Many textbooks assign phase changes to be a physical change, and yet ionic substances melting involves the breaking of ionic bonds. This is confusing to students who are not yet sure what bonds are and are presented with inconsistent correlation between bond breaking and chemical/physical designation. Bending a piece of metal involves the breaking of metallic bonds but is a physical change. Burning fuel involves bonding changes but is a chemical change. These later can interfere with students understanding of bonding. The majority of chemical reactions involve chemical bonds being formed and broken. There are a few where bonds are either only formed or only broken (e.g., $Cl_2\ (g) \rightarrow 2Cl\ (g)$). But most examples of physical changes shown to students are when intermolecular forces are changed. Teachers are more likely to present ice melting rather than lead melting.

Some characterize chemical changes by saying that a new substance is formed. This is problematic because different states of matter of a substance can vary wildly. Liquid zinc chloride conducts electricity while solid does not. Steam is invisible, but tiny drops of liquid water form clouds that can be seen. Even surface area within the same phase can have a dramatic impact on the reactivity of a substance. Steel wool burns with a shower of sparks emanating from it. But large pieces of iron rust at a slow rate that is hardly visible. Both steel wool and large pieces of iron have the same property of reacting with oxygen while they are in the solid state. But is it accurate to claim that their chemical properties are identical?

Teachers might give general guidelines for macroscopic observations that can help distinguish the two. If light is emitted, a color change occurs, or a gas is produced are signs of a chemical change. But these are inconsistent. Many physical changes produce light, change color, or produce a gas. Many chemical reactions do none of these. Some use the guidelines of reversibility to distinguish the cases that are not obvious. But again, many reactions are reversible, and many acid-base reactions in particular are easily reversible within a short time frame. Not all physical changes are easily reversible.

Chapter 1 – Chemistry Basics

Some teachers will claim that these objections are the entire basis for this unit. They want to produce these discrepancies so that students can think hard and learn. But I would contend then that these discussions fall prematurely. Typically, this unit comes early in the year, before students are ready to have discussions that are productive and reach a landing point. A student that does not know what an acid, a dispersion force, or a covalent bond is will struggle to produce an analysis that is fruitful. *Can students connect their ideas at a particle level?*

Furthermore, there is no underlying objective based on the original topic. If you want to discuss the differences between steel wool and iron rusting, do we need the chemical change distinction to have that conversation? If we want to look at the differences between melting and combustion at the particle level, do we need to distinguish them as physical and chemical first? I would argue that there is much more important prerequisite information required to have a productive discussion on these that have nothing to do with categorizing by chemical or physical.

Part of what drives the controversy behind this unit is that teachers enjoy teaching it and students like to learn it. There are some incredibly fun demonstrations that can be done. Experiments for students might be incredibly engaging. But is this unit needed to do those experiments, discussions, or demonstrations? I doubt it. Perhaps asking students if a tattoo is a chemical or physical change results in high levels of engagement. Why not revise the question to what do you think getting a tattoo looks like at the particle level?

If you are intent on teaching chemical and physical changes, poison can be a unifying theme. Hypothetically, you have a vial of poison, but you can make one change to it before you consume it. What would you do? Would you melt the poison and drink it? Would you cut the poison into two piles and consume the two piles? Would you dissolve it in water? What if you burned the poison and breathed in the gases produced? The physical changes are not going to help you. The chemical changes might, or they might not.

Then you can work through some examples. Hydrogen peroxide is toxic. But if we add a catalyst or UV light to it, the hydrogen peroxide decomposes into water and oxygen. This can be demonstrated using the elephant toothpaste demonstration.

$$2H_2O_2\ (l) \rightarrow H_2O\ (l) + O_2\ (g)$$

Methanol is an interesting poison to talk about because of the prohibition era. During prohibition bootleggers would steal methanol to use in place of ethanol. Both ethanol and methanol are toxic[*], but methanol is a bit worse. But the government would add poisons to methanol knowing that it would be stolen and used for drinks as a means to punish those who knowingly broke the rules.[14] Many were killed. If I had to consume some methanol, I could react it with oxygen first (burn it) and change it into CO_2 and H_2O that would be substantially less toxic to consume.

Even heavy metals such as lead, or mercury can be altered in their toxicity. Mercury ions (e.g., from Mercury (II) nitrate) are absorbed at much higher levels than mercury metal. When mercury has an organic group attached the impact on the

---

[*] *It is good to call alcohol a poison when talking to young students. The impact on their brain development is larger at their age and efforts should be made to limit their alcohol consumption until they are older. Marketing often presents alcohol as fun while downplaying the negative pieces like addiction, brain damage, or death.*

Chapter 1 – Chemistry Basics

body changes again. Consuming seafood will often help your brain with omega fatty acids, but it will also harm your brain with mercury. The net tradeoff is in your favor, and you can make this even better by being selective about which type of fish you consume. Larger fish such as tuna have higher concentrations of mercury. The correlation with fish size and mercury concentration is due to the fact that mercury bioaccumulates and that large fish must consume more food to survive. More food leads to building up higher mercury levels.

There are many other poisons that would work well for this purpose. The key is that you are showing examples where a chemical change alters the toxicity. If I melted mercury (II) nitrate the toxicity would not be affected much. If a student digs deeper, they will likely note that phase changes do result in changes in the LD_{50} (the amount per kilogram of body weight that gives a 50% chance of death). But the use of poison will give a more consistent distinction between chemical and physical changes will leading to higher engagement of the students.

Student Struggles

1. "I have no idea what I'm doing!" When students are lost, they often do not know how to describe what they need to progress forward. They often utter a phrase similar to this one as a means of expressing either their frustration or their apathy. When we encounter this phrase, we always want to encourage students to find something that they know and understand. Students are seeking to offload their discomfort, but an explanation is rarely the appropriate solution. Their prior knowledge is the beginning of all learning. Every "Student Struggles" will begin with this comment.

For this unit, the most likely issue that would lead to such a statement is rounding significant figures. We expect discomfort for students using significant figures. The student should only look at significant figures at the start and end of the problem. Next, we want them to put effort into distinguishing measurements as ranges, from numbers as points. We want them to see that 5.3 cm is different than 5.3 the number. 5.3 is a specific point. 5.3 cm is a range such as 5.2-5.4 cm. How those differences manifest in calculation will come with time if we continue to differentiate measurements from numbers.

2. "It makes sense when you do it, but I can't figure it out on my own."
This is a common refrain that chemistry teachers hear. When a teacher works out a solution for students, the teacher has a strong knowledge of the pieces that allow them to make decisions on how to start, critical components of the solution, and the sequence of calculations to find a solution. When a student struggles with that process, it is common to direct them to watch the solution again. But the student is going to view what the teacher presented differently than how the teacher does. The student is a novice, and the teacher is an expert. The teacher sees the sequence differently because of their wide range of content knowledge that supports how they view the algorithms they use.

To help a student with this, they should work on identifying variables and connecting them to a concrete example. Can a student identify approximately how much 5.2 g of something is? Do they know how big a millimeter or a cm^3 is? Students should also verbalize their understanding of units. The "For every" statements can provide a strong framework for students for units of a density like 13.2 g/mL.

Chapter 1 – Chemistry Basics

3. If a student has erratic math shown. It is common for teachers to blame a lack of math skills on an inability to do chemistry. But research is clear that "skills" tend not to translate between content areas. If a student is doing well with mathematics, that can often mean that the student is not learning the chemistry, but instead is treating the chemical components of problems as an abstract concept to use in a math problem. When a student struggles with organizing the mathematical details, they usually have a lack of understanding of the chemical concepts. A student that multiplies mass and density to get volume is more likely to not understand what density is than to not know how to do an algebraic manipulation.

4. *"How do I write a measurement that is exactly on the line?"* A student uses a ruler that has markings at 9.9, 10.0, and 10.1. What should they write down if the measurement happens exactly on the 10.0 line? Two things can help this student. The first is that the precision of an instrument is constant. The instrument should always have the same final decimal point. You cannot get measurements of 9.99 and 10.0 from the same instrument. The second helpful tip is to pretend the measurement is slightly off. If it's halfway between the 10.0 and 10.1 mark, we would write down that the measurement is 10.05 cm. Since 10.05 has two decimal places, being exactly on 10.0 should be recorded with two decimal places as well (10.00 cm).

5. *"I can't remember the rules for scientific notation."* If a student is adding $2x10^4 + 3x10^3$ they should get $2.3x10^4$. Multiplying the two values would give $6x10^7$. The patterns are clear when we have the answer. But when a student starts the problem later, they might forget what those patterns or rules were. Students should not just be memorizing rules for manipulating scientific notation. They should also understand that $2x10^4$ is equivalent to $2x10,000$ or $20,000$. Students know how to add $20,000 + 3,000$. As teachers, we should be using their math knowledge that is strong and connecting scientific notation to the rules they are excessively familiar with.

6. A student draws the particle level of a solid, liquid, and a gas. The gas particles are spaced far apart, the solids are in a lattice structure, the liquid are in between the two. The evidence shows that particle spacing in liquids and solids is nearly identical. Solids, liquids, and gases should not be thought of as small, medium, and large spacing. Rather the solids and liquids should be considered one group (condensed states) while gases differ. The evidence for this are the relative densities. The density of liquid mercury is about 13.5 g/mL. The density for solid mercury is about 14.2 g/mL.

This misconception has been built by inadequate textbook drawings in earlier grades. It is common for students to see particle spacings that vary inappropriately. There is also some underlying complexity that teachers don't always see. For example, when we discuss this progression, we are often maintaining a single substance. But how does the spacing of liquid mercury compare to liquid water? Until we tie this into the evidence provided by densities, the student has a gap in knowledge. These gaps cause hesitancy that allows misconceptions to fester. We'll see a similar misconception soon where students think that gases are fast, liquids are slow, and solids are either very slow or stationary. Water particles move at higher

Chapter 1 – Chemistry Basics

speeds in the liquid state than air particles in the gaseous state at a given temperature (e.g., room temperature).

7. "The solid starts to melt at this temperature, then when the temperature goes up even higher, it melts faster." Students struggles with melting point stem from a different issue. Students are taught that mixtures can be solids and liquids. Ice cream does not transition from a solid to a liquid. It's a mixture. It is inappropriate to describe a heterogeneous mixture as a single phase. A block of wood is not a solid. It has a variety of components including gases mixed in. In chemistry we want to correct this so that students can understand why ice cream does not melt at a given temperature. Ice cream has a variety of components that produce a range of temperatures where the viscosity varies. That doesn't happen for a pure substance (element or compound). There are a plethora of memes of cats being described as liquids that can help initiate this discussion. Does a cat have liquids? Yes. Does a cat have gases? Yes. Does a cat contain solids? Yes. The cat is a mixture of those things, and it is inappropriate to reduce that complex mixture to a single phase.

8. "I don't get why melting zinc chloride is a physical change." There could be underlying issues with misconceptions about melting here, but more generally students struggle to identify changes made to chemicals as physical changes at times. Part of this stems from being exposed to a lot of chemophobic propaganda. There is deceptive advertising from grocery stores, restaurants, cosmetics, environmental activists, schools, and many other trustworthy entities that students have experienced throughout their lives. One push is that chemicals are bad, and natural things are good. This can easily be reinforced because there are obvious anecdotes to both of those statements. This can be an opportunity to begin helping students understand that not all chemicals are harmful, that the amount of chemical matters, and everything is made out of chemicals (except light and a few other irrelevant technicalities). The Poison Squad is an excellent book and story that can help us understand why we went in this direction and also provide a pathway to help people understand more and fear less.[15]

Phenomena

1. Heavy-duty aluminum - Students can use the density of aluminum to determine thickness of the foil. They can compare regular duty and heavy-duty foils. This allows them to speculate what the differences between the two foils might look like at the particle level.
2. 1 mL = 1 cm³ - There are plastic cubes that can be purchased that have side lengths of 1 dm or 10 cm. These would therefore be 1 dm³ or 1000 cm³. A graduated cylinder with 1000 mL will fill the cube precisely to the brim without spilling. The cylinder appears much larger to students, which makes for a dramatic discrepant event.
3. Will it float? - A wooden block with holes in it will float just like a typical wooden block. In many classes, the predictions for what will happen to the wooden block with holes will nearly split down the middle between float or sink (Fig. 1-3).
4. Speed limits - When discussing significant figures ask students what the implications for significant figures would be for speed limits. A 70-mph speed limit

Chapter 1 – Chemistry Basics

could have a wide range of reasonable minimum speeds that could be enforceably ticketed. Could a ticket be written for 70.1 mph? Is everything under 71 mph acceptable? Or should the 70-mph limit be considered a measurement in which case the tens place provides us with a wide range of values that could be within the limits range of measurements? Disclaimer, do not field test this question. But if anyone knows the law behind this, I'd love to know what basis significant figures are given according to the text of the law.

5. Noise! - Have the students make an imprecise measurement. Maybe assign one wall of the classroom to be 0 and the other to be 1000 units. Then have the students individually write down what they think a point in between (but not exactly in the middle) would be. While most students will be incorrect when measured precisely, the average of the students' responses should be similar to the actual value. Each individual response contains what we call "noise" and that random noise balances as more and more values are taken.[16] This is a reality in measurement whether the measurement applies to science or not. This could produce a thought-provoking question for students, "How could we communicate the amount of noise in a measurement when we write down the final average result?" This can help students understand that measurements differ from numbers in that measurements are representative of a range and not a single point. This can help students establish that significant figures are critical for properly communicating measurements in a science classroom.

6. Which is heavier? - Have a student hold a large polystyrene sphere in one hand and a marble in the other. Ask which feels heavier. If you choose properly, you can get the student to sense the marble as being heavier even when the polystyrene sphere is much larger in mass. This showcases that our perceptions are not always in alignment with reality. In this instance, the polystyrene sphere has more area of contact which reduces the pressure. The student's sensation is based on pressure, not weight. Therefore, they experience a discrepancy with reality.

7. Number line - Put a clothesline up in your room and place notecards on it that have a 0, 1, and 10. Space them appropriately. Task students to put up 10^0, 10^1, 10^{-1} on the number line where they feel is appropriate.

8. Sig fig boards - Create significant figure boards (Fig. 1-2). Take three wooden boards and put the same colorful dot at the same location (not in the middle, but near the middle) on each. For board A, mark 0 and 100 cm with lines and nothing else. For board B, mark every 10 cm. For board C, mark every 1 cm. Have students measure the dot using board A first, then board B, then board C. Note how their measurements change as more lines are introduced.

9. Best conductor? - Silver is the best conductor for elemental metals.[17] Copper is nearly as good but is much cheaper. But we compare conductivity for metals by using common radii for the wires used. Aluminum has a conductivity that is about 56% of that of silver and 59% of copper. But how would aluminum's conductivity compare on an equal mass basis instead of equal volumes? Aluminum has a density of 2.7 g/mL while silver is 10.5 g/mL and copper is 9.0 g/mL.

10. More liquid - Show students samples of liquids that are not water or solutions with water. A surprising number of students will not be able to generate non-water liquids when pressed. An interesting anecdote is that once my children were making an ice candle at a science museum. Molten wax was poured into a cup of ice that resulted in a candle with holes where the ice was. When asked to describe what was

Chapter 1 – Chemistry Basics

happening, a young child responded that the water was turning into wax. The presenter assumed that the child was referring to the ice changing into wax but watching the child's reaction it was clear that they instead thought that the molten wax was water. They did not have a concept of non-water-based liquids. Bromine liquid can sometimes be purchased in a sealed glass vessel. Gallium can easily be melted using body heat. Organic compounds can be shown if safe in the classroom.

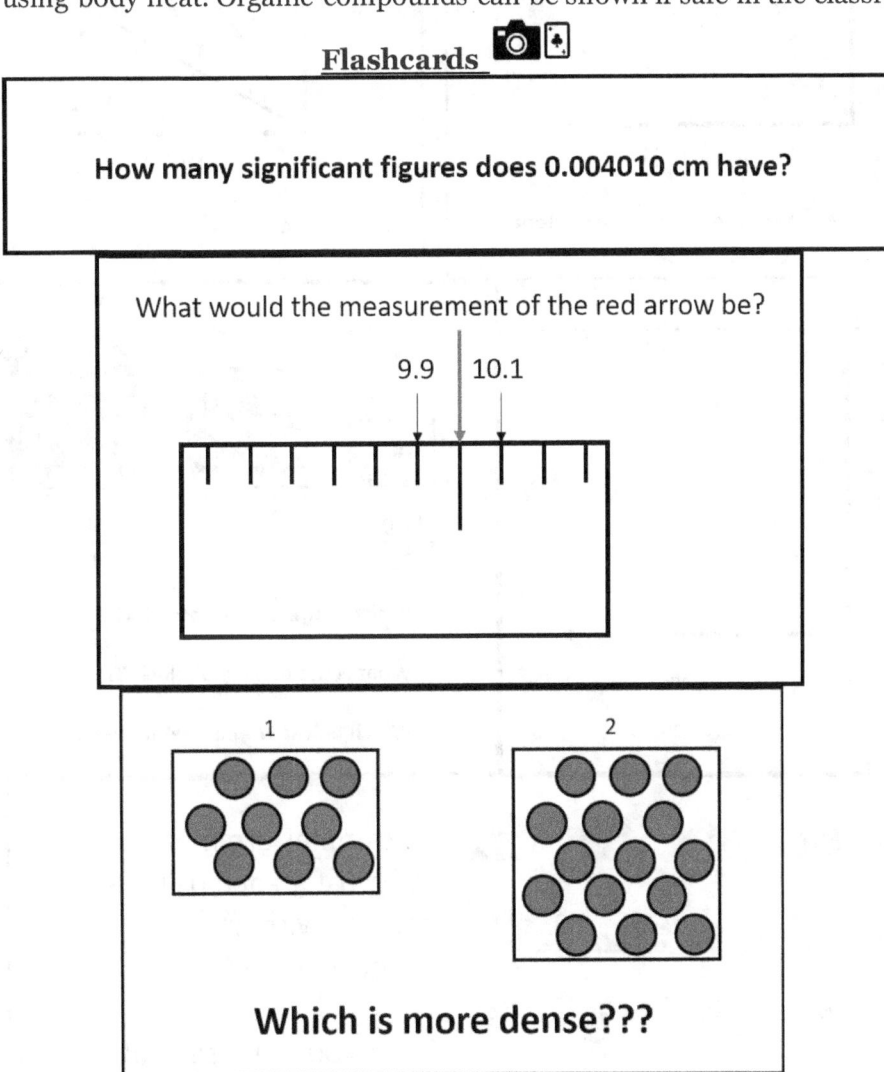

Chapter 1 – Chemistry Basics

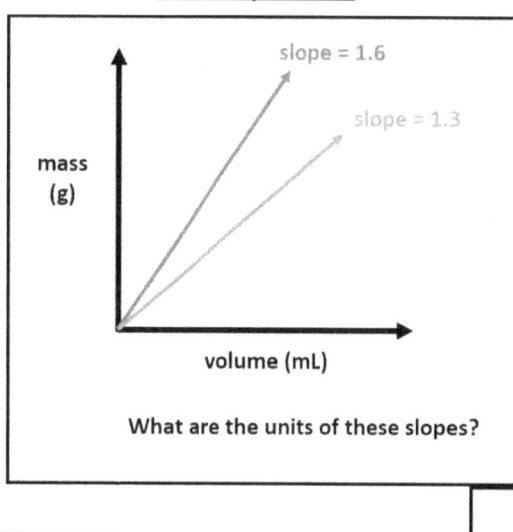

What are the units of these slopes?

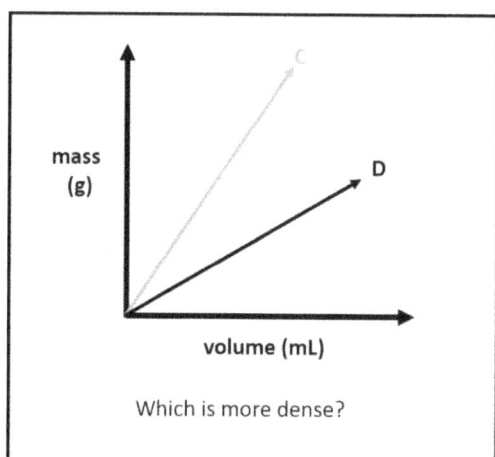

Which is more dense?

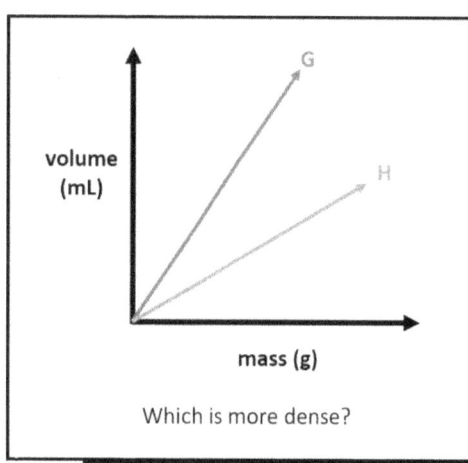

Which is more dense?

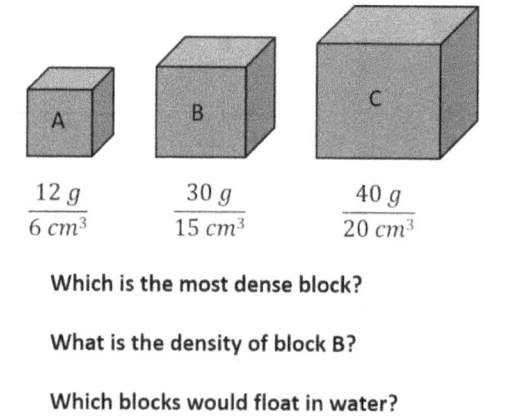

Which is the most dense block?

What is the density of block B?

Which blocks would float in water?

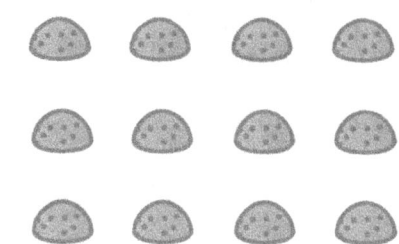

50 samples are measured to be 0.1 g each using a balance with an uncertainty of 0.1 g.

Why would the sum of these measurements not be 5.0 g?

Chapter 1 – Chemistry Basics

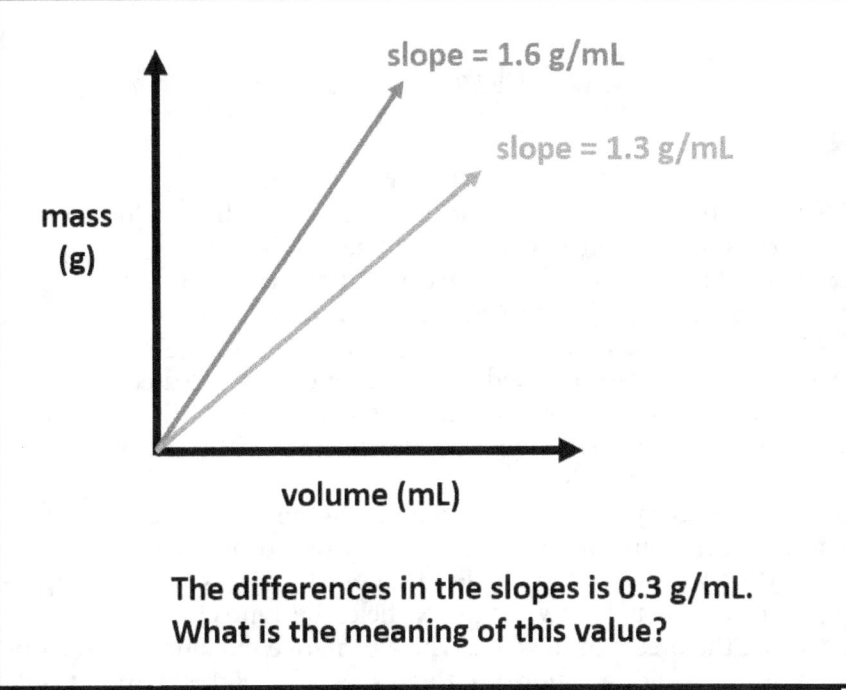

The differences in the slopes is 0.3 g/mL. What is the meaning of this value?

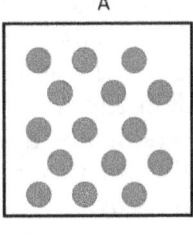

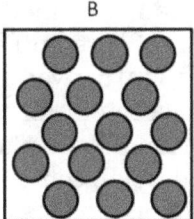

Will the results of these two calculations be the same or different?

Why?

A　　　　　B

Which is more dense???

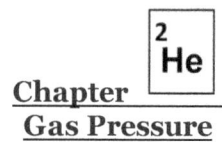

Chapter
Gas Pressure

History

Jan Baptist van Helmont (1580-1644) was the first to focus experiments on gases. He even coined the term gas.[1] His studies lacked precise quantitative details, but his work is the emergence differentiating "types of air." Priestley, Scheele, Lavoisier, Cavendish, and more used these "types of airs" (such as fire air, fixed air, etc.) to uncover hidden aspects of chemistry.

Robert Boyle (1627-1691) used Van Helmont's work to do quantitative research about the pressure of gases as changes were made. Boyle was the youngest child of 14 children, and his parents were extensively wealthy. His wealth allowed him to access lab equipment and education.

Boyle studied gas pressure as the volume of the gas varied. The relationship he found was that pressure and volume are inversely proportional just as the law that carries his name now states. Isaac Newton proposed that gas particles repelled each other using his inverse square law. As the particles became closer the repulsion increased causing the pressure to go up. The alternative explanation was that the particles were moving and in a more confined space the particles would collide more frequently.[2]

Jacque Charles concluded (but never published) a directly proportional relationship between volume and temperature of a gas at constant pressure (1780). This was over a century after Boyle's law in the. Charles's work was confirmed (and published by) Joseph Louis Gay-Lussac (1802). Gay Lussac published about pressure and temperature (1808).

Gay Lussac also published experiments showing that volumes of gases combine in simple whole number ratios, but it was Amedeo Avogadro that hypothesized (1811) that there would be equal numbers of gas molecules in equal volumes (at constant pressure and temperature). André-Marie Ampère published the same thing (1814) but had a considerably larger scientific presence that led to him receiving credit for the hypothesis for some time.

Notice how the progression of our understanding was sequenced. First scientists would compare two variables while holding the others constant. Then Benoît Paul Émil Clapeyron combined these into a single equation (1834). The ideal gas law, $PV = nRT$, was first based off empirical data. Finally, August Krönig (1856) and Rudolph Clausius (1857) derived the ideal gas law from physics principles.

The sequence we see in history works well in education for the same reason. We begin by helping students understand the relationships between each variable and pressure. Next, we develop connections for the kinetic molecular theory with the particle level representations and experimental data. Eventually we build to the point where we can describe a single state of a gas instead of just predicting how changes will correlate proportionally.

You should also note how the progression of dates corresponds with the qualitative descriptions of gases. Joseph Priestley's experiments to determine the "types of air" were mostly in the 1770s and 1780s. Antoine Lavoisier expanded on the experimental evidence and developed a systematic nomenclature system in 1787.[3]

Chapter 2 – Gas Pressure

After these shifts in knowledge and thinking, we see a more rapid progression forward.

Proportional Relationships, PVnT

Gas laws can appear to be a simple topic for many students. There is frequent success with doing calculations. The last few years I have been teaching gas pressure differently, and I would recommend the following change even if you are currently satisfied. Remove all of the equations and teach gas pressure using proportional reasoning.

The first thing you might do in this unit is introduce students to the variety of units that pressure can have. Ultimately pressure is a force on a given area. I talk about when I held a job as a phlebotomist and the implications of using a blunt needle (large area) vs. a sharp needle (smaller area). They get the idea pretty quickly.

If you have sensors that change units, have them measure standard pressure in the room in the various units. They will likely produce something such as the following:

1 atm = 100 kPa (formerly known as 101.3 kPa) = 100,000 Pa = 14.7 psi = 760 mm Hg = 760 torr = 1 bar

Trim that list down for students. They don't need 6 or 7 different units. It's worthwhile to give them some context. Otherwise, you'll find some unusual interpretations about what they think mm Hg is. In America you can give a more concrete connection to 14.7 psi since some students are more familiar with pounds and inches than Newtons and meters. Some students will have had experience with tire pressure, they can calculate pressure from ice skates on ice, and you can even purchase a 14.7-pound square steel bar that has 1-inch by 1-inch dimensions. The mm of Hg unit is the historical units since Hg was used in manometers and barometers. Atmospheres are the favorite for many students because of the simplicity of the number 1. Pascals are the standard unit (N/m^2 = 1 Pa).

From there we start our unit by doing experiments that seek relationships between pressure and volume, pressure and temperature, pressure and amount (we do not cover moles until later), and volume and temperature. In the experiments, we create whiteboards that display a particle level, a graph of the data, the data points, and a line of best fit equation. We use the line of best fit to determine if the variables are directly or inversely proportional. We want to stress what is being held constant in each experiment.

Pressure and temperature have a direct relationship but are not directly proportional. We know this because at 0 °C, there is still a large amount of pressure. This means that pressure increases as temperature increases, but that if we double the temperature from 20 °C to 40 °C, we won't double the pressure.

Chapter 2 – Gas Pressure

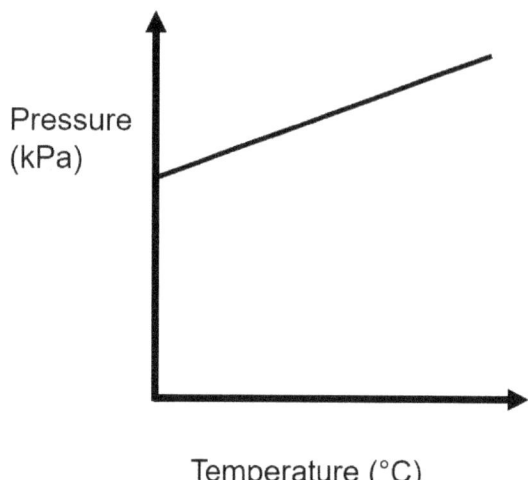

Figure 2-1: Pressure and temperature (°C) have a direct, but not directly proportional relationship

If we were to go to negative temperatures the pressure would continue to decrease in the same linear fashion as long as the gas does not condense. The x-intercept would be -273 °C. If we start a new set of axes at this intercept, we could construct a different temperature scale (Kelvins) that would have pressure and temperature as directly proportional (Fig. 2-2).

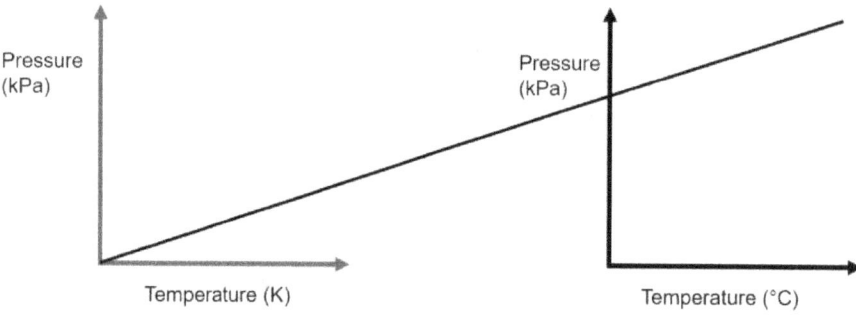

Figure 2-2: At a temperature of -273 °C the pressure is 0. If we start a new temperature scale here, temperature and pressure will be directly proportional.

What I am proposing is that we change -273 °C to be 0 K. In fact, any temperature in Celsius can be converted to Kelvins by adding 273. This would leave us with the same slope, but now when we double the temperature from 20 K to 40 K, the pressure would double. If we go back to 20 °C changing to 40 °C, this would be a change in Kelvin from 293 K to 313 K. This is an increase by a proportion of $\frac{313}{293}$ or 1.0683 and we would expect the pressure to increase by that same proportion.

The reason why we use the Kelvin temperature scale with gases is because 0 K has 0 pressure. Since pressure derives from collisions, this absolute zero implies zero particle motion (ignoring technicalities for the moment). By providing students with a justification for using the Kelvin scale, they will be more likely to remember to use it. The Kelvin temperature scale greatly simplifies the analysis of gas pressure, because it makes pressure and temperature proportional.

Chapter 2 – Gas Pressure

We do not discover what a mole is until later, so we use "puffs" of air to find the relationship between amount of gas and pressure. A puff represents a set number of particles where we often do not know the exact number of particles in each puff. We inject 5 mL puffs of air into a flask. Our flask starts with about 100 kPa of pressure. Every 5 mL air puff added causes a small increase in pressure. If we add two puffs the pressure increase is double than the increase from one puff. When we add three puffs the pressure increase is triple. When we find our line of best fit, we find that the number of 5 mL puffs of air that we would have to remove is approximately 27 or 28. This is about 135-140 mL of air, which happens to be the volume of our flask. From this we conclude that amount of gas and pressure are also directly proportional.

Next, we measure the pressure of a sealed syringe at a variety of volumes. This experiment involves some math. As the volume of trapped air gets smaller, the pressure gets bigger. This is an inverse relationship. *But how can we tell if it is an inversely proportional relationship?* That would mean that when we double the volume, the pressure halves. When we change the volume from 9 mL to 3 mL, the pressure would triple. The data values show the inverse relationship. We also get line of best fit equations that are somewhat close to P = 800 V^{-1}. This is volume raised to the negative one power, or the inverse of volume.

$$V^{-1} = \frac{1}{V} \text{ or } 800 \, V^{-1} = \frac{800}{V}$$

A good exercise to pursue is to plot 1 divided by volume vs. pressure. If the volume is 10 mL, plot 1/10, or 0.1, instead. If the relationship between P and V is inversely proportional, the relationship between P and $\frac{1}{V}$ should be directly proportional (we should have a linear function with a y-intercept of 0).

Trial	Pressure (kPa)	Volume (mL)	1/V (mL^{-1})
1	400	3	0.333
2	200	6	0.167
3	100	12	0.083
4	66.7	18	0.056
5	50	24	0.417

Table 2-1: Pressure vs. volume vs. $\frac{1}{V}$ data for a gas sample

The last experiment is with volume and temperature. Students might not have realized this earlier, but now it is worth pointing out that in each of these four experiments, two things were varied and two were constant. So, if we are looking for the relationship between volume and temperature, the pressure and amount of gas must be constant. If you're low on time you can save a little bit by just providing data or doing a demonstration for the 4th experiment.

To demonstrate take a flask and put a one-hole stopper in. Then heat the flask in a boiling water bath. The air inside the flask can move in and out of the hole. Let the temperature stabilize. Then plug the hole and invert the flask into a source of cool water. Some water will enter the flask. The amount of water that enters can be used to

Chapter 2 – Gas Pressure

figure out the final volume of the air. As long as the temperatures are converted into the Kelvin scale, the volume and temperature will be directly proportional.

$$P \propto T \text{ (in Kelvins)}$$
$$P \propto n$$
$$P \propto \frac{1}{V}$$
$$V \propto T$$

We do not need equations to determine how these variables will change. We can use proportional reasoning. The students do not need Charles's Law, and they do not need to learn about Robert Boyle (at least for equation purposes). Instead, we use charts to set up proportions. Let's say that you have a gas with a pressure of 1.2 atm and a volume of 28 mL. If the volume changes to 84 mL, what will the new pressure be (assume other variables are constant)?

For our students, they would identify that the volume has increased. Pressure is inversely proportional to volume according to our experiments. This means that the pressure must have decreased. It also must have decreased by the same proportion as the volume change ($\frac{84}{28}$ or 3). If the volume increased by a factor of 3, the pressure must have decreased by a factor of 3. Our new pressure must be 0.40 atm.

The algorithm our students use is to multiply the initial value of the unknown variable (1.2 atm) by the proportion of the variable that has two quantities (28/84).

$$1.2 \; atm * \frac{28 \; mL}{84 \; mL} = 0.40 \; atm$$

Note that we put the smaller volume on top, since the effect on the pressure of volume getting bigger is that the pressure must get smaller. Because P, V, n, and T are all proportional, the only decision students must make is whether to make the proportion bigger or smaller than one. We use IFE or PVnT charts to organize this.

	P	V	n	T
Initial	1.2 atm	28 mL		
Final	?	84 mL		
Effect		↓		

Figure 2-3: PVnT (or IFE) chart for the change in pressure when a gas changes volume from 28 to 84 mL

The PVnT chart is used to organize the different amounts of pressure, volume, amount, and temperature. The temperature must be in Kelvins, but the other variables just need to have matching units to form proportions. The initial and final values are organized first. After the values are set, any value that changes must have the effect of the change determined by the student. For the effect under the volume, the student is determining whether the volume going up will cause the unknown

Chapter 2 – Gas Pressure

pressure variable to increase or decrease. Because these two variables are inversely proportional, the volume change will affect the pressure by causing the pressure to decrease. An arrow is placed that points down.

When the chart is completed, the arrows pointing down form proportions with the smaller value as the numerator. The arrows pointing up form proportions with the larger value as the numerator. The initial value of the unknown variable is then adjusted using those proportions.

A second example would be that a gas has an initial pressure of 1.84 atm in a volume of 22.0 L at a temperature of 299 K. What would the final volume be if the pressure changes to 1.22 atm and the temperature drops to standard temperature (273 K)? Assume the amount of gas is constant.

	P	V	n	T
Initial	1.84 atm	22.0 L		299 K
Final	1.22 atm	?		273 K
Effect	↑			↓

Figure 2-4: A PVnT chart that organizes the values from the example two above

The pressure is decreasing (Fig. 2-4), but the effect of that pressure decrease on volume must be an increase. Thus, the arrow points up and we would use $\frac{1.84}{1.22}$ as our proportion. The temperature decreases and the effect of that on the unknown volume would be that the volume must decrease. The arrow points down and we would use $\frac{273}{299}$ as our proportion. Our calculation would then be:

$$22.0\ L * \frac{273}{299} * \frac{1.84}{1.22} = 30.3\ L$$

The benefits of using this method are that the method avoids equation memorization, and it forces students to identify the relationships between the variables. They are using proportional reasoning which builds number sense. Even though the algorithm is simpler than using the equations, the student must have the knowledge of the relationships to use this algorithm. Using proportional reasoning minimizes the focus on math so that students can focus on the relationships between the variables.

One interesting problem that some students have with the preceding analysis is that they struggle with three variables changing at once. Many students ask how it is possible that the temperature went down but the pressure got larger (Fig. 2-4). They don't carry the sense of proportionality from just two variables to realizing that the set of three become proportional when all three change. Initially you want to clarify that they should focus on just how each individual variable will affect the unknown variable. This can also be clarified by showing how the PV/T set is proportional to the new PV/T set. It might also be helpful to identify that the PV/T

Chapter 2 – Gas Pressure

values will equal a constant as long as the amount of gas (n) is constant. Just be wary that if you go down that rabbit hole you might have students emerge with the formulas.

$$\frac{PV}{T} = k = \frac{PV}{T}$$

I once had a student inform me that they were helping a student from another chemistry class that does gases much later in the year than we do. Their perception was that the student had a bunch of equations, but they had no conceptual understanding of how the variables affect each other and why. The student was just seeking how to plug into the equations without any knowledge of what the equations represented. Many students are capable of solving equations or manipulating units without knowing the meaning of their solution.

Measurement

How do we measure all of this stuff? Currently, we can use a smart phone to measure pressure. But what did they do way back when they figured out these relationships before electricity? How did they measure the pressure of a gas?

The first of two key measuring apparatus for pressure is a barometer (Fig. 2-5). A barometer is a long tube filled with mercury and sealed at the top. The tube is in a container filled with liquid mercury at the bottom so the mercury can flow in and out of the tube. The barometer measures the air pressure in a room. The more air pressure there is, the higher the mercury will rise in the tube. The typical pressure causes the mercury to rise about 760 mm above the surface of the mercury in the dish. The larger the pressure of the air in the room, the more weight of mercury can be supported.

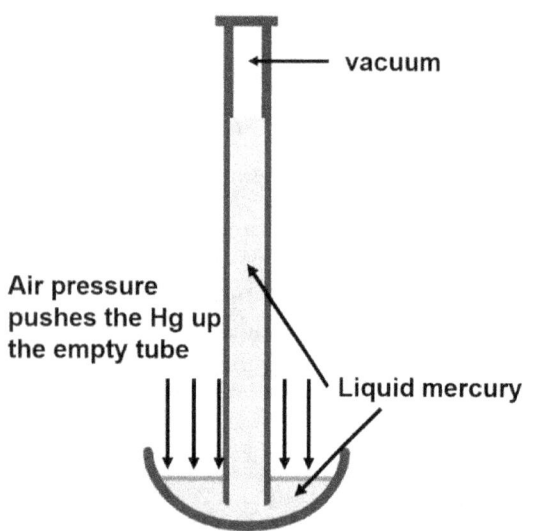

Figure 2-5: A barometer has a tube filled with mercury (Hg) inverted into a pool of Hg.

The second apparatus helps us to understand the relevance of the barometer. A manometer is a set of glass tubing with mercury in it. The tubing is open to the air in the room on one side while the other can connect to a gas sample. The gas sample pushes on the mercury from one direction, the air in the room pushes from the other. When the two sides push unequally, the mercury will rise up higher on the side with lower pressure. If we happen to know the pressure of one side, we can determine the pressure of the other side by looking at the mercury height difference.

Chapter 2 – Gas Pressure

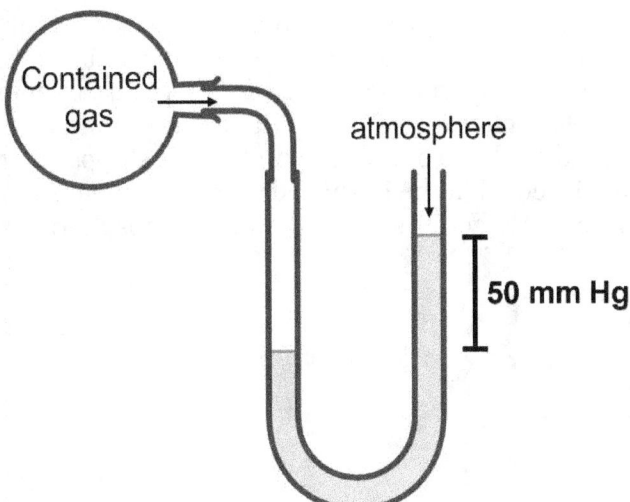

Figure 2-6: Manometer where Hg is 50 mm higher on side open to the air in the room

The pressure of the contained gas is pushing more than the air in the room (Fig 2-6). If the pressure of the air in the room is 760 mm Hg, then we know that the pressure of the contained gas is 810 mm Hg since it pushes enough to support 50 mm Hg in addition to the pressure of the air. Between the barometer and the manometer, we can now measure gas pressure under a variety of conditions. It might not be easy to control conditions, but it is possible.

Students have a wide variety of issues with both tools. For a barometer, students do not understand why there must be a vacuum at the top of the container. They struggle with differentiating between gas pressure, and the pressure of the liquid mercury. Many have experience of being underwater and feeling the pressure from the weight of the water on them. They do not have an easy time visualizing how changes in the shape of the tube do not impact the height of the mercury. If a tube with a larger diameter is used the amount of mercury supported will increase and so students might expect the height to lower.

Manometers are even worse. If you ask students to draw a manometer, some will produce gibberish.
- Some students do not understand that the mercury in the manometer can flow back and forth.
- Many will not understand the difference in height tells the difference in pressure between the two gases.
- Some will not understand that the amount of mercury is constant and remains the same length in the tubing.

We can address these through hypotheticals where something changes.

What would happen to the manometer in Fig. 2-6 if 10 mL of mercury were added without allowing air to leave the container? Let's assume that 1 cm of tubing will hold 1 mL of mercury. The mercury could rise on both sides by 5 mL, but this would lead the gas in the container being compressed. This could increase the pressure of the contained gas. The mercury would add a little less than 5 cm to the container side, and a little more than 5 cm would go to the open-air side.

What would happen if the container were heated? We would see the pressure increase in the container causing more mercury to flow from the container side to the open-air side. If the pressure increased by 20 mm Hg, then the container

Chapter 2 – Gas Pressure

side would drop by 10 mm and the open-air side would go up by 10 mm. This would lead to a 20 mm larger difference with the amount of mercury remaining unchanged.

If you have a U-shaped glass tube, you can help students visualize a manometer better. Put some water (food coloring helps make it more visible) in the tube, tip the tube to one side, put a rubber stopper on one of the sides. The two levels will not be equal in the U-shaped tube. This means that the pressure in the closed side is either larger or smaller than the atmospheric pressure in the room.

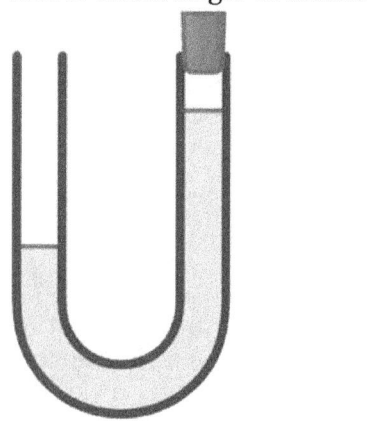 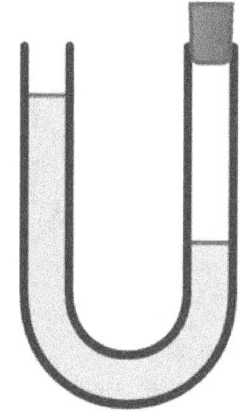

Figure 2-7: Two U-shaped tubes containing water with a rubber stopper on the left side

Because the fluid is water, the pressure differences are smaller for similar height differentials. Mercury's density is an enormous 13.6 g/mL. If you've ever lifted a small container of mercury before you know that the weight is shocking. But we can stipulate that in the first U-shaped tube (Fig. 2-7) the trapped gas has lower pressure than the atmospheric pressure. The atmosphere pushes as hard as the trapped gas, and additionally is able to support the weight of the excess water on the other side. If you go to the trouble to show this, let the students see the water shift back and forth when you remove the stopper. They struggle to see the inside as a fluid even when you use water.

If you ask them to draw a manometer you will realize that some students do not easily translate these images to a real-life object. Many students can follow the algorithm of adding or subtracting the height difference and the lower level of Hg has the higher pressure. But they don't know why that works, how this measures pressure, or why the apparatus was needed.

The next chapter will focus on temperature and heat. In this unit students need to have some basic concepts for temperature. They should know that as a substance increases in temperature that the particles of the substance move faster. They may have a few incorrect ideas that are going to be challenged over the next two units. Many students think that all gases have faster particles than all solids. Liquid water has a smaller relative mass than nitrogen or oxygen, so as long as air and water are at a similar temperature the liquid water particles move faster on average than the air particles.

Students will not have had a lot of time to think about variance in speeds of particles. If you have a sample of air, not all particles move the same. Additionally, the particles change their speeds as they collide with one another. There are many simulations that will allow students to note the straight-line motion of gas particles, the variance in speed in a sample, and how the speeds change as temperature is modified.

Chapter 2 – Gas Pressure

A mercury thermometer relies on particle motion to make measurements. As the mercury changes temperature, the mercury particles move faster which leads to greater spacing between the particles. Ethanol expands consistently like mercury but has a much lower boiling point (78 °C).

Put a long, thin, glass tube into a one-hole stopper. Then use the stopper to plug a test tube filled (overflowing!) with ethanol and dye. As the ethanol is heated the ethanol will expand up the tubing at somewhat regular intervals just like a thermometer.

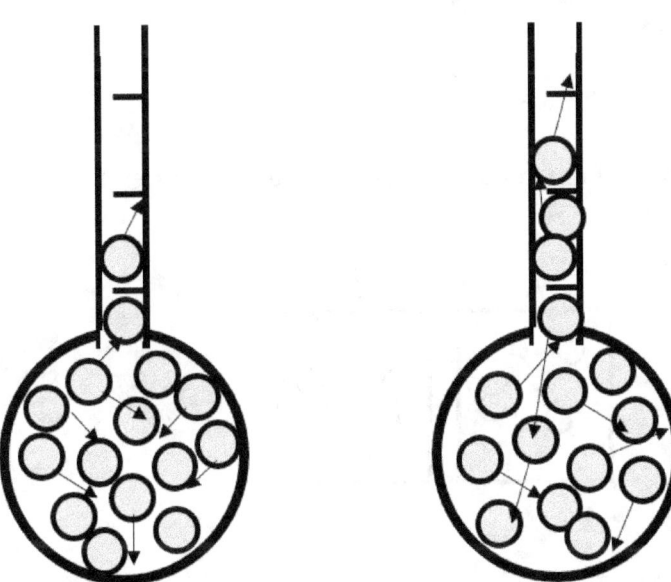

Figure 2-8: A particle level view of how expanding liquid in a thermometer allows measurement. Note that the number of particles is constant.

The way a thermometer measures temperature is that the spacing of the liquid particles increases as the particles move faster and collide harder (Fig. 2-8). This causes the particles to move apart from each other and with nowhere to go but up the tube the level rises on the macroscopic level at regular intervals. Expansion of a solid or liquid will yield some creative interpretations from students.

Kinetic Molecular Theory

If we return to the original experiments, we can learn a lot about kinetic molecular theory (KMT) by analyzing the collisions in each experiment. Why did pressure increase when temperature increased? The temperature increase we know leads to the particles moving faster. This increase in speed leads to more frequent collisions, but the collisions also are more forceful (Fig. 2-9). If you run a simulation of a gas that can alter particle speed, have the students count collisions with one surface at low and high temperature.

Chapter 2 – Gas Pressure

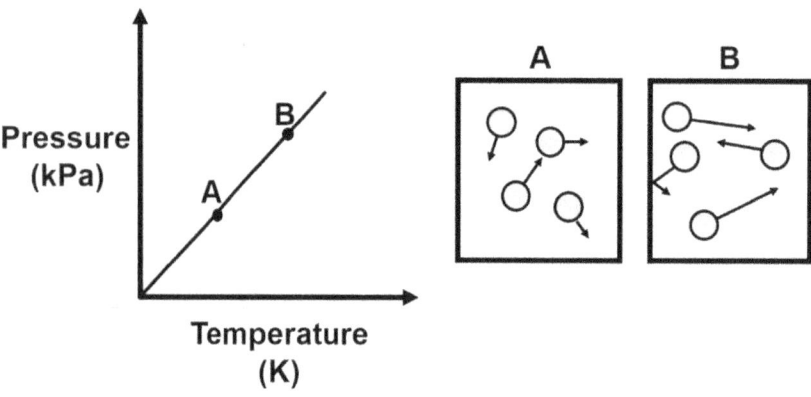

Figure 2-9: One sample of gas at low temperature (A) and high temperature (B)

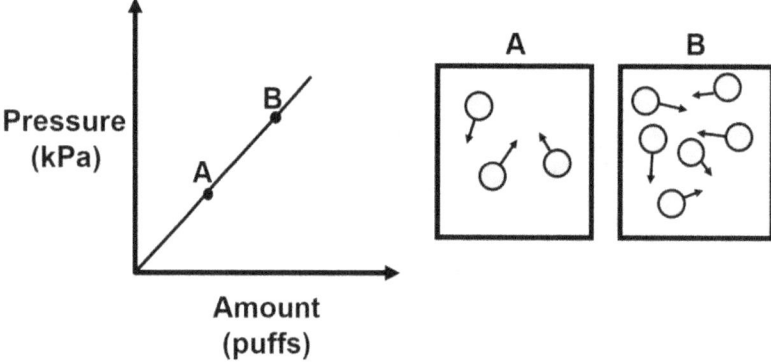

Figure 2-10: A container with one puff of air (A) and a container with three puffs of air (B)

When more particles are added to a container the collision frequency increases (Fig. 2-10). Some students will blend concepts together here. Because the pressure increases, they will assume that the temperature must have gone up. Make them be very clear about the arrow sizes so that it is clear that the temperature remains constant. That isn't to say that the temperature can't also change. But we are assuming in the experimental methods that a constant temperature is maintained.

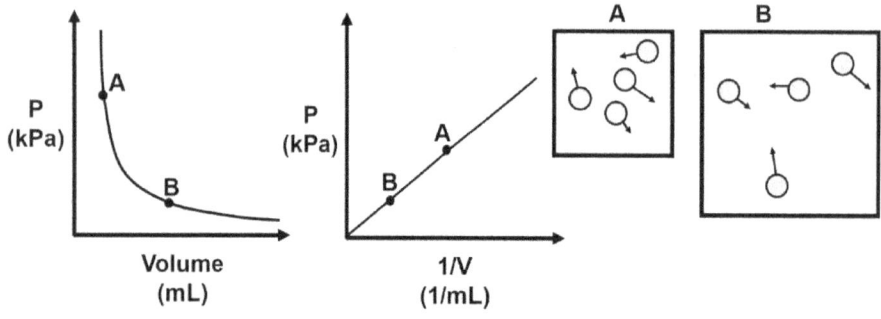

Figure 2-11: A sample of gas starts in a small volume (A) that is expanded (B)

Chapter 2 – Gas Pressure

Changing the volume changes nothing except for the space available to the gas particles. But the increased space means that collisions will be less frequent. This is easy for students to see, but difficult for them to verbalize. The students will often draw inconsistent arrow lengths where it is ambiguous if they intended for temperature to be changed. The changing container size affects their scale. But some students will claim that the pressure in A is higher because the particles move faster due to the constrained space. They may have had an experience where they have seen that work done on a gas can increase its temperature. Here we are controlling the temperature or at least assuming changes are negligible, but it is plausible for the temperature to change as a gas expands or contracts against atmospheric pressure.

Students may describe the particles as wanting to expand. They are taking the abstract phrases from elementary education where a gas fills its container and assigning human properties to the gas particles. **Push back on this!** We want them to describe what the particles do without assigning emotions to the particles. Otherwise, they will not fill in all of the details properly if they can rely on this crutch of anthropomorphism.

Have the students push on a syringe until they can no longer compress the plunger anymore. What stops them from compressing the syringe further? Have the gas particles run out of room to compress? What would happen if we pushed even harder? It is difficult for the students to understand how much force the air can push on us with. Especially for such a small sample. Even though we see tornados destroy homes, we struggle to connect the immense force of a surge of wind with static air.

Through these experiments we see two factors that influence pressure. If the frequency of collisions increases, pressure will increase. If the force (veracity) of the collisions increases, the pressure will increase.

Most text introduce Kinetic Molecular Theory (KMT) as a set of rules. But the objective of KMT is for students to have a visual picture of how gas particles move. We want students to understand that gas particles move in a straight line until they collide with something. When gas particles collide a force exists between the gas particle and whatever it collides with. We also make a couple of assumptions about the particles to give us greater predictive power.

The first assumption is that particles do not stick together at all. This can be worded in a couple of manners. The first way is *"all collisions are elastic."* An inelastic collision is when the kinetic energy is not conserved in the collision. This happens when things stick. If two cars collide, it's going to be an inelastic collision. They aren't going to bounce off of each other and continue on their merry way. The next way this can be described is by stating *"particles are assumed to have no intermolecular forces."* Since students do not know what these are yet, I prefer to just say that the particles do not stick when they collide.

The second assumption is that the particles are points. If a set of particles is in a 500 mL container, we assume that the entire 500 mL is available to the particles. If the particles are not points, they would take up some of that available space due to the volume of the particles themselves.

Neither of these are true assumptions. But they are close to being right a lot of the time. So even though they are technically not true, they are good assumptions. They are quite helpful at simplifying our ability to predict.

Chapter 2 – Gas Pressure

The first assumption falls apart when the temperature of the gas is close to the boiling point, the gas particles stick so much that the pressure is affected. If you have a container of steam at 104 °C our proportional reasoning will not function well.

The second assumption that disrupts our proportional reasoning is when the gas is stored at a high pressure. At high pressures the particles are closer together so the assumption that the particles are points starts to fall apart. As the particles become closer, they take up a substantial portion of the container and so the volume available to the particles decreases.

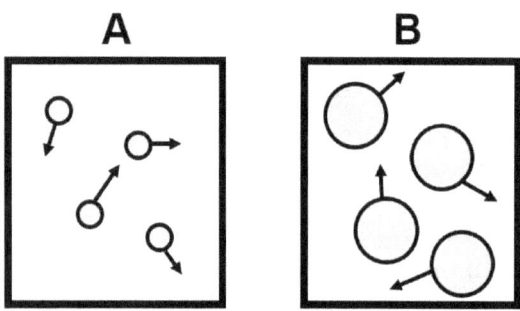

Figure 2-12: The particles in container A take up a negligible amount of space. The particles in container B reduce the available volume to well below the 10 mL.

Here's the thing about deviations from ideal gas behavior. Most students understand them in isolation, but they don't see any purpose to them. They can see how lots of particles occupy more space than a few particles. They have some intuition about how stickiness leads to weaker collisions. But they don't see any purpose or application to this knowledge.

When we make these 2 assumptions, all gases become identical. It does not matter whether we are talking about neon gas, or air, or a really smelly gas. They all start to work the same (except for the smell part). That's a big deal and we can do quite a bit of reasoning with that idea.

For example, if we make these assumptions, then 10 mL of any gases that are at the same temperature and pressure will have the same number of particles. So, if 10 mL of hydrogen gas reacts with 5 mL of oxygen gas, then I now can reasonably assume that the formula of the product is H_2O.

If I react 10 mL of hydrogen and 10 mL of chlorine, I will get 20 mL of HCl. This is surprising because if the H and Cl just combine to make HCl we would only expect 10 mL of product (Fig. 2-13). By looking at a particle level, this could potentially be explained that each hydrogen and chlorine particle actually consists of two particles (H_2 and Cl_2, Fig. 2-14).

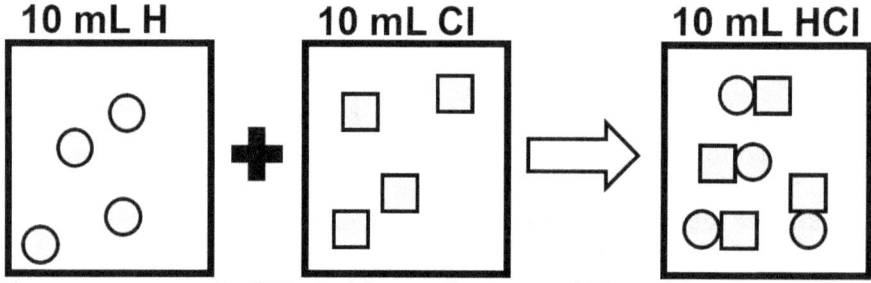

Figure 2-13: 10 mL of H particles and 10 mL of Cl particles do not form 10 mL of HCl

Chapter 2 – Gas Pressure

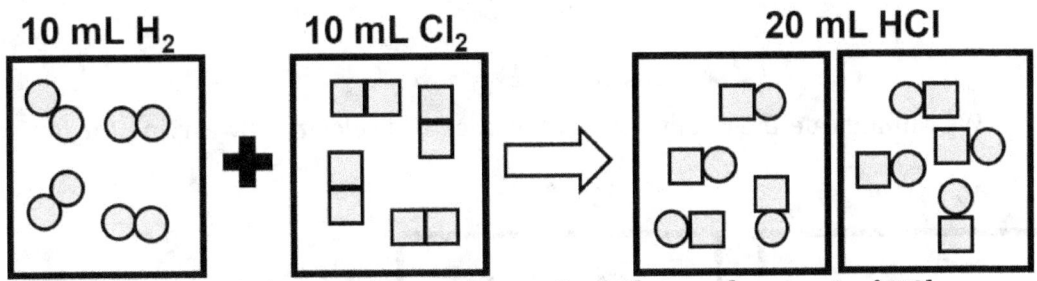

Figure 2-14: 10 mL of H₂ combine with 10 mL of Cl₂ to make 20 mL of HCl

This is a great concept for some of your advanced students to explore. The variables P, V, n, and T all align when we make these two assumptions regardless of the gas. If you have two samples of gases at the same P, V, and T then the amount of particles is the same for both. It doesn't matter what the two gases are, they will have the same number.

But there are other variables that are not consistent. If He and Ne have the same temperature, they do not have the same average particle speed. The variations in speed can be formalized using effusion rates, diffusion rates, or root mean square speed. The densities of different gases vary even when P, V, n, and T are identical.

A simple demonstration for students to see that gases move at different speeds involves putting Q-tips into a straw. One Q-tip should have a drop of concentrated HCl while the other has a drop of concentrated NH_3. The two gases will diffuse through the straw, but they do not meet in the center. A thin white line of NH_4Cl forms closer to the HCl starting point. Be wary of leaving these Q-tips in the garbage or you may frighten your custodian as they empty the trash. I'm not saying that I've done this, but I'm not going to deny it either.

If you're short on time you don't need the straw. You can just hold the open containers near each other and the white smoke forms closer to the HCl container. But exercise caution breathing near either of these containers while open. Neither is pleasant for the lungs.

The difference in effusion rates is the key behind the first method of uranium enrichment. Uranium ore is converted to uranium hexafluoride which is a gas. This gas flows into a container with filters that have tiny holes. The uranium hexafluoride molecules that have the 235 isotope will effuse through those holes slightly more frequently than the 238 isotope. This is repeated many times until the amount of 235 isotope increases from the initial 0.7% to somewhere between 3-5%. Uranium-235 is the isotope used for nuclear fuel rods (and bombs) while uranium-238 is the more stable isotope. The enrichment of uranium is a formidable chemical engineering challenge.

If you do compare rates of diffusion or effusion with the class, it works well to set up a derivation for a gas with large mass (M) and small mass (m). I use large blue particles (M_b) and small green particles (m_g) to give better clarity. At the same temperature, the blue and green particles have similar kinetic energies. Having different masses but the same kinetic energy would imply that the velocities are different.

Chapter 2 – Gas Pressure

$$KE_b = KE_g$$

$$\frac{1}{2} * m_b * v_b^2 = \frac{1}{2} * m_g * v_g^2$$

We can indicate the relative sizes of masses and velocities by varying the fonts.

½ m*v²   ½ m*v²

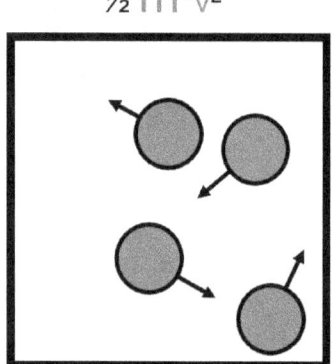 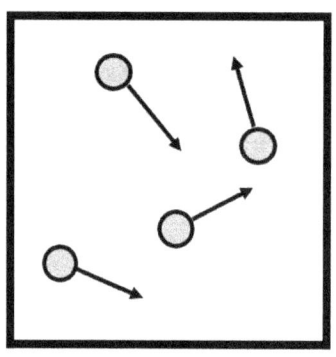

Figure 2-15: Blue particles (m) and green particles (m) at the same temperature and pressure

$$\frac{1}{2} * m_b * v_b^2 = \frac{1}{2} * m_g * v_g^2$$

We can rearrange this to get the ratio of velocities.

$$\frac{v_g}{v_b} = \sqrt{\frac{m_b}{m_g}}$$

If the blue particles have four times the mass, the velocities of the blue particles tend to be about twice as large. Methane and helium have relative masses of 16.05 and 4.00. At room temperature (293 K) the root mean square speed of methane particles is 675 m/s while helium particles are 1350 m/s. The high speed of helium particles is why helium balloons will deflate quickly. The He particles collide more frequently and some of those "collisions" find a gap in the balloon material. The use of mylar coatings on balloons helps to reduce the ability of helium to exit the tiny gaps in the polymer structure and this helps contain the helium for longer.

As you start to discuss particles moving at different speeds based on mass some students might notice that this complicates the idea that all gases exert the same pressure under the same conditions. The collisions of smaller particles impart less force per collision. The smaller particles collide more frequently due to the increased speeds. Overall, the resulting pressure is identical which might be tough for someone to wrap their head around.

Note that throughout this section I never assigned gas particles human qualities. Students will be quick to say, "the gas expands because the particles want to fill the container." Often students use desires and wants for particles because they want to assign a rationale but don't have one. It is far better to not have a reason why than to anthropomorphize a particle. The particles filled the container. That's all you need. The particles wanted to fill the container to reduce the loneliness they felt in their souls. Now you've ruined it.

Chapter 2 – Gas Pressure
Gas Pressure Demos

When I first started teaching, I used to do ten different exciting demonstrations for teaching gas pressure. I now do four and I approach them very differently. When I first started teaching, I was using demonstrations to engage students and get them excited. We would talk briefly about them, but we did not develop them in sufficient detail. Maybe I would point out the direction the pressure was acting, or that the surrounding pressure was relevant. Now I am trying to use them to push their thinking as much as I can.

The first demo is to put balloons into a vacuum chamber. More detail about how a vacuum chamber works can be found in the chapter 16 on entropy. For now, I tell students that the vacuum removes air by an engine that pushes air out, but does not let it back in. Over time the number of air particle decrease.

For the best effect, draw faces on the balloons. As the amount of air inside of the chamber decreases, the balloons get larger. Sometimes they even explode. If they do, you're in for a treat because they make very little noise due to the lack of air (the very thick glass might play a role too). When they explode you can inform students about how when balloons rise high into the atmosphere that the reduction in air pressure at high altitudes causes a similar expansion of the balloon followed by the material failing as it stretches too thin. At that point the balloon falls back to the ground where it can cause problems for small creatures. Note that I said the material fails as it stretches. The students are used to balloons popping due to a large pressure stretching the material and that is not what is happening here. We want to differentiate the two scenarios for them.

The key question to start the discussion after is, "how has the pressure inside of the balloons changed?" Many students will have an intuition that the pressure has increased. This is because they just saw the balloon expand and normally that happens because of an increase in pressure. Other students will push back. If not, ask the students how the pressure inside and outside compare to each other. Besides the stress of the material stretching, the pressure inside and outside should be similar if the balloon is not expanding or contracting. Once the students have had some discussion, have them draw a series three or more cartoon pictures of the balloon at different times.

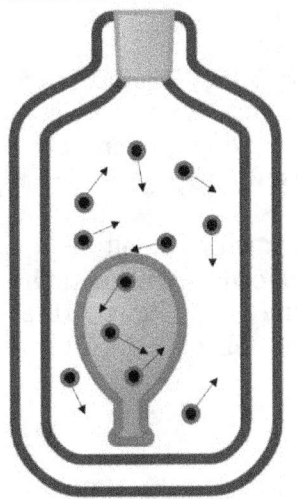
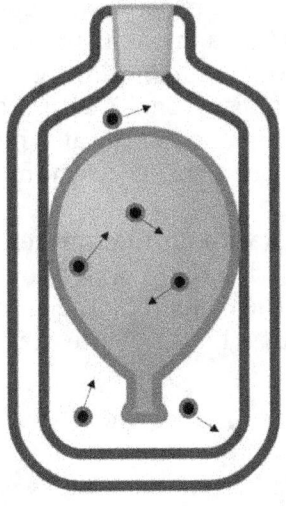

Figure 2-16: Balloons in a vacuum flask

Chapter 2 – Gas Pressure

The second demonstration is related, but slightly more advanced. Take a mason jar that has the screw on lid, but the metal disc that goes into the screw on lid has been replaced by a piece of screen from a window screen (the screen is optional, but it is worth it). I fill the mason jar with water from the sink (while students watch) and then place a notecard on top. The key to pushing the students' thinking on this demo is that when you flip over the mason jar, make sure that students note that a small amount of water comes out. Then the air pressure outside can balance the forces on the notecard internally (trapped gas pressure + weight of the water). If you have the screen in place, you can usually remove the notecard completely without much water coming out.

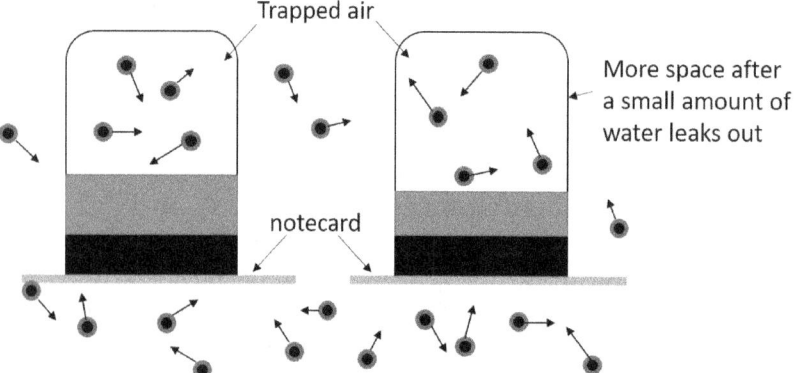

Figure 2-17: A notecard is held in place by the external atmospheric pressure

While the students are dumbfounded, the first question to ask is, "How does the pressure inside of the container compare to the pressure outside?" They are not equal! If they were equal, the water would come rushing out because there is a gravitational force between the water and the earth. The pressure outside must be greater. The air trapped in the mason jar must have somehow had its pressure drop. The key is the water that left. When you flip the mason jar upside down, there is always a small amount of water that comes out. Because that water comes out, the volume available to the trapped air increases. That increase in volume gives more space between the gas particles and thus there are less frequent collisions.

The third and fourth demonstrations are similar. But the conceptual development is challenging. Having two demonstrations gives you and the students two opportunities to work out what is happening. The third demonstration involves taking an empty pop can, putting a small amount of water in it, heating the can until the water boils, then finally flipping the can upside down into ice water. The can is quickly crushed.

The fourth demonstration starts with an Erlenmeyer flask with a small amount of water in it. The flask is heated until the water boils. Shortly after the water begins boiling, take a balloon and put the balloon on the lid of the flask.[†] Remove the flask from the hot plate either right after or right before the balloon is placed. As the

---

[†] Be careful, the rim will be hot. It helps to put some cool water on your hands first.

flask cools back down, the balloon gets smaller until it inverts and "blows up" backwards into the flask.

Keep the expectations high on these two. This is not a simple pressure and temperature variation. When the water in the pop can gets hotter, more water vaporizes. As the steam fills the can, air gets pushed out. At the point where condensed vapor becomes visible, there are now fewer particles in the can and most of them are steam. The pressure in the can is still similar to the pressure in the room because the can is open allowing matter to flow in and out of the can. When the can is flipped into the ice water, the gases cool down. During the cooling, much of the steam turns into liquid droplets. This causes two changes inside the can. The pressure decreases because of less frequent collisions and smaller collisions. The pressure also decreases because there are fewer gas particles as the steam condenses. There is a rapid decrease in pressure. The pressure outside of the can is constant, and thus crushes the can.

- Did the water or air crush the can?
- What was the pressure in the can when it was heated?
- What caused the lower pressure in the can?
- How much lower did that pressure get?

Again, be guarded against students using "wants" in their responses. Particles do not "want" to do anything.

In order for students to put all of that together they will need some guidance. They need to observe that some of the gas leaves and the cloud of liquid water drops can help make that visible. They need to think about all of the changes that have happened inside the can as the water began to boil. They need to realize there is more going on to decrease the pressure than just a temperature change. The temperature change of the can is pretty small on the Kelvin scale. Even from boiling to freezing the reduction in temperature is 273/373 or about 73%. But the can crushes immediately when placed in the ice water and before the temperature reaches that equilibrium.

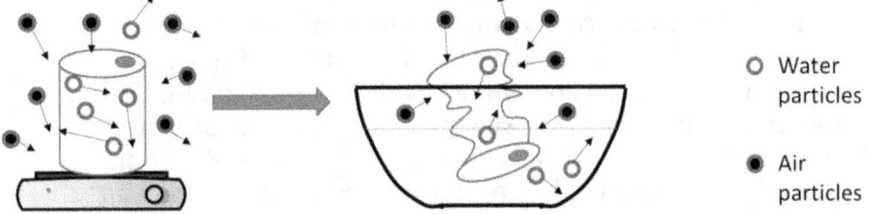

○ Water particles

● Air particles

Figure 2-18: A can is crushed by the external atmospheric pressure as the internal pressure rapidly decreases.

The balloon in flask reveals some of the hidden details because the flask is clear. When the water boils, we again see evidence that the gases in the flask are leaving. Once the balloon is secured to the flask the system becomes closed for matter. As the flask cools upon being removed from the hot plate, observant students will see quite a bit of water condensing on the sides of the flask. The reduction in the number of gas particles and the reduction in temperature both cause the pressure in the flask to decrease. The surroundings still exert the same constant amount of pressure. Thus, the balloon is pushed into the flask and blows up inside out until the pressure inside and outside become similar.

Chapter 2 – Gas Pressure

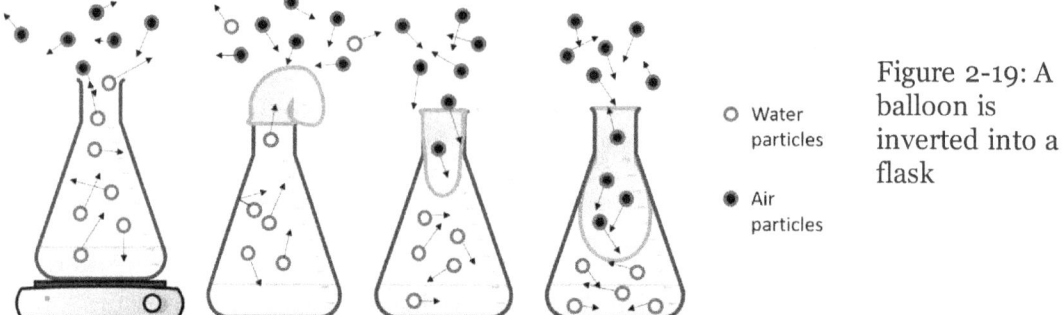

Figure 2-19: A balloon is inverted into a flask

There are other demonstrations that can be done in place of these. But focus on the nature of particles in gas demonstrations to guide students to critical thinking that they do not always do. Make sure that their explanations depend on multiple changes occurring at multiple time intervals. And then let them explore the ideas they have. Let them draw particle level representations at those different time intervals that show what they think is happening. You want them to reflect and revise after they discuss their initial ideas.

Student Struggles

1. "I have no idea what I'm doing." When students struggle solving PVnT problems, the first place to start is determining if they know what the units are. Start by asking them what mm Hg means and see if they know to read it as millimeters of mercury. From there check if they have any understanding of how mm Hg functions as a pressure unit via barometers or manometers. Next ask them to write down units for PVnT where they can choose any unit for each. Most students are not struggling with the mathematics, they are struggling with identifying the components of the problem.

2. "Do I need to change the units to match?" When students are using the IFE tables, the units for each variable should be the same for initial and final. If not, the proportionality is not going to hold. If a student uses mL for initial, and L for final, the result will be off by 1000. They can easily see this if they include units in the proportions. Students also forget to convert from Celsius to Kelvins. Providing students with the proportional reasoning justification helps them construct an understanding of why Kelvin temperature scale is used for gas pressure calculations. They will still need occasional reminders.

3. Student struggles explaining the sequence of changes in demonstrations. We have a perception that struggle is a sign of inappropriate pace for learning but struggle truly is indicative of a student putting in effort and productively learning. We want students to struggle, we just don't want them to become overwhelmed to the point where motivation suffers. Explaining gas pressure demonstrations will lead to productive struggle for most students. This is a good thing because you are challenging your top students in a way that can help them become more proficient in how to apply their knowledge and thinking.

To help students struggle through these, encourage them to draw particle level diagrams that showcase multiple points during the demonstration. Have them

Chapter 2 – Gas Pressure

consider whether the amount, volume, temperature, and pressure was changing or constant at each point. Some will be unknown due to two different plausible possibilities (e.g., pressure could remain constant or increase). Have them utilize collisions in their explanations that they craft from their drawing and thinking.

4. Student struggles graphing pressure vs. volume. When students struggle with graphing, a common disruption is that they look at the whole graph all at once. It can really help to direct them to focus on just one point, or a set of points. When moving from this point to this point, how does pressure change? How does volume change? Next students need to learn to take focus on points and trace those points to where they intercept the axes. For pressure vs. volume, highlight how one halves when the other doubles.

5. Students struggle with basic mechanics of demonstrations. Here the potential issue is possibly going to be that the student is not considering the external pressure, or the atmospheric pressure. They can identify the pressure of the system and its changes, but because they aren't considering the surroundings they quickly fall apart.

6. *"I don't know if I should add or subtract for the manometer."*
Manometers can appear simple because of the simple mathematics. But many students do not fully grasp what happens with them. They don't easily connect the two competing pressures (we rarely draw particles for the trapped gas and external atmosphere). Often students will not know what the mercury is inside of the glass tubing either. You can help them by asking to draw how a manometer would change if the trapped gas were heated. They should draw the liquid mercury shifting down on the trapped gas's side, but they also should draw a constant amount of mercury. Many will not. Unfortunately, it is difficult to show this due to the toxicity of mercury, but you can demonstrate the idea with water in a U-shaped tube.

7. The student's particle diagrams are flawed. Often students will group changes. If volume decreases causing the pressure to increase, the student might have the particles speed up to help the pressure increase. Even though the temperature was constant, the student is associating multiple changes at once. To help students with this, be explicit about what is constant in particle diagrams. Count particles to show n is constant, look at arrow length to show T is constant, look at the size of the container to show V is constant.

Phenomena

1. Drinking from a straw. Students develop a misconception from straws that they can "suck" or exert suction. There is no such thing as suction. Gases cannot be used to pull on an object, they can only push. When we see the appearance of suction, we are seeing a pressure imbalance where one side is pushing harder than the other to create an illusion that the weaker pressure is pulling on the object. Drinking from a straw is a common example of this. There is an apparatus for purchase where the straw is sealed into the cup so that surrounding air cannot move into the cup. Students who

Chapter 2 – Gas Pressure

try to "pull" air in via suction will find that they cannot do so effectively. The sealed cup limits the ability of the air in the cup to push the liquid up the straw.

You can also purchase new and clear plastic tubing that can be used to show how straws are limited in how high the liquid can be pushed up. Much like how air pressure only pushes mercury up to a certain height in a barometer, the atmosphere can only push a liquid up a straw to a certain height as well. The height is larger than 760 mm so long as the liquid is less dense than Hg (most are!).

2. Fierce! One of the core ideas in gas pressure is that particles move. A critical idea that assists students in thinking deeply about particle motion is asking them to explore what makes particles move. Spraying cologne in the room allows students to see how the cologne particles spread out or diffuse. If we ask them to determine what makes the cologne particles move, they produce a lot of external components such as air in the room, people shifting, the pump on the cologne bottle, air currents from vents, etc.

But it is critical that students identify that motion is an intrinsic property of all matter. Particles are moving, they are moving in a variety of directions, and they are constantly changing directions as they bump into other particles. Nothing is required to cause the cologne to diffuse. The cologne was already moving in those directions to start. If a room had no air, the cologne would still diffuse throughout the space.

3. Take a deep breath - Breathing is something that we all do. Breathing can also help students construct faulty conceptions about suction. When you breath in, you do not pull in particles. There is no such thing as suction. The way that breathing works, is that you can expand muscles causing your lungs and diaphragm to increase in volume. The increase in volume causes a reduction in pressure. Some then say that air moves from high to low pressure. Don't do that here. This statement won't undermine the misconception.

Instead go back to before you took a deep breath. If you just sit with your mouth open, what happens? Air moves in and out of your mouth at equal rates. When you expand your lungs, the rate of air leaving your mouth drops. Thus, you experience a rush of air into your mouth. And it feels like you're pulling that air in. Try it right now! You are not exerting force on any air particles outside of your body. The only thing you can do is push on them with the surface of your lungs and airways. And by expanding your lungs, you push on them less for a bit.

4. Cartesian diver - Cut the thin part of a plastic pipet. To the remaining portion, secure a hex nut and fill partially with water. Place the "diver" into a plastic bottle filled with water (this works best if the water overflows from the top). When you squeeze the bottle, the "diver" will sink to the bottom. When you release the bottle, the "diver" floats back to the top. This can be done as a discrepant illusion if you practice your sleight of hand.

What happens is that squeezing the bottle causes the pressure to increase in the bottle. This pressure causes the air in the diver to compress into a smaller volume. This means more water enters the diver. The additional mass causes the density of the "diver" to increase and therefore sink. If the diver does not sink, more water must be added. If the diver does not float, less water is needed in the diver.

5. The mason jar - Take a mason jar that has the lid that screws on to secure a metal disk in place. Remove the metal disk and replace with a portion of the material for a window screen (don't take from a window, you can buy a roll of the material). Fill the mason jar partially with water. Place a notecard on top of the mason jar that

Chapter 2 – Gas Pressure

covers the top. When you flip the mason jar over, the notecard is now underneath and supporting the water from falling out. When you remove your hand from the notecard, it stays in place!

When you flip the mason jar, a small quantity of water leaves. This causes the air inside the jar to increase in volume with the additional space. The decrease in pressure means that the atmospheric pressure can now support the weight of the water and the pressure in the jar. If you remove the notecard, the screen will usually hold the water in place due to a combination of the cohesive forces between the water particles and the external pressure. This can make for an exciting conclusion.

6. Marshmallow in a syringe - Buy a package of mini-marshmallows. Give each student one and let them draw a face on them with a marker. Then distribute syringes. The student should remove the plunger, add the mini-marshmallow, and replace the plunger. The students should be able to cause the marshmallow to expand and contract by altering the volume of air in the syringe. This is a great visual for students to picture gas pressure. Be wary that some students may not follow instructions precisely and can clog a syringe with the marshmallow.

7. Airplanes - Airplanes can fly in the air. And they are very heavy. I can barely lift one of them myself. And yet the air can easily lift several planes at once. The air pushes down on the plane from the top, and the air pushes up on the bottom. An imbalance at high speeds is sufficient for the plane to rapidly accelerate away from the Earth. We often misjudge how powerful air pressure is, and flying is a great way to appreciate its strength.

8. Bed of nails - A small bed of nails can be used as a good introduction to pressure as a ratio of force to area. When a balloon is pushed onto a single nail, the balloon quickly pops. But when a balloon is pushed onto a large number of nails, the balloon can be compressed substantially without popping. The difference is pressure. With a single nail, the contact area is quite small. But with lots of nails, that area is proportionally higher. The larger area reduces the pressure at each contact point, and this keeps the pressure below the threshold required for piercing the balloon material.

9. Gas simulation - When teaching students about collisions and particle diagrams, it can be incredibly helpful to do a summary with a simulation of gas particles. Adding motion can allow the students to count collisions, see how faster collisions impart more force, and allow students to see how speed varies within a gas sample. When showing a simulation, try to track a single particle to see how often it collides. If the volume can be adjusted, have the students count how many particles collide with the top surface over 10 seconds. Then repeat with a large change in volume.[4]

10. Ghosts of chemistry - Open a container of concentrated ammonia close to a container of concentrated hydrochloric acid. A white smoke that is ammonium chloride crystals will form, but it will form closer to the HCl. The HCl has a molar mass of 36.5 g/mol, while ammonia has a molar mass of 17.0 g/mol. This means that the ammonia particles move at a faster average speed, and we observe much faster diffusion. This can be done on a smaller scale by placing a Q-tip with 1 drop of concentrated HCl in one end of a straw, and a Q-tip with 1 drop of concentrated NH_3 in the other end of the straw. A white line of ammonium chloride will form closer to the HCl end. The straw allows for quantitative measurements and analysis.

Chapter 2 – Gas Pressure

Flashcards

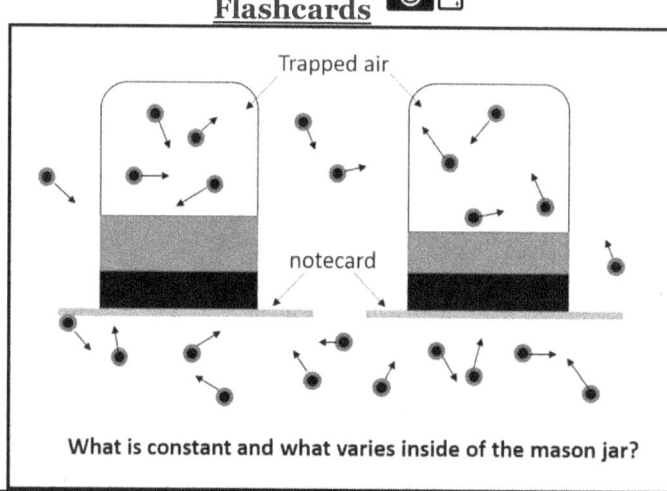

What is constant and what varies inside of the mason jar?

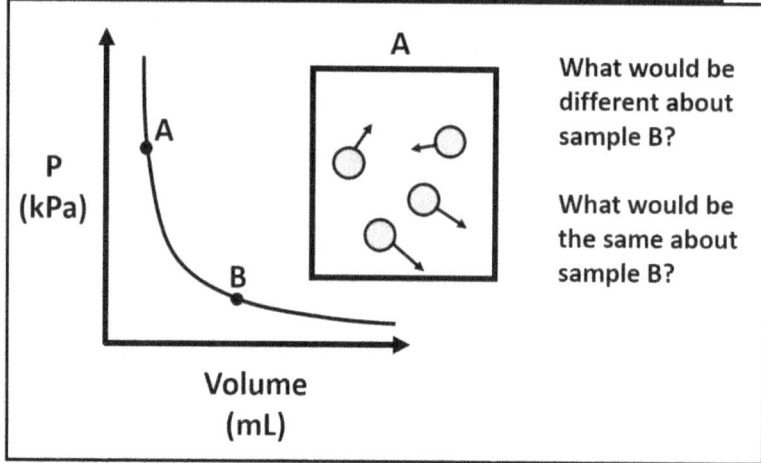

What would be different about sample B?

What would be the same about sample B?

The temperature of a gas increases from 25 °C to 75 °C, the pressure will _____ by a factor of _____

a. Increase, 3
b. Decrease, 3
c. Increase, something other than 3
d. Decrease, something other than 3

Chapter 2 – Gas Pressure

If the mercury level on the right shifted down by 10 mm what pressure would the trapped gas be?

a. Slightly less than 720 mm Hg
b. 720 mm Hg
c. Slightly more than 720 mm Hg
d. 730 mm
e. 750 mm
f. Slightly more than 750 mm Hg
g. 760 mm Hg
h. Slightly more than 760 mm Hg

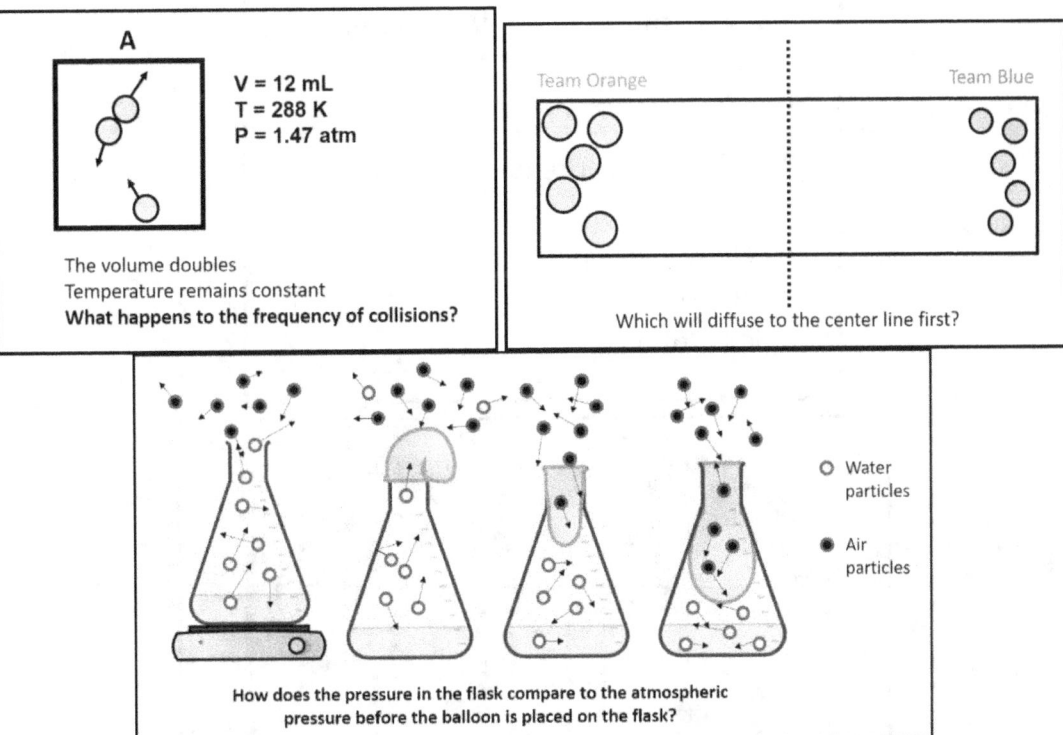

A

V = 12 mL
T = 288 K
P = 1.47 atm

The volume doubles
Temperature remains constant
What happens to the frequency of collisions?

Team Orange | Team Blue

Which will diffuse to the center line first?

○ Water particles
● Air particles

How does the pressure in the flask compare to the atmospheric pressure before the balloon is placed on the flask?

Chapter 2 – Gas Pressure

T = 20.0 °C, V = 20.0 mL

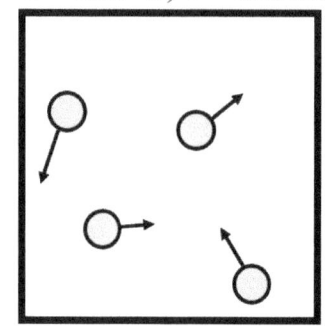

Which will increase the pressure the most?

A. Increasing T to 40.0 °C

B. Decreasing the volume to 15.0 mL

C. Increasing the volume to 40.0 mL

Initial
P = 1 atm
V = 22 mL
T = 450 K

Final
P = ?
V = 44 mL
T = 600 K

The final pressure is _____ 1 atm

a. greater than
b. equal to
c. less than

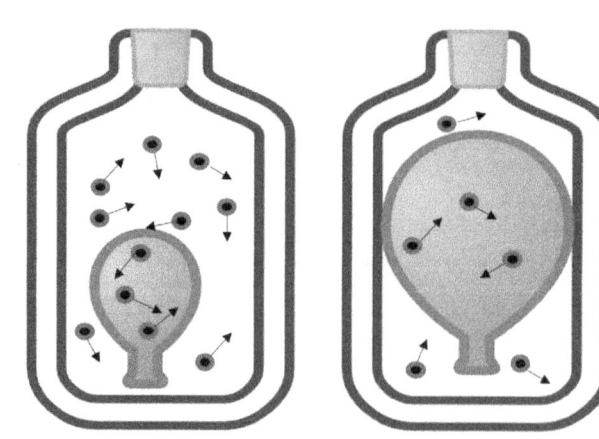

Why did the balloon get larger?

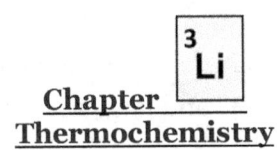

Chapter Thermochemistry

History

To this day fire is a captivating process. Why do some things burn, and others do not? What happens during burning? What happens to the fuel? Why do metals increase in mass when burned? What exactly is the flame? These questions were a struggle for scientists to explain. Along the way some models were developed that are no longer used.

One model that proved very convenient was the phlogiston theory. This theory stated that phlogiston moves into or out of substances as they burn which explained why the mass changed and why only some substances burned. Others that didn't burn lacked phlogiston. When insufficient oxygen was present, scientists proclaimed that the air had been saturated with phlogiston. They termed oxygen as dephlogisticated air since it could absorb so much phlogiston. When Lavoisier pushed back against the popular phlogiston theory, he had to explain what happened with the energy changes. For this he used the caloric theory.

It's important for teachers to understand how the caloric theory differs from the current kinetic molecular theory. Your students will blend between the two depending on the phrasing of the question they are tasked with. In caloric theory, when you burn a candle, caloric is released from the candle. Lavoisier included caloric and heat in his table of elements.[1,2] The caloric is a massless fluid that transfers to the surroundings causing a temperature change. In kinetic molecular theory (KMT), when you burn a candle, the particles change speeds because bonding forces change as particles relocate positions. Energy can be used to describe or predict those changes, but it is not a substance.

A big blow to the caloric theory came in 1847 when James Joule published his experimental results. James Joule was tutored by John Dalton (yep, that John Dalton) as a child. He would frequently play with electricity, and far too often attacked his peers with a jolt. But the most shocking (pun intended!) thing about Joule was his attention to precise measurement. His experiment involves creating a pulley system where he could drop a weight and the weight would cause a paddle wheel to mix water. As the weight fell, the paddle wheel would mix, and the temperature of the water went up. From this Joule was able to link between mechanical energy and thermal energy. He could measure and later predict if a weight of X was dropped a height H that the temperature of a sample of water would go up by T.[3]

His results showed that the mechanical energy changes for the weight were proportional to the temperature changes of the water. This meant that the change in mechanical energy (kinetic and potential energy) was equivalent to the change in thermal energy. *Energy doesn't change forms. Rather the same energy can be used for different things.*

When Dalton constructed his atomic masses by assuming formulas would also be in a 1:1 ratio, it was specific heat capacities that helped improve upon his error (Water is H_2O, not HO). It turns out that specific heat capacities of metals are nearly constant when considered on a per particle basis (It takes the same energy to heat up

Chapter 3 – Thermochemistry

1 million atoms of any metal). This means that specific heat capacities that are done per gram, will be inversely proportional to the atomic mass of each metal. Gold has a low specific heat capacity (0.129 $\frac{J}{g*°C}$) and large atomic mass (197 amu). Aluminum has a high specific heat capacity (0.900 $\frac{J}{g*°C}$) and small atomic mass (27 amu). Pierre Dulong and Alexis Petit (1819) used this relationship to improve upon the initial atomic masses which were often off by a factor of two.[4]

Temperature and Heat

Imagine the scientist James Joule as if he were about to go over Niagara Falls in a barrel. Inside of the barrel James is shaking a bottle of water. The water temperature increases a tiny amount. Why?

Why does mixing water make it hotter? Some might think of friction similar to rubbing one's hands together. But instead let's go back to that waterfall. As the water falls the water speeds up. When the water hits the bottom, it becomes static (the overall net motion). What happened to that motion is that the water particles went from all moving in the same direction to moving in a variety of directions. They are still faster, but their motion is no longer coordinated. Because of this increase in motion, the temperature has increased. Niagara Falls should have a temperature difference of about 0.1 °C between the top and bottom of the waterfall. *What exactly is temperature?*

Temperature is an underrated concept in science. Avoid giving students an abstract definition of temperature and move directly into converting between the Celsius and kelvins. They need to know more than "average kinetic energy." Students should link temperature with molecular motion and correlate faster motion with higher temperatures. *But students should also be pushed to think of temperature as a means to determine when molecular motion will change.*

If two substances with different temperatures come in contact, what happens? What do you know will happen without any additional information? What could you determine if you had more information?

- Will the temperatures change?
- Will each temperature get bigger or smaller?
- Does the temperature change depend on the chemical composition of the substances?
- Does the temperature change depend on the amount of each substance?
- What will happen if two substances with the same temperatures come in contact?

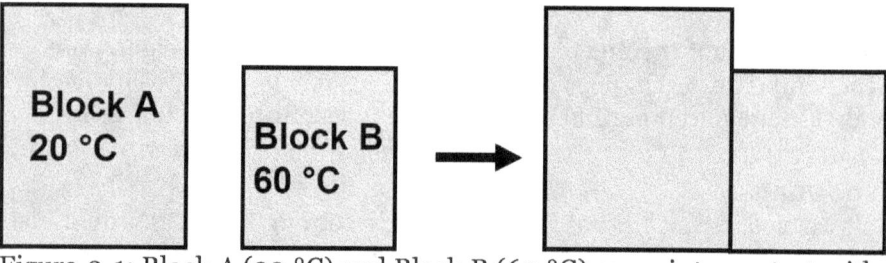

Figure 3-1: Block A (20 °C) and Block B (60 °C) come into contact with each other

Chapter 3 – Thermochemistry

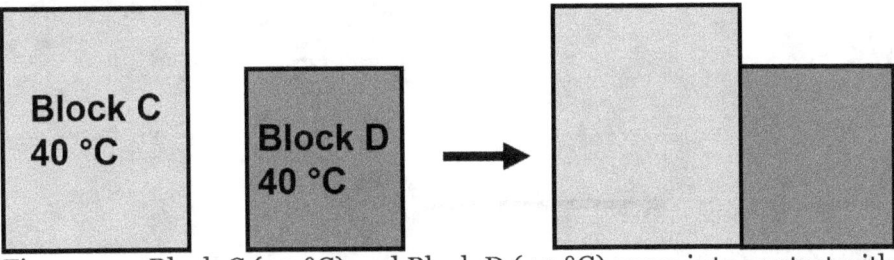

Figure 3-2: Block C (40 °C) and Block D (40 °C) come into contact with each other

What will happen when blocks A and B come into contact (Fig. 3-1)? What will happen when blocks C and D come into contact (Fig. 3-2)? The C and D scenario is quite interesting. Most students will respond for C and D that nothing will happen. But what about the fact that C is larger? Why doesn't that matter? How about the fact that we know nothing about what the two substances are? What if C is a liquid and D is a solid? Will that matter? What if C has massive particles and D has particles that are much less massive?

Think about that last question in particular. If the particles of C are much more massive relative to D particles, they cannot be moving at the same relative speeds (on average). And yet, if the temperatures remain constant, the slow-moving particles of C and the fast-moving particles of D remain at those speeds. What happens when a massive and slow particle of C collides with a fast but smaller particle of D? How does the C particle stay slow, and the D particle stay fast?

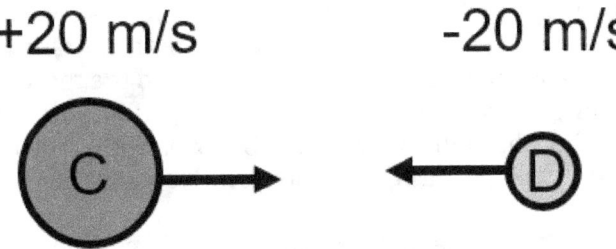

Figure 3-3: Particle C (mass 5 units) collides with particle D (mass 2 units) at equal speeds

If particles C and D collide while traveling the same speed the equal forces (during the collision) will result in particle C undergoing a smaller acceleration. Particle C (Fig. 3-3) will end moving slowly to the left with a velocity of -2.9 m/s. That is a change of -22.9 over the course of the collision. Particle D on the other hand ends up at a velocity of +37.2 m/s for a change of 57.2 m/s during the collision. Note that particle C experiences an acceleration that is ⅖ the acceleration of particle D since the duration of time when their velocities change is identical. *What would it require for both to leave at the same speeds that they collide with?*

Chapter 3 – Thermochemistry

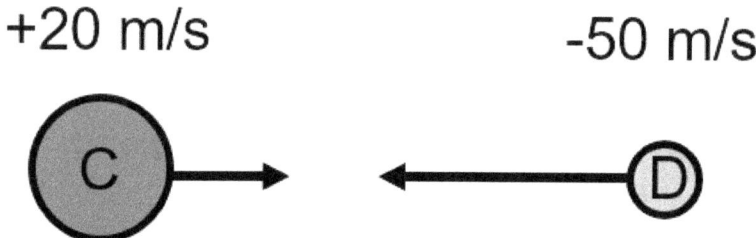

Figure 3-4: Massive particle C collides with the faster and smaller particle D

	C (Fig. 3-3)	D (Fig. 3-3)	C (Fig. 3-4)	D (Fig. 3-4)
initial velocity	+20 m/s	-20 m/s	+20 m/s	-50 m/s
final velocity	-2.9 m/s	+37.2 m/s	-20 m/s	+50 m/s
Δ velocity	22.9 m/s	57.2 m/s	40 m/s	100 m/s
average acceleration	22.9/time	57.2/time	40/time	100/time
mass (g)	5	2	5	2

Table 3-1: Data for initial and final conditions for Fig. 3-3 and Fig. 3-4

    Newton's third law stipulates that the force between a C particle and D particle will be equivalent on both particles while they collide. If the same force pushes the more massive particle C, the acceleration will be smaller than the less massive D particle. In Fig. 3-4 the smaller particle is moving faster than the more massive C particle. After the collision particle D moves with a velocity of +50 m/s for a change in velocity of +100 m/s during the collision. Particle C moves with a velocity of -20 m/s for a change in velocity of -40 m/s during the collision. Note that the ratio of accelerations is again 2:5.

    A liquid in contact with a gas at the same temperature results in collisions that do not transfer a net amount of motion. This is true even if the liquid is comprised of massive mercury particles (200.6 amu) and the gas above it is hydrogen (2.0 amu). At the same temperature, the collisions between the two substances will not result in either substance experiencing a net change in motion. This can become even more impressive when we take into account that not all particles move with the same speed. The speeds of the H_2 particles rapidly change as the particles collide with themselves. So, there will be some mercury particles that collide with the hydrogen that cause the hydrogen to speed up. There will be other mercury particles that collide with hydrogen and the mercury speeds up. But overall, the total transfer of motion fluctuates about zero.

    Most students have a misconception that all gases move faster than all liquids (which are faster than all solids). If you feel like giving the students some more opportunities for thinking, have them think about what temperature the air in the room is. How does the temperature of the air compare to the temperature of the solid objects in the room? For some students, they might need you to actually measure the

Chapter 3 – Thermochemistry

temperatures for them to believe that they are the same. But what would happen if they weren't the same?

Take a look away from the book for a moment. How many different chemicals in the room right this second are the same temperature? There are a few things that are alive or electrical and are therefore not the same as the others. But most are probably the same temperature. Even more interesting is that they might not feel that way. In Michigan the tile floor and the carpet floor do not feel like they are the same temperature during winter. *But they are, so why do they feel different?* The metal frame of the chairs in my classroom do not feel the same as the plastic seat. But they are the same temperature

Feeling warm or cold is about how quickly energy transfers from our body. Remember that we are constantly burning food, so we feel most comfortable with a net amount of energy leaving via heat transfer. When surrounded by air, about 70 °F feels comfortable while a pool at 70 °F is a different story. There's a faster transfer of energy to the pool than the air (at the same temperature gradient). Wind and humidity both affect how much energy transfer occurs.

When you let students wonder about these situations, you set them up for a lot of future topics that are tremendously abstract and difficult. Specific heat capacity is one that comes next. But entropy, enthalpy, kinetics, equilibrium, and more can all be improved by some deep thinking about temperature, collisions, molecular motion, and thermal equilibrium. We want them to have a strong model of particle motion to build those ideas on and not just abstract terms such as energy.

Specific Heat Capacity

Let's now go back to Fig. 3-1 and talk about A and B. A is at 20 °C and B is at 60 °C. When we bring them together what will the final temperature be for both? Will it be 40 °C because that's the average? Will the different sizes (let's assume the size correlates to mass here) matter? Will the materials make a difference?

With so many unknowns, it is tough to come up with a conclusion about the final temperature. But there are two things we know. One is that the final temperature will eventually be the same for A and B. The second is that the final temperature will be somewhere between 20 °C and 60 °C. It could be near the middle, or it could be close to either extreme. That depends on some factors we don't know. The big question that students need to wrestle with though, is how does the temperature change?

Energy can be a pitfall here. Some students might respond by stating that energy moves from B to A. Be warned that they may be using caloric theory here. While it is true that energy moves from B to A, a more effective model can be developed without energy. Instead, we want to look at how the particle motion begins, ends, and what changes had to have happened. Block A has particles start at some speed, and they end at a higher speed. They must have sped up. Block B has particles that start at some speed, and they end at a lower speed. They must have slowed down.

We want students to identify and use collisions as a key component of these motion changes. Most students at this level lack knowledge about the physics of collisions. But they really just need to know that when two particles collide that the speeds of both particles can change. We want them to explain cooling by stating that the particles slow down as a result of colliding with slower moving particles

Chapter 3 – Thermochemistry

(assuming similar masses), and that particles speed up as a result of colliding with faster moving particles (assuming similar masses).

    A good visual for students is a pool table. When a fast-moving pool ball collides with a slow-moving ball (or static ball), the fast-moving pool ball slows down while the other speeds up. It is also helpful to have students make simple particle drawings of substances that are changing temperature. After some time has been spent developing these concepts about temperature, we can finally introduce the term heat. If you want to add a fun collision, drop a tennis ball on top of a basketball. If you drop both from 6 feet, the tennis ball will bounce off of the basketball much higher than the six-foot mark you dropped them from.

    The technical definition of heat is a transfer of energy due to a temperature difference. Some take that definition, and they interpret heat as a noun that describes the energy. Others take that definition, and they interpret heat as a verb that describes the transfer. Both of these have their place in chemistry. This means that we must always be cautious when using the term heat that students will interpret what we say differently. I find it much better for heat to describe the transfer, and to use the term thermal energy to describe the amount of kinetic energy that particles have. I recommend removing the phrase ~~heat energy~~ from your classroom. Thermal energy should be thought of as kinetic energy of a set of particles where their motion is in a variety of directions. I prefer not to say random because the initial motions and positions do allow us to know exactly how they all move if we want to look hard enough. They may appear random to us, but they are not random.

*****

    Once a solid foundation of temperature leads us to this model of heat, we are now ready to look at specific heat capacity. As a physics and chemistry teacher, specific heat capacity had me stumped for years. I could explain specific heat capacity, I could show the algorithms, but I could never fully link what was happening at the particle level. It took me many years of thinking before it finally clicked.

    Let's assume we mix equal masses of hot metal and cool water. The hot metal cools down from 100.0 °C to 30.0 °C. The cool water warms up from 24.0 °C to 30.0 °C. At the particle level the water particles are speeding up a small amount. The metal particles on the other hand are slowing down by a much larger proportion. At the particle level, the rules of physics apply. Each collision between a water particle and a metal particle must involve the same force on both particles. *What happens in a collision where the water barely speeds up, but the metal particle slows down a lot?*

    Now just because the total masses of metal and water are the same doesn't mean that everything is the same. If the metal had been aluminum, the relative mass of each particle is 26.98 on average to the 18.02 relative mass of each water particle on average. So, the metal does have fewer particles given equal masses. But the ratio of particles is about 2 to 3. The ratio of temperature changes for aluminum to water would be almost 5 to 1 (specific heat capacities of 4.18 $\frac{J}{g*°C}$ and 0.900 $\frac{J}{g*°C}$). There is something else going on during these collisions.

    I was fortunate to find a master's program where I could take chemistry classes during evening hours. In my statistical mechanics course, we looked at the components that contribute to specific heat capacities. When you add energy to a particle, the energy can go towards the translational velocity, rotation, vibration, or

Chapter 3 – Thermochemistry

electronic (we ignore the electronic components usually because they aren't substantial contributions). A single atom does not have vibrational or rotational states and so a single atom will have a different specific heat capacity than a diatomic molecule that can vibrate (e.g., He vs. H_2). These roles made sense, but I still struggled to connect them to collisions between particles. A collision with a water particle causes more vibration, rotation and faster speeds. Therefore, it takes more energy to speed up the water.

But again, I felt like something was missing. Does the vibration not count as more motion? Does the rotation not affect the temperature? If I had water at 40 °C the water had those extra vibrations, but why then did the vibrations not cause metal at 40 °C to heat up when in contact with the water at 40 °C?

Something clicked. It felt for a long time as though sometimes vibrations counted (heat, energy) and sometimes they didn't (temperature). When there is not a temperature gradient, the vibrations and rotations do two things. Sometimes they cause a bigger collision. If the water particle is vibrating, the vibration can add motion to the collision. But it also can reduce the collision if the vibration is not in the same direction as the collision. If the water particle is moving to the right, but the hydrogen atom in the water is vibrating to the left, the collision will be reduced. At the same temperature, there are equal amounts of reduced collisions as stronger collisions.

But now we put water at a higher temperature. The water particles move faster translationally, but also vibrate and rotate faster. Because of the larger vibrations and rotations, the water is now more likely to have a larger collision due to the vibration or rotation. So now we see a great change in translational motion for the metal. Once the idea of having a temperature gradient causing the vibrational and rotational states to cause the collisions to differ made sense, my understanding of specific heat capacity increased. Furthermore, my confidence increased. In the past, I had been relegated to using simple abstract models. But now I felt comfortable with students questioning what was happening at the particle level. I needed that resolution to reduce my anxiety of being unprepared for questions.

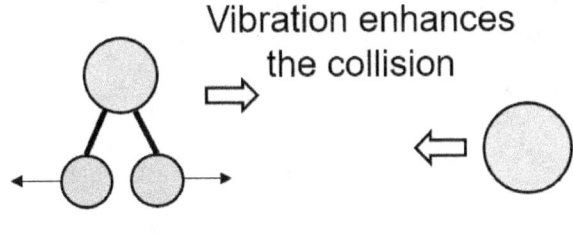

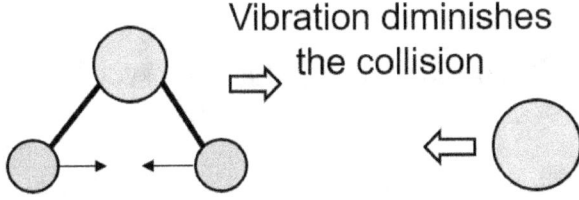

Figure 3-5: Water particle collides with a metal particle. The first collision is larger than the second due to the timing of the vibration.

Chapter 3 – Thermochemistry

The thinking I did about specific heat capacity is something I value. Teachers often repeat chemistry so frequently, that it becomes part of our system 1 thinking. We perceive it to be easy. But for students they are struggling through transitions between system 2 and system 1. It isn't easy for them. Being able to struggle through the content is so critical to being able to maintain a healthy perception of what chemistry looks like to novice students. Teachers fall victim to the "Curse of knowledge" mentioned in the introduction.

When students are first introduced to specific heat capacity, there are many difficulties. The units for the constant (C) have three different units combined. The concept of heat and the concept of temperature are both not developed. The collisions at the particle level are still abstract because students struggle to think of particles in a solid and a liquid. Students have limited knowledge of collisions, momentum, and other kinematics. Depending on sequencing, students might not have any prior knowledge of how particles differ between different substances such as water, metals and gases.

Starting specific heat capacity should focus on letting students observe and ask questions about the discrepancies in temperature changes. Having them draw a particle level representation can help them organize these questions. Why does the one set of particles change differently? The key is that they need to generate questions and make observations before they are introduced to algorithms. Once they see that there is an equation to manipulate, many students will lock into the equation and that will be all they see.

Before we get to equations, we want to get students to a common landing point. The resolution is that if we add the same amount of energy to different materials, the materials will not change temperature by the same amount. Even if students do not completely understand heat and temperature yet, this can be agreed upon because we have experimental evidence to support the claim.

*****

To transition to specific heat capacity calculations, proportional reasoning is the perfect transition. I put in 4.18 J of energy to 1 gram of water and the temperature changes by 1 °C. How much energy will it take to heat up 10 grams of water by 1 °C? How much energy will it take to heat up 10 grams of water by 10 °C? Will 1 gram of zinc metal heat up 1 °C if we add 4.18 J to it?

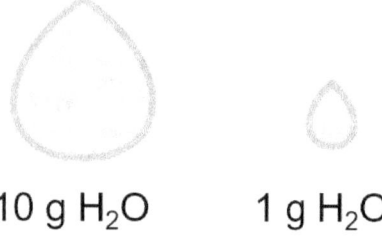

Figure 3-6: If it takes 41.8 J to heat up 1 g of water 10 °C, how much will it take to heat up 10 g?

When we reveal the constant for water of 4.18 J per gram per 1 °C up for students, they have models they can connect it with. Ideally at this point, they don't need an equation. If we know that 4.18 J will cause 1 g of water to heat up by 1 °C,

Chapter 3 – Thermochemistry

students should be able to use that information to determine how much the temperature will change when 387 J are added to 15.2 g of water. They just need the different constants for different materials. From the constants they can use units and proportional reasoning.

The introduction of the constants is also a good time for more student generated questions.
- Which substances have the biggest specific heat capacities?
- Does the mass of each particle matter?
- Do the states of matter have an impact on the values?
- Is the specific heat capacity of steam the same or different than the specific heat capacity of ice?
- What happens when you melt or freeze something?

The final landing point for specific heat capacity is to construct a "For every" statement. The students should do this on their own and ideally come up with something along the lines of, *"For every 4.18 J of energy added to 1 gram of water, the temperature will change by 1 °C."* If they struggle, have them do multiple "For every" statements using different specific heat capacities.

Now is a good time for students to try doing some problems. They have accrued many representations for heat, temperature, and specific heat capacity. Now it is time for them to deploy their models and see what they understand as well as what they do not understand yet. As they are doing these problems, you can direct them to things they have experienced such as metal and water changing temperature differently when heated.

If your students use equations to solve specific heat capacity problems, be ready for them to mix up heat with specific heat capacity. They do not have a strong concept of what either is and so the fact that both have J as part of the units leads to their brains connecting them. If this is the case, the "For every" statements can help the student. They also can be helped by comparing mystery substances to help give them a concrete example of specific heat capacity.

Mystery substance A has a lot of energy put in via heating. But the temperature change is small. Mystery substance B has a small amount of energy added and manages to change temperature a large amount. Assuming the masses of the two substances are equivalent, which has the larger specific heat capacity? What would a graph of Q (heat) vs. T look like? What would a graph of T vs. Q look like?

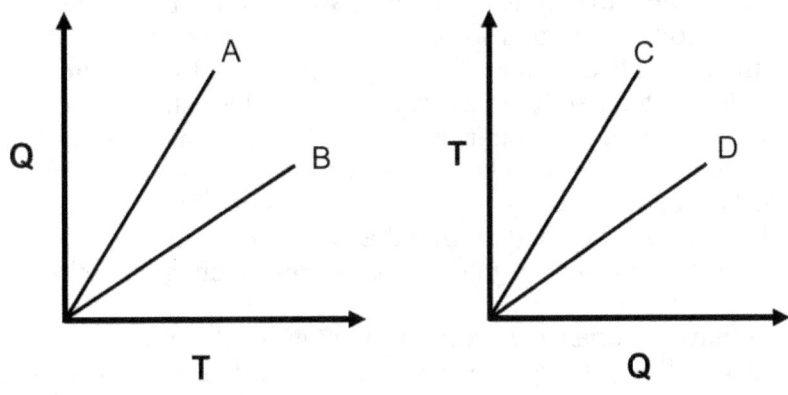

Chapter 3 – Thermochemistry

Figure 3-7: Which substances (A vs. B; C vs. D) have the larger specific heat capacities?

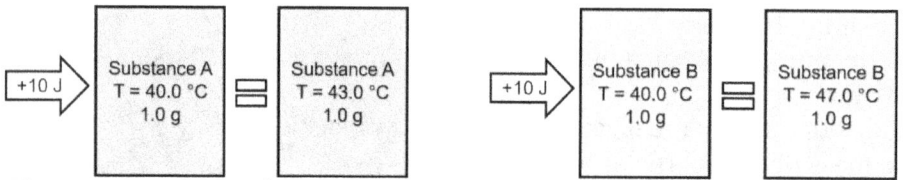

Figure 3-8: Which substance (A or B) has the larger specific heat capacity?

Specific heat capacity is a proportionality between energy change and temperature change for a gram of a given material. Giving students concrete examples that do not have numbers can help them learn to process the idea of scaling or proportionality that exists for a pure substance.

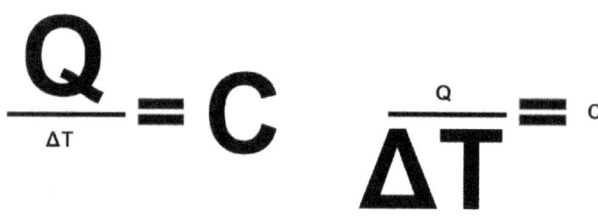

Figure 3-9: A large amount of heat with a small temperature change yields a large specific heat capacity. A small amount of heat with a larger temperature change yields a small specific heat capacity

There is something minor to be cautious of with your language used to teach specific heat capacity. We are constantly talking about the amount of energy added or removed. Be wary using rates of heating. If we add energy to metal and water, the same quantity of energy will cause different temperature changes in the two substances (assuming equal masses). *Which heats up faster? Which heats up more?* These two questions are similar, but the implications are different, and we want to be clear that we are using the latter when we pose questions to students. If we talk about the rate, we overlap thermal conductivity with specific heat capacity. This risks confusion and cognitive overload.

I am particularly vulnerable to confusing students when I talk about being burned. Boiling water is only 100 °C. But that water can cause a large burn. Metal at 100 °C will feel warm to the touch. When you add energy to metal and water, the metal experiences a much larger temperature change. It is helpful for students to process how metal can burn you, and how water burns you. When we heat metal in a flame, the metal does not stop changing temperature at 100 °C. The metal instead might come close to 1000 °C. If you have a quality flame and a metal that doesn't easily melt, the temperature can go even higher than that. Metal changing temperature by nearly a thousand degrees is going to cause some damage. Possibly more damage than water changing by 50 °C.

Some students will have experience cooking. When I grill salmon using aluminum foil, I am capable of lifting the corner of the aluminum foil safely while it is

Chapter 3 – Thermochemistry

on the grill. As long as I do not touch part of the foil that is too close to the fish, the aluminum is only about 200 °C and does not hurt to touch.

Another experience that is good to connect for students is unloading a dishwasher. When the ceramic dishes are heated, they require a lot of energy to reach those high temperatures. When the cycle is complete, they release that large quantity of energy causing the water to completely evaporate. Ceramic dishes come out dry. But plastic dishes do not absorb as much energy when they reach the temperatures of the dish cycle. So, when the cycle is complete, they are not able to evaporate all of the water on them. They often still have puddles of water and are irritatingly wet when they are removed.

Weather is a fantastic opportunity for students to make a concrete connection with heat capacity. Why does San Diego always have beautiful weather? Why if you drive east from San Diego to the desert does the temperature get much hotter with more variance? *How does the large body of water influence the temperature when the sun is raging and when the sun is gone?*

The final piece of specific heat capacity is calorimetry. The term calorimeter can be broken into *calor* (heat) and *meter* (measure). Heat cannot be directly measured, we can only measure temperature. But we can use temperature changes to calculate heat transfer or specific heat capacities. The bomb calorimeter is particularly fascinating to me. I'm convinced that some lucky person gets to explode food all day in a bomb calorimeter to measure the Calorie content for packaging. If that's not the actual job description, please don't ruin my dream of one day sitting in a lab and burning a hamburger in an oxygen rich container.

For me, I start specific heat capacity by looking at the phenomena of a hot metal being added to cool water. The water heats up a little bit, the metal cools down by a larger margin. But we can take that one step further and determine the specific heat capacity of the metal. And that starts with one big assertion. *All the energy that left the metal, went into the water.* There might have been a little bit of evaporation, a little bit of thermal energy going into the polystyrene cup, a little bit of thermal energy going into the air, but we ignore all of those as negligible for now.

If all the energy from the hot metal goes into the cool water, then we can set up a very simple equation.

$Q_{H2O} = -Q_{metal}$

The energy that leaves the metal during heating goes into the water. The negative sign here is about direction. Students get distressed with signs because they do not have clarity on directionality. The number is how much change happened; the sign tells us what direction. The metal decreases energy by Q while the water increases energy by Q. This equation is important for students because they might experience some cognitive overload when we add the 3 variables to each side. We want to stress now that the water is on one side and the metal is on the other.

$m_{H2O} \, C_{H2O} \, \Delta T_{H2O} = -m_{metal} \, C_{metal} \, \Delta T_{metal}$

$50 \text{ g} * (4.18 \frac{J}{g*°C}) * (30°C - 24°C) = -50 \text{ g} * C_{metal} * (30°C - 100°C)$

$C_{metal} = 0.358 \frac{J}{g*°C}$

Students will notice that the negative sign for Q is not needed if we treat the temperature changes as absolute values. This is true for this problem but does not work if the student is asked to find an initial or final temperature as the unknown

Chapter 3 – Thermochemistry

variable instead. It is helpful to them (upon completion of the first problem) to reiterate the Q = -Q initial set up. They can frame that connection better once they've seen the entire question and this will help them start out problems that have a large number of variables. Remember that students have a difficult time differentiating Q and C. Many students who struggle with the difference will write down odd units or no units for specific heat capacities.

Heating Curves and Energy Bar Charts (LOL Diagrams)

Even now I still get surprised with the difficulty students have with heating curves. There are numerous challenges students face with interpreting what information they have and making predictions from them. One issue is that students and teachers view melting points differently. Students rely on their experiences with water and ice to guide their thinking about melting and freezing. Teachers rely on a mixture of experiences and theory that students have not been exposed to yet. Ice and water are not a great foundation for melting and freezing. Water is not always pure and often large quantities are heated (or cooled) that prevent a consistent mixture at thermal equilibrium.

My definition of a melting point is based on an equilibrium between the solid and liquid forms. Because of this, I think of melting point as a single temperature. The melting point of lead metal is around 327 °C. To me, that means at 327 °C that solid and liquid lead metal co-exist in a dynamic equilibrium.

Many students think that lead is a solid until it gets to 327 °C and then it starts to melt. They believe the freezing point is somewhere above 327 °C. They can be led to believe that the range of melting is a small temperature range, but it is not a single point to them. They have too many experiences of mixtures melting such as butter or ice cream. They see melting and freezing as a process that depends on temperature changes.

If the temperature is higher, they think more will melt. If the temperature is lower, they think more will freeze. They struggle to separate temperature from the time it takes to achieve that temperature. The variability of speeds within a set of particles makes this extremely difficult to reconcile for students at the particle level.

When students are stuck, come back to what you can observe. Can you have liquid water at -1 °C? Can you have solid water at +1 °C? Can you have both at 0°C? Students may need clarification that the water itself must be at these temperatures, not the surroundings. You can put ice in a warm room (30 °C). That doesn't mean that the ice is now 30 °C. A fun side question to ask students is what happens when you heat up ice? They all say it melts, but they are only correct if the ice is at the melting point temperature. Otherwise, the ice gets hotter (what's cooler than being cool?).

Boiling point tends not to be as difficult. One reason is that students have more experience with a static temperature during boiling. They can sense that the temperature of boiling water for cooking is constant. They have an easier time rationalizing the impact of pressure on boiling point. Some students have noticed labels on food that cooking requires more time at high elevations like in Denver, CO. Others may have worked with a pressure cooker.

While the intuition is there for boiling point, the students can still be flummoxed if you question them directly. Boil some water and have them measure the temperature. Then ask them why the temperature has stopped changing. Many

Chapter 3 – Thermochemistry

will check to make sure the hot plate is still plugged in, and that the temperature sensor is working. They'll look to see if the temperature is going up a little bit or if the sensor isn't contacting the boiling water correctly. Some will hypothesize that the steam is what is getting hotter.

We want to use the liquid to gas vaporization phase change to help students differentiate thermal energy (E_{th}) with phase energy (E_{ph}). When we heat water, the heat source has fast-moving particles that collide with the water particles. The water particles increase speed as a result of the collisions. A typical sequence might be that the electricity causes the hot plate particles to move very fast, the hot plate particles collide with the glass particles in a beaker making them faster, the glass particles collide with water particles making them move faster. But what happens during boiling? The hot plate particles are moving fast, hit water particles (via the glass), but the water particles do not speed up. As we watch the temperature of boiling water maintain a constant thermal energy what would be happening? It is indeed perplexing. What does the collision look like?

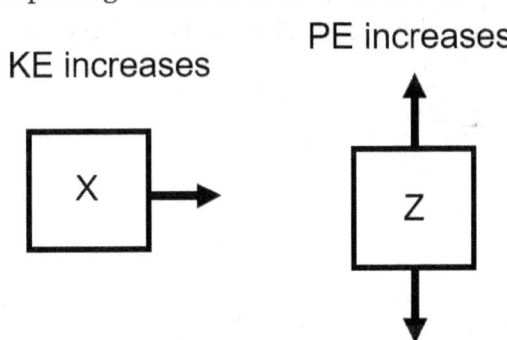

Figure 3-10: Box X increases in kinetic energy as a net force is applied. Box Z increases in potential energy as it moves upwards against the pull from Earth.

There are two things that can happen to an object when a force is applied. Picture a pen on a table. If I shove that pen horizontally the pen will accelerate and move faster because of the applied force. But if I lift the pen things are different. When I left the pen, I apply a force up and the earth pulls the pen down. So, I can apply a force without the pen changing its speed. The force can change the kinetic energy or potential energy depending on other forces present. If you think of examples of potential energy (springs, objects up high, magnets, etc.) you will find a counter force that causes that potential energy.

Now let's apply that thinking to heating water. When we heat liquid water, the particles move faster. The collisions from the hot plate cause the water particles to speed up. But when we reach boiling, the collisions now cause the water particles to separate. During this phase change the thermal energy is constant because the collision is working against the stickiness of the particles. Therefore, during the phase change we are experiencing a change in potential energy not a change in kinetic energy.

There is of course variability on the micro level that will confuse students (and teachers!). Some collisions will lead to an increase in thermal energy, but that thermal energy will quickly cause another particle to escape the stickiness of its neighbors. Students who think about this will realize that particles could have a bigger collision. Think of lifting the pen with such a force that it increases in potential

Chapter 3 – Thermochemistry

energy and kinetic energy. Get students to understand that these minor variations don't undermine the evidence of constant temperatures that they observe in the lab.

Phase energy is a potential energy based on the stickiness of particles. During phase changes the thermal energy is constant. When a particle does increase in kinetic energy, that energy is transferred quickly from other collisions to increase the spacing between particles. The increase in spacing is an increase in phase energy (except for water, which is frustratingly abnormal). A decrease in spacing is a decrease in phase energy.

The slope of the heating curve is also filled with important details. The rate of heating, the mass, and the specific heat capacity all come together to determine what the slope is. Those slopes should be different for each state of matter (solid, liquid, and gas). Temperatures during melting, boiling, freezing, and condensing are constant. *Can a student take a heating curve and modify it appropriately for the same situation but with half the mass? Can they modify it for when heat is applied three times as quickly? Can they do so for zinc instead of water?* For zinc the melting point, the boiling point, and all three specific heat capacities are different. How will each of those play a role in the new heating curve?

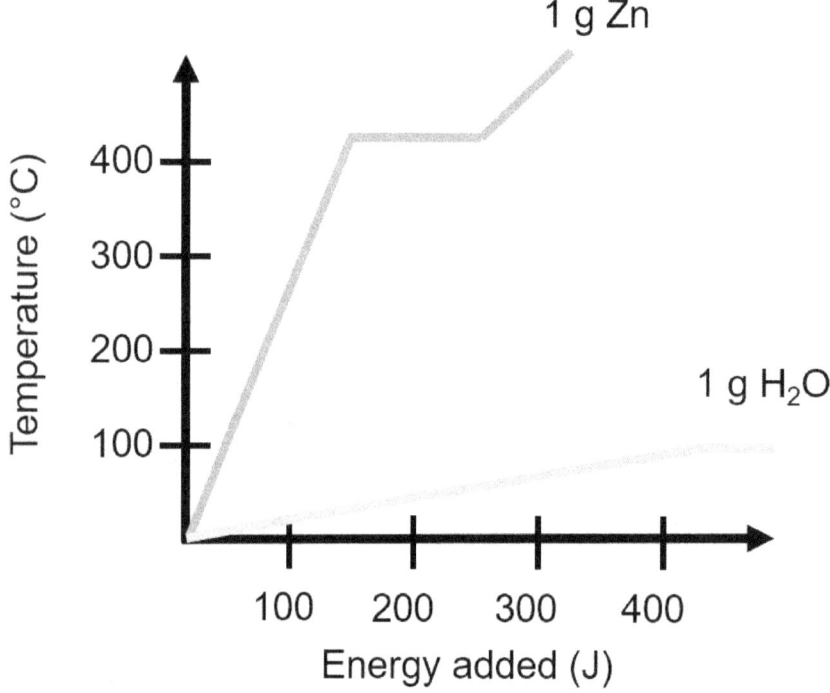

Figure 3-11: Heating 1 g of Zn and 1 g of H_2O. What similarities and differences would students notice?

Students understand better when they are shown a heating curve with something besides water. Lauric acid or BHT (butylated hydroxytoluene) are common examples. You can try using wax from a candle or crayons, but those are mixtures and may not function well. Freezing tends to give a more consistent transition, but there also can be issues with supercooling.

Chapter 3 – Thermochemistry

Part of me feels that it takes time for students to become accustomed to the large amount of information in a heating curve. Students will learn as they work out multi-step problems that involve heating a solid and melting it. Or cooling a gas and condensing it. I think back to my first years teaching and how quickly I would delve into those problems. Back then the problems were an opportunity to teach some of the concepts mentioned earlier. But by beginning with the algorithm, I lost a large number of students. Those students had too few experiences, too little discussion, and the unanswered questions led to cognitive overload. Many of the students who did learn to utilize the algorithm only did so with an abstract understanding. Students might know the number 4.18 is used for water, they might be able to solve a calorimetry problem, but they don't know what either truly represent at the macroscopic, mathematical, nor particle level. A very limited number of students were able to transition from the abstract understanding to concrete situations.

An "energy bar chart" (LOL diagram) is helpful for students to organize the energy components of the changes in a heating curve. Each L represents the phase and thermal energies at a specific moment and the O represents the system. If there is a change in energy from the initial to final moment, that change is indicated with the system. The number of bars for thermal and phase energies gives students a qualitative organization of the energy during a change.

<u>Phase Energy</u>
1 Bar - Solid
2 Bars - Liquid
4 Bars - Gas

<u>Thermal Energy</u> (again, these are not meant to be quantitative)
1 Bar - 0 ish °C
2 Bars - room temperature ish
3 Bars - between room temperature and 100 °C
4 Bars - 100 ish °C

A can of pop is placed in the freezer where it is forgotten. The next day someone finds the can has burst due to the pop freezing. Construct an LOL diagram.

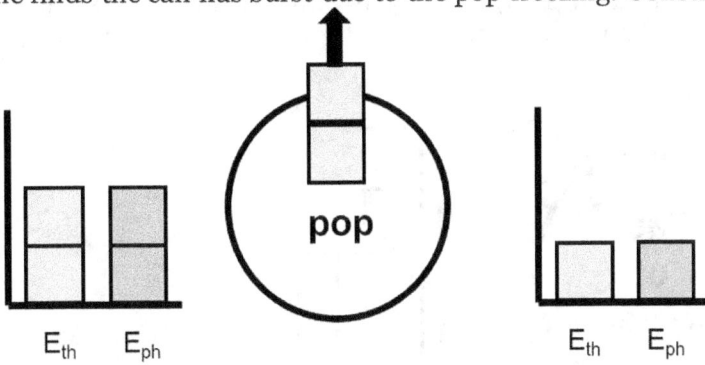

Figure 3-12: LOL diagram for a can of pop that was forgotten in the freezer

We are treating the pop as water here. The first L represents the initial conditions of the can before it goes into the freezer. The second L represents the final conditions. The O represents the energy changes from initial to final. The O specifically represents the system (can of pop) while the energy can transfer between the system and the surroundings. The pop cools down from room temperature to its freezing point. Next the pop freezes. Potentially the pop cooled down below the freezing point as we don't know the final temperature (e.g., it could be -5 °C)

This allows students to look at the initial and final states to determine what changes in energy have taken place. Our language should be precise here. The energy is always the same energy, but we're using that energy for two different things (thermal and phase). How much thermal energy depends on the temperature of the substance. The phase energy depends on the relative spacing of the particles and the state of matter. How sticky the particles are, the mass, and the average speed of the particles would push us further into knowing a quantitative amount of phase energy or thermal energy. For now, we are using the arbitrary number of bars to determine qualitative changes.

The LOL diagram helps us to organize what our calculations would involve if we wanted to know the actual energy changes that took place. We would do a calculation for the temperature change of the liquid. We would also do a calculation for the phase change. If the pop can cooled to a temperature below 0 °C we would require a third calculation.

The next important feature for an LOL diagram is that they highlight the system and the surroundings. The phrases endothermic and exothermic get thrown around casually by science teachers. It is important to note that any endothermic process is accompanied by an exothermic process. One for the surroundings and one for the system. There cannot be one without the other. And teachers have a clear image of what the system they are thinking of is. Students do not always see as clearly.

For example, students might hear a teacher inform them that melting ice is endothermic. They think endothermic means hot because you would need to heat ice to melt it. But those same students later might hear the teacher describe a chemical reaction where the solution gets hot as exothermic. To the student this could be very confusing because both processes involve heating. Yet one is endothermic, and the other is exothermic. But the ice cube is the system. The water in the solution is considered part of the surroundings. By specifically emphasizing the system and the surroundings we greatly enhance the clarity of our thoughts to the students.

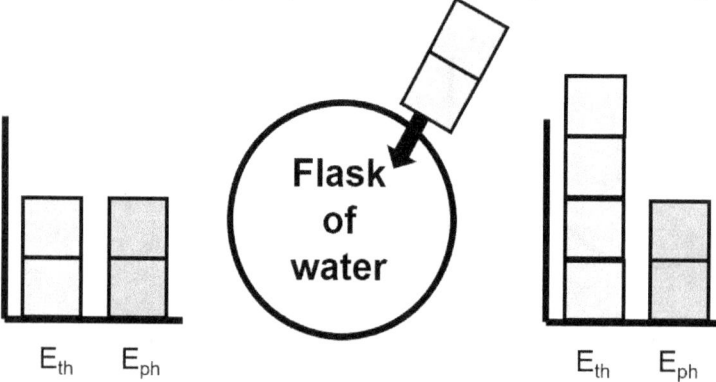

Figure 13-13: What would the flask of water look like before and after the energy changes?

Chapter 3 – Thermochemistry

One final note on LOL diagrams. There is a larger spacing for liquid to gas than solid to liquid. The experimental evidence shows that more heating is required to boil a sample of water than to melt ice. If it takes 5 minutes to melt ice, but then 40 minutes to vaporize all of it, then there was a much larger input of phase energy for boiling. If the students do not collect evidence, they can note this from the much larger enthalpy of vaporization 2260 $\frac{J}{g}$ relative to the enthalpy of fusion 334 $\frac{J}{g}$. This is the justification for having a larger bar increase (2 to 4) from liquid to gas than solid to liquid (1 to 2) for the phase energy. Either way you will also want to connect this to the heating curve to show that vaporization/condensation requires more heat (time of heating) than melting/freezing. The slope being zero makes this more important to address directly.

*****

The summary of the advice for teaching thermochemistry begins with energy. Energy is abstract and this can limit student thinking. Whenever possible expand beyond energy to include molecular motion and molecular positioning. Temperature is a critical tool to make this effective. Spend time emphasizing temperature beyond just a definition and conversions. Get students to develop some questions about temperature that deal with thermal equilibrium of different substances with different states of matter.

Do not start calculations too early. These build a false sense of mastery. Calculations should reinforce what a student learns conceptually. A good guide is to wait until students are confused, and then use calculations to clarify confusions that students identify.

When students are stuck or frustrated and you're not quite sure what the correct explanation is; go back to the experimental evidence. What did we observe? You might find some questions are too advanced for now and there is insufficient evidence, but maybe in a later chapter (Ch. 7 enthalpy, Ch. 16 entropy) we will acquire more observations that will then answer that question. An unresolved question is a good thing because it can push our thinking.

Along those lines, be patient. These are tough concepts that are very new. Let the students have some time to make connections, think hard, and figure them out. It took me half a decade to figure out why water and iron having different specific heat capacities didn't violate Newton's 3rd law. It's ok if it takes students 4 weeks to understand temperature. James Joule went over a waterfall for this chapter (he didn't really, I just made that part up), so the least you can do is take a deep breath and let students struggle through ideas.

Student Struggles

1. "I have no idea what I'm doing!" When students are calculating using specific heat capacity, they often confuse Q with C. This often implies the student does not have a strong understanding of what heat is. The student may not understand what it means to have three combined units (J per gram per °C). The student may not have any working concept of what specific heat capacity is. All of these issues have one thing in common. The approaches that the student is using are too abstract. They need more concrete models, experiences, and connections to be more successful. The solution to this is not just having the student identify that the thing with more units is C, and the thing with just J is Q. They need to use "for every" statements to get a

Chapter 3 – Thermochemistry

better sense of the units of specific heat capacity, explain proportional reasoning about the constant C (Fig. 3-6), and they need to be connecting the concepts of heat and specific heat to actual experiences with materials.

2. Students confuse a small slope of Q vs. T with having a large specific heat capacity (Fig. 3-6) - I have observed this to be a frequent issue where students have an intuition that is reversed for this particular relationship. I think that part of it stems from having seen heating curves in earlier courses where temperature is the y-axis and heat (or time) is on the x-axis. The best way to undermine this confusion is to have students construct "For every" statements for two slopes on the Q vs. T graph. Have them predict which would be water (high specific heat capacity) and which would be metal (low specific heat capacity). As they apply the "for every" statements for slope, some will start to notice the conflict.

3. Students struggle with calorimetry - This is another likely candidate for the "I have no idea what I'm doing" phase. The students should be directed to start with the idea that $Q_{sys} = -Q_{surr}$. Usually, the student is overwhelmed by the large amount of data and units that seem to overlap. They should begin the problem by labeling each data piece as belonging to either the metal or water (or other substances). They may need to produce constants that are expected to be memorized ($4.18 \frac{J}{g*°C}$). Often students miss these key starting points if the teacher works out an initial sample problem because they become engrossed in the solution. It can help to remind them at the end of the solution by calculating Q for both substances.

4. *"How many bars in the LOL diagram?"* - Students will hyperfocus on whether hot coffee should have 3 or 4 bars of thermal energy. Students will want to add fractional components (half-bars, or more divisions). But the purpose of the LOL diagram is to help organize the changes that are taking place. Whether a thermal change goes from 4 to 2 or 3 to 2 is not critical. What's important is that it is decreasing and this potentially means energy is moving from the system to the surroundings (exothermic). Don't miss out on the big ideas by making the bar assignments overly complex.

5. Heating curves - The level of struggle with heating curves always exceeds my expectations. To expose student confusions, ask them to determine states of matter along the heating curve during phase changes. If a substance is melting from 4-10 s on the x-axis, what is present at 9 s? Ask them to explain the slopes during temperature changes. Ask them to predict how phase change sections would change if more mass, or faster heating occurred? Have them assess on heating curves of substances that aren't water. During instruction you'll want to ask why the temperature does not change during boiling. Students tend to get stuck on this for melting and do not trust that the temperature of an ice/water mixture (at equilibrium!) will not change temperature.

Chapter 3 – Thermochemistry

6. Multiple-step calculations - These are difficult. If you start with 18.0 g of ice at -20.2 °C, and end with 8.0 g of steam and 10.0 g of boiling liquid water, how much energy did the water gain? Students have to be able to organize the various steps of this calculation. Heating curves and LOL diagrams both help with this. The most successful students with these problems are able to use the units of the constants to set up their calculation. For 4.18 $\frac{J}{g*°C}$ they know they need to multiply by mass and temperature change to get the energy. For 2260 $\frac{J}{g}$ they only need to multiply by mass (there is no temperature change). Successful students tend to notice patterns such as the H_v for water (2260 $\frac{J}{g}$) is much larger than H_f (334 $\frac{J}{g}$) and this correlates with the heating curve.

7. Particle nature of solids - Students do not know how close together solid particles are. Do they touch? Are there gaps? Do they move? Students know that gases move, they know that liquids move, but it can help them to provide evidence that solid particles are in motion. The best way to do this is to think about a solid substance that can cause a burn when hot. When I pull a baking pan out of the oven, the solid particles are moving so fast that the collisions with my skin particles cause damage. Ouch! There are models of solids where the particles are connected via springs. This can be helpful to giving students an idea of how solid particles can be close (condensed) in structure, but also able to have some spacing (solids expand when heated).

8. Particle nature of liquids - Students usually think that liquids are spaced much further apart than solids. They are not. You can compare densities to confirm this (density of Hg (l) is 13.6 g/mL vs. Hg (s) is 14.2 g/mL). The particles are similarly spaced in solids, but they can move to new locations relative to other particles. Ge, Ga, Bi, and Si have greater density as liquids than solids. Having students develop particle drawings will help them appreciate the similarity of particle spacing in condensed states.

9. Melting - Students think of melting as a process that spans over a temperature range. Melting is much more effectively communicated as an equilibrium between solid and liquid states that happens at a single temperature. If pressure changes then that temperature can fluctuate. The big issue with students is that they have experienced mixtures (butter, ice cream, crayons) melting over a range of temperatures, and they have experienced ice-water mixtures that are not thoroughly mixed. It is possible for the top portion of an ice-water mix to be at a higher temperature than the bottom. If thermal equilibrium was established in an ice-water mixture, then the temperature would be 0 °C.

Phenomena

1. Superheated steam - A flask with boiling water has a rubber seal that the steam escapes through copper tubing (Fig. 1-7). As the steam travels through a coil of the copper tubing, the steam is heated by a Bunsen burner causing the steam to increase in temperature. This demonstration helps students reconcile the fact that steam is

Chapter 3 – Thermochemistry

invisible with the clouds of liquid water droplets they see that many think is steam. This is also a unique opportunity to show the 5th section of a heating curve of water. You can use the superheated steam to char paper, ignite a matchstick, and ignite flash cotton. Thus, starting a fire using water.

2. Heating paper on a pipe - Wrap a piece of paper around a metal pipe. When the paper is placed into the flame of a Bunsen burner, the paper may char slightly, but it does not catch on fire. Students may think this is due to the pipe limiting the oxygen contact with the paper, but 50% of the collisions between the paper and oxygen are ongoing. Instead, what is happening is that the high-energy collisions are having their energy dissipated by the metal pipe particles.

3. Melting ice on copper and cork - Have students feel a block of copper metal and a piece of cork. Ask which is warmer? They will identify the cork as feeling warmer and the metal as feeling colder. When a chunk of ice is placed on both, the ice on the "colder" metal will melt faster than the ice on the "warmer" cork. This discrepant event can be explained easily once we define the systems and surroundings. Initially a person (temperature 37 °C) is in contact with the room temperature cork and metal (20 °C). Energy flows faster from the person to the conductor (copper), so the copper feels colder. But ice is at a lower temperature (0 °C). Energy flows faster from the conductor (copper) again, but this time energy is going from the copper to the ice instead of from the person to the copper. In both instances, the cork acts as an insulator.

4. Hot metal vs. cold water - This is such a powerful phenomenon to push student thinking. Hot metal (50 g) is placed into cool water (50 g), and the metal changes by a significantly larger temperature change. What does this look like at the particle level? How can we model these changes to predict future temperature changes? How do collisions result in one set of particles changing speed more than the others?
The data from this can later be used to determine the specific heat capacity of the metal as long as the temperature changes and masses are recorded.

5. Which is hotter? - A small flask (50 mL) of boiling water (100 °C) and a medium flask (250 mL) of hot water (60 °C) are presented. Which is hotter? Students can select the flask they think is hotter. The small flask is added to a liter beaker of water and the temperature increases a small amount. The medium flask is added to a different liter beaker of water and the temperature increases by a larger amount. *Which is hotter?*

An argument can be constructed for either flask being considered hotter. The smaller flask had the higher temperature. The medium flask caused the larger temperature change. We need a more precise question that replaces "hotter" with more specific terms. Which has the higher temperature? The small beaker. Which had more thermal energy? The medium beaker (relative to the large beakers).[5]

6. Food coloring in hot and cold water - Add 2 drops of food coloring to a beaker of warm water. Add 2 drops of food coloring to a beaker of cold water. The food coloring will diffuse quicker in the warm water. This demonstrates that as temperature goes up, particle motion increases. Be wary of the type of food coloring used. Blue food coloring is denser than water which can cause irregular patterns that distract from the purpose.

7. Sublimation of iodine - Iodine is a black solid that produces a purple vapor. Iodine actually does melt prior to vaporization. Many think that the triple point for

Chapter 3 – Thermochemistry

iodine is above atmospheric pressure, but it isn't. And liquid iodine does exist at standard pressure.

Place a small sample of solid iodine in a large test tube. Then use tape to secure a small test tube partially into the large test tube. Make sure the tape completely seals the gaps between the two test tubes so iodine cannot escape. Put some water in the small test tube and heat the large test tube gently. The iodine will vaporize, and solid iodine crystals will undergo deposition on the surface of the small test tube. The water helps absorb some of the energy.

8. Using a hot plate - At some point during your thermo unit your students will see a hot plate in use. The descriptions you use for the hot plate can make a large impact on what students see. I describe the hot plate as having particles that move faster due to electricity, then these particles bump into whatever is on the surface. Those collisions push those particles to move faster. Students that identify motion transfer via collisions will think more deeply than students who only state the thermal energy transfers during heating. Students will overapply this and refer to pressure from the hot plate.

9. Heating curve - Whether you have students observe a heating curve, or a cooling curve, you'll want them to observe a constant temperature during a phase change. Be aware of some of the consistency issues that might disrupt the constant temperature. If you are boiling a liquid, any dissolved solids will concentrate as the liquid vaporizes. The increase concentration will lead to a higher boiling point. A similar shift occurs as a mixture freezes. The remaining liquid tends to draw in more solute causing the melting point to drop.

Some liquids that are very pure will suffer from supercooling where the liquid does not crystal at the melting point. When the liquid does start to freeze, the temperature rises to the melting point.

But the most common issue with temperature changes during phase changes is that the energy changes too rapidly and thermal equilibrium is not present. When a hot plate is used, the surface near the hot plate is much warmer than the top surface. Mixing can help but does not completely fix the issue.

10. Dry ice - Dry ice (-78.5 °C) can cause burns. But when handled safely dry ice is a lot of fun for students to learn from. Dry ice is solid carbon dioxide. You can purchase it at some grocery stores, and some packages that must be kept cold come with it. If you place a chunk of dry ice on a table, it will often make a buzzing sound until the surface of the dry ice allows for escape of carbon dioxide gas. After the buzzing ceases, you can usually push the dry ice chunk and it will continue sliding with minimal friction. The "cloud" you see coming off of dry ice is actually liquid drops of water forming as steam condenses near the very cold surface of the dry ice. Gaseous carbon dioxide (like steam) is invisible. Try placing a small chunk in different liquids.[6]

Chapter 3 – Thermochemistry

Flashcards

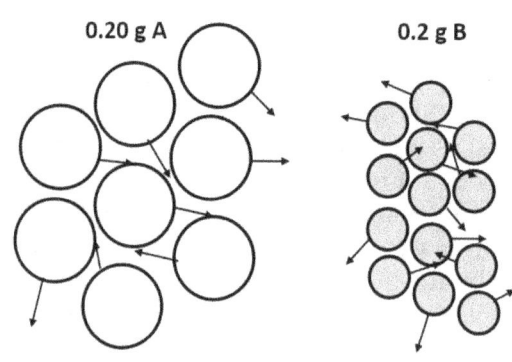

Substance A and B have particles moving at the same average speed. What will happen to those speeds when A and B come into contact?

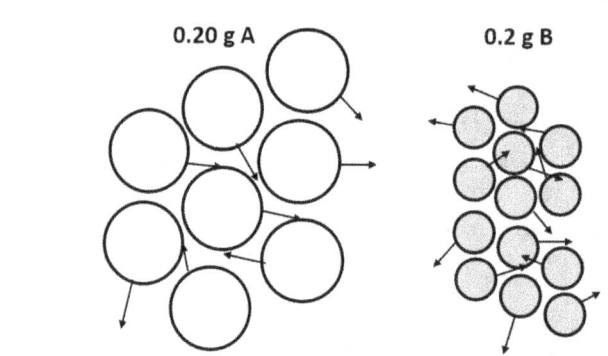

Substance A and B have particles moving at the same average speed. Which substance is at the higher temperature in Celsius?

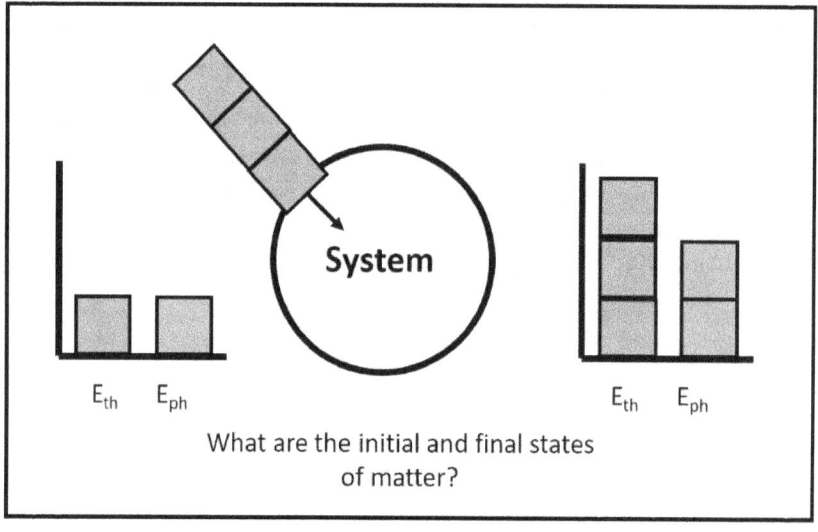

What are the initial and final states of matter?

Chapter 3 – Thermochemistry

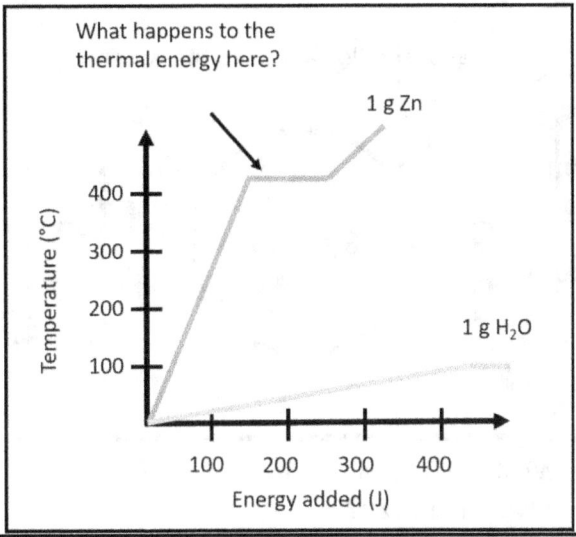

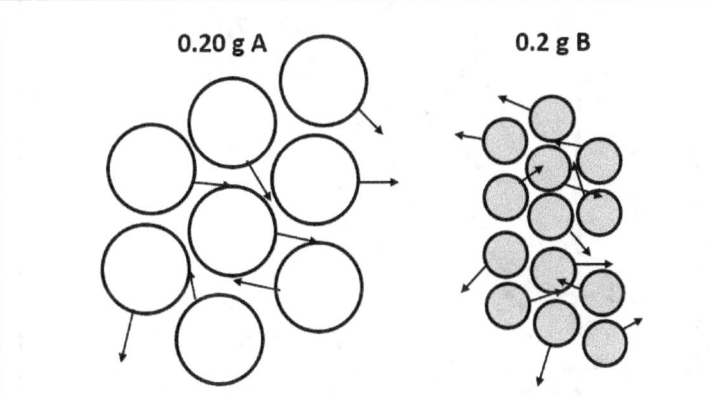

Substance A and B have particles moving at the same average speed. How could we tell which substance has the higher specific heat capacity?

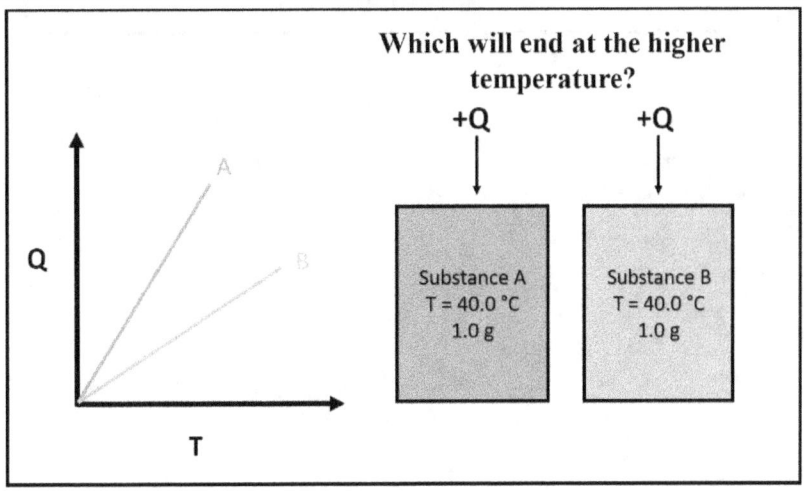

Chapter 3 – Thermochemistry

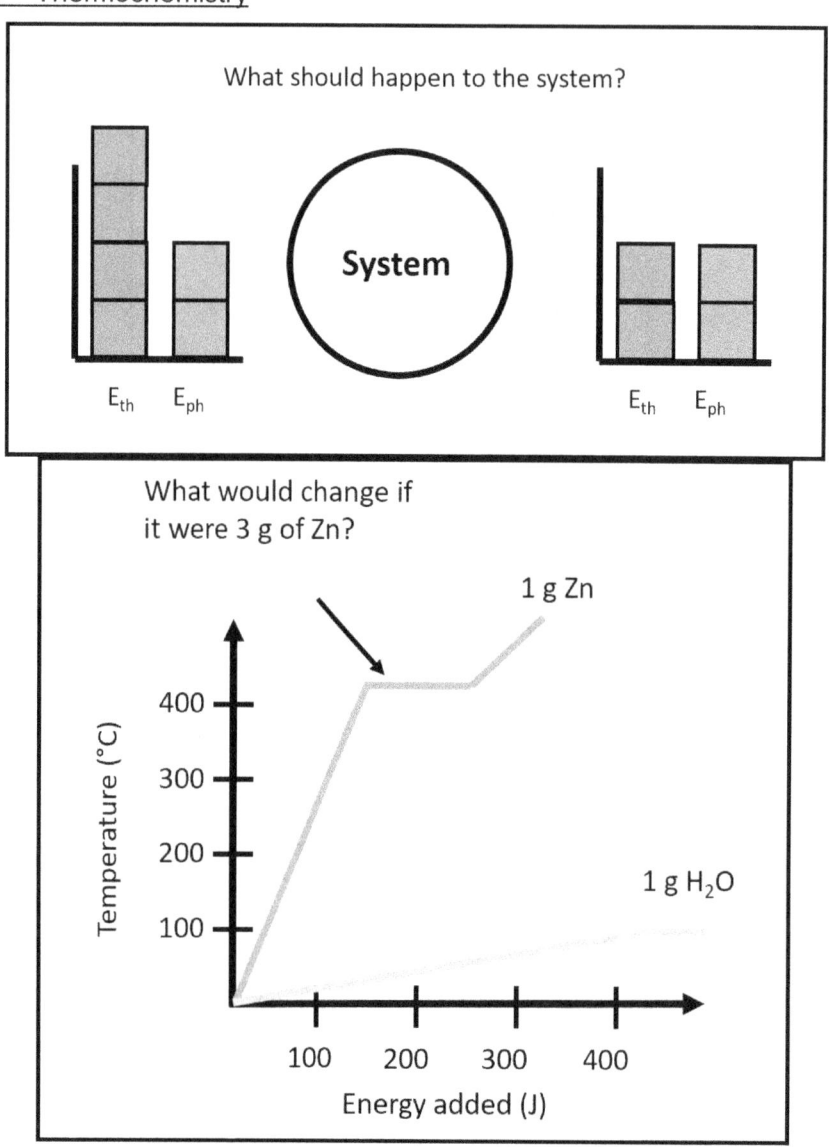

Chapter 3 – Thermochemistry

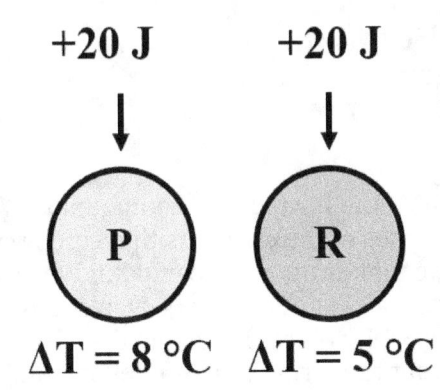

Which has the higher specific heat capacity?

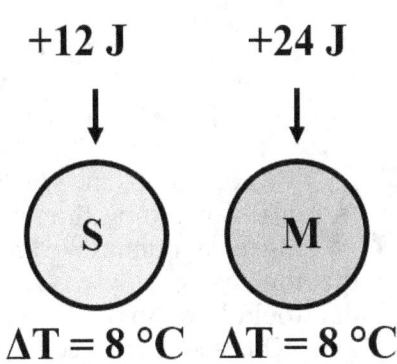

Which has the higher specific heat capacity?

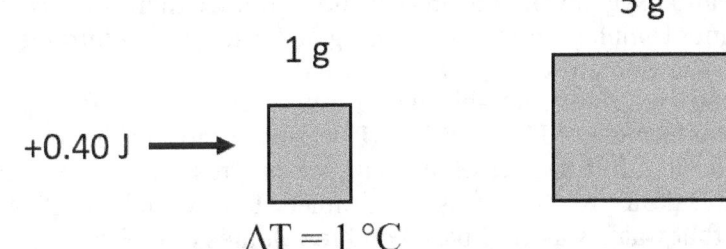

How much energy is needed to raise the temperature of the 5g chunk by 3 °C?

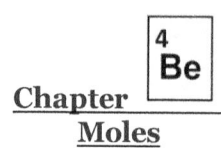

Chapter
Moles

History

We have a problem. We don't know the formulas of compounds. We also don't know the relative masses of the elements that comprise the compounds. We know if we make 100 g of a compound of sulfur and oxygen that there is 49.9 g of oxygen and 50.1 g of sulfur (there's even a 2nd compound that would have 60.0 g O and 40.0 g S). If you could tell me that the relative mass of S:O is about 2:1, I could figure out the formula. If you tell me the formula is SO_2, I can determine the relative masses. But with two unknowns, there's no way forward.

John Dalton finally cracked the code (1803). Just guess! Dalton had suspected that compounds have integer ratios of elements that do not exceed four particles. Dalton started progressing forward by assuming that compounds formed with oxygen and other elements would all be in a 1:1 ratio of particles. For element combinations that had multiple proportions, he used the limits of four as a guideline to estimate which compound would be a 1:1 ratio of the elements. Hydrogen and oxygen would form HO, carbon and oxygen would form CO, and so on. He then could compare how much of each element would combine with 100 g of oxygen and determine the relative masses of the elements.[1]

This worked except for a notable glaring issue. Hydrogen and oxygen form a compound with the formula of H_2O, not HO. This caused most of the relative masses to be off by a factor of 2. But in spite of this flaw, we progressed forward. Minor improvements on Dalton's relative masses continued. For example, in 1808 Gay Lussac proposed that water was H_2O based on the volumes of hydrogen and oxygen that combined. Dalton found issues with Gay Lussac's HCl because it would require hydrogen and chlorine particles to split in half and this must disprove the water formula. This would be resolved in 1811 by Amedeo Avogadro when he proposed diatomic molecules. But Dalton dissented again thinking that it would not be possible for two atoms of the same element to attract. It was a messy progression.[2]

Berzelius published more accurate atomic weights up until 1826. But the biggest shift forward for atomic masses came in 1858 when Stanislao Cannizzaro published a precise set of relative masses. He used Avogadro's hypothesis for equal volumes of gases having equal number of particles. He also used specific heat capacities to determine elements that did not easily vaporize. His work was distributed at a conference in 1860 and went on to have a profound influence on the first periodic tables.

As the relative masses are improved, there is also development towards establishing the number of particles within the atomic mass of a given element. The first attempt at determining Avogadro's number was done by Josef Loschmidt (1865) who estimated the number of particles in a given volume of gas. Michael Faraday knew the amount of charge in 1 mole of electrons (Faraday's constant, 96,500 C, 1834), but it wasn't until 1910 when Robert Millikan determined the charge of a single electron.[3]

Chapter 4 – Moles

Jean Perrin coined the terms "mole" and "Avogadro's number" (1909) when he used Brownian motion to improve upon the number of particles in the molar mass of an element. Perrin also switched from using oxygen as the standard to carbon. Originally Dalton had used hydrogen as the standard since it has the smallest relative mass of the elements. Wilhelm Ostwald (1898) proposed that 1/16th of an oxygen atom replace hydrogen as the standard. Perrin's definition of a mole being the number of carbon-12 atoms in 12 grams was recently altered (2019). Currently a mole is just defined to be $6.02214076 \times 10^{23}$ with no chemical standard.

Theodore Williams Richards was the first American recipient of the Nobel Prize in Chemistry (1914) based on his work improving the precision of atomic masses. He also found that measurements could vary depending on the source for lead. Lead from radioactive decay had a different atomic mass than lead obtained from nature.[4]

Avogadro's Number

Have you ever held a piece of metal in your hand and thought about how many atoms are in it? Not a tuba, but something small like a ring. When you take a deep breath, how many atoms are in your lungs? Are there more atoms in a grain of sand or grains of sand in the entire earth? If you didn't know how many atoms were in that piece of metal how could you figure it out? If you step on a LEGO with bare feet, how many atoms work to inflict maximum pain?

Teachers love to present students with examples of how large a mole is. If you had a mole of pennies, you could spend a billion dollars per second and still have money left when you die. That mole of pennies would stack a quarter mile high over the entire surface of the earth. Examples such as these are helpful to communicate that the mole is a large number. But it is important to take things one step further. The number is so big our brains can't understand how large it is.

Most people don't understand how big a billion dollars is. Try and think of sixty things you could buy that would cost a billion dollars each. An island, a skyscraper, a warehouse of expensive cars, or a professional sports team might work as options. Yet if you had a mole of pennies, you would be able to spend a billion dollars every second of your life and still have money left over after 100 years. If you list sixty things, that would only be one minute of spending. You would have years of spending a billion dollars every second. Then the space required for that mole of pennies would be immense. They would cover the entire surface of the earth to a depth of a quarter mile. The scale of the mole is too big for our brains to really comprehend.

I know you might think you get how big it is. You're a teacher. You went to college. The earth is big. A mole of pennies is big. But do you honestly understand how big the earth is? Our brains are stuck of limits of scale based on things we can observe. Even if you travel around the earth that doesn't mean that you understand the true scale of its size.

The flip side of this scale is critical. If a mole is incomprehensibly large, chemicals must be incomprehensibly small. A mole of pennies is too large to envision. But a mole of water molecules is a sip. So, if a sip of water has too many molecules to envision, each molecule must be incomprehensibly small. And that tiny size of atoms and molecules is relevant to understanding chemistry.

Chapter 4 – Moles

Just look at how absurd the number is. When we write it as 6.02×10^{23} it doesn't have the same impact as when we write it out.

602,000,000,000,000,000,000,000

It took me 83 seconds just to type this out and I had to correct two typos. People struggle to understand how much larger a billion is than a million. This number is bigger than a billion sets of a billion. Here's a mole compared to a billion.

602,000,000,000,000,000,000,000
1,000,000,000

The size of this 6.02×10^{23} is a problem. The first thing we need is an intermediate to help students approach the concept. That intermediate is a dozen. A dozen is not 12. A dozen is 12 of something. Likewise, a mole is not 6.02×10^{23}. It is 6.02×10^{23} of something. If you had 4 dozen donuts, how many donuts would you have? Students can easily shout out 48. If you had 4 moles of donuts, how would you figure out how many donuts you have?

Students will use moles throughout chemistry. We want students to move beyond just being able to calculate with moles. We want them to understand that moles give us a way to communicate about the number of particles without using absurd numbers. They need to have something more concrete than just being able to execute the algorithm.

Eventually we are going to use moles as a foundation for recipes in chemistry during stoichiometry. We want to establish that relationship now, but we expect students to struggle with the comprehension of the large quantities.

To expose student confusion, ask them to articulate how many atoms and molecules there are in a series of amounts that do not require a calculator.
- How many atoms in 1 mol of Fe?
- How many cookies in 1 mol of cookies?
- How many molecules in 2 moles of N_2 molecules?
- How many atoms in 1 mole of P_4 molecules?

Students give a wide range of responses to these questions. Some have said that we would need to know the molar mass of the cookies. Many would write 56 g for 2 moles of N_2 and 55.8 g for 1 mol of Fe. Those same students can successfully complete algorithms where they solve for grams and moles when they set up a conversion chart. But they don't know what it is they are doing. And that problem is going to undermine future content.

Molar Mass

Molar mass appears frequently in chemistry. Think of all the connections that students will make that involve molar mass as an initial step. We want to emphasize the relevance and to work on building student confidence.

There are grams, moles, and particles. We can't measure moles or particles directly. We can only directly measure mass for many substances. But our ability to predict and interpret lies in numbers. If you tell me that we have 4.5 pounds of pizza I have no idea how much that is. But if you tell me we have 3 large pieces with 10 slices

Chapter 4 – Moles

each I can start to identify how much each person can eat and whether we have enough.

The first thing we need to establish then is that in chemistry we measure mass, but must be able to convert the mass measurements into a number of particles. Knowing that you have 32 grams of oxygen, and 4 grams of hydrogen doesn't mean much. But knowing we have 2 hydrogen particles for every 1 oxygen particle means a lot. We can interpret and predict from that information.

Relative mass helps students connect moles and masses with the symbols. The molar mass of sulfur is 32.07 g/mol and the molar mass of oxygen is 16 g/mol. What is the relative mass of sulfur to oxygen? If we represent a larger mass by using a larger particle, how much bigger would a sulfur be relative to an oxygen particle?

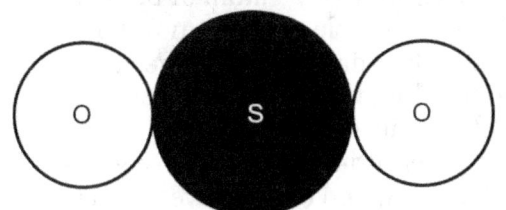

Figure 4-1: A particle of SO_2 where the sulfur is twice as large to represent its mass being about double that of each oxygen

Relative mass is about comparing all particle masses to a single standard. This has been done with three different elements in the past, hydrogen, carbon and oxygen. Hydrogen is the lightest element and therefore can be set to a relative mass of 1 where all other elements are compared to that. Oxygen was used by John Dalton when he worked to determine relative masses of elements that were not easily made into gases. Carbon is our current standard for relative mass, and now that we know about isotopes, Carbon-12 is our standard.

The historical path to finding those relative masses is impressive and helpful for students to see the importance of relative mass. Consider how difficult it was for Dalton to reach the conclusions he did with what we knew at the time. There was not a consensus that matter was made of atoms. There was no periodic table of elements. The relative masses of elements weren't known. There weren't stores to procure equipment from. There weren't electronic scales or electricity for that matter.

Dalton started progressing by making an assumption. He assumed that every element would combine with oxygen to form a compound that had a 1:1 ratio of particles. When C combines with O it forms CO. When Ag combines with O it forms AgO. When H combines with O it forms HO. We now know that some of these assumptions were correct, and others were off by a factor of 2.

If we assume that C and O form CO, we can now determine the relative masses of C and O. 3 grams of C are needed for every 4 grams of O. O has a relative mass of 1.33 relative to C. This means that every O particle has a mass 4/3 as large as every C particle. This matches our current relative masses where O is 16 and C is 12.

In order to find the relative masses, one needs to know what formulas the compounds form. But in order to know the formulas, the relative masses need to be known. It took an inordinate amount of creativity and persistence to come up with the relative masses. And just when you think you're making progress, diatomic elements throw a huge wrench in the problem.

Prior to Dalton, the simplest place to start making these assumptions was by reacting gases. Gases at the same pressure and temperature will have volume (V) and

Chapter 4 – Moles

amount (n) proportional. If you react 20 mL of gas A with 10 mL of gas B you had twice as many particles of A relative to B. The formula of the product is probably A_2B.

If we react 20 mL of hydrogen with 20 mL of oxygen, we will form water and there will be 10 mL of oxygen gas left over. This implies that the formula of water might be H_2O since 2 particles of hydrogen reacted with every 1 particle of oxygen. There is a video called *"Gases and How They Combine"* where George Pimental shows experimental evidence of gases reacting. Every year I sit on pins and needles hoping that somewhere a college professor who has the glassware capabilities reproduces this video with an updated camera.

One interesting sticking point in gases combining is that some elements are diatomics. Let's assume A is actually a diatomic A_2 in the example of A and B in the preceding paragraph. There are now 2 molecules of A_2 for every 1 atom of B. This would change our formula of the product to A_4B (or a multiple of that ratio). For water this still produces the same answer because both hydrogen and oxygen are diatomic. 2 particles of hydrogen are really 4 atoms while 1 particle of oxygen is 2 atoms. The ratio remains 2:1 even though it could be considered 4:2.

But Dalton went even further than gases by measuring how much oxygen gas reacts with several elements. From this he was able to expand our relative masses to include elements that are solids or liquids. He did not consider the possibility of diatomics and thus frequently ended with results that were off by a factor of 2 for relative masses.

If these past few paragraphs of the history of relative mass are confusing, then think of what a great opportunity they are to help students struggle through the concepts of relative mass and molar mass. Think of how much more effective students would be in later chemistry topics if the 16.00 g/mol was not just the molar mass of oxygen, but also indicative of the relative mass of an oxygen particle to a hydrogen particle.

This dual interpretation gives students a means to connect the particle level (relative mass) and symbolic level (molar mass). Without this, students are stuck in an abstract understanding where they must wait for future topics to make those connections.

The historical development of what a compound is can be used as a means to help student make molar mass and relative mass into concrete details. Give students data from Dalton's and Avogadro's experiments and see what they can make sense of. *If 100 g of a compound of these two elements has 80 g of the first element, and 20 g of the second element, what is the relative mass of the two elements?* They will struggle. But that struggle allows you to provide feedback and push them to understanding.

The tradeoff you want to consider is should the students struggle with this concept now or should they struggle through it later? If you've taught chemistry, you've undoubtedly noticed that even students who conduct these calculations easily have an apprehension about their understanding.

Without relative masses being connected to the particle representations students must instead use the abstract algorithm for molar masses of combining the numbers from the periodic table. A student might know that for Na_2O they must add 23 + 23 + 16 to get 62 g/mol (or 61.98 g/mol). Do they know what the result is though? Because this algorithm is so fundamental to so many things, we want

Chapter 4 – Moles

students to have something concrete to attach it to. Without that concrete connection, students can develop an underlying lack of confidence about their process. You may see this later when students do a problem correctly yet are convinced that they do not understand it.

The molar mass of calcium carbonate is 100.09 g/mol. The units of molar mass are not analyzed by many students. The biggest miss for many students is that "per" means "per one." There are 100.09 g of calcium carbonate per one mole of calcium carbonate. Unfortunately, we need students to also extend that one more step. There are 100.09 g of calcium carbonate per one mole of calcium carbonate, which is 6.02×10^{23} particles (formula units) of calcium carbonate. You want to ask students to articulate what the molar mass represents frequently.

When students struggle with mole calculations it is imperative that the teacher checks their use of units in calculations. Many struggling students omit units, and this leads to erratic calculations. When students have to include units, the processing becomes more difficult initially, but this initial difficulty forces the student to grapple with the meaning. I recommend students include both units and chemical formulae whenever they are having a difficult time.

The inclusion of units is also relevant to feedback. A student that forces themself to include units will get feedback along each step of the problem. Do I know which units here? Do I know which chemical? But the student that omits units can quickly throw down numbers and operations. They do not get feedback until they reach the final answer. Because their engagement is lower, they learn less and show more frustration.

Many teachers have students use dimensional analysis for mole conversions. This is fine since it is simple, and the algorithm is easy to follow. But if students struggle, using proportions can be much more effective for helping students uncover meaning. For calcium carbonate, the molar mass can be thought of as claiming that 100.09 g of calcium carbonate is equivalent to 1 mole of calcium carbonate. If students write with 18.21 g of sidewalk chalk that we assume is composed of pure calcium carbonate, we can then set up a proportion.

$$\frac{1 \ mol \ CaCO_3}{100.09 \ CaCO_3} = \frac{x \ mol \ CaCO_3}{18.21 \ g \ CaCO_3}$$

One benefit of using proportional reasoning is that this allows students to easily connect between the symbolic level and the macroscopic levels. If we draw a pile of chalk under the left side (100.09 g $CaCO_3$), we know that the pile on the right (18.21 g $CaCO_3$) must be smaller due to the smaller mass. This means that x must be smaller than 1 mole as well. Because the conversions involve multiplication or division, we know that the solution to x would be found by dividing 18.21 g by 100.09 g/mol (Fig. 4-2).

Chapter 4 – Moles

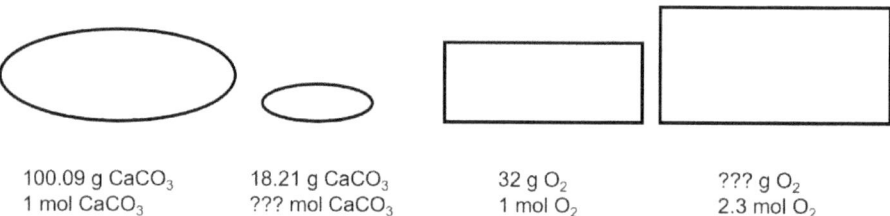

| 100.09 g CaCO₃ | 18.21 g CaCO₃ | 32 g O₂ | ??? g O₂ |
| 1 mol CaCO₃ | ??? mol CaCO₃ | 1 mol O₂ | 2.3 mol O₂ |

Figure 4-2: Proportions of $CaCO_3$ and O_2. Is there more or less than 1 mol in 18.21 g $CaCO_3$? Is there more or less than 32 g in 2.3 mol O_2?

The proportional reasoning is more concrete. The algorithm for dimensional analysis can be faster to follow, but it is more abstract. Picture the difference between a student trying to replicate a dimensional analysis algorithm and a student who has never seen an example trying to determine how many moles 27.4 g of H_2O is. What goes through each of their heads in order to successfully compute the moles?

The student following the sample algorithm is trying to match pieces. They see 27.4 g H_2O as the given and they try and fit that given into the same location in the sample algorithm. The 1 mole and molar mass of the sample chemical they see came from the periodic table. So, they find the molar mass of H_2O from the periodic table and put a 1 mole of H_2O in the denominator. It is likely that many students will not think critically about the relevance of the 1 mole nor its equivalence with the relative mass. Even if asked about the equivalency by the teacher afterwards, they are likely to process the question separately. They plug the numbers into their calculator and obtain the result. They did not have to think about what their prior knowledge was. They did not have to make any concrete connections. They did not have to know what any of the variables were.

The second student is told that 18.02 g of H_2O is 1 mole and asked to determine what 27.4 g of H_2O would be. This student is more likely to search their prior knowledge to figure out what the 1 mole of H_2O means. Ideally, they realize that 1 mole of H_2O tells us how many molecules there are. The student might then ascertain that 27.4 g of H_2O is more than 18.02 g of H_2O. This should tell them there is more than 1 mole of H_2O present and hence more than 6.02×10^{23} molecules as well. Now they need to find a method to solve the problem. They might use units to do a conversion. Or they might set up a proportion.

$$\frac{x \; mol \; H_2O}{27.4 \; g \; H_2O} = \frac{1 \; mol \; H_2O}{18.02 \; g \; H_2O}$$

Notice how the proportion allows students to see that there are more grams (27.4) on the left side. This means that there must be a proportionally larger number of moles as well (1.52 mol H_2O).

This ties back to some of the lessons in the cognitive science chapter. The more students must think, the better they will learn as long as we can avoid frustration. Many students will come in with an expectation that you must show them how to do everything and that they are helpless. This is a dangerous expectation that the teacher should reject. They are capable to set up a proportion, and they need minimal guidance to do so. But they will resist because it involves them thinking and many students have been trained to be fearful of that risk taking. How lucky are us

Chapter 4 – Moles

chemistry teachers that we will have so many opportunities to teach them how to do just that?

This is also a time to reflect on what your goals are as a teacher. Are you aiming to have students follow mathematical analyses of chemistry problems, or are you teaching chemistry? There is a difference, and we are at a point in education where both options are available to most teachers. What is your aim? I would also note that it is challenging for teachers to make a philosophical shift because most chemistry teachers had such a positive experience with doing mathematical analyses of chemistry problems during their own education. But in my experience the new approaches in teaching are sufficiently more rewarding and I am convinced I would have enjoyed them as a student just as I enjoyed my initial experiences.

% Composition and Empirical Formula

When students get to chemistry, they have a mostly complete fluency in calculating with percentages. The one thing I find useful to add is that all percentages in chemistry compare a part with a total. When we need the percent for nitrogen of ammonium nitrate (NH_4NO_3) by mass, the student must know to find the part that is nitrogen and the total molar mass of the compound. Once that comparison is made and adjusted to be out of one hundred parts, most students have an easy time working with percentages.

Students will initially have a couple of questions. For a chemical such as N_2O_4, they will wonder whether the percent of nitrogen should incorporate the 2 subscript or not (it should). The part is all of the nitrogen and there are two of them for every molecule.

This might also be one of the first exposures students have to parentheses for chemical formulas. Most students are prepared to distribute the subscript outside of the parentheses from their math class, but not all students will translate that to a particle level unless directed. For $Mg(NO_3)_2$ the students will know there are 2 nitrogens and 6 oxygens, but they won't necessarily process that there are two groups of NO_3 particles. They might draw a formula unit by putting a chain of Mg-N-N-O-O-O-O-O-O which indicates they are not considering this as two groups of NO_3. We have time to help students with polyatomic ions in later units, but we want to set that direction now if possible.

You may find it helpful to have two problems that compare percent composition of different compounds that share the same empirical formula. *Both C_2H_4 and C_4H_8 have 85.7% carbon. Ask students why that is?*

You might also consider including percentages that are not based off of compounds. Percent purity is another fun discussion. When high purity is required, chemicals tend to be more expensive since a more challenging separation may have been required. Silica is mostly used for concrete and glass. But a growing use of silica is in computer chips and solar cells. The purity of silicon used for computer chips is 99.99999%. Special sand is used for this, and that supply will not last forever.[5]

Cognitive science research supports having students try problems they've never seen before on their own. When we present a sample problem, students will work at mimicking the sample problem. But when we give students the opportunity to try a new problem that they've never seen before they have to follow an important sequence. They have to determine what is known, what is unknown, and what tools they have to use the knowns to find the unknown. Empirical formula is a fantastic

Chapter 4 – Moles

topic to give students the opportunity to learn how to learn. Reacting zinc with hydrochloric acid is a reliable experiment that is easy for students to set up their own empirical formula calculation. But it is even easier to give them some data where they must find the empirical formula from a set of masses or percentages.

Let's assume students are told that 75% of a compound is carbon by mass, and the remainder is hydrogen. The number of possible methods students might use to seek the formula is intriguing. They might figure out the ratio of moles, but they also might find the ratio of atoms. *If there are four times as many moles of hydrogen, does that imply that there must also be four times as many hydrogen atoms? Does the fact that hydrogen is diatomic matter?*

If you give students the initial opportunity to solve a new problem, they will have to identify the parts they know, connect to prior knowledge, make a plan to solve, evaluate the plan to solve, and reflect on the process. If you show them the algorithm, they only have to try and mimic what you did. The typical algorithm for finding an empirical formula from % composition is three steps.

1. Assume you have 100 grams of the compound
2. Change the grams of each element into moles
3. Divide by the smallest number of moles to obtain integers

Let's assume we give that algorithm to students, show an example, and then have them try it. Initially things will work out well. The students will solve their problem. Most of them will be successful, which will increase their motivation. But the solution will require little thinking. Little thinking means that their brains aren't changing. When we come back to this later, they are less likely to remember. When we make a modification, many students won't know what to do. If we provide a new problem that starts with the number of moles of each element, some students will promptly change the moles to grams. They do this because they lack familiarity with the content, and they haven't had a chance to address that yet.

There are a few simple labs that allow students to find the empirical formula from experimental data. Burning magnesium to find the empirical formula for magnesium oxide is a fun one. The Mg does not always burn completely, and nitrogen reacts in addition to oxygen. Of course, the students must be prepared to not watch the reaction as the light produced can cause damage to students' eyes. Do not just tell them not to look directly at the reaction, physically show them your head turned away with your eyes pointed away from the Mg where your only visibility is from your peripheral. A perk to the magnesium experiment is the curious observation that the mass increases. This can be used to invoke wonder about what happened in terms of system and surroundings. If the mass of increased, our system must have gained particles from the surroundings.

The increase in mass of burning metals was a historical conundrum that can add some personal narrative to the reaction. Antoine Lavoisier and Joseph Priestley were two scientists that explored oxygen in a time when the theories about burning were quite interesting. Burning was thought to release phlogiston from a substance, but this did not easily translate to metals that increased in mass when burned.

Priestley would mix NO (he called it nitrous air) with air over water. The NO would combine with O_2 to make NO_2 which then dissolves in the water. He would

Chapter 4 – Moles

then measure the volume changes of the gas mixture above the water to collect data.[6] Lavoisier was inspired by Priestley, repeated his experiments and developed conclusions from them. Priestley then produced oxygen gas by decomposing HgO (calx) and he tested how long mice could breathe in this new mystery gas. Lavoisier would use this experiment to undermine the phlogiston theory and replace it with oxygen gas. But Lavoisier was a very wealthy tax collector that fell victim to the guillotine during the French revolution.

Zinc reacts with hydrochloric acid to form zinc chloride ($ZnCl_2$). The excess hydrochloric acid and water can be evaporated away to leave behind only the compound. This is another reaction that works well. The issues in this experiment are that the zinc chloride is hygroscopic and absorbs water from the air, the reaction can splatter a bit as the hydrogen evolves, and the evaporation of the water takes time. Perks of using zinc chloride are that many students don't know what charge zinc forms yet so they can't predict the formula from charges. The reaction also produces consistent results.

Once the students have had an initial opportunity to determine an empirical formula, you'll want to provide them with a variety of problems. They should be able to determine an empirical formula from % composition, masses of each element, moles of each element, or volumes of gases. If you always do the same style of given information the students might not truly understand, and you won't know what they know.

Finally, you'll want to differentiate between an empirical formula and a molecular formula. The empirical formula has the subscripts in the smallest ratios. The reason why we start by determining the empirical formula is because the molecular and empirical formula give the exact same % compositions. CH_2 has 85.6% carbon. C_2H_4 has 85.6% carbon. C_3H_6 has 85.6% carbon. All have 14.4% H. When we start with the percentages (or masses of each element), there is no way to determine which we have. We need more information.

When students get to working out problems where they find the molar mass, it is worth prepping them on how to deal with the problem in two steps. Because the molar mass is a given many students will try to use that molar mass.

A compound is 85.6% carbon and 14.4% H with a molar mass of 56.12 g/mol. Determine the empirical and molecular formula.

The student needs to know that the 56.12 g/mol does not get used until the empirical formula is found. Students will sometimes start with the molar mass and things quickly devolve from there. Instead, the student should first find the empirical formula. They must compare the molar mass of the empirical formula with the molar mass of the molecular formula. CH_2 is the empirical formula which has a molar mass of 14.03 g/mol. 56.12 g/mol is four times that amount. That means the molecular formula has four times as much stuff as CH_2 does. The molecular formula is C_4H_8.

You can use the definition of a compound to enhance the idea of relative mass effectively if you follow the historical progression in this unit. A compound differs from a mixture because of how the particles interact. The law of definite proportions claims that a compound always has the same proportions between the elements. It's not just that sodium and sulfur stick together, it's that they always have 1.434 times the mass of sodium as they do sulfur. If there is 1 gram of sulfur, there will be 1.434

Chapter 4 – Moles

grams of sodium. If there is 32.07 grams of sulfur, there will be 45.98 grams of sodium.

Note that the relative mass of sodium is 0.717 to 1 for sulfur. Because there are 2 sodium particles for every 1 sulfur, our proportion for the compound is double that relative mass (1.434). When the first chemists were deciphering the relative masses of elements, they did so by precisely measuring the masses of elements that combined to make compounds. In chemical education it helps to start with that same puzzle to let students see how the concepts of relative mass, molar mass, and compounds can be used to strengthen student understanding. Note how this advances students' abilities to calculate compositional stoichiometry.

Student Struggles

1. "I have no idea what I'm doing!" Students who are lost with moles are often assumed to not have sufficient math skills. The scientific notation used can be daunting. But a component of what most students struggle with is the sheer scale being used. Because students can't connect with how numerous 10^{23} particles is, they struggle to create concrete representations. One method that can help with this is by making sure students connect the macroscopic image of a given mass of substance with a proportional relationship to the number of moles. If 83.2 g of mystery substance X has 1.0 mol, then having 13.4 g of X will be less than 1.0 mol. A visual connection for that proportional relationship (Fig. 4-2) is helpful for addressing this.

2. "I don't know whether to multiply or divide." If you hear something to this effect and see that the student work is disorganized, you need to provide the student more structure in solving the problem. This doesn't come by setting up an algorithm for them, it stems from having students identify the components of the problem. They must include units and chemical symbols for all quantities. This means that the given should include units (g or mol) along with the symbol for what chemical it is (1.2 g $MgCl_2$ or 2.3 mol N_2O_4). The molar mass in particular cannot be written as g/mol. Instead, the number 1 must accompany the mol along with symbols for both units (18.02 g H_2O/1 mol H_2O). This will test the student initially, but ultimately will provide them with enough structure to find a way to use the mathematics they are familiar with to solve the problems.

3. "I got the empirical formula correct, but I don't know what to do from there." Here the student doesn't get how to use the ratio of the molar mass for the empirical formula and the molar mass of the molecular formula. We always want to start by making clear that an empirical formula of CH_2 means the formula could be any multiple of that (C_2H_4, C_3H_6, C_4H_8, $C_{10}H_{20}$, etc.). All those formulas have identical percent compositions (85.6% C, 14.4% H). Determine what the empirical formula's molar mass would be. When we compare that to the molar mass of the molecular formula, we will find a whole number ratio. If that ratio is 3, that means the molecular formula has 3 times as much stuff as the empirical formula. Triple each coefficient and you'll have your molecular formula. It can help for students to see examples. CH_2 has a molar mass of about 14 g/mol. C_2H_4 would be 28 g/mol. C_3H_6 would be 42 g/mol. If we know the molecular formula has a molar mass of 70 g/mol, that's 5 times the molar mass of CH_2, therefore our formula must be C_5H_{10}.

Chapter 4 – Moles

I think students also have some trouble understanding why we don't just know the molecular formula in the first place. Why do we have to go through all these comparisons? It is easy to determine the percent composition for a compound. There are also methods available to determine the molar mass. But neither of those is sufficient to produce a molecular formula in isolation. Only when we have both pieces of information can we deduce the molecular formula. Students at this point haven't seen the many ways we can find the molar mass of a compound yet.

4. "*I didn't know how to start the problem besides the given.*" Once a student has written down that they have 24.3 g KBr, they have to know to use the periodic table next to determine the molar mass. This can be a break from how students typically are able to find all the information within the problem. It can take mentioning this several times and being explicit about finding the molar mass from the periodic table.

5. "*I can't figure out the number of molecules.*" Some students struggle with the scientific notation. Ideally, they have a calculator that can do the scientific notation for them, but some will fail to use parentheses correctly. If they end up with 10^{46} moles of a substance, they forgot parentheses in their calculation.

6. "*How many atoms are there in 2.0 moles of O_2?*" This is a question students get stuck on because the units don't match. O_2 are molecules. Each molecule of O_2 has 2 oxygen atoms. 2.0 moles will have 1.204×10^{24} molecules of O_2, which will be 2.408×10^{24} atoms of O.

7. "*How should I round my molar mass?*" Whatever you pick, be consistent and make it simple. For me, we round from the periodic table to 2 decimal places. Each H is 1.01 g/mol. Each Fe is 55.85 g/mol. But I choose that because that's how students that progress into my IB chemistry class are going to be expected to round on their IB exam. If your students are on a different track, it might make more sense to round to one decimal place. Or you might want better precision for answers, so you want to have at least 4 significant places for the molar mass. This has implications for how students round their final results.

Phenomena

1. Sidewalk chalk - Have student mass a piece of sidewalk chalk. Then take them outside to write their names or draw a school appropriate picture. Have the students measure the remaining mass of chalk. From the change in mass, they should be able to determine the number of moles of chalk on the sidewalk, the number of formula units, the number of ions, and the number of atoms. Chalk is either $CaCO_3$ or $CaSO_4$ depending on what type you use.

2. Individually wrapped chocolates - There is a type of chocolate that is wrapped in aluminum foil. Have the students measure the total mass, then eat the chocolate, then mass the aluminum wrapper. They can now determine the moles of aluminum in the wrapper and the moles of sugar in the chocolate (assume it's all sucrose or fructose).

Chapter 4 – Moles

3. Guess how many marbles! Have a container filled with an excessively large number of marbles (or other small items). Have each student write down their guess of how many are inside. The average of a large collection of guesses should be very close to the actual amount as long as each student makes their prediction before hearing what others predict. Then ask the students how they could determine the exact number without having to count all of them. They should propose an idea that you could remove one of the objects, get the mass of it, and use the total mass of the objects to figure out how many.

4. Show the students some samples of metals that are all 1.0 mole. Put them into containers to save year to year. You can purchase samples of 1 mole. If you show the masses, emphasize that all of the metal samples have the same number of atoms. This works as a great opportunity to bring up relative mass. The mole of Al atoms is about 27 g, and the mole of Au atoms (I wish!) is about 197 g. Each gold atom must be 7.3 times as massive as each aluminum atom.

5. How big is 1.0 mole of gas? It turns out that gases tend to be about the same volume for 1.0 mole of any gas. That volume is 22.4 (22.7 for the purists) liters (dm^3). For me, I can usually pull out 11 empty 2-L bottles to demonstrate how big that is (I drink a fair amount of diet pop). But you might also have a cube with a side length of 0.281 meters to show a 22.4 dm^3 space. Note how much more space that occupies than your 27 g bar of aluminum from earlier.

6. Chewing gum for percent composition. The concept that students don't translate from math is that a percent compares a part and a total. They can apply this seamlessly when it's about saving money, but struggle to do so in a chemical context. Chewing gum presents a second concrete example to emphasize the abstract idea of percentages. Have students mass a stick of chewing gum using the wrapper to protect the gum from the balance. They should then chew the gum for 10-20 minutes or until the flavor is lost. Ideally, they can let the gum dry a bit, but even without that the mass of the gum will have decreased substantially. Have them determine what percent of the gum was sugar (and flavoring) that left the system during chewing.

7. React zinc metal (get the mass first) with hydrochloric acid. Hydrogen gas will be evolved. If the HCl is in excess, you'll be left with dissolved zinc chloride. Heat the solution gently (in a fume hood) to evaporate the water and excess HCl. You can measure the zinc chloride mass and have students determine the relative quantities of zinc and chlorine in moles. Ask them to determine the empirical formula of the zinc chloride compound. When they get $ZnCl_2$, ask them what molecular formulas they might have? Note that Zn_2Cl_4, Zn_3Cl_6, etc. would all have identical percent composition. (We're assuming that students would not know about ionic compound composition at this point)

8. Coke reduction of malachite - If you have a Meker burner, you can heat malachite ($Cu_2CO_3(OH)_2$) in the presence of charcoal. This "coke reduction" technique is how we used to produce metals when temperatures of ovens were limited. The malachite will turn into CuO and eventually Cu.[7] Be wary of the carbon monoxide produced. Students can use this as an opportunity to apply percent composition to ores that have relevant economics.

9. Mole Day - Either October 23rd or June 2nd, many teachers participate in mole day. Students might sew moles, do relay races with 18.02 mL of water, or listen to music about moles. There is a #molympics that had participation from 31 states and 13 countries in 2022. Mole day can help create strong memories about moles.

Chapter 4 – Moles

Flashcards

Sulfur molar mass: 32 g/mol

Oxygen molar mass: 16 g/mol

How will 1 oxygen atom compare to 1 sulfur atom?

Empirical Formula: C_2H_5
Molar mass C_2H_5: 29 g/mol

Molar mass: 116 g/mol
Molecular formula: ???

---

1 mol XY = 25.6 g XY

3 mol XY = ??? g XY

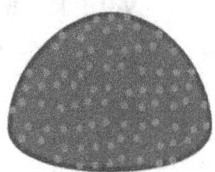

Chapter 4 – Moles

Molecular formula: N_2O_4

Empirical Formula: ???

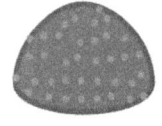
1 mol XZ = 52.0 g XZ

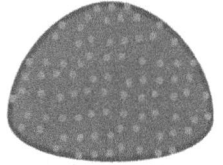
??? mol XZ = 156.0 g XZ

SO_2 SO_3

Which has the greater percentage of sulfur by mass???

Chapter 4 – Moles

Explain how you would find the molar mass of Cu(NO$_3$)$_2$

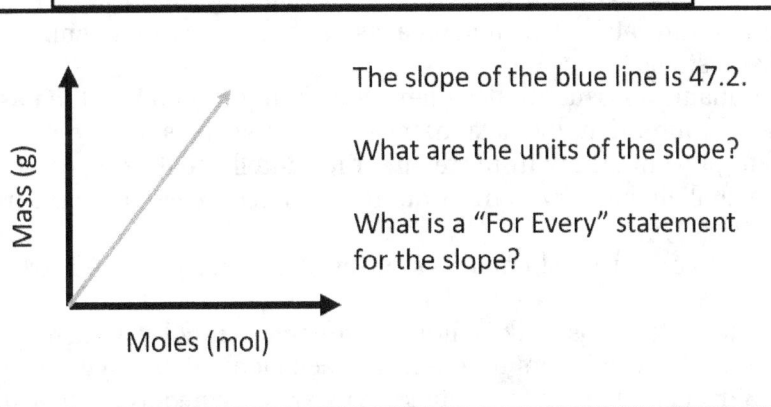

The slope of the blue line is 47.2.

What are the units of the slope?

What is a "For Every" statement for the slope?

C$_2$H$_2$ C$_2$H$_6$

Which has the greater percentage of carbon by mass???

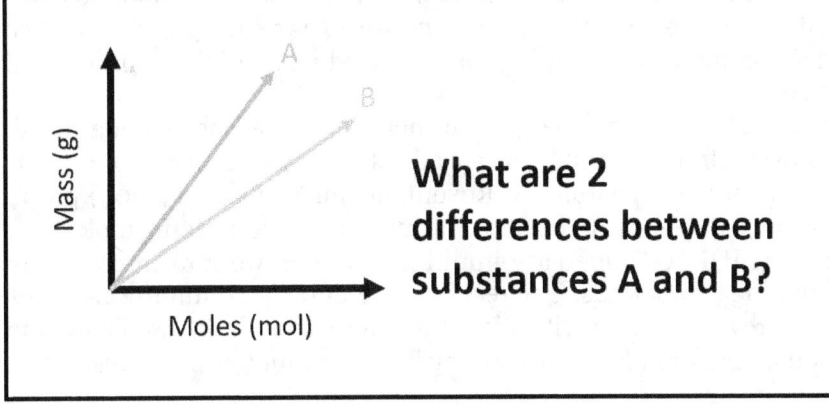

What are 2 differences between substances A and B?

Chapter 5 B
Naming and Formula Writing

History

Prior to Antoine Lavoisier's publishing of Method of Chemical Nomenclature (1787, along with Berthollet, Fourcroy, and de Morveau), there wasn't a universal system for naming chemicals. This means that scientists had to translate between languages as well as varying chemical nomenclature systems.

This remains an issue for deciphering alchemy research of the past. Some alchemists would intentionally make names obscure such as "divine water" or "bile of the serpent" to prevent others from stealing their intellectual property. Even common names can be challenging. Try to translate the chemical reaction below from 1612:

stibnite + corrosive sublimate → butter of antimony + cinnabar[1]

Jons Jacob Berzelius took Dalton's confusing symbols[2] and created a systematic method to use letters to represent each element as a symbol (1813). Nonmetals (Berzelius called these metalloids) were assigned to be their first letter of their names unless that letter was taken by an earlier symbol. Metals were assigned the first two letters unless there were two such metals in which case the second metal would go to the first consonant that differed.

A few of Berzelius's element symbols have since been altered (Ch/Cr, Tn/W, Cl/Nb, Pl/Pt, Pa/Pd, Ma/Mn, Gl/Be, Ms/Mg, So/Na, Po/K, M/Cl). Initially there was pushback to using these symbols, including from Dalton.[3] But in 1834 a recommendation from the British Association for the Advancement of Science recommended their use and they slowly gained favor from there.

Humphry Davy made use of the zinc-copper voltaic pile to improve our knowledge of chemicals. He isolated several elements including sodium, potassium, and calcium. Jons Jacob Berzelius had done similar experimentation (1803) and proposed that some elements were positive (+) in nature and others were negative (-). This new "dualism" model worked well for explaining some features in chemistry. Ionic compounds in particular fit the model well. Molecular compounds had some issues. Eventually Gilbert Lewis (1916) along with Irving Langmuir (1919) constructed a valence electron model for bonding that resolved many of the conflicts of Berzelius's dualism model.

As the 1900s start we emerge with increasing needs for standardization of naming. Bohuslav Brauner (1902) proposed a suffix system (a = 1, o = 2, i = 3...) as a standard system of nomenclature. A. Rosenheim and I. Koppel (1909) propose a numeral system 3-iron 4-oxide (Fe_3O_4).[4] Neither of these systems took hold. After World War I, the IUPAC formed a committee (1938) to work towards a common method of naming compounds. The results were initially communicated in 1940, and after 8 years of revision and clarification they were settled. The Stock system was the result, named after Alfred Stock, who contributed a number of suggestions that were utilized.

Chapter 5 – Naming and Formula Writing

Despite these efforts, common names persist to this day. They will continue to persist in the future. Some organic compounds have much simpler common names than the official IUPAC version. Other common names (water, table salt, etc.) are entrenched in everyday language. Some prefer to know both the traditional names used as well as the Stock names.

Even before we knew the exact formulas of compounds, we did know that some had more oxygen than others. Thus, some of the acid nomenclature we still use was from Antoine Lavoisier back in the late 1700s (Sulfuric and sulfurous acid). Lavoisier used the name acid, which means begets oxygen, thinking that all acids contain oxygen. This was already countered by Berthollet showing that hydrocyanic acid was comprised only of H, C, and N (1787). Scheele had shown that hydrosulfuric acid was just H and S (1777). And later Humphry Davy confirmed that hydrochloric acid was just H and Cl (1808).

Names are selected with intention in chemistry. Learn some of the linguistic and geographical connections because they provide students with a feeling of comfort when they can bridge chemistry knowledge with other subject material. Methyl alcohol or methanol has Greek origins from tree spirits. Fulminic acid comes from the Latin for lightning. Nickel's name came from kupfernickel or the devil's copper since it often contaminated copper ores.[5]

Naming and Formula Writing

In 1787 Louis-Bernard Guyton de Morveau teamed up with Antoine Lavoisier, Claude Louis Bertholet, and Antoine François de Fourcroy to publish Méthode de Nomenclature Chimique. This book set out to establish a systematic means of naming chemicals that would improve communication. Since then, the scope of complexity and knowledge of chemicals has grown considerably, yet our objective remains the same. *How do we effectively communicate with other chemists?*

High school students do not share in this desired objective. They seek a system of simplicity. But chemical nomenclature is about providing the maximum amount of necessary information in the shortest communication possible. *The system is designed for experts to use.* The omission of obvious components is easy for experts, but novice students can struggle immensely.

Differentiating elements and compounds should be more of a journey than a set of phrases. At the start nomenclature most students know that elements are one type of atom and that compounds have two or more stuck together. It is imperative that as we teach naming, we expand their understanding of elements and compounds to include the charges of the particles in both forms (elemental and compounded).

When elemental iron reacts with elemental sulfur, they form a compound with completely different properties. When metallic magnesium combines brilliantly with gaseous oxygen, they form a white salt. Yet these compounds are made of the very things that we started with. Why do their properties differ so much?

Naming is our opportunity to begin students down the path to understanding how atoms change properties when they combine into compounds. The fundamental difference is the amount of electron density or charge. As frequently as possible we want to identify charge with symbols and with particle representations.

But why is the symbol Na for sodium? Shouldn't it be So (originally it was!)? Even if students aren't asking this, they are wondering why that's the case. You might

Chapter 5 – Naming and Formula Writing

enjoy telling some jokes about Hey You (A U), gimme some gold. Or the classic Ahhh, gee, (A G) it's just made of silver. But it's also good to give them some closure and provide the students with a little bit of Latin and rationale.

Many elements have been known for a long time. Their names and symbols weren't always the same as what we use now. This is an ongoing problem in chemistry. When we try to develop a systematic naming system we run into problems with the existing names. Some of the original names won't fit easily into a new system. Should we push for people to only use the systematic names, or should we stick with water? The answer depends on how commonly the elements or compounds are used. The ongoing jokes about dihydrogen monoxide are clever, but we also aren't going to replace water with the IUPAC name anytime soon.

Element	Symbol	Latin basis
Gold	Au	Aurum
Sodium	Na	Natrium
Potassium	K	Kalium
Iron	Fe	Ferrum
Copper	Cu	Cuprum
Silver	Ag	Argentum
Tin	Sn	Stannum
Antimony	Sb	Stibium
Tungsten	W	Wolfram
Mercury	Hg	Hydrargyrum
Lead	Pb	Plumbum

Students tend to view naming from their lens as a secondary student in an academic setting. They see naming as a frustrating obstacle. It's better if they view naming as a necessity for chemists to be able to communicate. Students will ask why there are so many different rules because they want this to be simplified. What they don't understand initially is that naming has been simplified down, but that task is more difficult than they expect because of the sheer number and variety of chemical compounds. There is no one size fits all algorithm because the differences between ionic, molecular, and other compounds complicate the process.

Should students memorize the element symbols? I will do my best to present the arguments for and against. Regardless of which you choose I would encourage you to strongly consider the opposing argument. The rationale for memorizing the elements mostly stems from students being able to chunk better. When learning new things, our short-term memory can only hold 7 +/- 2 items at a time. A 7-digit number could maybe be remembered for 30 seconds or so. But if some of those 7 digits form a single chunk then it becomes much easier. 23,571,113 could be quite formidable to memorize. But if I point out that it is the prime numbers from 2 to 13, it becomes a single item to remember and much easier. In fact, I will now type the number again from memory. 23,571,113.

When a student knows that Na is sodium, they are better able to chunk information. Na_2O is just 2 sodiums and one oxygen. But a student that does not know what Na is has to hold the Na symbol in their short-term memory while they also work out the algorithm for naming. A student learning Spanish might translate "arbol" to tree and then they visualize a tree. But a native speaker hears the word

Chapter 5 – Naming and Formula Writing

"arbol" and they picture the tree directly. That lack of extra step is an advantage cognitively. We want the student to have greater capacity for connecting representations in chemistry.

The argument against memorizing the elements is that they are abstract. The student that knows that Na is sodium still might not know what either of those is because they do not have concrete details to attach them to. The student can't chunk information because both are abstract. A tuba is like a big trumpet. A trumpet is like a little tuba. If you don't know what either is, you haven't progressed by associating them together.

Some claim that students will learn the relevant element symbols as they practice the naming and formula writing. In this event, I do not see the harm in giving students some specific focus and attention on the memorization since inevitably not all students will memorize the elements without that. But I also think that a key here is that students just learning how to name simple compounds probably won't need to know the symbols or names for terbium or iridium.

I personally am in favor of having students dedicate effort to memorize some common elements. But I do so while giving them concrete details about the elements. We look at pictures of elements, they hear stories about them, and I do demonstrations with them. There are even some small collections of elements that are surprisingly affordable to purchase.

Two years ago, I tried not having students memorize elements. I did not find a substantial difference in their test scores at the end of the unit. But I do worry that may have only been true for the average score. Part of my reasoning for having students memorize the elements is because I frequently see struggling students unable to easily match the element symbols correctly throughout the year. I have concerns that some students have had more exposure to chemistry prior to high school than others and I want to level out that playing field. I am intentional to make sure that grading around element symbols is reflective of final mastery and not how long it took to get there.

Another perk for memorizing is that you can demonstrate quality study techniques. You can do spaced practice by starting each day with a few minutes of element symbol retrieval practice. You can teach about dual coding and have students pair imagery with the element or symbol. You could even have students create a memory palace to memorize the order if they were interested.

Element memorization should take place during class in order to demonstrate effective study tools such as retrieval practice. (S is written on the board)
- What element is this? What does it look like?
- Is it a metal or a nonmetal?
- When it is charged, what charge does it tend to form?
- Which symbol is used for nitrogen, N or Ni?
- Which symbol is used for copper, Co or Cu? What the heck is W?

Notice how these questions are a mix of abstract and concrete.

Later in the unit you're going to need students to use polyatomic ions. It is more important that they understand what they are, but again you'll need to decide if you want students to memorize some of them or not. But first they need to know a few things. Polyatomic ion has three features in the name. "Poly" means many, "atomic" means atoms and "ion" means that we're dealing with something charged.

Chapter 5 – Naming and Formula Writing

We have a group of atoms that have a net charge. The charge part is slightly new, but the group part is surprisingly more difficult for students to grasp.

Don't forget that you the teacher know about bonds, different compounds, intermolecular forces, the scale of molecules, dissolving, and many more concepts that help you understand what a group of atoms is and is not. The students have limited understanding of those and therefore can't reliably use them as models to easily break down what SO_4^{2-} means. Many of the struggles in this unit and in reaction types stem from a lack of understanding of polyatomic ions.

What you can provide them is a particle diagram. Give them the particle representation and formula for one polyatomic ion. Ask them to determine what the formula would be for a second particle representation (Fig. 5-1). Then ask them to draw NH_4^+. You can deal with charges by using a plum pudding model. This forces students to focus on subscripts and charges. If memory is the residue of thought, this helps students acquire the skills to be successful in formula writing. If you put charges like in Fig. 5-1 students may even look ahead by wondering why the charges go with the oxygens and not the sulfur or chlorines.

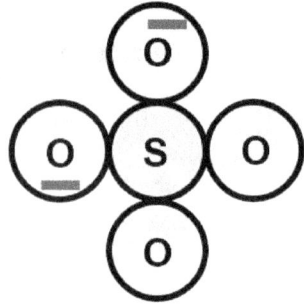 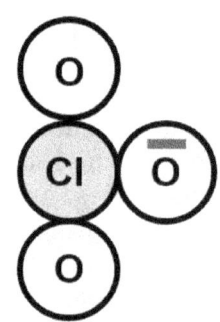

Figure 5-1: Particle representation of sulfate and chlorate with a - for the excess charges

Polyatomic ions are semipermanent. They tend to remain intact for many reactions. But they do not always do this, so be cautious about how you describe them to students. In the following unit they will see things like
$CaCO_3$ (s) → CaO (s) + CO_2 (g) where the polyatomic ion does not stay intact. Leave that possibility alive in their models for now.

Just as with elements, be sure to give students some experiences that they can build their memories of polyatomic ions from. There are a lot of interesting chemicals that contain polyatomic ions. Mercury (II) thiocyanate (Pharaoh's serpent) is something else when lit on fire (super dangerous, show a video!). Chlorates are explosive when heated in the presence of something organic. Gummy bears in particular are popular, but the chlorate will quickly eat through your skin, so again a video is preferable. Students may have consumed barium sulfate prior to an X-ray since the barium absorbs X-rays well. The sulfate makes the barium insoluble which is pretty important for when you consume it since barium is toxic. Try and think a few compounds that local businesses might use.

As you work towards helping them understand the concept of polyatomic ions, be sure to include some memorization techniques as well. When I was in high school, we had to memorize about 20 of them. I still remember that there were 4 ate/ite pairs we had to know because I organized them alphabetically (ClO_2^-/ClO_3^-, NO_2^-/NO_3^-, PO_3^{3-}/PO_4^{3-}, SO_3^{2-}/SO_4^{2-}) and then noted some trends. The first two sets have 2 and 3 oxygens while the last two have 3 and 4. The "ate" form always has 1 more oxygen

Chapter 5 – Naming and Formula Writing

(the ate, "ate" more oxygen). And the charges are small (-1) for the first two sets and larger for the last two sets (3-/2-). The third one is the only one that had a charge of 3-.

By seeking patterns students are likely to develop chunks (these 3 sets of polyatomic ions have a 2- charge). This helps with memory and processing speed. There are a lot of trends in polyatomic ions that students can use. The series of ClO^-/ClO_2^-/ClO_3^-/ClO_4^- goes hypochlorite, chlorite, chlorate, perchlorate. What do you notice that might help a student remember something? Permanganate is MnO_4^-, how does that compare to perchlorate? If a student can't remember $C_2H_3O_2^-$, maybe they can remember cha coo (CH_3COO^-) or maybe they can remember that sodium acetate spells out *nacho* ($NaC_2H_3O_2$). Just remember that memorizing the formulas alone is going to leave students without understanding of what these polyatomic ions are.

Even if you do not give a formal assessment or quiz, if you give the students opportunities to do some retrieval practice at the beginning of the unit it will help your students. The fear is that you do not want to cause frustration. I find elements to be my time to shine. When I first taught, I had a lot of methods that did not work well. But I was a storyteller, and I could capture attention. Telling the stories and tales of the different elements takes me back to those days. If I show pictures to accompany these stories, I am incorporating dual coding that justifies some of my showmanship.

After we've taught for several years, we start to forget about all the interesting stories that most students have never heard of. Don't assume that they know about the Hindenburg blimp igniting in mid-air. Don't assume that they know about aluminum being such a valuable metal that Napoleon used to serve his top guests with aluminum silverware instead of gold that was used for the lesser visitors.[6] Don't assume anything above iron on the periodic table was formed by a star exploding (or neutron stars colliding). Don't assume they know that nearly 1% of the air they breathe is composed of argon. Don't assume they know about gallium melting just above room temperature and how useful it can be for pranks.

Ionic and Molecular Compounds

Traditionally chemistry classes provide students with the rules for naming and then proceed to have students practice applying those rules. This works for some students and falls apart for others. Think about the ones that this doesn't work well for. What goes wrong for them? *When a student calls $CaBr_2$ calcium dibromide, what went wrong and how do we fix it?*

Naming systems are about communication. And communicating about organic, ionic, and molecular compounds is too complex for a single set of rules to function well. Try to envision the characteristics that we would want of a systematic naming system. We want to be able to communicate the exact formula with a minimal amount of information.

The problem in chemistry is that some compounds are so different than others that this is best accomplished through having multiple naming systems. But how exactly do ionic and molecular compounds differ that leads to the different rules for naming? And how can students uncover those differences?

For an ionic compound you only need to know the elements involved. Because the charges dictate the formula you don't need to inform me that there are two

Chapter 5 – Naming and Formula Writing

bromides for every magnesium ion. I already knew that just because of the elements involved. There is no compound with magnesium and bromide besides $MgBr_2$. But for a molecular compound more information is needed because two elements can combine to form many different proportionalities. Nitrogen and oxygen can form NO, NO_2, N_2O, N_2O_3, N_2O_4, and N_2O_5 in addition to some polyatomic ions. We therefore need a different naming system for molecular and ionic compounds. The first thing we want students to observe is how many elements and what kind (metal or nonmetal) in order to characterize the type of compound.

One thing that is extremely helpful is to give the students a single idea to focus on first. I use charge. Students will not always differentiate that elements exist in both a charged form and a neutral one unless explicitly informed (multiple times). Furthermore, students rarely connect on their own how elements differ based on their current charge. They can be easily confused by statements such as "sodium is a metal." They use sodium chloride, they know the formula for salt, but they don't experience anything as a metal when they eat it. But if we stress the idea that sodium can be in a neutral metallic state as an element, or it can have a positive charge in an ionic compound, then we can begin to strengthen the ability of the student to distinguish the rules involved in naming. *"But Mr. Milam, why don't bananas explode when I eat them if they have potassium in them?"*

If a student understands that ionic compounds are neutral overall but have charged ions in them; that student can then understand why naming ionic compounds is so simple. You don't need a prefix to tell me how many bromides are with a calcium because I already know that based on their relative charges. Emphasizing that the ions in the compound are charged even though we do not write those charges can also pay dividends when students must write double replacement reactions later. We write NaCl so it is not surprising that they fail to see those elements as charged. Especially if you use language such as balance or cancel the charges when we talk about how to determine subscripts.

Many students will ask questions about molecular compounds such as N_2O_4 where they ask how can the nitrogen be 3- charged and balanced by four 2- charged oxides? It takes them time and multiple concrete examples for them to be able to distinguish between the reasons for the differences in nomenclature rules. By repeatedly emphasizing the charges of the elements we give them an additional concrete example that applies to all compounds.

How can we draw students' attention to charge? A great way to connect charge in with phenomena is to show students that putting ionic compounds in water increases the reading on a conductivity meter. Adding molecular compounds typically does not cause changes. When I do this lesson, I put the formulas of the substances on the board. Then I ask the students to group the compounds by which ones conduct electricity in water, and those that do not conduct. Then we look for patterns.

Chapter 5 – Naming and Formula Writing

Formula	Effect on conductivity of water
NaCl	increases
KI	increases
$Al(NO_3)_3$	increases
C_2H_5OH	none
$C_{12}H_{22}O_{11}$	none
$HC_2H_3O_2$	Slight increase

Table 5-1: How compounds impact conductivity when dissolved in water

What patterns will a student see? At first, they will look for patterns in numbers. Many will quickly notice the compounds with C, H, and O do not impact the conductivity. But they will also notice the $HC_2H_3O_2$ does a little and the $Al(NO_3)_3$ does.

The location of the elements relative to the staircase is the key. When a metal and nonmetal element are present, the compound increases conductivity. When only nonmetals are present there is no change, or the change is minimal for the compound with H in two different spots.

This means that the components of compounds that have metallic and nonmetallic elements are charged. We call these ionic compounds and the pieces that comprise them are ions. The other type of compounds only has nonmetals, and these are neutral molecules with neutral atoms that don't ionize in solution.

If you include acids in this activity, I would analyze those separately after they've found that compounds with metals and nonmetals conduct, and that compounds with only nonmetals do not conduct. The acids being analyzed at the end is interesting because many students will visualize hydrogen as being a hybrid somewhat of metals and nonmetals based on its position in the periodic table. They sometimes see the staircase as continuing up to border hydrogen in the first period. The fact that some acids conduct a little bit, and some conduct a lot makes sense.

While we are talking about charge, polyatomic ions are a struggle for students. Students often encounter these for the first time when they are just learning about the intricacies of subscripts and charge balancing. For students that think critically, polyatomic ions do not make sense for many of the charges they are learning about. For example, cyanide has a 1- charge. Many students would see the nitrogen and think of the 3- charge it typically has. This would mean the carbon would be a 2+ charge. If they think of molecular compounds, then neither element would be charged so where did the extra negative charge come from? Further complicating the issue is that many polyatomic ions are inconsistent to a novice student. Ammonium has a nitrogen with a 3- charge and four 1+ charged hydrogens in their heads. Hence the 1+ charge overall. Nitrate and nitrite have N with a 5+ and 3+ charge which seems

Chapter 5 – Naming and Formula Writing

plausible based on its group position. Chlorate is not going to fit into their developing scheme. There's something else going on.

To address the complexity of polyatomic ions there are two reliable strategies. The first is to draw polyatomic ions as particle diagrams. For a nitrite, draw two oxygen particles connected to a nitrogen and add one additional electron for the negative charge. This is the perfect time to introduce parentheses in formulae. Emphasize the differences in subscripts. In $Ca(NO_3)_2$, what does the 3 represent? What does the 2 represent? How many nitrogen, oxygen and total atoms are in a single formula unit?

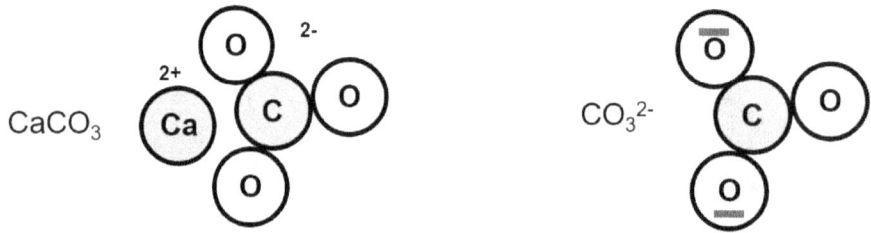

Figure 5-2: Particle representations of $CaCO_3$ and CO_3^{2-}

Do you notice how the symbolic representation for the compound omits the charge of the calcium and carbonate? What would have to change to include those? The use of lines for excess electrons would not work to show positive charge unless we assumed we started with a certain amount. For example, if calcium normally has 20 electrons when neutral, it will have 18 in the 2+ state. Having these discussions along with the particle representations provides clarity to the students by giving them concrete details to understand the abstract symbols.

The second tool for polyatomic ions is to have students count how many elements there are. Na_2SO_4 has three elements, so it probably has a polyatomic ion. They need to be on the lookout for peroxides (Na_2O_2), and a few molecular compounds, but it's a good thing for students to start by making observations about how many elements are in a compound. From there they can take the next step to classify those elements as metals and nonmetals.

Working from the name to the formula is easier for students in most instances. The prefixes used for molecular compounds make this very simple. Dinitrogen trioxide (N_2O_3) is straightforward for every student. Many students are already familiar with CO and CO_2 and so it is easy for them to learn the norm to omit the prefix mon- on the first element.

Writing ionic formulas is harder with more pitfalls. The first thing that students do not understand is charge balancing. There are many creative ways to address this, but they are all much more effective if you address the underlying concern. Students don't understand why you can't have more of one charge than the other. Who cares if CaCl has a 2+ and 1- charge? It's close enough, isn't it?

The reason why the charges must be balanced is because of the large number of pieces. When we talk about calcium chloride, we aren't discussing three ions. We're talking about trillions of ions for even a small speck. And if you have an extra positive charge a trillion times the amount of force is beyond the nuclear scale. When we write a formula, we aren't talking about a single formula unit. We're talking about moles of them potentially. And the formula describes the entire lot. Within a crystal of calcium

Chapter 5 – Naming and Formula Writing

chloride, you are likely to have defects where you might have a missing ion in the structure. This is confusing if you consider the formula $CaCl_2$ to represent discrete molecules. If we consider the formula to be the ratio of the number of total ions in a crystal this is less of a concern.

The reason why students don't realize that the formula is describing a large number of pieces is because we frequently represent particle diagrams with a minimal number of ions. We draw two electrons leaving a single calcium to go to two chlorine particles (Fig. 5-2). And the symbols themselves seem reasonable to represent a small number of particles or even 1 particle. The language we use to describe the relative numbers of particles could easily be interpreted that way. There are twice as many chloride ions as calcium ions could be heard by a student as there are 2 chlorides and 1 calcium. Try to work in images that have multiple formula units for ionic compounds as part of your instruction.

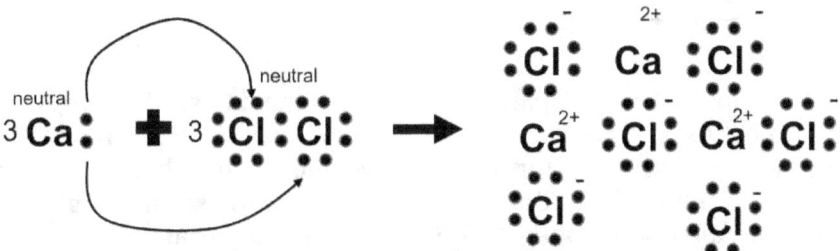

Figure 5-3: chlorine particles take electrons from calcium to form calcium and chloride ions

Using subscripts to balance charges has many functional algorithms. You potentially don't need any of them. Just ask the students how many of each ion you would need to balance the charge to neutral. Say neutral instead of zero.

You can also have students swap the charges. Al^{3+} and S^{2-} becomes Al_2S_3. You'll need to be wary of compounds with like charges for the ions. Ca^{2+} and O^{2-} make CaO not Ca_2O_2. Cutout activities (Fig. 5-4) are common and can help students grasp the connections and differences between subscripts and charges. Students struggle because subscripts and charges correlate, but sometimes more visually than others. One perk to the cutout activity is that this really can help students differentiate the charges from subscripts for polyatomic ions.

Chapter 5 – Naming and Formula Writing

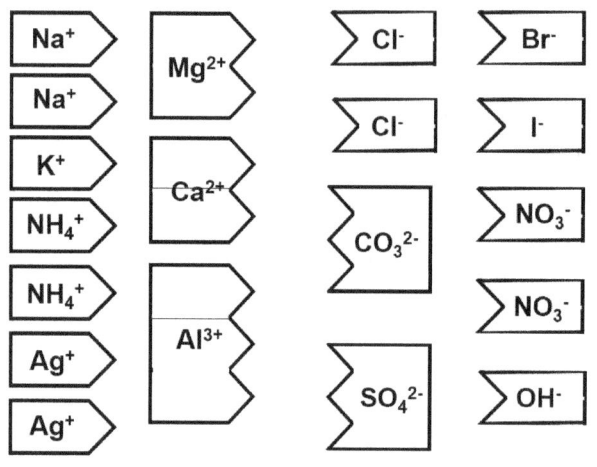

Figure 5-4: Cutout activity to help students balance charges in ionic compound formulas

   Be prepared for some issues with parentheses when students start writing formulas with polyatomic ions. You'll want to stress that the ion is a single group. If there are multiple of those groups in the formula, then you need parentheses. Copper (II) nitrate is $Cu(NO_3)_2$ and not CuN_2O_6. But we don't need parentheses for single polyatomic ions nor for multiple ions. Both $(Cu)_2S$ and $(NH_4)Cl$ are examples of incorrect representations that should not have parentheses.

   Speaking of roman numerals, some elements have multiple charges possible. Many transition metals can have multiple charges (Fe, Co, Cu, etc.) while others have only one charge besides neutral (Ag^+, Zn^{2+}, etc.). Students will mistake the subscript of the anion as the cause of the roman numerals. This is their system 1 noticing the link between the two. The first example you want to use is one where the subscripts do not match the charge of the metal.

   FeO is perfect for this. Iron (II) is used in a common redox titration, students are more familiar with iron and rust than some of the other transition metals, and it spells out ugly when translated to Spanish. But most importantly the charge of the iron is 2+ but there is not a 2 subscript anywhere. Students will wonder how you knew the charge. They need to identify that we know the total charge is neutral and we know the anion charge. This is easier for them to do with something like $FeCl_3$, but we want to avoid the first example having matching roman numerals and anion subscripts. The next issue that arises is that students will overuse roman numerals just as they might do for parentheses. Magnesium (II) bromide or carbon (IV) chloride are overgeneralized responses that students will produce.

   In my classroom the students get a list of common polyatomic ions to use, and it includes all of the elements with multiple charges that I will potentially use on a test or a quiz. I allow them to use the traditional names or the stock names of these ions. For FeO they can name it as iron (II) oxide or ferrous oxide. A chemist will understand either.

   Before you start your naming unit, you'll want to consider what you want students to learn, what do they need to memorize, and what mistakes are acceptable. Do you want them to follow the rules correctly or do they need to know the exceptions? Is dinitrogen tetraoxide going to be ok? What about phosphic acid or sulfous acid? These demonstrate an understanding of how to apply the algorithms

Chapter 5 – Naming and Formula Writing

correctly. Do they need to have chromate or hydroxide memorized? Can they use the name hydrogen chloride, or do they need to show they can write hydrochloric acid? Will the students get marked down for putting parentheses around a single polyatomic ion such as Na(NO$_2$)?

The best time to make such a list is directly after grading the test. If this will be your first year, consider giving a mid-unit quiz and tracking the questionable marks. You're going to need to communicate with the students about what your expectations are because there is a lot of gray area where names are wrong, but the rules were followed correctly. If there's something ambiguous in your rubric, they will find it. Trust me. There are also many minor errors that you may find acceptable for students since those that will need to know the difference later in their chemistry careers will probably figure it out as they gain experience.

Complex ions can be used to push student thinking further on charge. A complex ion is a combination of metals and anions that carries a net charge. These are stable in solution such as [Al(OH)$_6$]$^{3-}$. [Zn(OH)$_4$]$^{2-}$ is an example of a complex ion that can be used to deposit a layer of zinc on a penny using electrolysis in order to form a brass penny that looks gold. Compounds can be made with cations and complex ions such as potassium ferricyanide K$_3$Fe(CN)$_6$ which dissolves to make the ferricyanide complex ion [Fe(CN)$_6$]$^{3-}$.

Acids

Naming acids will be a struggle. Students are still learning polyatomic ions which slows recognition. It can be helpful to begin with an example of multiple similar acids. HCl, HClO$_2$ and HClO$_3$ work well. We need to have three different names for these three different compounds so that we can communicate consistently with other chemists. The anions of these three acids are chloride, chlorite, and chlorate. The difference between these anion stems are one or two letters. So, there is little room for error.

If the anion name ends in ide (chloride), we use hydro_____ic acid to name the acid. In this case, HCl is hydrochloric acid. I often have several students who have heard that name before, perhaps even in my class. Their system 1 recognizes that they are familiar with that name.

If the anion ends in ite (chlorite), we use _____ous acid. There is no hydro prefix here. The hydro prefix is tough for students. When we write the name chlorous acid (HClO$_2$), they often ask what they should add to indicate the H portion of the molecule. They see the chlorite become chlorous, but they fail to connect that the "acid" of "chlorous acid" tells us that we have hydrogen present.

If the anion ends in ate (chlorate), we use _____ic acid. For HClO$_3$ we name it as chloric acid. Showing students the similarities and differences between hydrochloric acid and chloric acid is valuable and might take them some time to process. A POGIL is a good way to introduce acid naming because it gives students the chance to process some minor details that are likely to be filtered by their system one if we use traditional instruction. The POGIL will force them to make observations about the elements involved, the subscripts, and the patterns of prefixes and suffixes.

Constructing the formulas of acids is simple. The students use the name to determine the anion. Then they balance the charge of that anion with the same number of hydrogen ions. Phosphoric acid must have phosphate as the anion. The

Chapter 5 – Naming and Formula Writing

phosphate is balanced by 3H⁺ to make H_3PO_4. When you get to acetic (ethanoic) acid the students will be curious if they should combine the hydrogen ions with the hydrogen atoms in the acetate ion. This is a great opportunity to let them know that some hydrogens are acidic (whatever that means!) based on what part of the molecule they're connected to. Others are not and so we typically do not write them together ($HC_2H_3O_2$ vs. $C_2H_4O_2$). You can take this one step further by showing them the form CH_3COOH that some students may see later.

Common errors students make

Formula	Incorrect name
H_2SO_4	hydrosulfuric acid
HCl	hydrogen chloride
CBr_4	Carbon (IV) bromide
$CaCO_3$	Calcium monocarbon trioxide
$MgBr_2$	Magnesium dibromide
NO_2	Nitrite
NO	Nitrogen oxide

Name	Incorrect formula
Copper (I) oxide	CuO
Aluminum oxide	$Al_2(O)_3$
Sodium bicarbonate	$Na(CO_3)_2$
Magnesium perchlorate	$Mg(ClO)_4$

What these errors all have in common is they show the student mixing rules. This means that the rules are too abstract for the students. They are experiencing cognitive overload, or they have overlearned some of the naming algorithms. Too many new things are going on at once and they can't track them all. What a teacher hopes for in this unit is that the student learns how to organize the decisions like a flowchart. The student should be looking at formulas for the number of elements and types of elements. Then deciding whether the compound is ionic, molecular or an acid. If the compound is ionic, they should be looking to see whether the cation has multiple possible charges or if a polyatomic ion is present. If the compound is an acid, they should be determining what the end of the anion name is. From there they should be applying the correct nomenclature rule.

Practice doing retrieval with students in steps. Go through these questions with multiple examples with different students answering each question.
1. How many elements are there?
2. Are the elements metals, nonmetals, or H?
3. What type of compound is it (ionic, molecular, or acid)?
4. If ionic, do we need parentheses, is there a polyatomic ion, do we need a roman numeral?
5. If acid, what is the anion ending?
6. Are the elements charged or neutral, if charged, what are the charges?
7. Can you draw a particle representation of the compound?

Interleaving is very important to consider when teaching naming and formula writing. This is something that should be considered every year how you will balance

Chapter 5 – Naming and Formula Writing

the introduction of naming systems with the practice. If you teach all of the systems and then have students practice a variety of compounds to name, they will learn better. Unless this is too much for them to do and they become frustrated. The extreme alternative is a sequence such as: introduce binary ionic compounds, practice, introduce molecular compounds, practice, introduce acids, practice, etc. In this alternative the practice would be on the topic just introduced.

When students practice naming only ionic compounds, they are much less likely to observe details about the elements and formulas. This leads to them becoming overconfident because the practice is easier. When they get to a test and must differentiate the type of compound, they are unprepared.

A big part of teaching is to monitor the students in order to make an informed decision about how much interleaving you can incorporate without disrupting the student's motivation and confidence. There is no perfect balance, but you should aim to have an optimal balance.

When assessing naming and formula writing you want to interleave as much as possible. It is more valuable to have students name NaCl, SF_4, and HBr than it is to give a quiz with only molecular naming and formula writing. A big component of naming is to identify the type of compound from the formula or name. I also prompt students to use their quiz. If they forget when a roman numeral should be used, look at the names on the quiz to decipher when roman numbers are used.

At the beginning of the unit emphasize charges, retrieval of element symbols, and retrieval of polyatomic ions. In the middle of the unit emphasize interleaving and giving structure to how students analyze a formula or name. Be prepared to counter student frustration. Yes, some of the rules are complicated, but the systems we used are highly functional. Enjoy the challenge of learning them, even if it doesn't appear to be designed for optimal learning of high school students.

Student Struggles

1. "I have no idea what I'm doing!" With naming, most struggling students just write down gibberish instead of verbally proclaiming defeat. A student struggling with all names and formulas has two problems. First is that they likely don't know their element symbols or polyatomic ions. Even if these are provided, not knowing enough of them will cause students to be unable to process the procedures. The second issue is that students are being too abstract with the names and symbols. The first issue is fixed through spaced practice, retrieval practice, and seeing some of the chemicals in action to construct memorable experiences in the brain. The second issue can be helped by emphasizing the few compounds students were previously familiar with. Many students already know NaCl is sodium chloride and CO_2 is carbon dioxide. Have them start there and then connect to similar compounds. Students will also benefit from seeing actual chemicals. They will benefit from drawing particle representations. Particle representations help students slow down and connect why subscripts are what they are. And finally, have students emphasize whether the elements are charged or neutral as they process formulas.

2. "Is it hydrosulfuric or sulfuric acid?" Sometimes we overcomplicate acids by using terms like binary and ternary acids. The ending of the anion tells you which naming system to use. If the end is "ide", use hydro_____ic acid with the original stem. If the anion ends in "ate" it becomes _____ic acid. It helps students to see both

Chapter 5 – Naming and Formula Writing

examples side by side. The chlorine set of polyatomic ions works great for this. HCl, HClO, $HClO_2$, $HClO_3$, and $HClO_4$ are hydrochloric acid, hypochlorous acid, chlorous acid, chloric acid, and perchloric acid.

3. "When do I use roman numerals?" This is tricky because the answer is when a cation has multiple charges possible. All elements have multiple charges, but there are usually two and one is neutral. A lot of transition elements have multiple charges, but not all do (Zn, Ag, etc.). There are also elements like tin and lead that have multiple charges. By now you see how a student is going to need assistance with this. Perhaps they have a list somewhere of which elements you expect them to use roman numerals for.

4. "Why are there parentheses around some parts of the formulas?" This student does not understand what a polyatomic ion is yet. You need to address that and not just verbally describe the symbolic formula. Instead draw some particle diagrams. You're going to have a charge for the polyatomic ion, but that charge isn't always localized somewhere.

5. "I just can't remember all of these polyatomic ions." Polyatomic ions have a lot going on in a tiny space for symbols. To help students memorize them, draw the Lewis structures. Show why chromate has 4 oxygen atoms, and dichromate has 7 oxygen atoms since one of the oxygens serves as a bridge between the chromiums. Help them notice how charge is constant within a group (hypochlorite, chlorite, chlorate, and perchlorate are all 1- charged). Have them search for patterns with the numbers of oxygen and charge. After they've done some sort of thinking that involves transfer between STM and LTM you want them to use retrieval practice at spaced intervals. Show them flashcards or quiz the class verbally for a few minutes during class that week. If some students don't participate because the others are too fast, have them write their prediction down before students shout out responses.

6. "Is this calcium dibromide?" The students that mix rules have not taken on a conceptual understanding of the types of compounds. You can address this quickly by telling them the answer, but it won't last long, and they'll end up in a worse spot for reaction types. Students need to have a checklist when they see a formula. How many elements are there, are they metals or nonmetals, are there any polyatomic ions, what charges do all the elements have, and what type of compound is it? When students skip those steps, they fail to lock in the critical details distinguishing the compounds from each other. They also fail to make observations that are going to help them in the next chapter on chemical reactions. You have to get them to slow down, engage with the material, and have some phenomena besides "naming rules" to attach that thinking to.

7. "Is FeO iron (I) oxide?" Here the student thinks that the roman numeral is the subscript used on the anion. They think this because too many examples have had that be the case. $FeCl_3$ is iron (III) chloride. $CuBr_2$ is copper (II) bromide. It is critical that within the first few samples, they see an example such as PbO as lead (II) oxide and you point out the roman numeral being two.

Chapter 5 – Naming and Formula Writing

8. *"I don't know where you get the subscripts from."* Showing sets of names and formulas first helps students decipher rules while observing patterns. When we start with the rules, we overload the short-term memory of students struggling to make sense of abstract details about a topic that students have limited concrete examples of. Switching the initial order can have a substantial impact on how many students remain cognitively engaged. Charges are also helpful here. Ask the student what they picture when you say sulfur has a 2- charge.

Phenomena

1. Colorful precipitates - Create equal concentration solutions of copper (II) nitrate, iron (III) nitrate, and sodium hydroxide (e.g., all are 0.10 M). Add 10 drops of $Cu(NO_3)_2$ solution to a test tube followed by 2 drops of phenolphthalein. Swirl the test tube to mix the contents and the color should be blue. Then add 20 drops of the NaOH. The color should become pink in parts of the test tube, but after a swirl everything returns to blue. As the 21st drop is added, the color shifts to pink. For every 1 drop of $Cu(NO_3)_2$ solution added, you will need 2 drops of NaOH to change the color back. This helps to emphasize that the formula of the compound being formed is $Cu(OH)_2$. For every 1 Cu^{2+} ion, you need 2 OH^- ions. When there is more Cu^{2+} the solution is blue. When there is more OH^- the solution is pink.

Repeat with the $Fe(NO_3)_3$ solution but ask students to predict how many drops of hydroxide solution are needed to turn the 10 drops of $Fe(NO_3)_3$ solution from yellow to pink. If the drops are consistent, you'll see the change happen at 31 drops.

2. Sticky tape - Adhere two long pieces of clear tape to a table. It helps to make tabs on the end of both and have the top tape be a little shorter so it's easy to remember which is which. Rip the two pieces of tape off the table. Then rip the two pieces of tape off each other. They should now have opposing charges. Your students can explore what the two pieces of tape attract and repel. This can give them a functional model of charge. Charge is the deficit or surplus of a type of particle (electron) that causes attraction or repulsion.

The tapes will attract to neutral items. This can help lead students towards the concept of polarization. A neutral item should repel and attract the tape. In order for the attraction to be greater, there has to be a shift in charge that increases the distance between particles that repel. This shifting of charge within a substance is called polarization and is a critical idea for the bonding unit. Insulators will experience shifting within particles, and conductors will experience shifting between particles for a greater polarization.

3. Polarization ruler - Carefully balance a ruler on top of a watch glass so the ruler can freely rotate. Use a pool noodle and rabbit fur or any other method of charging an object (hair and balloon works well). Hold the charged pool noodle near the ruler and the ruler will begin to spin in a circle. The ruler is neutral. How is a neutral object attracted to the charged pool noodle? This is a dramatic presentation that sets up the concept of polarization and can help students in their physics classes when we use the noodle to slow the rule gradually.

4. Conductivity of solutions - Use a conductivity meter to measure how conductivity varies as additional quantities of chemicals are dissolved in water. Use a variety of substances that are ionic and molecular. At the end you can include acids as a 3rd type of compound that shares properties of both. You should show how

Chapter 5 – Naming and Formula Writing

molecular compounds (multiple nonmetals) do not typically impact the conductivity of the solution. Ionic compounds will cause the conductivity to increase in proportion with the amount of ionic compound added. This shows that the ionic compounds are composed of charged particles that can move once dissolved. The molecular compounds (even when dissolved) do not conduct. This means that their particles are neutral. This is a big component of why different naming systems are required for the two compounds. Charged ions combine in predictable ratios. Neutral nonmetal atoms combine in multiple combinations necessitating a prefix description for subscripts.

5. Electrolysis of $CuCl_2$ - Elemental copper is produced on the negative terminal and the odor of chlorine is produced at the positive terminal. This can help guide students to the distinction between metals and nonmetals as it did for Humphry Davy. We're seeing evidence that metals have an affinity for negative charge while nonmetals have an attraction to positive charge. This "dualism model" still holds for ionic compounds today. It also highlights that copper can exist in a charged state (blue-green, dissolves in water) or a neutral state (looks like a penny, metallic). Chlorine also has distinguishing characteristics whether it is neutral or charged.

6. Burning Mg - Magnesium produces an excessively large amount of light when burned. The metal can react with nitrogen, but for classroom purposes let's assume the reaction is isolated to oxygen and magnesium. The magnesium starts as a neutral metal. The oxygen starts as neutral molecules of a nonmetal. Yet the final product is not a metal, nor a gas. The final product is a crumbly (brittle) white salt. The MgO product contains charged components as a transfer of electrons from the magnesium metal to the oxygen nonmetal has occurred.

7. Splint tests - This is around the time where you'll want students to be aware of splint testing for gases. If a gas is produced during a reaction, a burning splint can be used to distinguish between three common gas products. If the gas is carbon dioxide, the splint will be extinguished. If the gas is oxygen, the splint will burst into a vigorous flame (and the splint will reignite if a hot ember if placed into the oxygen environment). If the gas is hydrogen, a small explosion will be heard (pop!).

These tests help us to provide evidence for the composition of compounds. To generate carbon dioxide, mix an acid (vinegar) with a bicarbonate or carbonate (baking soda). To generate oxygen gas, add a catalyst (MnO_2, KI, others) to hydrogen peroxide. To generate hydrogen gas, either mix zinc metal and hydrochloric acid, or add calcium metal to water. There are other alternatives to produce any of these gases, but many of them are difficult or have increased risk and hazard.

8. Cement cycle - The world's population is approaching a plateau around 11 billion people. Most of those people are going to live in cities. In order to house that many people, large quantities of concrete are needed for infrastructure in those cities. Teach students about the cycle for cement (limestone → quicklime → slaked lime → limestone; or $CaCO_3$ → CaO → $Ca(OH)_2$ → $CaCO_3$). The environmental impact of finding the best sand to use for concrete will likely have implications for where you live regardless of where you are. Along the way toward learning environmental chemistry, students can apply some of their naming concepts.

9. Cigarette in a test tube - Place a cigarette in a test tube. Cover the test tube with wet paper towel held in place by a rubber band. Clamp the test tube at an angle over a bunsen burner. The cigarette will burn as the Bunsen burner flame is applied even inside of the glass container. The residual tar and ooze will make a memorable impression on students. Then have them look up what some of the compounds are

Chapter 5 – Naming and Formula Writing

that are used in cigarettes and find which ones they know how to write the formulas for.

10. Teacher favorites - Any demonstration where students use retrieval practice after an experience is a good thing. Perform any of your favorite reactions, stories from history, or showcase interesting molecules in this unit to help keep the material fresh for students.

Flashcards

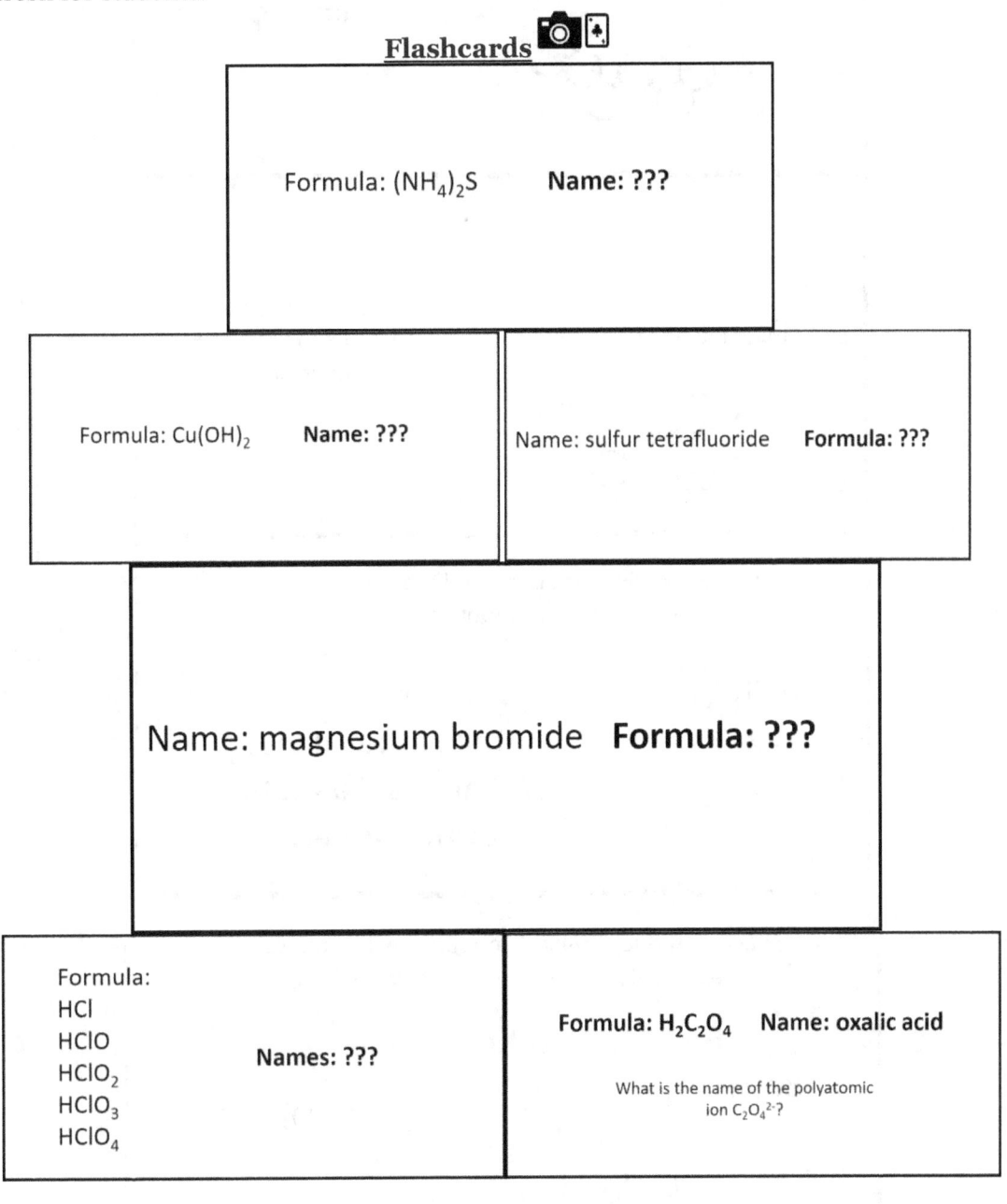

Formula: $(NH_4)_2S$ Name: ???

Formula: $Cu(OH)_2$ Name: ???

Name: sulfur tetrafluoride   Formula: ???

Name: magnesium bromide   Formula: ???

Formula:
HCl
HClO
HClO$_2$
HClO$_3$
HClO$_4$

Names: ???

Formula: $H_2C_2O_4$ Name: oxalic acid

What is the name of the polyatomic ion $C_2O_4^{2-}$?

Chapter 5 – Naming and Formula Writing

Formula: Ca(ClO$_3$)$_2$

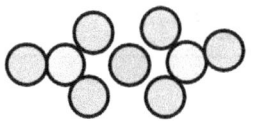

Which elements are charged and which are neutral?

Formula: N$_2$O$_4$

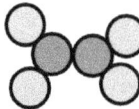

Which elements are charged and which are neutral?

A new element X is discovered that forms the following compounds:

XCl$_2$
XO
X(NO$_3$)$_2$

What is the charge of X and is X a metal or nonmetal?

Which of the following compounds will conduct electricity when dissolved in water?

KBr

Ba(NO$_3$)$_2$

H$_2$SO$_4$

C$_2$H$_5$OH

Chapter 6C
Reaction Types

History

When we burn hydrocarbon fuel, we observe a mass decrease. When we burn a metal, we observe a mass increase. Today we know this is based on particles moving into or out of the system. Scientists in the 1700s could not easily discern lab observations by translating to the particulate level. Different gases produced in chemical reactions had a variety of names. Inflammable air (hydrogen), fixed air (carbon dioxide), dephlogisticated air (hydrogen or oxygen), and other names served as imprecise descriptions of what was being observed.

Enter phlogiston. Phlogiston theory served as a catch all for discrepant observations. It worked well in this regard. When something was burned, phlogiston transferred from the substance to the air. Phlogiston could be added to metals when they were reduced from calx. Because of the functionality, the phlogiston model limited the advance of chemistry for some time. Antoine Lavoisier was fundamental in undermining the phlogiston model.

If we come back to combustion, one perk to phlogiston theory is that it did help explain why some substances burn and not others. Why would oil and metal both burn? Such different substances seem unlikely to share a commonality that would explain their mutual flammability. Phlogiston serves as a simple unifying theory. But is this phlogiston in the substances, the air, or both? Such a question begins to reveal the limitations of phlogiston theory. Phlogiston can't predict future events. It only produced partial and inconsistent explanations of phenomena.

Here is an interesting breakdown using phlogiston theory that was debated in the late 1700s.[1] If zinc metal is mixed with sulfuric acid, inflammable air (hydrogen) is produced as the zinc dissolves in the acid. If the water is evaporated, a substance remains (zinc sulfate) that upon heating releases the acid (chemists at the time thought the gas was the acid) and a zinc calx remains. But if zinc calx is mixed with sulfuric acid, the same cycle can be produced with no inflammable air evolved.

Reaction 1: $Zn + H_2SO_4 \rightarrow H_2 + ZnSO_4$
Phlogiston version: $(calx + \phi) + acid \rightarrow (calx + acid) + \phi$

Reaction 2: $ZnO + H_2SO_4 \rightarrow H_2O + ZnSO_4$
Phlogiston version: $calx + acid \rightarrow (calx + acid)$

This was easy to explain using phlogiston theory. The zinc metal was zinc calx with phlogiston. The zinc calx was zinc without phlogiston. Henry Cavendish would refer to hydrogen gas as phlogiston since it would be released as the zinc converted to calx.

To explain the difference without phlogiston, a chemist would have to know that the oxide portion of calx reacts with the acid to produce water. The water is already present in the acid solution, so the impact of more water being produced is negligible.

Chapter 6 – Reaction Types

Interestingly, it was Cavendish (a phlogiston believer to the very end) who showed that inflammable air (H_2) combines with flammable air (regular air) to form water. But his initial response to this new information was to consider inflammable air as water combined with phlogiston.[2]

When the change in mass conflicted with phlogiston theory, they shifted which chemical the phlogiston pertained to. When that didn't work, they assumed that the inflammable air was contained in the metal and not the acid. No matter what problems arose, the scientists adjusted phlogiston to address the issue. Many of these adjustments can be quite convincing to those doing the analysis.

It was the combination of Lavoisier's conclusions and Joseph Priestley's experiments that eventually led to a shift away from phlogiston theory.[3] There was intensive pushback initially, but Lavoisier compromised by adding caloric theory. The caloric theory treated energy as a fluid, and this allowed people to adjust away from phlogiston by providing an explanation of why some chemicals were more reactive than others.

This transition led to a better understanding of the components of gases, determination of chemical formulas, more systematic nomenclature, and other information that allowed us to see the patterns among chemical reactions. In Chapter 5 we saw the first double replacement reaction written: stibnite + corrosive sublimate → butter of antimony + cinnabar. This was published in 1612 by Jean Beguin. In 1617 Angelus Sala reacted copper metal with a series of reactions that eventually led to the precipitation of copper metal again. This common recycling copper lab was a historic breakthrough the effectively challenged the four elements theory from ancient times. But when Robert Boyle observed that burning a metal causes the mass to increase, these developments became stuck in the mysticism known as phlogiston.

Synthesis and Decomposition

Students struggle with predicting products for chemical reactions well beyond their introductory chemistry course. This makes sense from the sheer volume of algorithms, combinations, and exceptions that they encounter. But when students have success in writing reactions, it builds a strong foundation for future chemistry.

At this point students are not able to classify into the three more general reaction types (redox, precipitation, acid-base). These three classifications show that chemical reactions always involve the movement of charged particles. Electrons move in redox reactions. Ions move in precipitation reactions (double replacement). either protons or electron pairs move in acid-base reactions, depending on the particular definition used.

At the introductory level, students are instead provided with a set of reaction types such as synthesis, decomposition, single replacement, double replacement and combustion. These classifications have patterns that are more superficial, which allows students a more gradual introduction. Perhaps a 6th type (e.g., neutralization) is added or one is omitted for the time being. When you watch students begin the unit you will notice that they gravitate very heavily towards the generic forms for the different types. They are substantially more likely to represent a synthesis reaction as A + B → AB than using an actual example such as $MgO\ (s) + CO_2\ (g) → MgCO_3\ (s)$. For synthesis and decomposition in particular students struggle to make sense of patterns with formulas. But they do connect the meaning of synthesis as two things combine to make one and decomposition as one thing breaking apart.

Chapter 6 – Reaction Types

Part of the issue is that there are not many consistent patterns for synthesis and decomposition reactions. Some are acid-base reactions, some are redox reactions, some are two elements making a compound, some use metals, some use nonmetals, some have acids, etc. One pattern that holds up frequently is that decomposition reactions generally produce a recognizable gas (CO_2, O_2, H_2O). Chlorates produce O_2. Carbonates produce CO_2. Hydroxides produce H_2O. Many of the exceptions to this involve binary compounds like KI that are relatively simple for students to predict.

Having students search for gases gives them a good place to start for decomposition reactions. This can also help avoid the part where students just chop a chemical into random pieces. An incorrect student prediction is $CaCO_3$ (s) → Ca (s) + CO_3 (g). That is not a simple task to convince a student that this is not possible. They do not know enough about bonding to exclude CO_3 as a possible chemical yet. They don't know that Ca is unlikely to form through heating because it is reactive. They see this as simple, and what is simple is often correct. There are two tactics to help students see why this is incorrect.

The first is to show actual reactions where some of the products are observable. When hydrogen peroxide decomposes it is easy to see the water produced as it forms tiny droplets of liquid water to make a cloud. The oxygen is also easily detectable using a splint test. If you've never done the oxygen splint test, do it. It is so fun to reignite a splint by dunking it into a nearly pure oxygen environment. I could be entertained by that for hours.

The reason why hydrogen peroxide works well is the products are easily observable, but also, they don't fit the schema for students. If you asked students to predict what would form from H_2O_2 decomposing, many would guess that it would be H_2O_2 (aq) → H_2 (g) + O_2 (g). Instead, it's H_2O_2 (aq) → H_2O (g) + O_2 (g). Students immediately are confused because where the heck did that additional oxygen come from? "Mr. Milam are you sure you went to college for this stuff?"

And they should be confused. I know that it seems easy for teachers to immediately jump in and start using coefficients to balance the reaction. But there's something that you may not realize when you do that. Many of the students are trying to picture this at a blend of particle level and symbols.

Reactions with lots of pieces are confusing. These reactions often occur in multiple steps, but students don't know that yet. For the decomposition of hydrogen peroxide students are trying to picture how an oxygen atom could start in one molecule, fly through space to another molecule, and end up as a diatomic oxygen molecule. The visualization is much more complicated to them for some of the actual products formed than it would be for just chopping the molecule down the middle. How a solvent fits into this picture is probably still a way off.

The second tactic is to practice having students predict the gas produced first. Just put up a single compound and ask them to identify the gas produced.

$Ba(OH)_2$ → which gas? (H_2O)
Li_2O → which gas? (O_2)
KCl → which gas? (Cl_2)
Na_2CO_3 → which gas? (CO_2)
$KClO_3$ → which gas? (O_2)

Chapter 6 – Reaction Types

This helps them start. The second step is to put together the second compound (or element). When barium hydroxide decomposes you end up with a compound of barium and oxygen. Because many of these reactions balance without coefficients (or at least without many), students will often just put all the remaining pieces together. You want them to instead go through and figure out the compound using charges. The barium ion is 2+ charged and oxide ion is 2- charged. The formula is BaO.

Otherwise, when you get to something like potassium chlorate there will be issues. They'll start making something like KClO. That's not the end of the world (technically a small amount of potassium hypochlorite is possible), but students that take the time to write the ionic compounds correctly fare better when we get into the replacement reactions.

Synthesis and decomposition should be taught in tandem. Any synthesis reaction can be reversed into a decomposition reaction. Metal carbonates decompose into carbon dioxide and a metal oxide (decomposition). A metal oxide combines with carbon dioxide to form a metal carbonate (synthesis).

$CaCO_3$ (s) → CaO (s) + CO_2 (g)
CaO (s) + CO_2 (g) → $CaCO_3$ (s)

One set of pairings is water and oxides. If you combine a metal oxide and water, you'll produce the metal hydroxide (base). If you combine a nonmetal oxide and water, you'll produce an acid. The decomposition of the acids and bases is also possible.

SO_2 (g) + H_2O (l) → H_2SO_3 (aq)
Na_2O (s) + H_2O (l) → 2NaOH (aq)

These reactions are confusing for students who have learned double replacement. They view these as two compounds reacting, but when they go to swap the cations (or anions) they find that they end up with the products being the same as the reactants. They are swapping oxygen for oxygen.

Ideally this sticking point can be identified as a signal by the teacher early on. When you go to a double replacement and it doesn't work, check to see if it's something with oxygen being added to water. If so, this is a synthesis reaction even though the reactants are two compounds.

If the reaction forms an acid, the student must think through what polyatomic ions are possible based on the nonmetal element. For the reaction SO_2 (g) + H_2O (l) → H_2SO_3 (aq), we had two options. The polyatomic anion of the acid was either going to be sulfite or sulfate. The student has no means of knowing which it will be so they should just guess one. If the reaction balances easily, they guessed correctly. If they guess incorrectly, (SO_2 + H_2O → H_2SO_4) they will find the reaction is impossible to balance.

For N_2O_3 + H_2O → ?, we will either produce HNO_2 or HNO_3. Which one balances easily? The correct product is HNO_2 and the balanced reaction is N_2O_3 + H_2O → $2HNO_2$. Of the common nonmetal oxides, the most difficult one to balance is the reaction P_4O_{10} + $6H_2O$ → $4H_3PO_4$. These reactions can be revisited in periodic trends, and in the acid-base unit.

Combustion reactions are simple to predict, and the demonstrations can be quite lively. Combustion is a good reaction type to work on balancing with students. Two general guidelines for balancing are to do oxygen last, and if you get stuck try

Chapter 6 – Reaction Types

doubling everything. You can always reduce coefficients at the end if you shouldn't have.

C_2H_6 (g) + O_2 (g) → CO_2 (g) + H_2O (g)

First we balance the carbons and hydrogens.
C_2H_6 (g) + O_2 (g) → $2CO_2$ (g) + $3H_2O$ (g)

Note that the number of oxygens on the right side is 7. This cannot be balanced unless we use non-integer values for the oxygen gas reactant. So, we double everything.

$2C_2H_6$ (g) + O_2 (g) → $4CO_2$ (g) + $6H_2O$ (g)

Now we can add a 7 to the reactant O_2 to balance.
$2C_2H_6$ (g) + $7O_2$ (g) → $4CO_2$ (g) + $6H_2O$ (g)

Students have a substantial amount of experience with combustion. They've seen lit candles, sat near a campfire, cooked with fire, and probably seen some form of fire in their science classes. They have questions about fire that we aren't going to be able to answer in this unit. When students ask what fire is, they aren't asking what chemical reaction takes place. They are asking what the flame is. The flame is colorful, hot, and several students have noticed they can run their finger through a small flame if they are speedy.

When something burns, the products of the reaction move at high speeds. These fast-moving particles collide with air particles (or other particles). These collisions can lead to electrons being removed or promoted to a higher energy state. When the electrons return back light is emitted. When we claim that energy is produced during combustion, we don't really address the explanations that students are seeking. And since students aren't ready currently to talk about electrons and energy states it is best to defer while acknowledging that this is an intriguing question.

A very honest response would let students know that we don't currently know enough to give a complete description of what a flame is, but that we can make observations about the flame now that might help us figure it out later. The flame is colorful, the colors can change, the flame emits light, the flame has a high temperature, and the flame is not a condensed state.

If you get into incomplete combustion where carbon monoxide and other products are produced, you should look into getting some flash cotton. Flash cotton was invented on accident. If you soak cotton in a mixture of concentrated nitric and sulfuric acids the cellulose in cotton gets nitro groups (-NO_2) added. Those nitro groups are a large source of oxygen making flash cotton easy to ignite. Once the cotton is lit it burns completely (in a flash!). When oxygen is limited carbon monoxide and carbon can be produced instead of CO_2. Students may be aware of this from watching a show where an automobile runs in a garage, and anyone nearby suffers from carbon monoxide poisoning. They can also see the production of carbon if you run an inefficient Bunsen burner flame under glassware. A black solid will collect on the glass which is carbon/coal.

Synthesis, decomposition, and combustion all involve product prediction that are not algorithmic in nature. The student must rely on the teacher telling them what

Chapter 6 – Reaction Types

to do to predict. This leaves a lot of discretion to the teacher on what is most important. For my class I might choose to minimize the number of possibilities to help students focus on formula writing. Others might want to include complex ions or net ionics if students will see them later in the course.

Students should be given a framework for how to best practice writing reactions. The first step is to identify how many compounds and elements are present. Then what type of compounds and elements. From these two pieces of information the student should be able to identify the reaction type, and then predict the product. After this the student can balance, add in states of matter, and draw a particle representation that includes charges.

$$KClO_3 \rightarrow ?$$

Here we have 1 compound. The compound is ionic. Because there is only one thing, this must be a decomposition reaction. Chlorates produce oxygen gas.

$2KClO_3$ (s) $\rightarrow 2KCl$ (s) $+ 3O_2$ (g)

The potassium has a 1+ charge, chlorate has a 1- charge, chloride has a 1- charge, and oxygen is neutral.

$$K_2O + H_2O \rightarrow ?$$

- How many elements and compounds are there?
- What types of elements and what kinds of compounds?
- What is the reaction type?
- What is the product?
- Does the full reaction require balancing?
- What are the charges of the atoms, ions, and compounds?

Single Replacement and Double Replacement

Single replacement reactions are the best reaction type to get students to think about what happens with electrons and charge. Start with a demonstration where one reaction occurs, and the other does not. I use the following:

1. Zn (s) + $CuCl_2$ (aq) $\rightarrow$ Cu (s) + $ZnCl_2$ (aq)
2. Cu (s) + $ZnCl_2$ (aq) $\rightarrow$ no reaction (NR)

This is where you really want to hammer in that elements are neutral, but in an ionic compound those same elements will be charged (even though we don't indicate it in the formula). The initial zinc metal in #1 is neutral and the Zn in $ZnCl_2$ (aq) is 2+ charged. What has to happen for the zinc to go from neutral to 2+ charged? The zinc must lose 2 electrons per atom. This prompts an interesting question. Where did those electrons go?

The copper in reaction #1 starts out as 2+ charged and finishes as neutral Cu metal. How does the copper go from 2+ to neutral? Each copper (II) ion must add 2 electrons. Where did those electrons come from? They came from the zinc. What is happening in this electron is that copper (II) ions are pulling electrons away from the zinc. The chloride is mostly irrelevant. If we check the chloride it starts with 1- charge and ends with 1- charge. It may or may not be helpful to show students that the chloride ions in solution are separated from the copper (II) or zinc ions if you intend for them to later identify them as spectator ions.

Chapter 6 – Reaction Types

Either way now is a great chance to work on connecting the symbols just discussed with the particle level representation. But we want to include electrons with the particles to show what happens when the charge changes.

Figure 6-1: Particle representation of a single replacement reaction between $CuCl_2$ and Zn

We pretend not to know the actual number of electrons in Cu and Zn for now (mostly because I don't want to draw 27-30 electrons for each). But what we are showing here is that the copper (II) ions take two of the electrons from the neutral zinc. It is easy to show students the different color of neutral copper and copper (II) by showing them a penny and a solution of copper (II) chloride. If you run electricity through a solution with copper (II) you will see the brownish orange copper metal form on the electrode connected to the - terminal.

Something that I wish was emphasized more in chemistry is that when elements change the number of electrons, their properties change. Whenever someone puts a chunk of potassium into water, there's always someone who asks, "Why don't bananas blow up then?" It's important for us to emphasize that potassium ions are not the same as potassium atoms. The calcium in your bones is not calcium metal. Students get confused from that when we show students the common charges that elements form based on their periodic table column without mentioning the neutral form as well. Sodium can have a 1+ charge or it can exist in the neutral state. Sulfur can be 2- charged or exist in the neutral state. We often describe metallic elements as metals whether they are neutral or ions which can lead students to think of them as having a single state.

Some teachers use dance partners as a means for students to predict products for single replacement and double replacement reactions. If you do this, please be thoughtful about making sure you are not normalizing heterosexuality at the expense of others. In single replacement reactions a heterosexual couple (Adam and Bella) goes to a homecoming dance. There is a single male (Chris) at the dance who has a superior personality. The female leaves to go with the other male.

AB + C → CB + A

At a second dance (in Michigan we have snowcoming) Bella and Chris go together and this time Adam is there on his own. Will Bella leave Chris for Adam? Of course not!

CB + A → NR

What is particularly nice about this example is that you can use the males to represent metals and females to represent nonmetals. This will help students for halogen displacement reactions. Chris is at snowcoming with Bella and Daria is there without a date. Daria has a superior personality (in spite of her lackluster volleyball efforts) in Chris's opinion and so he ditches Bella at the dance.

CB + D → CD + B

Chapter 6 – Reaction Types

When you use the dance analogy the students will take a strong hold to it. They will be particularly struck by it if you inject yourself as a high school student into it. But if you do not quickly transfer that towards actual reactions, they will mostly just remember the story about betrayal and not the chemistry. Using charges allows them to connect the story to chemistry much better.

F_2 (g) + 2NaBr (aq) → Br_2 (l) + 2NaF (aq)

It is important to note that the dance analogy produces a misconception that the reaction is about who ends up with the partner. Most single replacement reactions do not form that partnership since the reactions are aqueous. For example, the NaBr (aq) above could also be represented as Na^+ (aq) + Br^- (aq). The sodium ions aren't bonded to the bromide or fluoride as long as they remain in solution. Furthermore, the dance analogy has the superior element ending up as the charged ion in the compound (as opposed to a neutral element). But for the metals, the metal that ends as the element was actually better at pulling on electrons. For nonmetals this is reversed and the nonmetal that pulls on electrons better ends up charged and in the compound form. If you make the choice to use a dance partners analogy, give some indication that there are limitations to the analogy that fall apart later.

Double replacement reactions do not have charges change. The reaction here is not predicated on electrons moving around, rather it is based on ions changing locations.

$Ba(NO_3)_2$ (aq) + Na_2SO_4 (aq) → $2NaNO_3$ (aq) + $BaSO_4$ (s)

Get ready for students to put parentheses everywhere!

There is a lot going on with symbols when we transition to double replacement. One simple idea that students miss out on is that the reactant solutions are being mixed. It is surprisingly enlightening to show them that the barium nitrate starts out in one container and the sodium sulfate in a second container before they are mixed. That helps them visualize that the barium ions from the one container meet up with the sulfate ions from the second container only after mixing. The plus sign in the reaction means they are being mixed. Once they mix, those two ions stick together to form the insoluble solid. I prefer to start this reaction type by mixing two solutions and then drawing particles on the board.

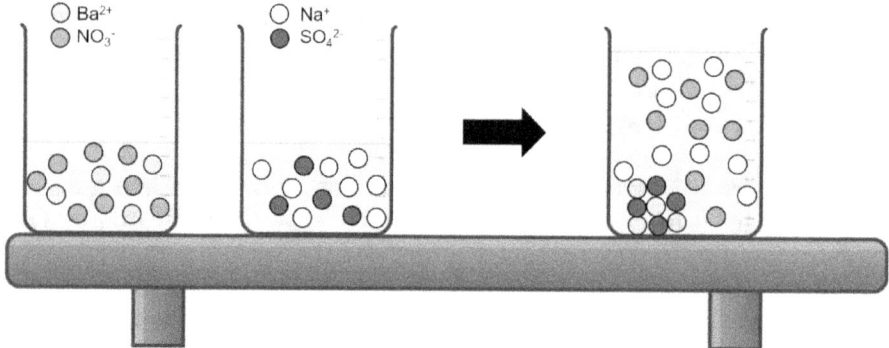

Figure 6-2: Double replacement reaction between $Ba(NO_3)_2$ and Na_2SO_4

The barium sulfate precipitate that forms will settle to the bottom. A great follow up question after is to ask the students what a test tube would look like that had equal amounts of copper (II) sulfate mixed with barium nitrate. The white

Chapter 6 – Reaction Types

barium sulfate will precipitate and settle to the bottom, but the copper (II) nitrate remains in solution. The test tube will have a white solid at the bottom and a blue solution above.

Because double replacement frequently focuses on precipitate formation, many teachers put in guidelines about solubility. Very little is gained by having students memorize solubility trends, but it can be helpful to point out that ions with more charge are less likely to be soluble. Some solubility trends that teachers know are that nitrates, chlorates, acetates, halides (Cl^-, Br^-, I^-), alkali metals, and ammonium ions tend to be soluble. Note that all of those have a 1+ or 1- charge.

When we explain why something is soluble, we compare how strong the ionic bonds are with how strong the interactions between water and the ions would be. Ions with a lot of charge will have stronger ionic bonds and are more likely to be insoluble. There are exceptions to this because more charge also means stronger interactions with the water. Smaller ions have stronger ionic bonding, but again also stronger interactions with water.

Now it's time for students to practice. You give them a prompt and they write:

$CaCl_2$ + $NaNO_3$ → $NaCl_2$ + $CaNO_3$
What do you do next?

The student has taken in the idea of double replacement as an algorithm but is not applying the rules of charge balancing. To start we should think about why they might not be transferring over that skill from the previous unit. When a skill doesn't transfer, the student is not recognizing the need for that skill in the new context. And for this problem, that means that the student is seeing the reaction in the abstract as symbols only.

This can be remedied by incorporating the particle level while including charges. If they add in the charges, they will quickly notice that $NaCl_2$ does not make sense, and of course it doesn't. Sodium chloride is NaCl. They know that, yet they didn't recognize their error in writing.

You give the student some feedback and for the next problem they write:

$3Mg^{2+}$ + $2AlCl_3$ → $3MgCl_2$ + $2Al^{\beta+}$

They're a little uncertain about the charges, but they aren't sure what to write besides that. Again, they have applied the algorithm correctly, but are now including charges with neutral elements. They do not understand that elements can exist with different charges. When elements are not in compounds, they are neutral.

You move on to another student who has written down:

Li + Cl_2 → Cl + Li_2

This student is lost somewhere further back. They are not identifying the algorithm correctly, which implies they don't understand how to identify how many elements and compound and what type of each. They could use some experience with macroscopic chemicals to get a better sense of metal elements, nonmetal elements, ionic compounds, and molecular compounds. This student should practice identifying the components of the reaction, and then progress to writing down the type. Once they master those, they can resume practicing writing out products.

Chapter 6 – Reaction Types

Hopefully I have convinced you by this point that reaction writing is a lot for students to take in at once. Once you have introduced the four reaction types it is time for the students to do some practice. I recommend holding off on adding any special reactions until they have had some practice doing this on their own.

For example, you could add neutralization reactions with double replacement reactions. The overall structure is the same except you end up producing water as a product. You could add in double replacement reactions that produce H_2CO_3, H_2SO_3 or NH_4OH. These products all decompose upon being formed.

1. $CaCO_3$ (s) + 2HCl (aq) → $CaCl_2$ (aq) + H_2CO_3 (aq)
2. $CaCO_3$ (s) + 2HCl (aq) → $CaCl_2$ (aq) + H_2O (l) + CO_2 (g)

Reaction #1 is what we would have predicted if the carbonic acid product did not decompose. Reaction #2 is what occurs instead. The carbon dioxide can easily be observed due to the bubbling that occurs when carbonates and bicarbonates mix with acid. This is a minor addition to double replacement, but it is best to give students an opportunity to interleave the four types before adding in examples such as this one. The gains of interleaving are that the student must search the information harder to make connections, find patterns and make sense of those patterns. But that can only happen if we don't hit cognitive overload.

When starting the first practice sets, have students first work to identify what the reactants are. Are they elements, metals, nonmetals, ionic compounds, molecular compounds, and how many of each are there? From that information can we determine if the reaction is a synthesis, decomposition, single replacement, or double replacement?

While single replacement (SR) and double replacement (DR) have a general algorithm to predict products there are the following exceptions:
- *SR can have the element as a metal or nonmetal*
- *SR can involve H from acids*
- *SR some metals can displace H from water*
- *SR some products have multiple charges possible*
- *SR half of the reactions don't happen*
- *DR can produce a product that decomposes producing a gas*
- *DR can produce water through neutralization*
- *Some reactions begin with two compounds that look like DR but are actually synthesis*

1. 2Na (s) + 2H_2O (l) → 2NaOH (aq) + H_2 (g)
2. 2Na (s) + H_2O (l) → Na_2O (aq) + H_2 (g)
3. Na_2O (aq) + H_2O (l) → 2NaOH (aq)

During the single replacement reactions where hydrogen is displaced from water, the student might find it helpful to consider the oxide product reacting with water to form a hydroxide. Reactions 2 and 3 combine to be equivalent to reaction 1. But reaction 2 might make more sense for a student who is still largely using abstract reasoning with the symbols. Reaction 3 will come back later in periodic trends and acid-base chemistry.

Chapter 6 – Reaction Types

In addition to the product predictions, many students have little understanding of what being dissolved entails at the particle level. Students struggle to differentiate whether ionic compounds should be represented as liquids or in the aqueous state. Students often don't understand that the + sign in the reaction indicates the mixing of the two substances. This leads to them asking questions about if it matters which order things are in (is AgCl + NaNO$_3$ the same as NaNO$_3$ + AgCl). Prior to predicting reaction types, balancing reactions is introduced. While many students can quickly master the idea of balancing reactions, interleaving balancing with all of the new ideas in predicting products can add another layer of confusion.

There are two keys for having students be successful in learning how to predict reactions. *The first key is again to have students focus on charges.* They should know that elements are neutral, ionic compounds have charged ions but are neutral overall, and molecular compounds are neutral overall with neutral components. When students assign charges to the reactants, they start to note other details that are relevant into making predictions. They notice how many elements and compounds there are. They notice whether the elements are nonmetals or metals, and they notice what types of compounds there are. This also opens them up to some key changes in replacement reactions. In single replacement reactions, electrons are transferred so charges change. During double replacement reactions, charges remain constant indicating electrons have not moved between ions.

The second key for predicting reaction types is to have students search for evidence of what products formed in observable experiments. Try holding a piece of glassware over a burning candle to show the condensation of water that occurs. Show students how a single replacement reaction occurs, but the reverse reaction does not. Show the formation of white barium sulfate precipitate and green nickel (II) nitrate solution when barium nitrate and nickel (II) sulfate are mixed. Without evidence students will devolve to using their intuition from learning how to read when they get overwhelmed.

When students become overwhelmed, they write reactions like the following:
MgBr$_2$ + NaClO$_3$ → MgClO$_3$ + NaBr$_2$

They should be aware that there is no such thing as NaBr$_2$ (or MgClO$_3$ for that matter). But that incorrect option shows the student has given up on detangling all the algorithms and exceptions when they write such a response. They aren't looking at charges correctly or at all. But we force them to do so when they identify the chemicals being produced from observation and evidence instead of from the formulas of the reactants. They know water is H$_2$O and oxygen is O$_2$. When we decompose hydrogen peroxide and show evidence from splint tests and a cloud forming, they write the reaction as H$_2$O$_2$ → H$_2$O + O$_2$ instead of something incorrect like H$_2$O$_2$ → H$_2$ + O$_2$.

When students do problems and see examples you want to make sure that they see variations where the patterns of subscripts force them to consider charges. If my first example is AgNO$_3$ (aq) + NaCl (aq) → NaNO$_3$ (aq) + AgCl (s), the students might learn to not worry about altering subscripts in future examples.

At the conclusion of double replacement, you may wish to include some neutralization reactions. Balancing a neutralization reaction is quite simple if you

Chapter 6 – Reaction Types

know a trick. If you balance the number of H^+, OH^-, and H_2O particles, the entire reaction will be balanced.

$$\_\_\_H_3PO_4 + \_\_\_Mg(OH)_2 \rightarrow \_\_\_H_2O + \_\_\_Mg_3(PO_4)_2$$

Currently there are 3 H^+, 2 OH^-, and 1 H_2O particles in the skeleton reaction. To balance we make all of those 6 particles using the following coefficients:

$$2H_3PO_4 + 3Mg(OH)_2 \rightarrow 6H_2O + Mg_3(PO_4)_2$$

Some students will use pushback on mistakes in this unit to challenge their developing conceptions of particle diagrams. A lot of students will ask what the difference is between 2N and N_2. They will ask if the 2 subscript in $Ca(NO_3)_2$ applies to the oxygens or both the nitrogen and oxygen. Sometimes they will stop in the middle of asking the question because they start to realize the answer and feel embarrassed. It is not good that they feel embarrassed, but the realizations and thinking they are doing is important and should be celebrated.

A tough decision to make during this unit is how much interleaving to use. The concept of interleaving is that it is more effective for learners to switch between different types of problems rather than repeat similar ones. For example, a teacher could introduce just single replacement reactions and have students complete ten single replacement reactions. Or a teacher could introduce single and double replacement and have students work out ten reactions that are a mixture of both. The cognitive demand of the second arrangement is higher, which would generally lead to more struggle, but also more learning. The difficulty is that at some point there is too much information, and the short-term memory is incapable of encoding so much information so quickly without experiencing cognitive overload. How students fare in naming and formula writing can be very helpful with deciding how much is the appropriate level of challenge when starting to predict reactions.

- How should you start reaction writing?
- How should students practice?
- How should you scaffold?

These are the three questions that teachers will take time and experience to develop. If you choose to start with synthesis and decomposition you want to focus on providing evidence of the products. If you choose to start with single replacement and double replacement, you'll want to focus on charges.

When students practice you want them to articulate the information that guides their decisions. How many elements and compounds are there? Are the elements metals or nonmetals? Are the compounds molecular, ionic, or acids? After they become experienced with this, they should quickly learn to identify the reaction type and predict the products.

Anticipate that students will struggle to write formulas correctly when they first start predicting products. Charges help and so do particle representations. Students struggle when everything is abstract, so adding in pictures and charges gives them more opportunities to identify patterns.

Chapter 6 – Reaction Types

The more you can interleave, the more the students will remember. Unless students become overwhelmed with cognitive overload. This chapter can be rough for students so how you scaffold will shift over time as you gain experience with teaching. Focus on adding in as many core reaction types and save states of matter and exceptions for later. A brief introduction to the basics of single replacement, double replacement, synthesis, and decomposition can be done without adding in neutralization, combustion, products that decompose, or other unusual cases. The minor details can be added once students have had a chance to struggle through the basics.

Student Struggles

1. "I have no idea what I'm doing!" Predicting chemical reactions involves a lot of parts that work together. Students have to dissect what types of compounds, how many elements and compounds there are, what types of elements, which reaction type seems likely, and the individual patterns for each type. It's a lot. Start by giving them one thing to focus on, charge. For students that are lost, have them assign charges (including neutral) to elements, ions, and molecules in the reaction. This will help them because they'll be forced to identify the number of elements, compounds, and the types of each. Once they start, their practice will help them progress through the latter components, but they need a consistent starting point to predicting products just like students start algebraic manipulations by writing out the givens, units, and equation.

2. "I don't know where to start when balancing." The key here again is to start. You can make mistakes while balancing and fix them. But if you don't start, you won't progress. Students should begin by picking an element.

"But which one?" they'll press. It doesn't matter (avoid oxygen though). Start with any element. Is it balanced? If no, then use coefficients to balance that element. If it is balanced, then pick another element.

3. "I can start balancing, but I get stuck in the middle." Sometimes balancing part of a reaction creates imbalance in others. For $H_2O \rightarrow H_2 + O_2$, the hydrogen begins as balanced. But to balance the oxygen, we make the hydrogen unbalanced ($2H_2O \rightarrow H_2 + O_2$) until we add another coefficient to the elemental hydrogen ($2H_2O \rightarrow 2H_2 + O_2$). One helpful tip is when stuck to double everything. Then at the end if you need to reduce the coefficients you can. This tends to come in handy for combustion reactions.

4. "Why do I need the activity series for single replacement, but not double replacement reactions?" This student is seeing all of the reactions in abstract mode. They aren't seeing how charges vary. In double replacement reactions the ions are changing position, but charges are constant. Single replacement reactions involve a transfer of electrons that is directional. Electrons will move from zinc to copper (II) ions, but not from copper atoms to zinc ions.

5. "How do you tell what products are for decomposition, do I just need to memorize them?" Yes. A high school chemistry student does not have the knowledge to predict products from decomposition reactions. The exception to that

Chapter 6 – Reaction Types

rule is that many decomposition reactions produce a small common molecule (CO_2, H_2O, O_2) and that binary compounds usually decompose into their elements (not hydrogen peroxide though).

6. "$BaCl_2 + Na_2SO_4 \rightarrow BaSO_4 + Na_2Cl_2$" Here the student understands the pattern for identifying and predicting a double replacement reaction. What they're missing is that they don't understand how to construct the product compounds using charge balancing. I try to make this point with NaCl because students know that formula by now. Sodium chloride isn't Na_2Cl_2 or $NaCl_2$. It's always NaCl. You can't just change the formula of a compound for convenience in balancing a reaction. But this idea takes time to grab hold. There is value in having students identify why a response is wrong.

Phenomena

1. Aluminum + iodine - Mix powdered aluminum and solid iodine in a heat resistant dish. Add a couple of drops of water. The water reacts with the aluminum initially which causes enough thermal motion to initiate the aluminum and iodine reacting. An impressive flame results with large quantities of iodine vapor being produced. The end result is a white solid showing another example of neutral elements (metal + nonmetal) combining into charged ions for an ionic compound.

2. Water softeners - Zeolites are chemicals that have a cage like structure. Sodium cations are embedded within the structure, but calcium and magnesium ions can replace them. If you add zeolites to a solution with calcium or magnesium dissolved, you'll note a considerable reduction in precipitate formed when carbonate is added. Students may have hard water issues where they live.

3. Flame discharge of an electroscope - An electroscope can show repulsion of like charges. If you hold a flame near a charged electroscope, the vane will quickly return to contact with the other metal surface. We tell students that a flame is a combustion reaction, or that a flame is plasma, but neither of those helps a student to really understand what the flame is. But the evidence from the electroscope does help us. What happens is that as the fuel reacts, the product particles speed up. They speed up so much that they can dislodge electrons from air molecules during collisions (hence the plasma term). As those electrons find their way back to the ionized particles, light is emitted (hence the flame). The electroscope confirms the presence of these charged particles since they can neutralize the charge on the electroscope. A wire gauze will block the flame from neutralizing the electroscope.

4. Alkali metals plus water - Some schools ban the use of sodium and potassium in the classroom. But for those who can safely run these reactions, they help students. The reaction $2Na + 2H_2O \rightarrow 2NaOH + H_2$ conflicts with other single replacement patterns. This can be rationalized by thinking of this as two reactions, or the teacher can use the phenomena to help students remember the products. The two-reaction pathway is:

$2Na + H_2O \rightarrow Na_2O + H_2$
$Na_2O + H_2O \rightarrow 2NaOH$

Alternatively, the teacher can add sodium metal to water with phenolphthalein. Two observations are evident. There is a small explosion indicating hydrogen. And the indicator turns pink indicating the presence of a base (NaOH). The reaction can be

Chapter 6 – Reaction Types

done in a layered mixture of oil and water. The sodium slowly sinks from the oil to the water where partial reaction occurs. As more hydrogen gas is produced the sodium starts to rise into the oil. This slows the reaction down considerably.

5. Light a candle - As the candle burns, have students hold a watch glass upside down over the flame. They should see water condense on the cooler glass surface showing that water is a product of combustion.

6. The fake candle - This doesn't serve much purpose educationally, but it's worth knowing. Take a banana, candle holder, almond sliver, and a lighter. Make a fake candle using the straightest part of the banana and put the almond sliver into the top. The almond sliver will burn just like a wick. You can even darken the sliver before you light the "fake candle" in front of students. Then pretend to use the candle to read a story or some other distracting event. Finally blow out the candle and casually take a big bite out of it. Then never mention it again.[4]

7. MOM - Put some milk of magnesia (MOM, $Mg(OH)_2$) into a large beaker with a stir bar. The $Mg(OH)_2$ does not dissolve much. But the little bit that does will turn universal indicator purple. Add a small amount of HCl to the mixture. A flash of pink will occur and then quick transitions through orange, yellow and green. After green the mixture will slowly fade to blue and eventually purple again.

8. Firetube - Obtain a long glass tube and two rubber stoppers. My glass tube is about 2 feet in length and has a diameter of about one inch. Place a small amount of ethanol or propan-2-ol in the tube and swirl it around to coat the inside walls. Remove both stoppers and place the tube over a lit candle. In the dark, a visible flame will dance around inside the tube and audible haunting noises can be heard if students are quiet.

9. The vapor ramp - DO NOT USE METHANOL for this. Take a metal trough and a flask with a volatile fuel in it. Heptane works well (hexane does too, but heptane is less toxic to you). Hold the flask so the heptane vapor (NOT LIQUID!) runs down the ramp towards a lit candle. After several seconds, remove the flask. When the vapor reaches the candle, a trail of flame will run back up the ramp. The fuel must be volatile, dense, and flammable.

10. $NiSO_4$ **(aq) +** $BaCl_2$ **(aq)** → $BaSO_4$ **(s) +** $NiCl_2$ **(aq)** - Do at least one double replacement reaction that involves colorful solutions. In this example, students will see a colorless solution mixed with a green solution. The products will be opaque at first, but the suspension of barium sulfate precipitate will settle over time. When it does the nickel ions will remain in solution and the green color will remain. This can be done with copper (II) as well. The barium sulfate is also used to enhance X-ray contrast. It's possible you will have students who have consumed this chalky drink as part of a medical examination.

Chapter 6 – Reaction Types

Flashcards

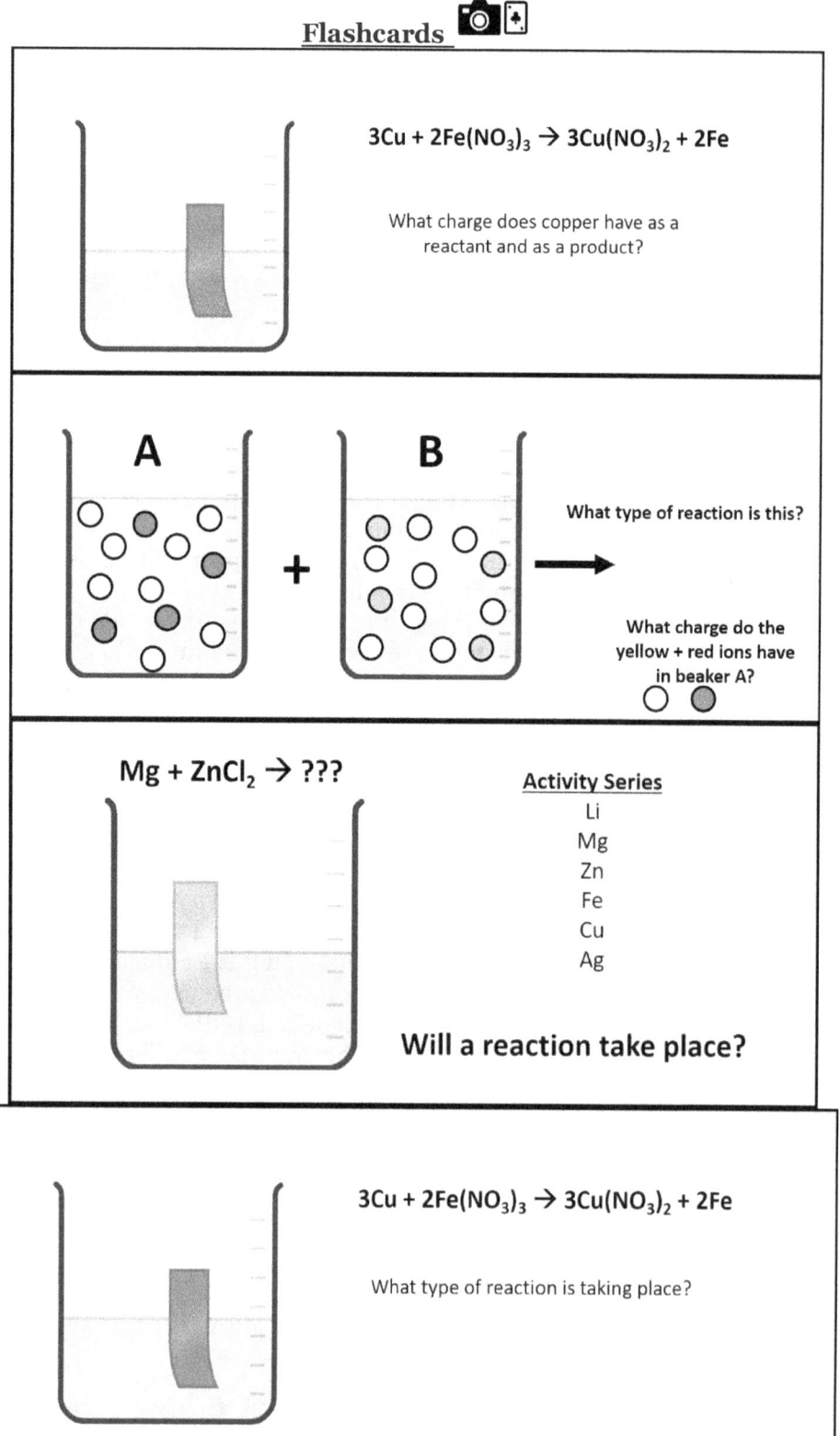

$3Cu + 2Fe(NO_3)_3 \rightarrow 3Cu(NO_3)_2 + 2Fe$

What charge does copper have as a reactant and as a product?

What type of reaction is this?

What charge do the yellow + red ions have in beaker A?

$Mg + ZnCl_2 \rightarrow ???$

Activity Series
Li
Mg
Zn
Fe
Cu
Ag

Will a reaction take place?

$3Cu + 2Fe(NO_3)_3 \rightarrow 3Cu(NO_3)_2 + 2Fe$

What type of reaction is taking place?

Chapter 6 – Reaction Types

Which reaction types can have an element and a compound as reactants?

Synthesis
Decomposition
Single Replacement
Double Replacement
Combustion

___Al + ___HCl → ___AlCl$_3$ + ___H$_2$

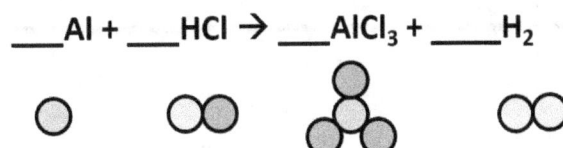

What coefficients are needed to balance the reaction?

What are the 2 products for combustion reactions?

What is a product that can form during incomplete combustion?

Chapter 6 – Reaction Types

K_2CO_3 → _____ + _____

What products are formed?

Which elements are diatomic?

Do you have to have 2 of a diatomic element when the element is part of a compound?

Na_2SO_3 + $SrCl_2$ → $NaCl$ + ???

+ → + ???

What is a reaction you have memorized?

Chapter
Stoichiometry

History

Chemistry teachers often miss the magic of stoichiometry. You can easily construct a functional recipe for chemical reactions when you have a periodic table, valence bond theory, knowledge of charge, and a particle model of matter. But historically, we didn't know any of those things. Regardless, chemists were able to conjure up valid conclusions using minimal theory and evidence.

Stoichiometry began with confusions about mass changes. What happens when we burn something, and it seemingly disappears? Why does burning a metal result in an increase in mass? Why do compounds maintain constant composition? We began to resolve these confusions by establishing that gases combine to make compounds in simple whole number ratios. Avogadro's hypothesis (1811) that equal volumes of a gas would contain equal numbers of particles was a big step toward definite proportions and recipes.

Because of the relationships between combining gases, we were able to build confidence in our understanding of solid and liquid element combinations. Prior to Avogadro's hypothesis, Priestley and Lavoisier compiled a large amount of experimental evidence that helped clarify some of the mass changes observed. They did this by controlling the system and surroundings. Indeed, here we first see the divergence from phlogiston theory by Lavoisier. Soon after that, Dalton assumes that a set of binary compounds with oxygen were all formed in a 1:1 particle ratio (1803). From this an initial set of relative masses for elements is organized. Berzelius improves upon these by 1826, but not everyone uses his data. Relative masses became much more precise and consistent in 1860 when Stanislao Cannizzaro's data became widespread.

All these developments came with flaws. Lavoisier replaced phlogiston with oxygen but added caloric theory. Dalton assumed water was HO, and that diatomic elements were impossible. There was skepticism for Cannizzaro's data since some elements seemed to be out of order (I and Te). And yet, through all of this, our understanding of the recipes for chemical reaction gradually and persistently improved.

Stoichiometry was able to test these developments in the field of organic chemistry. Organic chemistry was used for industry and medicine. But how could we decipher such a large number of compounds when their combinations of carbon and hydrogen were so variable?

One method that provided the necessary precision was to analyze combustion products to determine the relative quantities of carbon and hydrogen. Justus Liebig (1837, studied under Gay Lussac) improved upon the methods initiated by Lavoisier and Berzelius using glass apparatus. The organic compound was burned, and the products were passed through a drying agent to absorb water followed by a solution of potash (KOH) to absorb carbon dioxide (forming K_2CO_3). By merely measuring the gains in mass for each, data could be collected that could determine empirical formula of various compounds.[1] Liebig's glass instrument with 3 bulbs is called a kaliapparat and is part of the ACS symbol.

Chapter 7 – Stoichiometry + Extensions

The components of stoichiometry are found in many of the earlier chapters. In chapter 1 we see measurement established. Advances to precise measurement were critical in being able to track mass changes in chemical reactions. In chapter 2 we see that gas particles combine in simple, whole-number ratios. Our knowledge of pressure, volume, amount, and temperature are needed to interpret these combinations. In chapter 4 we find the conflict of having a two unknown problem for relative masses and chemical formulas. Yet we still managed to make those determinations through a cycle of assumptions, reflections, and revisions. This culminates here with stoichiometry and chemical reactions where all these factors are deciphered simultaneously.

BCA Tables

Stoichiometry is predicated upon proportional relationships between amounts of reactants consumed in a reaction and amounts of products formed. The recipe in this case is the balanced chemical reaction. We should anticipate that students will have constructed ideas about recipes that are easy to connect to, but that those recipes do not easily connect. Our recipes are most easily deciphered when units of moles or molecules are used, but the majority of our measurements use mass.

The fundamental concept in stoichiometry is that the amount of chemicals that react are proportional. But those proportions are most apparent when comparing molecules or moles. 45.98 g of Na react with every 70.9 g of Cl_2. Alternatively, we can say that 1 mole of chlorine gas reacts with 2 moles of sodium atoms. From this starting point we want to build the idea that if we doubled the amount of sodium, we could require twice the amount of chlorine as well. If we had 6.34 moles of sodium, we would require 3.17 moles of chlorine to complete the reaction. There is a recipe, and it must be followed.

Perhaps the best introduction to stoichiometry embraces the obscurity of mass-mass relationships. Provide the students with a chart of various starting amounts of reactants along with the theoretical amount of product produced. The reaction is $4Na + O_2 \rightarrow 2Na_2O$. What do you notice? Could we construct a for every statement based on the data?

Trial	Amount of Na (22.99 g/mol)	Amount of O_2 (32.00 g/mol)	Amount of Na_2O produced (61.98 g/mol)
1	22.99 g	32.00 g	30.99
2	45.98 g	32.00 g	61.98 g
3	91.96 g	32.00 g	123.96 g
4	183.92 g	32.00 g	123.96 g
5	183.92 g	64.00 g	247.92 g

Table 7-1: Initial amounts of Na and O_2 along with the final amount of Na_2O product

Chapter 7 – Stoichiometry + Extensions

For every 91.96 g of Na, 32.00 g O_2 react to form 123.96 g of Na_2O. It doesn't quite roll off the tongue. We could enhance the depth of observations by adding additional columns.

Trial	Amount of Na (22.99 g/mol)	Amount of O_2 (32.00 g/mol)	Amount of Na_2O Produced (61.98 g/mol)	Leftover reactant
1	22.99 g	32.00 g	30.99	24.00 g O_2
2	45.98 g	32.00 g	61.98 g	16.00 g O_2
3	91.96 g	32.00 g	123.96 g	none
4	183.92 g	32.00 g	123.96 g	91.96 g Na
5	183.92 g	64.00 g	247.92 g	none

Table 7-2: Initial amounts of Na and O_2 along with the final amount of Na_2O product and excess reagent

If we had reacted 19.14 g of Na, could we predict the maximum amount of Na_2O produced? What is a method we could go about this? By starting with students making observations of patterns they are better able to connect with their prior knowledge. If instead we had started by showing students a dimensional analysis set up, we would have limited their ability to make concrete connections. This leads to them being able to do the calculations and find a correct solution with little understanding of what they did.

$$\frac{19.14 \text{ g Na}}{} \cdot \frac{1 \text{ mol Na}}{22.99 \text{ g Na}} \cdot \frac{2 \text{ mol Na}_2\text{O}}{4 \text{ mol Na}} \cdot \frac{61.98 \text{ g Na}_2\text{O}}{1 \text{ mol Na}_2\text{O}} = 25.8 \text{ g Na}_2\text{O}$$

The instructions are simple and redundant. The tradeoff is that we are obscuring the "why." Why are we doing what we are doing? A student can be proficient in finding answers without understanding the algorithm. As teachers we should be particularly concerned that a student that sees the algorithm first without any understanding will struggle to shift towards a more comprehensive model.

Students understand the concept of recipes well even if they struggle to apply recipes with chemicals. At any point that students are stuck, it is helpful to use a simple cooking recipe analogy. I like to use a sandwich recipe where four pieces of salami (4), one piece of stringed cheese (C), and two pieces of bread (B) combine to make one sandwich (Sw, or S_4CB_2).

$$4S + C + 2B \rightarrow Sw$$

- If we have 20 pieces of salami how many sandwiches can I make?
- How many slices of bread would I need?
- Would 8 pieces of cheese work?
- Would I have extra?

These questions are simple for students to answer. They understand proportionality and they have experienced the idea of scaling up food to prepare. But stoichiometry can often be a struggle because that recipe and proportional thinking is

Chapter 7 – Stoichiometry + Extensions

not readily apparent. Students struggle to transfer the concepts that are easy for food over to chemistry.

There are two common methods used for doing stoichiometry calculations. The first is more common and has been used for longer. It involves conversions using dimensional analysis. The second method uses BCA tables.

$$\frac{9.00 \text{ g Na}}{} \bigg| \frac{1 \text{ mol Na}}{22.99 \text{ g Na}} \bigg| \frac{2 \text{ mol NaCl}}{2 \text{ mol Na}} \bigg| \frac{58.44 \text{ g NaCl}}{1 \text{ mol NaCl}} = 22.9 \text{ g NaCl}$$

A BCA table has the reaction and underneath each chemical species we will in how much there is Before, how much the chemical amount Changes, and how much is left After the reaction is complete. Everything in a BCA table must be in units of moles (molecules, formula units, and atoms would work as well). The changes must be proportional to the balanced coefficients.

Typically, a BCA table starts with 0 product before the reaction and ends with 0 of at least one reactant after. BCA tables take longer to fill out than dimensional analysis but require the student to connect the concepts within the problem.

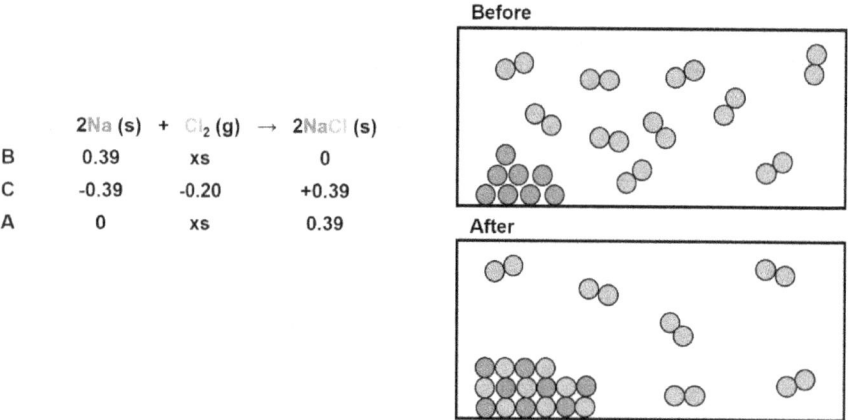

Figure 7-1: BCA table with accompanying particle diagram for 2Na + Cl₂ → 2NaCl

9.0 g of Na is 0.39 moles of Na. All the sodium reacts so the final "After" value is 0. Note that 9.0 is smaller than 22.99 g (relative mass of Na) and therefore we have less than one mole of Na. The chlorine is in excess (xs). That means that we have enough chlorine that we will not run out. We might have 0.21 moles, or we might have a billion moles. Because we have extra, we do not need to be concerned with the impact on the reaction. The sodium that does run out will determine (or limit) how much product we make.

Students may try to incorporate the 2 coefficient into the mole conversion. They will end with a result of 0.20 moles of Na because they figured that they need twice as many Na. Stressing the per 1 mole nature of the molar mass can be helpful, but it does take them some time to process why the change portion is the only time when coefficients are utilized. You can begin by mixing any 2 amounts of reactants, but what reacts is limited by the recipe.

Something hidden in stoichiometry is that students do not easily translate the recipe because moles are not the same as molecules. It can be helpful for them to be

Chapter 7 – Stoichiometry + Extensions

walked through a comparison. If we gave students a BCA table like the sulfur and oxygen reaction below, they are much more successful in determining how many oxygen molecules are needed to complete the reaction and how many sulfur trioxide molecules would be formed.

	2S (s) +	3O$_2$ (g)	→	2SO$_3$ (g)
B	2 atoms	? molecules		0 molecules
C				
A	0 atoms			? molecules ?

    The difference is that they can readily connect 2 atoms with a particle level representation. But 2 moles has that extra step with a large number in the way. If we show them an example like this and ask how we would change each piece into moles, they can see why moles also works. To change 2 atoms of S into moles we would divide by 6.02 x 10^{23}. To change 3 molecules of O$_2$ into moles we would divide by 6.02 x 10^{23}. To change 2 molecules of SO$_3$ into moles we would divide by 6.02 x 10^{23}. Because the conversion to moles from particles is always going to be relative to Avogadro's number, moles work the same as particles.
    This would be a good time to compare moles to dozens again. If 2 atoms of sulfur react with 3 molecules of oxygen, how many dozens of molecules would react with 2 dozen S atoms? Students can translate dozens easier because the number 12 is simpler than 602 sextillion. They are also more familiar from experiences with food.
    The BCA table method is more effective for instruction because it provides students with more information and structure. Both methods work similarly mathematically, but when you watch students struggle with stoichiometry calculations a few commonalities appear frequently. Students who struggle with stoichiometry initially will often put a variety of numbers into a dimensional analysis table, but the numbers will be without units and chemical labels. This is usually because they don't know what they are doing, and they are just trying to mimic the algorithm they saw the teacher perform. It is more difficult to connect dimensional analysis to visual representations of the reaction.
    Have students fill in the units of grams, moles, and the chemical formulas. The student will be forced to think more about what they are doing. They will also notice more patterns and relationships that help them understand. You want them to ask questions such as, "Should I put a 2 or a 1 with the moles of Na over the 22.99 grams of Na?" This is more difficult initially but pays off later as students are forced to evaluate their processing.
    A BCA table not only addresses issues of why, they also simplify more advanced questions that involve excess reagents and equilibrium calculations. The change portion of the BCA tables makes the mole ratio much more concrete for students than seeing the mole ratio in the midst of a conversion.
    Another perk to using a BCA Table is that they connect to a graphical representation of stoichiometry. If we start with 32 particles of Li and 11 particles of O$_2$ 16 particles of Li$_2$O will be formed before we run out of Li (Fig. 7-2, Fig. 7-3). As the particles react to form Li$_2$O the reaction progress is tracked using a graph (Fig. 7-4).

Chapter 7 – Stoichiometry + Extensions

	4 Li (s)	+ O$_2$ (g)	→ 2 Li$_2$O (s)
B	32	11	0
C			
A			

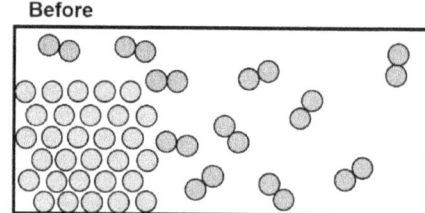

Figure 7-2: BCA table and particle model for lithium reacting with oxygen before reaction starts

	4 Li (s)	+ O$_2$ (g)	→ 2 Li$_2$O (s)
B	32	11	0
C	-32	-8	+16
A	0	3	16

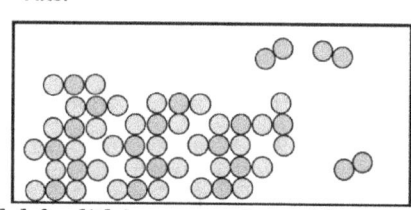

Figure 7-3: BCA table and particle model for lithium reacting with oxygen once the reaction is complete

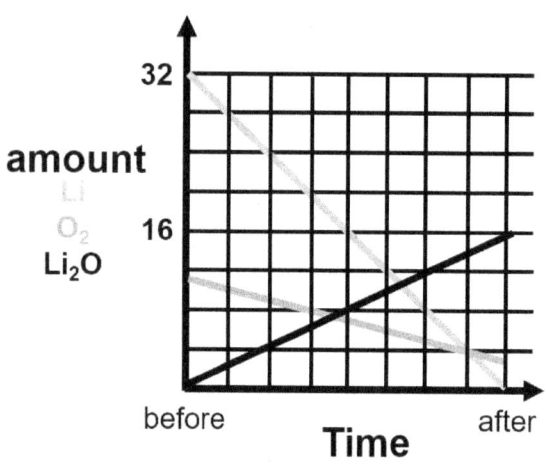

Figure 7-4: BCA table and particle diagrams for lithium and oxygen reaction

    Construct the graph by following the recipe one set at a time. After 1 time interval, 4 Li particles and 1 O$_2$ particle combine to make 2 Li$_2$O particles. Then another set reacts, and another. Adding the element of a gradual transition over time helps students to visualize the process with what they observe macroscopically. When a candle burns, the entire candle doesn't convert instantly into all CO$_2$ and H$_2$O. Rather part of the wax burns, then another part, and another.

    The graphical representation allows students to connect pieces such as collision theory for kinetics, equilibrium ICE charts, relative rates, and limiting reagent. The collision theory can be seen when the oxygen molecules react only with the surface particles of Lithium in the particle model. The relative rates can be seen by comparing the slopes of the lines in the graph model. The BCA table and graph both help develop concepts that are important for ICE charts in equilibrium. In stoichiometry we assume reaction completion. In equilibrium we reach a point where there is still some of each chemical. The products are colliding and forming reactants at the same rate as the reactants are forming products. Highlight this by drawing a

Chapter 7 – Stoichiometry + Extensions

line on the graph in a new color before the reaction completes to talk about reversible reactions.

We can connect particle diagrams, BCA tables and the graphical analysis to limiting reagent. When we run out of lithium particles in the particle model the reaction stops. But we have excess oxygen particles that could react if we added more lithium. On the graph we see that the lithium has reached zero, but we could extrapolate the amount of lithium oxide that we could form if we had used up all of the oxygen instead.

Limiting and Excess Reagent

When we mix two chemicals the most likely scenario is that there will be extra of one when the other runs out. A BCA table is helpful to seeing which is which. But if you are committed to using dimensional analysis there are some options for how students can determine the limiting reagent, amount of excess reagent, and the amount of product produced.

The simplest method for doing all three is to start by doing two full stoichiometry calculations. Let's expand on the earlier example and say that 9.00 g of Na reacts with 24.3 g of Cl_2. There are three questions that can be asked.
1. What is the limiting reagent?
2. How much excess reagent is leftover?
3. How much product is made?

We begin by doing two separate calculations for each reagent:
$$2Na\ (s) + Cl_2\ (g) \rightarrow 2NaCl\ (s)$$

9.00 g Na	1 mol Na	2 mol NaCl	58.44 g NaCl	= 22.9 g NaCl
	22.99 g Na	2 mol Na	1 mol NaCl	

24.3 g Cl_2	1 mol Cl_2	2 mol NaCl	58.44 g NaCl	= 40.1 g NaCl
	70.90 g Cl_2	1 mol Cl_2	1 mol NaCl	

Just by doing the two calculations three things become immediately apparent. *The sodium is the limiting reagent. The chlorine is in excess. There will be 22.9 g of product formed.* We can see that as the sodium chloride is produced that the sodium will run out once 22.9 g of NaCl is formed. Chlorine would be capable of continuing the reaction until 40.1 g of NaCl is formed. The lesser of the two product amounts is how much product will form and identifies the limiting reagent. If students do not see this, work through some salami sandwich examples.

The only thing left to determine is how much excess remains. There is an equation for this calculation.

(1-ratio of answers)*original quantity of excess reagent = amount of excess

For our example that would be:

$$(1 - \frac{22.9}{40.1})*24.3\ g\ Cl_2 = 10.4\ g\ Cl_2\ remaining$$

Chapter 7 – Stoichiometry + Extensions

The ratio of answers should be thought of entirely in context of the excess reagent. The reaction stops when 22.9 g of NaCl are made, but there is enough chlorine to produce 40.1 g. This means that 22.9/40.1 is proportional to the amount of chlorine used. If we subtract that from 1 that would be proportional to the amount of leftover Cl_2.

It might be helpful to think of this in percentage terms for students. 22.9/40.1 is 57.1%. That means that we are using up 57.1% of the chlorine. 42.9% is what will be leftover. Since you have enough chlorine to make about 40 grams of NaCl, when you get to 20 grams of NaCl you would have used about half of the chlorine. When we get to 22.9 grams of NaCl we would be a little bit past that. Hence the 57.1% of chlorine used.

What I like about this method is that you get most of the calculation done separately from the analysis. There are ways for students to work out how much one reactant would make and then work backwards from that amount to find how much of the other reactant would be needed. But here the student does 2 stoichiometry problems. This is easy because the second half of the second calculation can be copied from the first problem. So, the student repeats the algorithm and is likely to develop a system 1 transfer where they can set up stoichiometry problems with minimal thinking.

But once the stoichiometry calculations are done, the analysis can be done where the student does the rationalization of what will get made and what will be left over. I prefer the BCA tables, but this is a good alternative. If you use it, spend some time helping students make sense of the equation to find the amount of excess reagent leftover.

	2Al	+	6HCl	→	3H$_2$	+	2AlCl$_3$
B	0.040		0.090		0		0
C							
A							

Limiting and excess reagent problems also fit seamlessly when using BCA tables. Here we see a question where we have 0.040 moles of Al reacting with 0.090 moles of HCl. We assume that one of these reactants ends up with 0 moles and the other will have some leftover excess.

If we assume the Al goes to 0 moles, the change for Al is -0.040 moles. The HCl must change by a 6:2 proportion, so HCl would change by -0.120 mol. This, of course, is not possible as we would end up with a negative quantity of HCl.

Chapter 7 – Stoichiometry + Extensions

	2Al	+ 6HCl	→ 3H₂	+ 2AlCl₃
B	0.040	0.090	0	0
C	-0.040	-0.120	+0.060	+0.040
A	0	-0.030	0.060	0.040

If we instead assume that the HCl goes to 0 moles, the change for HCl is -0.090 moles. The Al would need to decrease by a 2:6 proportion, or by -0.030 moles. This would mean we would end with 0.010 moles of excess Al.

	2Al	+ 6HCl	→ 3H₂	+ 2AlCl₃
B	0.040	0.090	0	0
C	-0.030	-0.090	+0.045	+0.030
A	0.010	0	0.045	0.030

For BCA tables the student can arbitrarily guess which reactant runs out. If they guess incorrectly, they will end up with a negative amount of the other reactant. If they guess correctly, they'll end up with a positive excess (or 0 if the amounts were perfectly balanced). Students can frequently correctly predict by doing some mental math prior to selecting the limiting reactant as well.

Once the BCA table has been correctly completed, the student can easily determine the amount of product formed, the amount of excess reagent, and the identity of the limiting reagent.

Stoichiometry is easy to construct a variety of practice problems. The students will need practice and feedback on that practice. But don't lose sight of the concept. We want students to understand the recipe concept behind how chemicals react. At some point that means we want them to be constructing particle diagrams. We want them to articulate the recipe concept and how mole ratios are used in their calculations.

There is a demonstration where you put increasing amounts of baking soda into balloons. Then the balloons are fixed to five bottles with equal amounts of vinegar. When 1 gram of baking soda is emptied into the vinegar the first balloon blows up a tiny amount. When 2 grams of baking soda is emptied into the vinegar, the second balloon gets a little bigger than the 1st. When 3 grams of baking soda is emptied into the vinegar, the third balloon gets even bigger. But when 4 grams is added the balloon becomes the same size as the 3rd balloon. And the 5 grams of baking soda again produce the same sized balloon (Fig. 7-5). *What is happening?*

Chapter 7 – Stoichiometry + Extensions

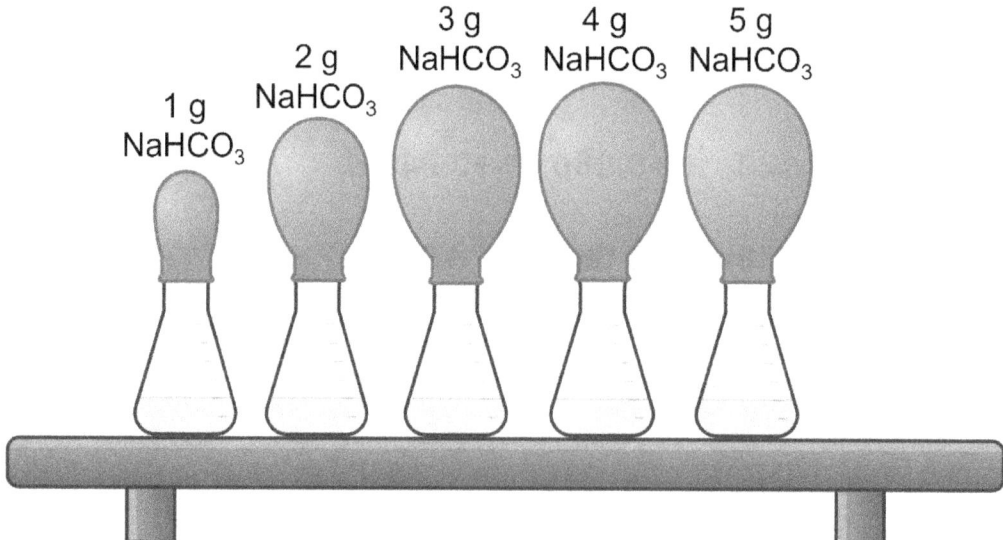

Figure 7-5: 1g to 5g of baking soda are mixed with equal quantities of vinegar in bottles with balloons attached

- Why did the 4th and 5th balloons remain the same as the 3rd?
- What is the limiting reagent in the 1st balloon and the 5th balloon?
- Can you draw a particle diagram of each solution and balloon after the reaction is completed?

A particularly good question to bring up focuses on the 3rd balloon. The assumption is that in the 3rd balloon that the two chemicals are evenly balanced, and both are used up completely. But that isn't necessarily true. I could have added a little extra of one. Which one would I have added a little bit extra of? A hint is that the 3rd balloon is the same as the 4th and 5th. What was extra in those?

    Stoichiometry is a critical subject. Make sure that students move beyond just finding solutions to understanding what they are doing. BCA tables can help with this, but even if you stick with dimensional analysis ask probing questions to help students identify key components of the algorithms used. Whenever possible, avoid starting with the algorithm and let students determine patterns first. This forces them to utilize their knowledge and working models first, which builds a stronger cognitive memory. The concept of a recipe should be present constantly, and the obscuring of that recipe via relative mass can get you the initial confusion to help students observe carefully.

    Some teachers group stoichiometry problems with compositional stoichiometry (percent composition and empirical formula). Instead, I propose immediately shifting into topics that are all predicated upon proportions of the balanced chemical reaction recipe. Concentration, enthalpy, and the ideal gas law are all connected to amounts of chemical reaction taking place. By teaching these extensions to BCA problems now we form a strong model for stoichiometry by linking in a variety of applications. This is how experts would solve problems involving titrations, or mass to volume of a gas calculations. Students will require time, feedback, and reflection to be successful. But there is a large upside at play here for students who will go on to higher levels of chemistry instruction.

Chapter 7 – Stoichiometry + Extensions

Molarity

Students have a large amount of experience with sugary beverages. Use that connection. Imagine the sweet taste of your favorite pop (in Michigan we say pop). My favorite is Dr. Pepper because I am an intellectual. A 20 oz. Dr. Pepper has 64 g of sugar in it. Let's assume that sugar is fructose ($C_6H_{12}O_6$, 180.18 g/mol). In the morning I drink some of the Dr. Pepper until there are 15 oz. left. How much sugar is there now? What about at 10 oz. and 5 oz.? Have students figure these out. Then have them plot the values but with a couple of quick changes. Instead of plotting grams of sugar, we want moles. And instead of fluid ounces of pop, we want liters. The conversion factors are 1 L = 33.814 oz., and the students can figure out the molar mass of fructose from the formula.

We want them to plot the four values, find the line of best fit in y = mx + b format, and draw particle representations of the 2-4 of the amounts. You'll have to excuse my significant figures because I am not sure how precise Dr. Pepper measurements are. But students should produce an equation that is somewhere close to y = 0.601x.

Fluid ounces	Liters	Grams of fructose	Moles of fructose
20	0.591	64	0.355
15	0.444	48	0.266
10	0.296	32	0.178
5	0.148	16	0.089

Table 7-3: Volumes, masses and moles for various amounts of Dr. Pepper

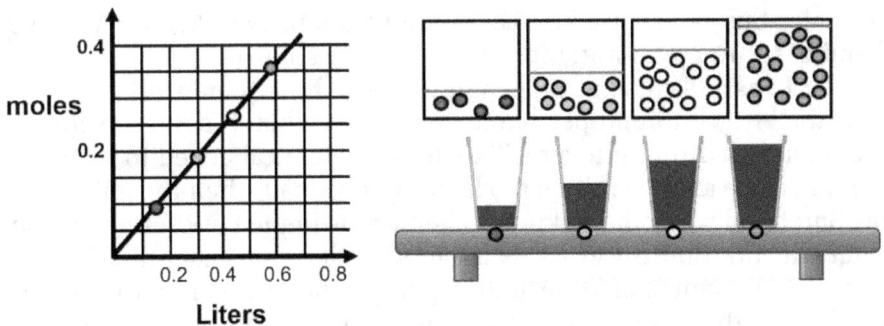

Figure 7-6: Graph and particle representations of data from Table 7-3

Note that students should find the volume in liters to be proportional to the moles and grams of sugar. Our y-intercept is 0. The y-intercept makes sense that if there is no pop, there would be no sugar. The slope provides a flood of good questions.
- What is the "For every" statement for the slope of 0.601?
- What are the units for the slope?
- What is the equation for the slope?
- What is the physical interpretation of the slope?

Chapter 7 – Stoichiometry + Extensions

- Is the slope a density?
- What would a large slope taste like? What would a small slope taste like?
- Why is the slope constant?

As before, don't let the students use vocab terms like concentration. Have them explain with simpler terms what the slope represents. Interestingly they tend to use density. The more sugar dissolved is a denser solution. This makes for a compelling discussion on what the difference between density and concentration is. A quick comparison of units shows that g/mL and mol/L are quite similar. Both are amounts per unit volume. Concentration tends to be used to compare two components in a mixture whereas density can describe a single substance. But they are very similar concepts.

After some discussion of the slope, the students are ready to chunk the information. The slope is in units of molarity. Molarity can be represented as M or mol/L (or mol*dm^{-3}). We should improve the units by writing out the chemical symbol after moles and L of solution. That maintains our connection to the macroscopic and symbolic representations because it makes the numbers specify what they represent.

If we go back to our y = 0.601 x equation, we can plug in moles of fructose for y and liters of solution for x to get:

*Moles of fructose = 0.601 M * liters of Dr. Pepper*

We can write this more generally as:

*moles = M * volume*

It is worth stressing that a large number of calculations in chemistry begin by finding the moles of a substance. Knowing that the concentration multiplied by the volume gives moles can be critical. This can be further enhanced by looking at why dilutions can use the formula $M_1V_1 = M_2V_2$ since the moles of solute do not change during a dilution. Later this can be used for acid-base titrations with $M_aV_a = M_bV_b$ so long as the acid and base are monoprotic (or react in a 1:1 fashion).

A good means to assess student understanding is to have them work out the direction of systematic error for improper buret cleaning. If a buret is rinsed with water, the small amounts of drops leftover will dilute the chemical added to it. Ask students to determine if the answer will be too large or too small when the unknown chemical is in the buret and when the unknown chemical is in the flask for a titration.

Let's assume the titration is $KMnO_4$ being used to titrate a solution with iron (II) ions. If the iron (II) amount is unknown, the improper buret cleaning will cause the $KMnO_4$ to take longer than expected to reach the end point. The iron (II) amount will therefore be overestimated. But if the $KMnO_4$ is unknown, the larger volume due to dilution will cause the concentration of permanganate to be underestimated.

When students struggle doing calculations of molarity it is usually because they fail to automatically change grams of a chemical into moles. Also many students who get stuck will show work without sufficient labels. Having them write the formula of the chemical they are analyzing is a key step to them assigning concrete meaning to the numbers they are manipulating.

Proportionality can be used in place of an equation. For every 1 liter of solution there are 2.30 moles of $CuBr_2$ (MM = 223.35 g/mol). How many grams are needed to make 75.0 mL?

Chapter 7 – Stoichiometry + Extensions

2.30 moles of CuBr$_2$ is equivalent to 514 g of CuBr$_2$ and 1 liter is 1000 mL.

$$\frac{514 \text{ g CuBr}_2}{1000 \text{ mL}} = \frac{? \text{ g CuBr}_2}{75.0 \text{ mL}}$$

Obviously, we should have less than 514 g of CuBr$_2$ in order for this proportion to hold. Note that the student must do two calculations prior to the proportionality though. They must make the volumes have equivalent units and change the moles to grams (before or after).

Many teachers will clarify the difference between adding 1 liter of water and preparing 1 liter of solution. Ethanol and water make for a good demonstration that volumes are not additive. A fun way to demonstrate this is to add yellow dye to ethanol and blue dye to water. Then fill a long tube halfway with the water. Then add ethanol until the ethanol so that it slightly overflows when the tube is closed. As you invert the tube to mix the two substances the students will expect the color to become green. But a bubble will also appear. That bubble gradually gets larger as the two solutions mix more. What happens is that the ethanol and water occupy the interstitial spaces causing the total volume to reduce. Since the tube is closed a bubble forms due to the liquids not occupying all the space in the tube. If the tube is 100 mL in volume, the 50 mL of ethanol and 50 mL of water will not occupy 100 mL when mixed. They will occupy about 95 mL with a 5 mL bubble present. A fun question to ask the students is what is the bubble made of?

You can assess students' ability to prepare a solution quickly by filling a plastic pipet with shot and calibrating it with lines. If the solution causes the pipet to float between the target lines, they get an A for the assignment. The next set of lines would earn a B and so forth. In Figure 7-7 the student transfers the solution to a beaker and would get a B grade.

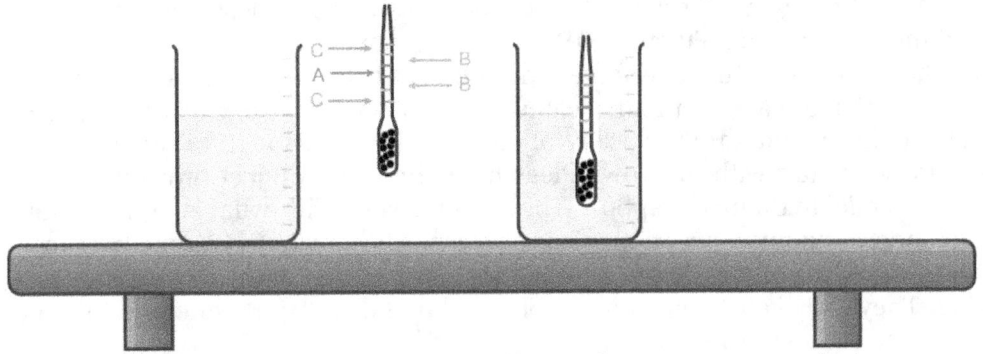

Figure 7-7: Quick assessment of how well a student prepares a solution

The phrase *"when in doubt, find moles"* can pay dividends for students who are lost at the start of a problem. Concentration can start to overwhelm students struggling to connect the recipe concept to a growing number of extensions. The emphasis on M*V = moles can help students have a simple tool to use in getting started. The slope analysis and proportional reasoning will help push student thinking to the point of mastery.

Chapter 7 – Stoichiometry + Extensions

Enthalpy

Like many topics in chemistry, enthalpy can be presented in a technical or simplified version. For introductory students start with three keys for enthalpy.

1. Provide students with a mechanism of why chemical reactions result in a net change in energy. Using force and motion is critical for students to move past misconceptions.
2. Use an energy organizational tool to emphasize systems and surroundings.
3. Help students appreciate the concept that energy produced or consumed in a reaction is proportional to the amount of chemicals reacted.

After those three keys have been established you can move on to the technical side of enthalpy and the supporting calculations. Some do this by teaching enthalpy later (e.g., with Chapter 16).

We start by having students make some basic observations about a chemical reaction. A convenient phenomenon for students to observe is to have students mix vinegar and baking soda in a plastic baggie. They will expect the bag to blow up, but instead you want to draw their attention to the fact that the chemicals become colder as they mix and react. If you discuss this, you will find that they have very few conceptual understandings that help explain this. I recommend having them begin by agreeing that the particles in the mixture have slowed down relative to their speeds before reacting. This is a starting point where you can begin with consensus.

Next you want them to identify that the particles are all still there, but they have changed their arrangements. For some reactions, the rearrangements during the reaction lead to a net increase in motion (burning/combustion). For others, the reaction leads to a net decrease in motion. This implies that there are two things happening during a reaction. When bonds, or the stickiness between the particles, are broken the particles change speeds. When bonds are formed, the particles change speeds the other way. It is critical not to dictate to them which one is which at first. Let them discuss, write, or draw which one they think is which. Push them on their ideas and see what thoughts they have. Some students will carry ideas about this from biology and the changing of ATP and ADP. Others will have physical ideas about how the charges have stronger attractive forces when the particles are closer. Others will be using models that are abstract and vocabulary based such as "bonds store energy." But you need to have them think critically about what it means to "store" energy.

If they don't start with their own ideas here, then you will just provide them an alternative model to their conception. They need to start with what is in their brain so they can undermine misconceptions. It is also critical that you take some time to see what they think so you can address some flaws in their reasoning through questioning. They will likely be uncomfortable, and that discomfort is helpful to undermine some of their wrong ideas.

Do particles speed up or slow down during a reaction?

Some students have an idea that a bond breaking is accompanied by a minor explosion. They think that the bond explodes or something to that effect. It is important to now specify that when a bond breaks, all that is happening is the particles are moving farther away from each other. When particles move apart, they have an attractive force pulling them together and they are moving in opposition to this force. Therefore, they will slow down.

Chapter 7 – Stoichiometry + Extensions

⇨ Direction of force
→ Direction of motion

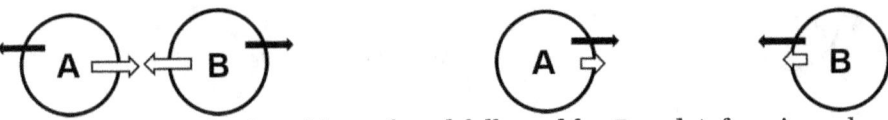

Figure 7-8: A and B breaking a bond followed by B and A forming a bond

The attractive force pulls A and B together (Fig. 7-8) but A and B are moving apart. Therefore, the particles slow down. When moving left and experiencing a net force to the right, speed will decrease. In the second part of Fig. 7-8, the attractive force pulls in the same directions as the motion. This causes the particles to move faster. When moving left and experiencing a net force to the left, speed increases. When the bond is formed the molecules have an increase in motion. If someone is riding a bike east and a gust of wind blows east, they will speed up. If the wind blows westward the bike will slow.

When an object falls towards the Earth, the object is moving and experiencing a net force downwards. Therefore, the object speeds up. When the object is moving away from the Earth, the net force and motion are in opposition and thus the object slows down. If this seems like overkill, you have to understand that the student has had sensations with magnets that undermine their ability to process this. They think of magnets as close as being high in energy because the force is stronger when they are close. Since "storing energy" or "releasing energy" are vague, it is easy for students to fill any a wide array of explanations for their experiences.

Many texts also omit whether the energy of particles is based on kinetic, potential or both. If we just provide students with abstract models of energy in bonding, they will not be able to replace the faulty ideas that they begin with. It can be extremely helpful to run through an example where gravitational potential energy is converted to kinetic prior to these exercises. Holding an object higher or lower, you are focusing on the larger separation correlating to a faster speed prior to impact. If you hold a bowling ball up high, it has more gravitational potential energy. When you release it and it has fallen halfway, it now has some potential and is moving. But if you had just released it from that middle height, it would have only had that amount of potential and no motion (relative to the Earth).

Chapter 7 – Stoichiometry + Extensions

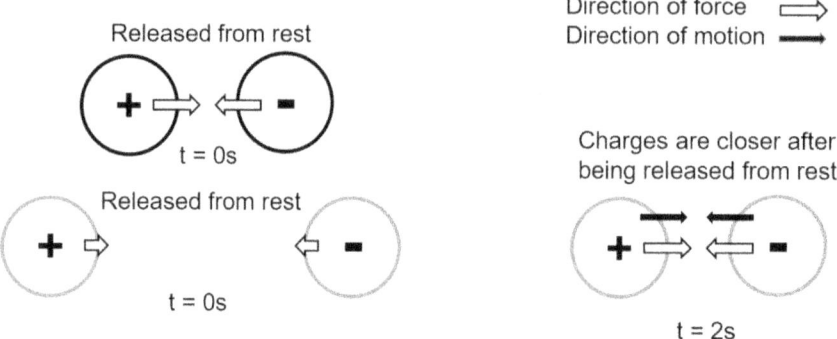

Figure 7-9: The forces and movement between oppositely charged particles before released from distance d, before released from distance 2d, and after being released from distance 2d but at distance d.

When the particles are released in the first arrangement, they move closer together (Fig. 7-9). The force and speed both increase as the particles move. In the second arrangement the particles experience less initial force, but that force gets larger as the particles approach. The third arrangement is just the same as the second 2 seconds later. The particles are now at the same position as they were in the first arrangement, but because they were released further apart, they are now moving instead of being stationary. So now only do they experience the same force as the initial set, they have motion in addition. Therefore the 2nd and 3rd arrangements had more energy than the first set.

Fig. 7-9 is the type of simple model that we hope to build consensus with for our students. *We want them to think through positions and relative speeds for charged particles in order to have a meaningful model for why larger separation equates to more energy.* They do not need mathematical support for these models. Instead, we are trying to understand why a larger separation has a greater electrical potential energy for charged particles. From this we can reach the point where students have a stronger understanding of why breaking a bond requires energy and why forming a bond causes particles to speed up and transfer thermal energy to the surroundings.

Once students have a better working understanding of energy changes, we want to organize the forms of energy as changes occur. Energy organizational tools accomplish two things. The first is that they validate energy conservation for students. The second is that they help students track changes in the form of energy during changes by drawing attention to system and surroundings.

Chapter 7 – Stoichiometry + Extensions

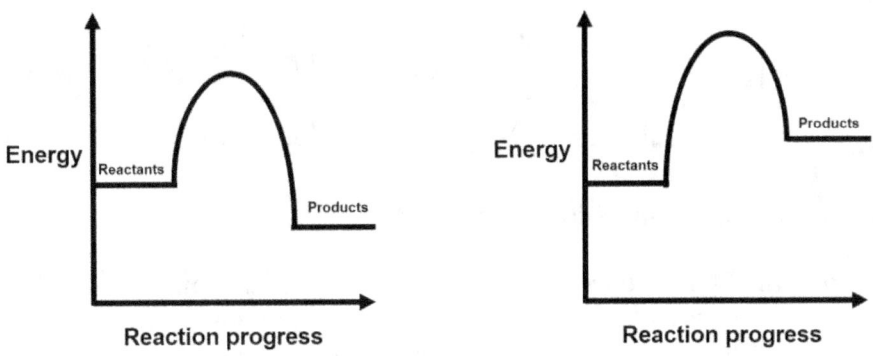

Figure 7-10: A reaction energy diagram for an endothermic and exothermic reaction

My first attempts at teaching reaction energy diagrams were both successful and ineffective. The cognitive load was minimal, and the algorithms were simple. If the energy goes up, energy went into (sounds just like endo). If the energy goes down, energy exits (sounds like exo). And that was it. Most students could easily recall which diagram was which. They could repeat what I had stated. But into what exactly did that energy leave? What type of energy? Why did the energy change? Why was the shape of the curve the way that it is? What does energy change look like at the particle level? What happens with kinetic energy? Does the reaction energy diagram represent a single set of molecules or the average of a lot of molecules?

There are an overwhelming number of questions that I never gave the students a chance to wonder and think about. This was yet another missed opportunity for students to think. Instead, I was helping them to avoid that thinking.

At some level I grew frustrated with energy as I taught chemistry. Since then, I've started to avoid its use because I find that energy is common to avoiding thinking. Recently I've begun to use organizational tools called energy bar charts that are better equipped to help students classify energy changes without destroying their curiosity in the process. These energy bar charts are affectionately called "LOL diagrams" or "LOLOL diagrams."

LOL and LOLOL diagrams are productive scaffolds for teachers and students. Each L contains bars that represent two different forms of energy within a system. The bars are meant to be qualitative more than quantitative. We're looking for changes more than specific amounts. LOL diagrams can be introduced along with specific heat capacity (Chapter 3). At that point the students do not know as much, and so we look at thermal (E_{th}) and phase (E_{ph}) energies. When we get to enthalpy and LOLOL diagrams, chemical energy (E_{ch}) replaces phase energy.

The number of thermal energy bars is determined by the temperature. The number of phase energy bars is determined by the state of matter. The number of chemical energy bars relies on other changes we see taking place. For example, if we know that no energy has been exchanged with the surroundings, but the system is now hotter, the chemical energy must have decreased to offset the increase in thermal energy.

Chapter 7 – Stoichiometry + Extensions

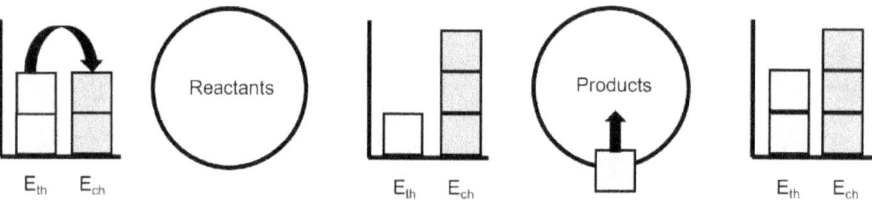

Figure 7-11: LOLOL diagram for an endothermic reaction

Fig. 7-11 shows the LOLOL diagram for an endothermic reaction. The first thing that happens in the reaction is that the reaction occurs. For some endothermic reactions, heating is the first step. In this one as soon as the chemicals are mixed, they react. As the reaction proceeds, the chemicals get colder. The particles are now moving slower. Because of this cooling, there is now a temperature gradient between the surroundings and chemical system. Energy flows into the system through heating (indicated as a bar moving into the circle), so the thermal energy increases back to the point where the temperature reaches the initial starting point. Overall, there is a net influx of energy to the system. Note that there are only 3 variations of LOLOL diagrams. There are endothermic with initial heating, endothermic without heating, and exothermic reactions.

    LOLOL diagrams give students a framework to organize the various uses and changes that energy undergoes during a chemical reaction. Note how many more distinctions must be made relative to a reaction energy diagram. The teacher might also use LOLOL diagrams to emphasize the collision nature of heat transfer and chemical reactions. The exchange of energy between system and surroundings occurs through particles speeding and slowing. The rearrangement of particles occurs when the chemicals in the system collide with each other fast enough to break bonds and rearrange.

    Let's use our initial example of vinegar and baking soda. At room temperature, these two chemicals collide with sufficient relative speeds to break apart bonds. Energy that was previously thermal is now potential or chemical energy. Then particles form new bonds in a new arrangement. The energy is now being used more as thermal and less as chemical. The collisions between the particles is the mechanism by which these energy changes can occur.

    Overall, the particles have slowed down since the bonds being broken were stronger than those formed. The particles slowed down (breaking bonds) more than they sped up (forming bonds). Since the particles are moving slower, the surroundings and system experience heat transfer. Energy goes into the chemical system via collisions between particles of the room temperature surroundings with the cooler products of the reaction.

    The LOLOL diagrams help expose all the components of energy changes that happen during chemical reactions. But we also want to connect back to the idea of recipes. If I react 10 grams of baking soda, there will be a certain amount of chemical energy change. If I react 20 grams of baking soda, what will happen? Enthalpy is going to be our tool to quantitatively analyze energy changes in chemical reactions.

    There are four common methods that introductory students will calculate enthalpy. These methods include calorimetry, enthalpy of formation values (ΔH_f°), Hess's law, and bond enthalpies. You will want to provide organization to help students differentiate these four algorithms that have some process overlap.

Chapter 7 – Stoichiometry + Extensions

Method #1 Calorimetry

Calorimetry involves doing a chemical reaction in a vessel that can have its temperature change measured. If you mix a strip of magnesium with hydrochloric acid in a polystyrene foam cup, the temperature of the solution is easily measured before and after the reaction. The equation used here is $\Delta H = Q/mol$. There are three components to that equation and every single one of those three can trip up students. Not only is the precise definition of ΔH tricky, but the equation for it here is deceptive because ΔH refers to the chemical system while Q refers to the surroundings (everything except the chemical system). And that line is easily blurred. In the example, ΔH refers to the Mg and HCl changing to $MgCl_2$ and H_2. But the Q is being measured by taking the temperature of the water/solution that those chemicals are changing within.

You must be very clear about what is being evaluated. Otherwise, students are going to resort back to previous misconceptions about heat when they determine the sign for enthalpy. If a student had previously learned that melting ice was endothermic, they then associate endothermic as being hot. So later when we mix Mg and HCl and the solution warms up they might connect this with the melting ice. To help push students past this loose association you want to be clear that all changes where heating occurs involve endothermic and exothermic changes. If the system is exothermic, the surroundings are endothermic. When the ice melts, the system (ice) is undergoing an endothermic change while the surroundings are exothermic. When the Mg and HCl react, the chemical system is undergoing an exothermic change and the surroundings (water/solution) are undergoing an endothermic change. Because many examples they encounter will have the system and surroundings be so similar it becomes critical to have detailed clarity about the specific identifications being applied.

The heat (Q) suffers from the same issue as above. To add a little more clarity, the two sides are describing different things. Enthalpy describes the energy of the chemicals. But that cannot be directly measured. So instead, we watch everything else in the universe to see how the energy of the surroundings changes. The Q describes everything else. The ΔH describes the chemicals. When $\Delta H = +$, the reaction energy diagram shows the products with more chemical energy than the reactants had. But the surroundings lose energy (Q = -) to the system.

Students will additionally need to be able to calculate using specific heat capacity to determine Q. One issue students have a harder time with is that they often see two or more masses and they must decide what to use. If the reaction involves 0.40 g of Mg being combined with 100.0 mL of 0.10 M HCl, what is the mass for $mC\Delta T$? If the 100.0 mL of solution is 100.0 g, then the mass that should be plugged in is 100.4 g since that was what was heated. The 0.4 g can be considered negligible or part of H, so there are inconsistent presentations on what should and should not be included.

Per mole of what? Is it per mole of magnesium or per mole of hydrochloric acid? Since there are twice as many moles of HCl reacting, which is it? It is per mole of the reaction. You are trying to come up with a recipe for the reaction between Mg and HCl and you want a single enthalpy to describe the entire recipe. So, we say that it is per mole of reaction.

$$Mg\ (s) + 2HCl\ (aq) \rightarrow MgCl_2\ (aq) + H_2\ (g) \quad \Delta H = -500\ kJ/mol$$

Chapter 7 – Stoichiometry + Extensions

The 500 kJ/mol describes the entire reaction. When 1 mole of Mg reacts with 2 moles of HCl the enthalpy change is 500 kJ. If we react 3 moles of Mg with 6 moles of HCl there would be an enthalpy change of 1500 kJ. Keep in mind it is not possible for 3 moles of Mg to react without 6 moles of HCl reacting as well. You could have an excess of HCl, but the excess would not react. The 500 kJ/mol means both 500 kJ per 1 mol of Mg and 500 kJ per 2 mol of HCl (as well as per 1 mol of either product being formed).

Using Q/mol to find enthalpy changes is the experimental method. But some reactions are extremely inconvenient to do while measuring the energy exchanges with the surroundings. Fortunately, we have some tricks up our sleeves. Potential energies can be set to arbitrary amounts. If I have a 1.0 kg mass on a table, I can say that the mass has 0 gravitational potential energy (GPE) or 1,000,000 J of GPE. If I lower the mass to the floor the potential energy will change by the same amount regardless of what arbitrary value I select. If I had chosen 0, it would now have -10 J if the floor is about one meter below. If I had chosen 1,000,000 J the GPE would change to be 999,990 J. It doesn't matter what we pick because we can only use the changes effectively. For elements we assign the enthalpy for every element to be 0 as long as the element is in its standard form.

The beauty of this is that we can do experiments where we react elements to form compounds. Scientists have done a lot of these and organized the values into tables for us to use.

Method #2 Enthalpy of formation values

Using $\Delta H_f°$ (standard enthalpy of formation) can be much easier for students than understanding what it represents. They can execute the algorithm without knowing what it is they are doing. Whenever an algorithm is simpler than the concept, we want to be wary of student's perceptions of their understanding.

An enthalpy of formation is the enthalpy change that occurs to form something from the elements it is made of. 1 mole of CH_4 (g) can be made by combining 2 moles of hydrogen molecules and 1 mole of carbon atoms. The specifics matter here. The enthalpy of formation is different for CH_4 (g) than CH_4 (l). There is a different change in energy to get from carbon and hydrogen to methane as a gas than as a liquid. Carbon has two kinds of carbon that could be used, and their different structures means we have to pick one. We choose graphite to be the standard allotrope and diamond has an enthalpy of formation of +2 kJ/mol.

After a brief introduction to what a standard enthalpy of formation is, give the students some data and have them make some observations. What do you notice about the enthalpy of formation values? They should notice that the elements are 0, some compounds have negative, and others have positive values, and larger compounds tend to have larger enthalpy of formations. They might recognize some fuels and products of combustion in the table.

After they have had some time to wrap their heads around the concept, now they should learn how they can use them to determine the enthalpy of a reaction. One key idea behind this algorithm is that enthalpy is a state function. This means that it doesn't matter how you get from the beginning to the end, the change will be the same.

Once I climbed up a steep mountain called the Manitou Incline in Colorado. It increases about 2000 feet in just 1 mile. It took four adults about 90 minutes to lug

Chapter 7 – Stoichiometry + Extensions

four small kids up the incline. The return trip is down a separate trail that is 4 miles and much more gradual. If we go up the incline and down the path our change in elevation is 0. If we go up the path and down the incline (not recommended) our change in elevation is 0. If we just sit there for 3 hours our change in elevation is 0. Change in elevation is a state function. It doesn't matter what you did in the middle, all that matters is your initial and final position.

Now let's connect that idea to chemistry with the reaction CH_4 (g) + $2O_2$ (g) → CO_2 (g) + $2H_2O$ (g). All that matters for the enthalpy change in this reaction is the beginning and the end. I can do anything in the middle I want. I can take the reactants, turn them all into elements, and then turn them into products. The enthalpy change will be identical.

CH_4 (g) + $2O_2$ (g) → C(graphite) + $2H_2$ (g) + $2O_2$ (g) → CO_2 (g) + $2H_2O$ (g)

The first step of this is the opposite of an enthalpy of formation for the reactants.

Step 1 CH_4 (g) + $2O_2$ (g) → C(graphite) + $2H_2$ (g) + $2O_2$ (g)

This means that we could determine the changes from the enthalpy of formation values of the reactants, but each one being flipped from + to - or - to +. The second step is the enthalpy of formation values for the products.

Step 2 C(graphite) + $2H_2$ (g) + $2O_2$ (g) → CO_2 (g) + $2H_2O$ (g)

The end result is that we can add up the sum of all the enthalpy of formation values for the products. Then we subtract the enthalpy of formation values of all the reactants to get the total change in enthalpy for the reaction.

$$\Delta H°_{rxn} = \Sigma \Delta H°_{f,products} - \Sigma \Delta H°_{f,reactants}$$

Students will need to pay attention to states of matter and coefficients. But otherwise finding the enthalpy for a reaction is as simple as adding up all the product values and subtracting the sum of all of the reactant values.

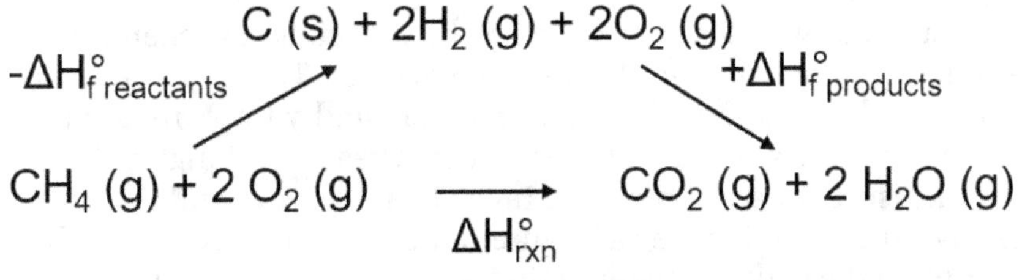

$$C\ (s) + 2H_2\ (g) + 2O_2\ (g)$$

$-\Delta H°_{f\ reactants}$ ↗   ↘ $+\Delta H°_{f\ products}$

$$CH_4\ (g) + 2\ O_2\ (g) \xrightarrow{\Delta H°_{rxn}} CO_2\ (g) + 2\ H_2O\ (g)$$

$$\Delta H°_{rxn} = \Sigma \Delta H°_{f\ products} - \Sigma \Delta H°_{f\ reactants}$$

Figure 7-12: A reaction can have reactants converted to elements, then the elements to products to demonstrate how enthalpy of formation values can be used to determine the overall enthalpy for a reaction.

Chapter 7 – Stoichiometry + Extensions

Method #3 Hess's Law

Hess's Law is fun even if you have no clue what you're doing. There are two rules for students to follow. If the reaction is reversed, we flip the sign of the enthalpy. If the reaction is scaled, the enthalpy is scaled as well (doubling the coefficients leads to double the enthalpy change).

Getting students to understand why these rules function is best done by connecting to the other models. Using a reaction energy diagram is helpful. If we reverse the reaction, we reverse the diagram. Now instead of increasing by +10 kJ/mol, we decrease by the same amount. The reaction energy diagram represents a single reaction. If that reaction occurs, one mole of times we get a certain amount of enthalpy change. If we now do the reaction two moles of times, we get double that enthalpy change.

When students do problems, you'll want some of those problems to either have reactions that aren't needed or are only needed to cancel out terms.

Unknown reaction A (g) + 2B (g) → X (g) + 2Y (g)	$\Delta H°_{rxn}$ = ?
Rxn 1: A (g) + E (g) → X (g) + Q (g)	$\Delta H°$ = +22 kJ/mol
Rxn 2: Y (g) → B (g)	$\Delta H°$ = +48 kJ/mol
Rxn 3: E (g) → Q (g)	$\Delta H°$ = +14 kJ/mol
Rxn 4: U (g) → R (g)	$\Delta H°$ = +86 kJ/mol

In order to construct the unknown reaction, we would require Rxn 1 as is, Rxn 2 flipped and x2, and Rxn 3 flipped. The enthalpy values in the same order would be +22 kJ/mol, -96 kJ/mol, and -14 kJ/mol for a total of -88 kJ/mol. Rxn 4 has no impact on the calculation.

Rxn 1:	A (g) + E (g) → X (g) + Q (g)	$\Delta H°$ = +22 kJ/mol
+Rxn 2 x2 reversed:	2B (g) → 2Y (g)	$\Delta H°$ = -96 kJ/mol
+Rxn 3 reversed:	Q (g) → E (g)	$\Delta H°$ = -14 kJ/mol
Sum of Rxns:	A (g) + 2B (g) + ~~E (g) + Q (g)~~ → X (g) + 2Y (g) + ~~Q (g) + E (g)~~	
Unk. Rxn:	A (g) + 2B (g) → X (g) + 2Y (g)	$\Delta H°_{rxn}$ = -88 kJ/mol

When students struggle with Hess's law direct them to circle the chemicals shared between the unknown reaction and the known reactions. This can give them an initial thing to look for. They also are likely not to understand why we're doing what we are. The reason for Hess's law is that the unknown reaction is highly inconvenient for us to measure for some reason. Maybe it is explosive and dangerous or maybe it is extremely expensive. But we can figure out the enthalpy indirectly by using 2-4 other reactions that can be combined into that unknown reaction. It can help to stress that we are using the known enthalpy reactions to figure out the enthalpy of the unknown reaction. To novice students the whole calculation can blend together.

Method #4 Bond enthalpies

Bond enthalpies are less precise than the other two methods. The energy change for the formation or breaking of a bond depends on what else there is nearby. The C-H bond in ethane is not the same as the C-H bond in methane. The data from tables is based on an average of multiple data points. This means that we should

Chapter 7 – Stoichiometry + Extensions

expect some minor differences when we use bond enthalpies instead of enthalpy of formation values or calorimetry.

In order to calculate with bond enthalpy changes you will need to have the Lewis structures of the reactants and products. After that you have two options. Option one is to break all the bonds that are not between the products and then forming all of the new bonds for products (Fig. 7-14). Option two is to break every single reactant bond and form every product bond (Fig. 7-15).

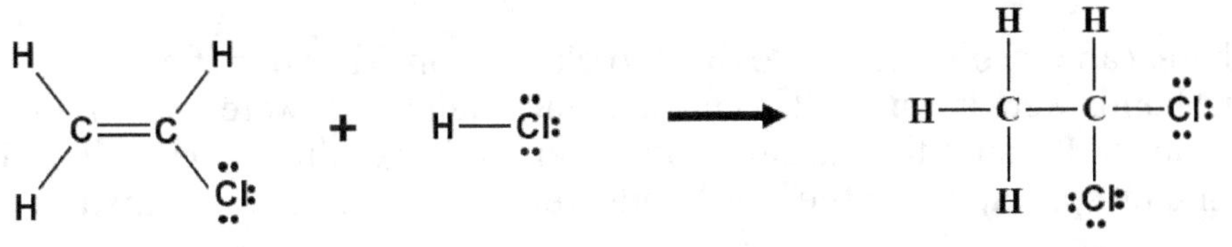

Figure 7-13: HCl reacts with chloroethene to form 1,1-dichloroethene.

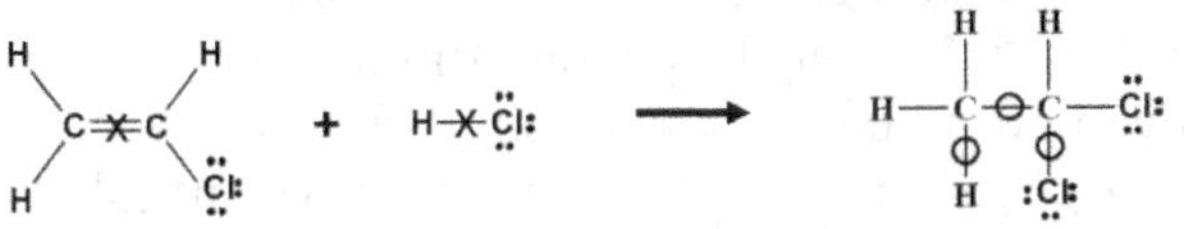

Figure 7-14: Two reactant bonds are broken, three product bonds are formed

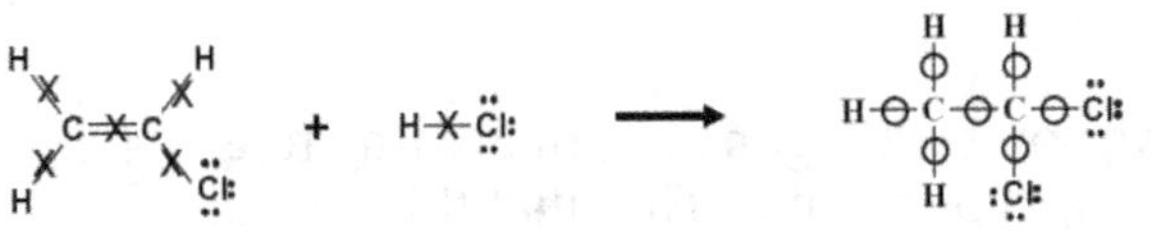

Figure 7-15: All reactant bonds broken, and all product bonds formed

Students will need to briefly think through that they cannot just break one of the double bonds between the carbons. Instead, the entire double bond is broken and then a single bond is formed. The pi bond between the carbons is not identical in energy to the sigma bond and so we cannot just subtract a single bond from a double bond to get a single bond.

Otherwise, both methods (7-14, 7-15) work. Students are more prone to error when they make fewer marks on their paper. It is helpful for them to mark the bonds they are going to break and make. It is also helpful for them to total all the broken bonds and the formed bonds in separate calculations.

Some students will be assisted by organizing their calculations. Otherwise, they will leave out a bond (or two). But I find it best to avoid organization that resembles the other enthalpy algorithms. Some use $\Delta H°_{rxn}$ = Σ Bond enthalpies (reactants) - Σ Bond enthalpies (products). I have two issues with using this organizational tool. The first issue is that it is too similar to the equation for enthalpy of formation values. So now students are going to have to remember one is products minus reactants and the other is reactants minus products. The second issue is that I want to challenge students' early conceptions about bond energy. It is a frequent problem that students think breaking bonds releases energy and making bonds requires it. Force them to identify those changes when they do calculations.

Chapter 7 – Stoichiometry + Extensions

I have students add up all the bonds that are broken. They assign these a + value since energy must go into separating the particles from each other. They then add up all the bonds that are made. They assign these a - value since potential energy decreases when particles form a new bond. Then they just add the values together. This is a great time to connect back to the introductory phenomena students observed when mixing vinegar and baking soda. Which bonds were stronger if the mixture got colder as the reaction progressed? Were the reactant bonds or product bonds stronger?

From there, can we connect that to the particle motion. The correct answer is that the reactant bonds were stronger. This means that when bonds were broken, the particles slowed down. Because the reactant bonds were stronger the particles slowed down more than they sped up when the new bonds were formed. Hence the mixture became cooler.

Once all four methods of enthalpy are introduced, now help students organize them into being able to recognize them. If students get a question that has temperature changes and experimental data, it is likely based on Q/mol (method #1). If they see a set of Lewis structures, they might need to look up bond enthalpies (method #4). If they get a large number of reactions where all but one have enthalpy values, that would be a Hess's law (method #3). If they get nothing but a reaction, they'll need to locate an enthalpy of formation data set (method #2). They should assume that they might need to look up data values without being prompted in the question.

What is enthalpy though?

Perhaps you've noticed that I have somewhat gone out of my way to give a definition of enthalpy to this point. That was intentional. The final thing I want to discuss is what enthalpy is exactly. Why is it not just the heat released during a chemical reaction?

Enthalpy is defined as the sum of internal energy and pressure-volume work.
$H = E + PV$

Where internal energy is:
$E = Q - W$

Here work is defined to be work done by the system on the surroundings, so a positive value for W means that the system is expending energy by pushing on the surroundings.

If you're unfamiliar with pressure volume work, think of opening a 2-liter bottle. Since the carbon dioxide inside is stored under high pressure the gas rapidly expands when the bottle is opened. But for the gas to expand it has to push the air around the bottle out of the way. This causes the "cxsh" sound, and it also causes the carbon dioxide gas to cool down. There is a little bit of steam mixed in with that carbon dioxide and as the steam cools you can sometimes see the steam condense into tiny liquid droplets inside the bottle.

In terms of particle motion, heat occurs when particles collide and the net amount of motion changes for the system and surroundings. Work is similar, but the change in motion is in a coordinated direction. If you sit in a cold room, the air in the

Chapter 7 – Stoichiometry + Extensions

room starts to move faster as your thermal energy transfers to the air from your body. But if you wave your hand, you can push a clump of air in mostly one direction. This would be you doing work (briefly) on the air. Alternatively in a windy situation, air could push you back where work is done on you.

When a chemical reaction occurs, there are typically small changes in volume. These changes require air to be pushed further away when volume expands (work done by the system), or air to compress the sample when volume decreases (work done on the system).

For reactions without gases there is very little change in volume. But things are different if you do a reaction where a gas is produced such as:

$$Mg\ (s) + 2HCl\ (aq) \rightarrow MgCl_2\ (aq) + H_2\ (g) \qquad \Delta E = -467.0\ kJ/mol$$

Now as the hydrogen gas produced expands, the hydrogen has to push the surrounding air away. This causes the hydrogen particles to slow down, and slower particles means that less heat is transferred to the surroundings.

Some of the internal energy change goes towards the work of expanding against the surroundings. Some goes toward heating the surroundings. If 1 mole of Mg reacts with 2 moles of HCl in a sealed container, then no expansion work will be done. The heat transferred to the surroundings will be 467.0 kJ. The coordinated pushing of air away is restricted by the rigid container and over time the coordinated motion results in increased thermal motion in all directions.

If 1.00 mole of Mg reacts with 2.00 moles of HCl in an open environment (assuming 1.00 atm), then 1.00 mole of hydrogen gas should expand until it occupies 22.4 L of space. That expansion involves 22.4 L * 1.00 atm = 22.4 L*atm of PV work. 22.4 L*atm is 2270 J or 2.27 kJ. That work is being done to the surroundings which means that it is negative for the chemical system. The total amount of heat released in this case could be 464.7 kJ.

If the same reaction occurs under double the pressure, the hydrogen gas will expand to half the volume (11.2 L) and the same amount of work will be done. We end up with -2.27 kJ of work done, and 464.7 kJ of thermal energy is transferred to the surroundings.

Enthalpy is defined as the change in internal energy minus the PV work done. As long as we are at a constant pressure, this value will be constant and equal in magnitude to the thermal energy transferred (464.7 kJ).

But wait a minute. Doesn't that mean that enthalpy changes are just equal to the heat released? Yes, if we are at constant pressure. But the highly technical way in which we defined enthalpy makes it a state function, whereas Q is not. Enthalpy is a state function, which means that the path it takes doesn't matter. Only the starting point and ending point matter. And that property allows things like Hess's law, and enthalpy of formation values to function. I don't think that students need the technical definition to understand the versatility of enthalpy. They really just need to understand that only the initial and final states matter.

If students have taken physics, they should have familiarity with this concept through distinguishing displacement and distance or speed and velocity. But even students without physics can easily understand the concept. The change in elevation example from the Manitou Incline is a simple example. If you have students that run, you can talk about how far you travel if you start at your house and end at your house.

Chapter 7 – Stoichiometry + Extensions

Did you run 0 miles? Your displacement was 0 even if your distance was 3 miles. Displacement is a state function, distance is not.

Understanding a precise definition of enthalpy is less important than knowing its versatility. Knowing the recipe component is the crucial detail for now. The more reaction that takes place, the more energy will change. If 1.0 moles of A reacting causes a change in 2875 kJ of energy, then 3.0 moles of A reacting causes a change in energy of 8625 kJ. Giving students the four methods to calculate enthalpy helps them identify how to start problems. Using LOLOL diagrams will help them translate those problems to the particle and macroscopic levels.

Ideal Gas Law, Partial Pressures

It is entirely reasonable to put ideal gas law with the other pressure calculations. The benefit of separating them is that you can space out the practice and connect moles (n) with BCA tables one additional time. When students finish a test on gas pressure, they immediately start forgetting the concepts. By relearning some of them in a later unit they will remember them much better. The tradeoff is time. It is faster to put this all into a single unit, but students will remember less. What will you lose out on by not combining them is something to consider when making choices between spaced practice and timing?

For me I like to look at midterm results. I am usually surprised at how poorly students do on midterm and final exams. The students will claim this is because the tests are too hard, but I am well aware of the relative difficulty between the questions on a midterm and on their regular assessments. The midterms and finals are much easier. Yet the scores are substantially worse. This is because students do not get enough spaced practice to demonstrate long-term retention. Therefore, I want to take opportunities to teach them how forgetting works and how spaced practice works whenever I can. Here we're using the stoichiometry concept with extensions for ideal gas law, enthalpy, and concentration as three ways to link to the moles needed to do stoichiometry analysis.

Historically we discovered the relationships between P, V, n, and T by comparing two variables at a time. In physics we can start with some basic assumptions about gas particle collisions and arrive at the same result combined into a single equation.

$$PV = nRT$$

There are two big new things with this equation. *We no longer need to compare two different sets of variables for a gas*. We aren't changing an initial state to a final state. We can now look at a single snapshot of a gas and determine meaningful information about it through calculation. *And there's an R in this equation. What is R?*

Students recognize P as pressure, V as volume, n as moles, and T as temperature. Hopefully they remember that temperature must be in Kelvins. But what is R? It is a constant. Its units are $L*atm*mol^{-1}*K^{-1}$. The units could also be $L*kPa*mol^{-1}*K^{-1}$ and they could be other things as well. And if the units change the value of R changes, even though it's a constant. Make sense?

Students are going to either ignore R or get stuck on R. Remember that students often struggle with units that have two units in them. At the beginning with density, we had to construct "For every" statements for grams per 1 mL. Now we are

Chapter 7 – Stoichiometry + Extensions

looking at a monstrous "For every" statement. For every 1 mol and 1 K there will be a pressure of 0.0821 atmospheres for each liter of gas. But any of those can be altered so that really isn't it. For every 1 product of moles times Kelvins, there are 0.0821 products of liters times atmospheres. Even with a functional "For every" statement, we're not clarifying anything.

Where does R come from in the first place? R is the Boltzmann constant k multiplied by Avogadro's number (6.02×10^{23}). There is an alternative version of the ideal gas law that is written as PV = NkT where N is the number of particles and k is the Boltzmann constant. To change n to N we multiple by Avogadro's number, and to change R to k we divide by Avogadro's number, so they are basically the same equation.

The Boltzmann constant relates temperature to kinetic energy. If you take a small set of particles, you can determine from their masses and speeds how much momentum change will occur when they collide with the walls of a container. From this you can determine the force during collisions and the frequency of collisions to get a pressure. But in order to write this in terms of temperature we require a constant that relates temperature to the original kinetic energy. Since we measure temperature instead of kinetic energy, we use a constant in PV = nRT.

But that isn't likely to help students understand the constant. So, give them a concrete connection to make. Have them either measure the constant in an experiment or show them how to use standard values to find the constant. During gas pressure calculations you likely used the phrase STP with some frequency. Standard temperature is a bit chilly and so later (in thermodynamics) it was set to 298 K. But in gas laws we are stuck at 273 K (brrr...) or 0 °C.

At STP 1 mole of any gas takes up 22.4 L (This was later changed to 22.7 L but very few use that value). The preceding sentence has four values in it. Standard temperature is 273 K, standard pressure is 1 atm, there is 1 mole of gas, and the volume is 22.4 L. Those values can be used to determine R.

$$R = \frac{PV}{nT} = \frac{1 \, atm * 22.4 \, L}{1 \, mol * 273 \, K} = 0.0821 \, \frac{atm * L}{mol * K}$$

$$R = \frac{PV}{nT} = \frac{101.3 \, kPa * 22.4 \, L}{1 \, mol * 273 \, K} = 8.31 \, \frac{kPa * L}{mol * K}$$

These calculations can help students understand how R is both a constant and can have different values for different units. These calculations provide a safety net for a student who forgets which R value is which during a test. They help add concrete examples for a student to understand the conceptual components. This has an added perk of helping students realize the units of R and the units of the other variables must match. Temperature must be in Kelvins, but pressure units have to match the units in R.

In order to measure the gas constant, we would need to know the other four variables (PVnT). This can be done by collecting hydrogen gas over water that is generated by reacting HCl and Mg. Another simple method is to bubble butane gas out of a lighter into a graduated cylinder in a bucket. Either way you are pushing students to understand that when any of $\frac{PV}{nT}$ change, the other components adjust as well to maintain the original constant.

Chapter 7 – Stoichiometry + Extensions

Collecting a gas over water was a big breakthrough historically. A lot of early chemical research by Priestley, Lavoisier, Boyle, Avogadro, and others was done by analyzing gases. Work by Dalton built upon the fundamentals they discovered. Collecting a gas over water comes with two potential systematic errors. The first is that some of the gas will dissolve in the water. Carbon dioxide, HCl, NH_3, and many other gases have substantial solubilities in water. But the other error we must consider is that some of the water is going to become a gas as well. This means that the pressure of the gas we collect is not entirely from the gas we collected.

Fortunately, the pressure of steam above water is only contingent upon the temperature of the water and allowing sufficient time for equilibrium to establish. In the open atmosphere that second part can be a problem, but for a sealed container the timeframe is nearly zero. Dalton's law of partial pressures is intuitively simple. The total pressure in a container is the sum of all of the partial pressures.

When the piston becomes static (Fig. 7-16) the forces are balanced on the right and left. Since the area exposed to both is the same, we can assume the total pressures are equivalent. This means that $P_B = P_A + P_C$. When collecting a gas over water, we create a similar dynamic except the equation would be $P_{ext} = P_{H2O} + P_{gas}$ where P_{ext} is the external pressure in the room and P_{gas} is the pressure of the gas being collected.

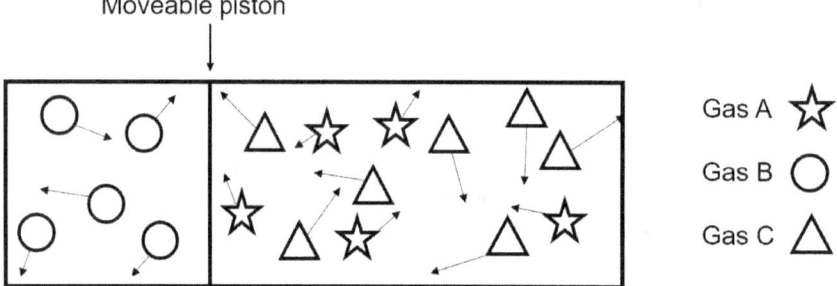

Figure 7-16: Gases A and C have a combined pressure equal to gas B

The convenience lies that the vapor pressure of water, P_{H2O}, is dependent on temperature and therefore easily determined. The external pressure can be measured using a barometer. Therefore, minimal glassware is needed to measure the pressure of the gas collected over the water. In the 1700s this led to new discoveries. In the current year this makes for reasonable experimentation in a minimally funded classroom.

Partial pressure is a great topic to develop some creative questions that challenge students mathematically. I am particularly fond of the questions where a reaction results in a net change in the number of gas molecules and only the initial and final pressures are known.

$$N_2 (g) + 3H_2 (g) \rightleftharpoons 2NH_3 (g)$$

In a 2.0 L container at 298 K, nitrogen has a partial pressure of 1 atm and hydrogen a partial pressure of 3.0 atm. After several minutes go by the total pressure remains constant at 3 atm. Determine the amount of each gas at this equilibrium.

Chapter 7 – Stoichiometry + Extensions

$$N_2\,(g) + 3H_2\,(g) \rightleftharpoons 2NH_3\,(g)$$

I	1 atm	3 atm	0
C	-x	-3x	+2x
E	1-x	3-3x	2x

Total pressure at equilibrium: 3 atm = (1-x) + (3-3x) + 2x = 4 - 2x
$$x = 0.5 \text{ atm}$$

Therefore, at equilibrium: $P_{N2} = 0.5$ atm, $P_{H2} = 1.5$ atm, and $P_{NH3} = 1.0$ atm

    We can combine partial pressure along with the ideal gas law to show that if the gases are in the same container the number of particles is proportional to their partial pressures. The same container will have the same values of P and T (and R!). Therefore, P is directly proportional to n. Note in the ammonia equilibrium problem how the changes are all proportional to the coefficients. We can treat pressure and amount interchangeably if the gases share the same container.

    We wouldn't be able to do that in Fig. 7-16 because gas B likely has a different volume than gases A and C. But gas A would have a partial pressure that is twice as much as gas C because gas C has twice as many particles.

    BCA tables connect well with these three extensions (molarity, enthalpy, and PV = nRT). They all involve moles and can be used to determine the relative amounts of chemicals before or after a reaction has taken place. In particular I like how the BCA table is the constant source for analysis of the recipe, while the chemistry that connects to the recipe can vary widely. This helps the students wrestle with these ideas, make connections, and find patterns. As students understand the conceptual components better, they inevitably become better at solving problems.

    Stoichiometry is a key unit and a student that can do these successfully is set up well to handle topics like equilibrium, kinetics, and electrochemistry. But there is a big difference between a student who can successfully calculate stoichiometry problems, and a student that knows what they are doing. Give space for your students to be knowledgeable about the methods they use.

Student Struggles

1. "I don't know anything!" If a student is writing down incoherent gibberish, they are missing the background tools. To do a stoichiometry problem, one must have basic understanding of naming, formula-writing, balancing reactions, predicting products, and changing grams to moles. The most relevant of those for students will be writing formulas correctly and changing grams to moles. You'll want to emphasize what molar mass is both as the number from the periodic table, and as a proportion of grams to moles.

    The best way to get students to move toward better understanding of the fundamentals is for them to include full units with chemical formula when showing their work. This forces them to match numbers within the algorithm with the appropriate application. When reading a student's work that is overwhelmed, you'll likely see erratic manipulations. By forcing them to put units and the chemical

Chapter 7 – Stoichiometry + Extensions

formula, you can begin to address inconsistencies so they can develop schema about the processes within the algorithm.

If you are not using BCA tables, the BCA tables split up these components so the student will have an easier time identifying what they can and can't do. If they can't change grams to moles, they'll have the incorrect information to start the BCA table. If they can't apply a mole ratio, they'll have errors in the change row.

A good assessment is to give students a problem where they start with the moles of one of the chemicals. Students who don't have a strong grasp will tend to convert the moles to grams before applying a mole ratio.

2. Students use the coefficients in the balanced reaction for the grams to mole conversions.

You mix 18.7 g of Na with an excess of O_2. The student calculates:

$$18.7 \text{ g Na} * \frac{4 \text{ mol Na}}{22.99 \text{ g Na}} = 3.25 \text{ mol Na}$$

The student is applying the 4 coefficient from $4Na + O_2 \rightarrow 2Na_2O$ inappropriately. This is a fantastic error for students, because it doesn't require intervention from the teacher. Feedback should inform the student they are incorrect, and then they can work independently to determine why. We can start a reaction with any quantity of chemical that we want. The reactants do not need to be mixed in specific proportions; they are limited to reacting in certain proportions. In this particular problem, the 18.7 g of Na tells us how much chemical we begin with, not necessarily the amount that reacts.

3. "I know what to do, but I don't know what I'm doing." Many students can follow the algorithms we provide to find stoichiometric solutions. But they fail to elicit the underlying concepts. Stoichiometry is about understanding how chemicals react in proportional manners. These proportions or recipes are predictable. If we know the reaction taking place and the starting amounts, we can predict the final products. In history, we often would use mass relationships to determine what products had formed when the reaction was the unknown.

4. What in titration? - Students will struggle with titrations. They're applying M*V for one chemical species, while the other is often determined by mass or moles. Because students are analyzing two chemicals with similar data, it is critical that the student organizes information so that the identity of each chemical is explicit throughout the calculation. A BCA table can help, but it is critical that students include labels with work (20.0 mL of 2.0 M HCl instead of 20.0 mL of 2.0 M).

One small factor with titrations is that students need to become acclimated with changing mL to L. It's not that they can't divide by 1000, but it helps to automatically see 50.0 mL as 0.0500 L. Try to work similar conversions into retrieval practice during sample problems or with flashcards to build their familiarity.

Be wary of technical details that may distract students. When we titrate iron (II) with permanganate, we might add a stabilizing agent that prevents the iron (II) from oxidizing to iron (III) in solution. Even the addition of water to the analyte can trip up a student.

Chapter 7 – Stoichiometry + Extensions

5. Enthalpy signs - It really depends on which type of enthalpy calculation is being performed for how sign errors manifest. But a common issue is when students calculate with $\Delta H = Q/mol$, they incorporate signs in their calculations that lead to errors. The better method is for them to calculate just with magnitudes and then at the conclusion determine whether the change was endothermic or exothermic with respect to the chemical system. This will help them focus on the system and surroundings. If the surrounding solution absorbs +340 J of energy from a reaction between 0.010 moles of reactant, the sign of ΔH is - since the chemical system must have lost that energy. This helps students clarify exothermic reactions lose energy to the surroundings without being oversimplified into "hot" or "cold" descriptions that students are drawn toward.

6. Ideal gas law - $PV = nRT$ calculations are pretty straightforward. The most common issue is not converting units correctly, but occasionally you'll get a student that just does not seem to know how to proceed. These students usually lack the knowledge of which units go with which variable. The "Curse of Knowledge" makes this challenging for the teacher to notice, but to a novice student they really are not familiar with units of pressure and sometimes the other variables as well. This makes it easy for them to become confused and shut down.

7. "I can never remember the gas constant!" When a student forgets the constant R, they should plug in standard values into $PV = nRT$. With 22.4 (22.7 for purists!) L being 1 mole at STP, the student has the four values. By plugging in 101.3 kPa, 1 atm, or any other pressure unit they can find the appropriate R constant while understanding how the units of R are determined.

Phenomena

1. Fizzy Kool-aid - If citric acid and baking soda are mixed in the correct proportions, you will produce carbon dioxide and tasteless sodium citrate. Provide the students with the following reaction:
$H_3C_6H_5O_7 + 3NaHCO_3 \rightarrow Na_3C_6H_5O_7 + 3H_2O + 3CO_2$
They should determine the amount of baking soda needed to completely react with 0.90 g of citric acid. Then they can mix those quantities into a cup before adding some type of juice that would taste good when carbonated.
If they are wrong, they will either end up with a very sour (excess citric acid) or very bitter (excess baking soda) drink. It does take a minute for the reaction to mix and take place so don't drink too quickly. If you have a lot of students, smaller cups will make the amount of juice needed more reasonable.

2. Limiting reagent balloons - Fill 5 flasks with vinegar and then add increasing amounts of baking soda into balloons that get stretched out over the flask (Fig. 7-5). The third flask should have equivalent amounts of baking soda and vinegar so that the first three balloons increase in size, but balloons 4 and 5 end up the same as balloon 3. The flasks transition from having baking soda as the limiting reagent (flasks 1+2) to having vinegar as the limiting reagent (flasks 3-5).

3. Mysterious overhead - If you're a younger teacher, you may have to locate a more senior teacher to explain what an overhead projector is. Take two beakers with a copper (II) sulfate (or other Cu^{2+}) solution and put it on the overhead so that the blue

Chapter 7 – Stoichiometry + Extensions

color is projected onto a screen (or the wall). Both solutions should have the same concentration and volume so the projected circle should be the same color. Ask students to predict what will happen to the projected circle of blue light when more water is added to one of the beakers. When the water is added, both circles remain the same color. Why? Both solutions have the same amount of copper (II) ions that the light must move through, leading to the same amount of absorbed light. The one now has a longer path to get there, but the concentration of the solution with that longer path has decreased. This is a good phenomenon to do prior to showing students that M*V = M*V (or before Beer-Lambert law).

4. Hot Glow Stick - This one makes a mess and involves broken glass so feel free to use a video instead. Activate a glow stick and then carefully cut open the glow stick from the top end so solution doesn't spill out (it's worthwhile to invest in some advanced scissors). Carefully pour the mixture from the glow stick into a test tube. Then heat the mixture using a Bunsen burner. The mixture will become very bright, but the glow will end abruptly after a shorter time interval. The glow stick contains a dye, catalyst, hydrogen peroxide, and diphenyl oxalate. The diphenyl oxalate will be the limiting reagent (especially if you add a bit of more concentrated peroxide). What's nice about this demonstration is that you're introducing ideas from kinetics. You can refer back later to show how stoichiometry still plays a role in kinetics and equilibrium.

5. Burning Mg - Burning magnesium produces light so intense it can damage your eyes. Do not watch directly (your face/eyes should not point toward the Mg). Burning metal is a good phenomenon because students will be surprised to see the mass increase (even if you tell them). Steel wool can also be burned with fewer safety hazards. Both risk a metallic splinter in your eye if you aren't careful. You'll also need to be sure to contain all the product. Burning steel wool will send small pieces flying and burning Mg will lead to a portion of the oxidized Mg strip falling away from the tongs. Both can be mitigated by using a ceramic dish underneath the reaction. Remember that burning Mg will react with nitrogen in addition to oxygen.

6. Reduction of malachite - Take small pieces of malachite ($Cu_2CO_3(OH)_2$) jewelry and heat using a Meker burner (these get hotter than Bunsen burners) while in contact with charcoal. The malachite will turn to black copper (II) oxide. With patience you'll end up with copper metal. Only use small amounts as some carbon monoxide is generated. This is a good demonstration for students to predict the amount of copper that should be there at the end. They can use stoichiometry or compositional stoichiometry to make that determination (they can apply the % composition).

7. Double the energy - Have students determine the enthalpy of reaction for Mg + 2HCl (aq) → $MgCl_2$ (aq) + H_2 (g) using a coffee cup calorimeter. Then have them repeat the process, but with double the amount of Mg (Mg should be the limiting reagent for both). The amount of energy change should double, but so do the number of moles reacting. The calculated enthalpy should be constant.

8. Vinegar and baking soda in a baggie - Have the students crease the bottom of a sandwich baggie so the two bottom corners have a rise in between them. Put vinegar in one bottom corner and baking soda in the other. Seal the baggie without letting them mix. Once sealed students should mix the two chemicals. They will feel the mixture get colder as the reaction proceeds. Ask them to articulate why they think the

Chapter 7 – Stoichiometry + Extensions

mixture gets colder. If they get stuck, ask them whether the particles have sped up or slowed down.

9. Mini-thermite spheres - Start with two steel balls (spheres!) that are rusty. Cover one in aluminum foil. When you strike the two balls together with a glancing blow, the thermite reaction can occur at a micro-level.

Al + Fe_2O_3 → Fe + Al_2O_3

This is a great demonstration to accompany a reaction energy diagram. There is a minimum energy for activation required (some blows will be insufficient). But once that minimum is accomplished, students can see an array of sparks as the reaction takes place. They will associate those sparks with the "release of energy" as the new bonds begin to form.

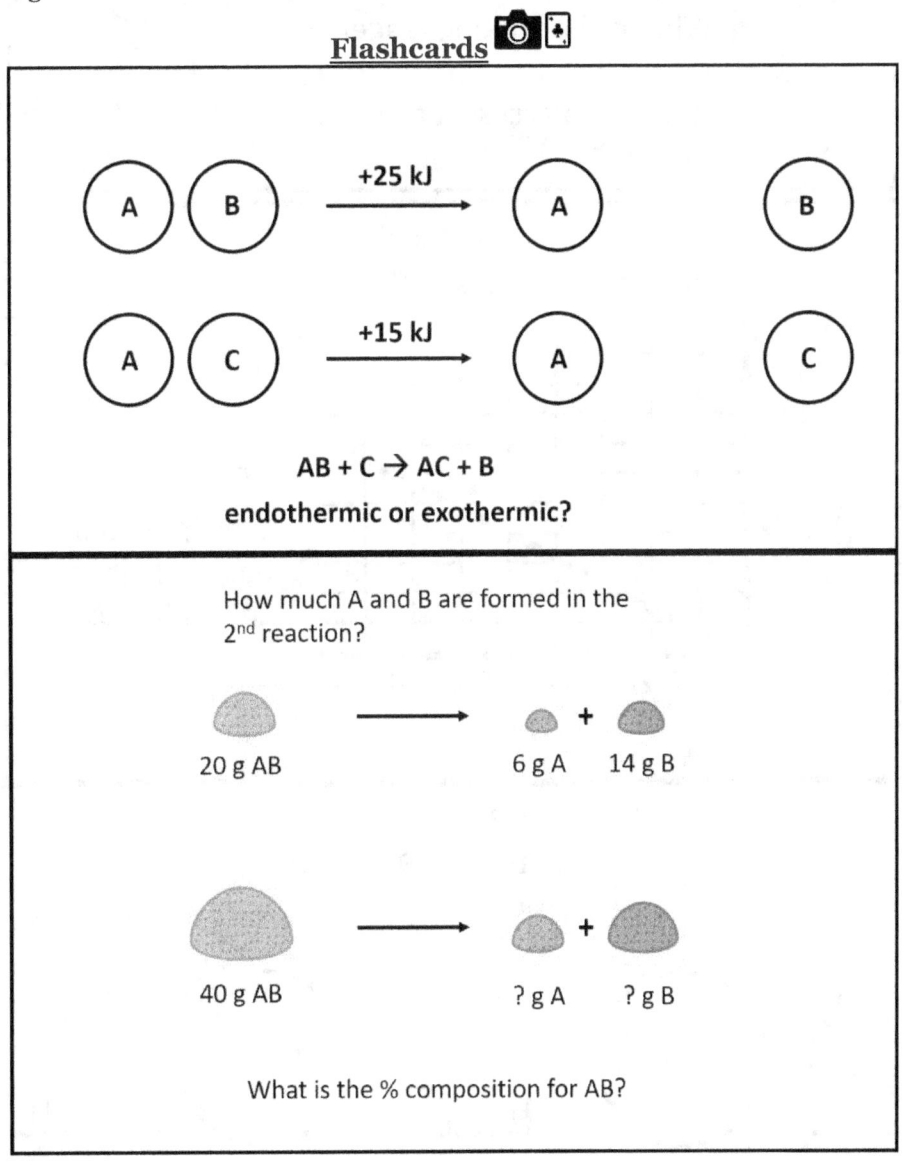

Chapter 7 – Stoichiometry + Extensions

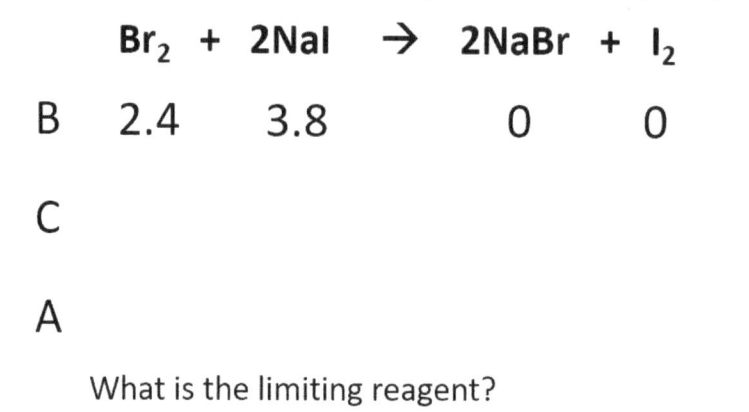

Br_2 + 2NaI → 2NaBr + I_2

	Br_2	2NaI	2NaBr	I_2
B	2.4	3.8	0	0
C				
A				

What is the limiting reagent?

How much excess reagent will remain?

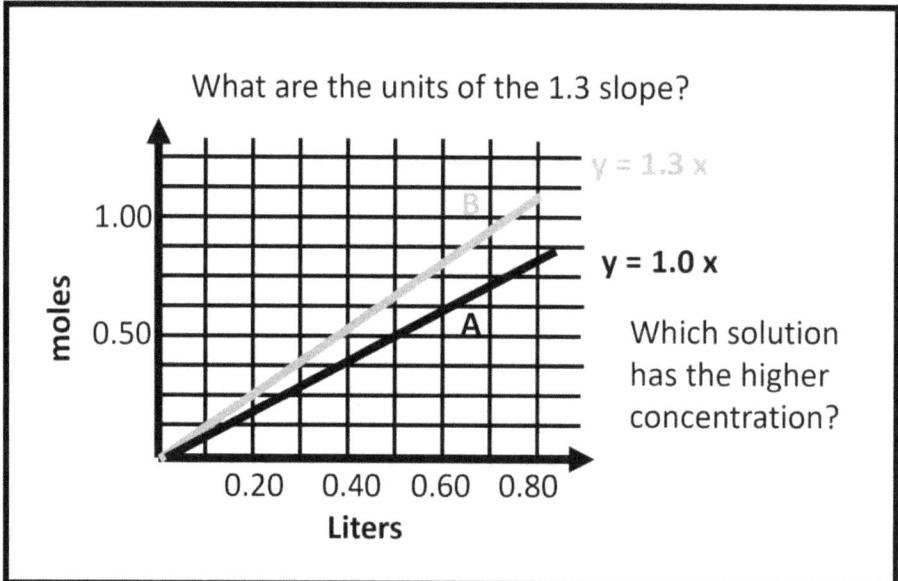

What are the units of the 1.3 slope?

y = 1.3 x

y = 1.0 x

Which solution has the higher concentration?

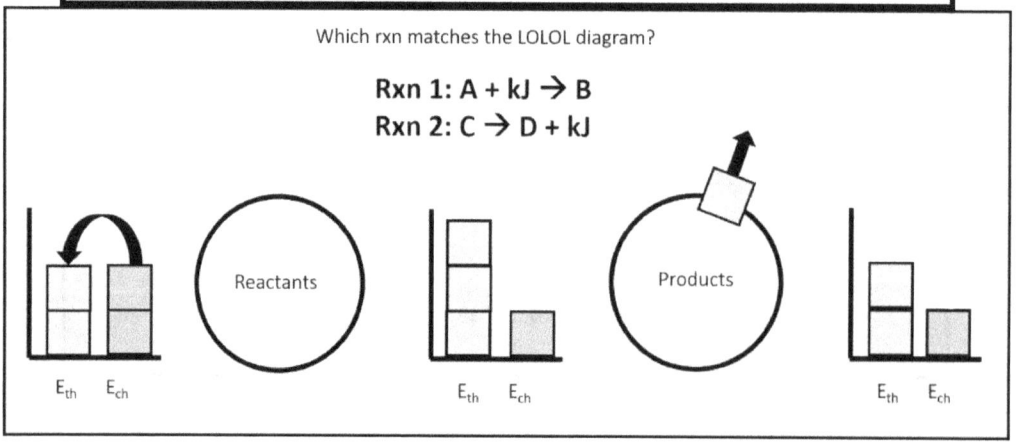

Which rxn matches the LOLOL diagram?

Rxn 1: A + kJ → B
Rxn 2: C → D + kJ

Chapter 7 – Stoichiometry + Extensions

Which beaker has more moles of $CuCl_2$?

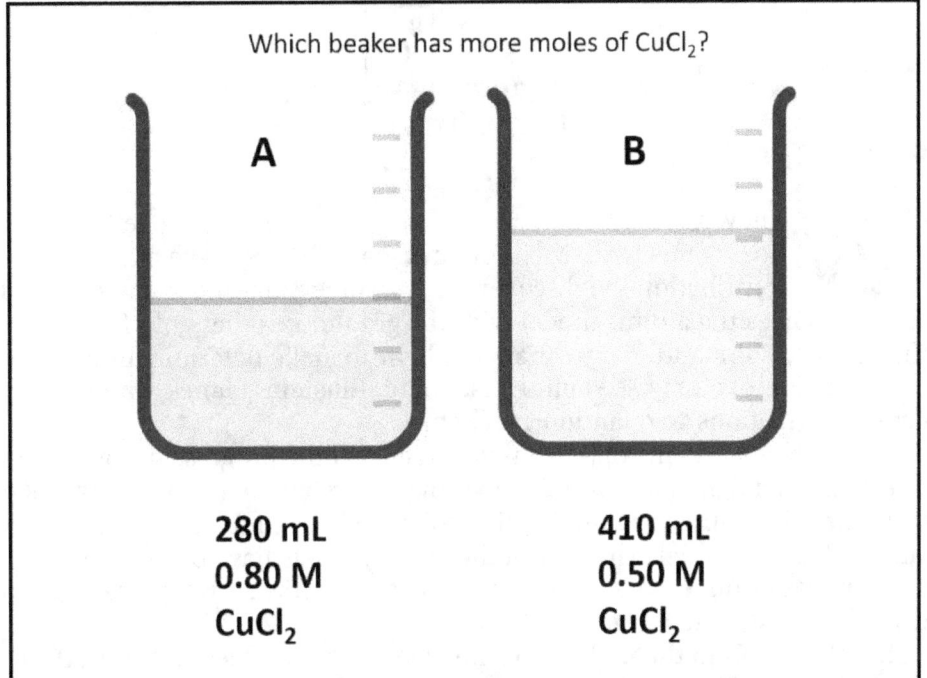

A	B
280 mL	410 mL
0.80 M	0.50 M
$CuCl_2$	$CuCl_2$

33.7 g Na reacts with excess water, how many grams of hydrogen gas are formed?

$$2Na + 2H_2O \rightarrow 2NaOH + H_2$$

33.7 g Na	2 mol Na	2 mol H_2	1.01 g H_2
	22.99 g Na	2 mol Na	1 mol H_2

Identify all of the errors in the calculation set up above

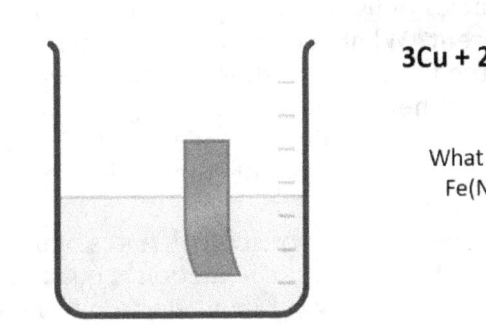

$$3Cu + 2Fe(NO_3)_3 \rightarrow 3Cu(NO_3)_2 + 2Fe$$

What will the beaker look like if the $Fe(NO_3)_3$ is the limiting reagent?

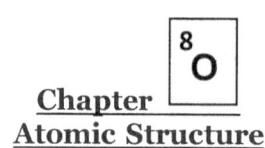

Chapter
Atomic Structure

History

We now shift to a time when all of the pieces come together. Becquerel, Roentgen, Curie, Meitner, and many others make major contributions to discovering and understanding radioactivity. Ernest Rutherford advances the atomic model with the groundbreaking gold foil experiment. Kirchoff, Bunsen, Bohr, and many others use light to make determinations about the electronic structure of atoms.[1,2] Pauli, Heisenberg, Einstein, Planck, and many others make large contributions to quantum mechanics.

Most teachers use this opportunity to focus on the progression of the atomic model. Perhaps we might start with Democritus before moving to Thompson's plum-pudding model, the Rutherford model, the Bohr model, and finally our current quantum mechanical model. The first of these two models have been used and discussed. The latter three involve far more collaboration than we typically give credit for. Let's do our best to make amends for that.

The late 1800's to the mid-1900's involved a chaotic mess of contradictory evidence and claims. Before we look for resolution, it's important to understand some of these discrepancies. If you read historical texts, you'll find chemists struggling to explain: paramagnetism[3], helium having four times the mass of hydrogen (but only twice the protons)[4], electrons don't collapse via a spiral into the nucleus, chlorine appears to be able to have positive or negative charge (radical theory vs. dualism),[5,6] pitchblende is more radioactive than uranium, and many more pieces of evidence that exposed holes in the current scientific models.

In the 1860s we have the first periodic tables and accurate relative atomic masses (courtesy of Cannizzaro). These provide organization to emerging theories and evidence. Just prior, Robert Bunsen and Gustav Kirchoff had constructed a spectroscope that they used to study light from sodium added to a flame. They soon discovered two new elements (Rb and Cs) by identifying new spectral lines which started a new trend of searching for missing elements using light.

The emission of only select frequencies of light proved problematic for some time. Why were so many colors of light absent? What determined which frequencies would be seen? Why were the colors absorbed the same as those emitted? Johann Balmer (who was a teacher) determined a mathematical pattern (1885) for the wavelengths of visible light emitted from hydrogen. This was generalized by Johannes Rydberg (1888) to include more types of light. Then Niels Bohr (1913) developed a model of the atom that involved a major leap.

In this new model of the atom, the electrons have restricted states of motion. They can move in a few different motions, but not others. The motion is quantized (not continuous). From this idea, we see a shift in thinking about electrons where a new branch of quantum mechanics is used to describe the physics of how electrons move. This shift is a critical change in chemistry that we should be emphasizing in our instruction. Quantum mechanics is also very difficult to teach because it is abstract and can have limited examples.

Chapter 8 – Atomic Structure

The Bohr model works unusually well for hydrogen and ions with a single electron (e.g., Li^{2+}). But the Bohr model treats electrons as particles orbiting the nucleus in defined orbits. We get better agreement with experimental evidence when we use wave mechanics to describe electrons in atoms and molecules. As teachers, we need to be cautious. Students will be drawn to the particulate model of the electron. It is simpler for them to visualize. But the wave model is superior. In later units (periodic trends, bonding, etc.) we use the particle model of electrons because it is simpler to represent. Therefore, it is important that we make the distinction now of the flaws in using the Bohr model in chemical education.

Right before Bohr was developing the details for his model of the atom, Ernest Rutherford had started his gold foil experiment. In 1911 Rutherford published his paper on the scattering of alpha particles by gold foil. The two conclusions from this paper were radical improvements on atomic theory. The first is that the vast majority of the mass of the atom is located in a small, dense, positively charged region. We call this the nucleus. The second conclusion was that the rest of the atom is primarily empty space occupied by electrons.

The new atomic structure and the use of light to determine the electronic structure of atoms and molecules led to significant improvements in our understanding of chemistry. Both should be focal points of this unit. Because the conclusions from them are both radical, it is critical to include experimental evidence and time for students to process. Students observe hydrogen to emit red and teal light, but not orange or yellow. What could they propose as that would explain this observation?

Light and Electrons

Get ready for a confusing mess! The atomic structure unit has too many new ideas that are all complex with new scientific models needed to explain highly unusual evidence. This makes the atomic structure an incredibly challenging unit to teach. Where should we begin?

Students in introductory chemistry have limited concepts about what light is. Everything about light is abstract.
- What is light made of?
- Where does it come from?
- Why does it always go the same speed no matter what?
- What happens when light hits a chemical?

If you start to answer some of these questions, it typically gets worse before it gets better. If I tell you light is an oscillating magnetic and electric field variance you might have several new questions. What is an electric field? What is a magnetic field? Do the fields move up and down or change their size? What would a field changing its size mean?

When you start digging into light, you find more questions than answers. The things previous teachers have told students about light are often unhelpful or confusing. When I start teaching about light in a chemistry class, I need to give them somewhere to start where we all can go to the same place. I aim to give them some information about fields, but I avoid using the term. Because there is so much student development needed, we want to begin with an anchor point.

I start with where light comes from. Light comes from a charged particle accelerating. Accelerating means that the velocity is changing, so this can be a

Chapter 8 – Atomic Structure

charged particle that is vibrating back and forth, moving in a circle, or it can be a charged particle that is speeding up in a single direction. If I shake an electron, light gets produced. If I shake it really fast or really slow, different types of light are made.

Unfortunately, electrons are so tiny that you can't just sit and watch them. We can't track them. As far as I know, even in the most technically developed experiments we cannot watch electrons in the same way we can watch a car drive down the road. But when we look at light coming from atoms, something very odd results.

If you have access to spectral bulbs, have students look through a spectroscope at the light that hydrogen emits. They should see a single red line, a teal line and two violet lines. Have the students draw what they think is happening to cause these different colors of light and why there is no orange light. Students will produce a wide variety of models to explain why we see red light but not orange. Some draw fast moving hydrogen particles as violet and slow ones as red. One group drew clumps of 4 H particles as red, clumps of 3 H particles as teal, and 2 H particles as violet. Others might incorporate the electrons, vibrations, or the electricity.

These models are wonderful even though they are incorrect. There is no orange! That is strange, and perhaps our brains need time to fully digest how unusual that is. How can there be no orange? One common initial model that students construct is that red H particles are slow and violet H particles are fast. If slow particles are red, and fast are violet, why is there no medium speed orange ones? Wouldn't a red and a violet collide at some point to make an orange one?

Let's reconnect back to the original starting point we want for our students. Light originates when a charged particle changes how it moves (accelerates). If we get red light from some motion changes, and violet from others, but we never get orange, what does that mean?

Potentially it means that electrons can only move in certain ways. This is especially troubling and weird because we can't track them, so we can't watch them move. But the fact that we get some colors, and not others, means that electrons can only move in certain ways. There are restrictions on how they move. The term we use is we say that their motions are quantized. There is no analogy to this in the macroscopic world.

Since we cannot track an electron's motion, the best we can do is predict what the motion will look like and try and find evidence to support or refute that claim. We have been doing this for over a century and found some surprising evidence. The first big leap was to have electrons in quantized orbits around the nucleus. Since then, we have found that describing the electron motion as a wave has produced far greater explanatory power. But it is extremely difficult to visualize wave mechanics, and this presents challenges for chemistry teachers.

Teachers will be tempted to say that electron configurations are difficult enough without understanding what they are. I would encourage you to try teaching more than just that the electron configuration of Na is $1s^2\ 2s^2\ 2p^6\ 3s^1$. The orbital model is useful for predictions and explains evidence that is not explained easily by particulate models of the electron.

To develop this picture, we must consider the electron as a standing wave bound within the atom. A standing wave on a slinky results when wave pulses move back and forth at a specific frequency. When we put more energy into the slinky by creating pulses at a higher frequency, we get more frequent nodes.

Chapter 8 – Atomic Structure

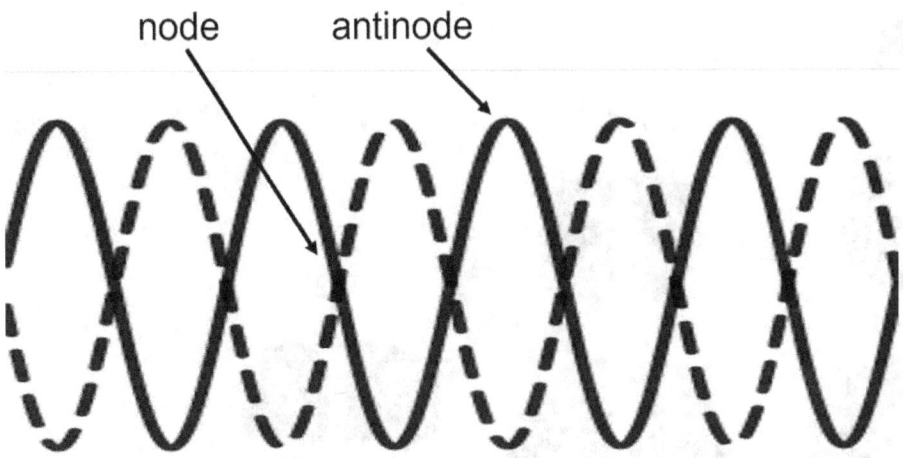

Figure 8-1: Standing wave that could be constructed using a slinky bound at both ends

For an electron bound to an atom, we are going to need 3-dimensional standing waves. In 3-dimensions we are going to find two types of nodes: radial and angular. Radial nodes occur when the electron does not exist at a given radius from the nucleus. Angular nodes occur when the electron does not exist at a given angle. See if you can determine how many radial and angular nodes the following cross section of an electron has:

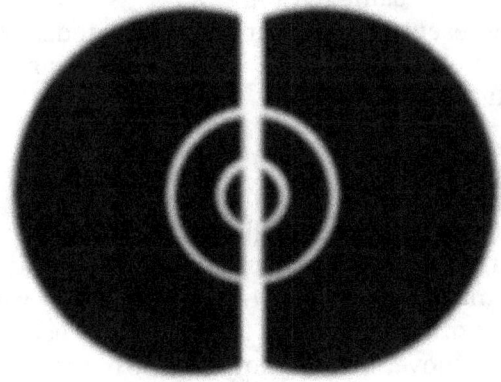

Figure 8-2: An electron that has 3 total (1 angular and 2 radial) nodes

As we increase the energy of the electron there are more nodes. For all electrons there are nodes at the nucleus and at infinity. We'll ignore those two nodes for a bit. For the first energy level (n=1), we have no other nodes. When we get to the second energy level, we can either have a radial node (2s orbital) or an angular node (2p orbital).

Chapter 8 – Atomic Structure

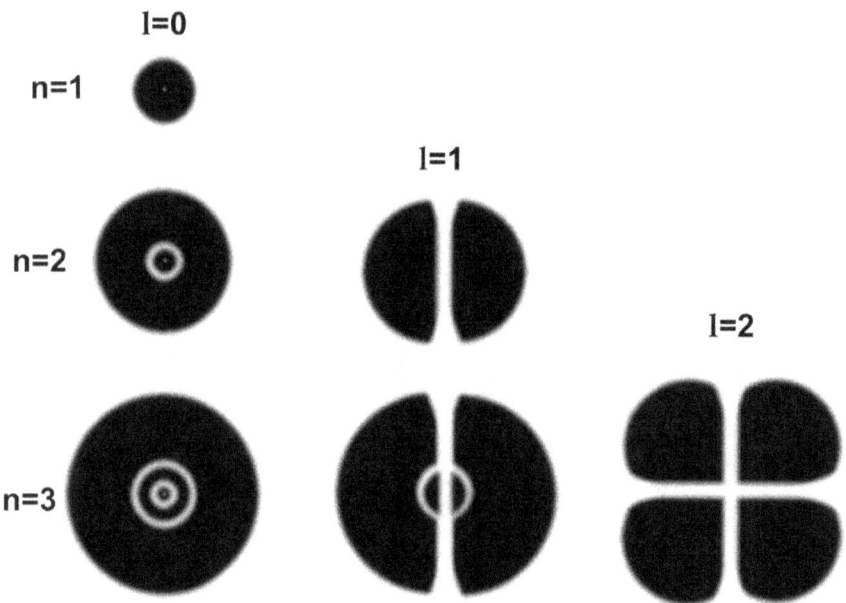

Figure 8-3: Images representing electrons using waves[7]

When we get to the third energy level, we can now have 2 radial nodes (3s), 1 angular and 1 radial node (3p), or 2 angular nodes (3d). This is why there is no such thing as a 2d orbital as you cannot have that many angular nodes in the 2nd energy level. Likewise, there are no 1p orbitals for the same reason.

The orbital types s, p, d, and f represent the number of angular nodes. We use the letter l to describe these. If l = 0, we have an s-orbital that has no angular nodes. If l = 1, have a p-orbital that has 1 angular node.

*****

It is difficult as a teacher to summarize what an orbital is. You are trying to balance between an accurate description and simple enough terms for a novice student. I would describe an orbital as a mathematical description of the position and motion of the electron that allows us to make predictions. Light is often how we confirm or discard those predictions. The orbital is related to where the electron tends to be, and how the electron tends to move, but it is not just one of those things. I have heard that the orbital is the electron itself. I would argue that the orbital is a representation of the electron, but that might not clarify the statement any better.

An orbital is not a container. It looks like one to students. Representations that are composed of dots are better than enclosed shapes. I once asked an expert if the orbital is a sum of all of the dots you would get by localizing the electron with light. He was thrilled with the question. He informed us that the orbital is absolutely not that. His analogy he used to explain was pretend that there is a universe where you could not see cantaloupes. They exist, but you just can't see them for some reason. The only thing you can do is smash them with a hammer. When you smash them with a hammer you can see a piece of the cantaloupe that's been smashed. In this analogy, the invisible cantaloupe is the orbital. The smashed piece is the electron

Chapter 8 – Atomic Structure

after being hit with a high energy light wave. The electron is almost completely different afterwards and bears little resemblance.

Our instincts are to try to fill in that gap. We can't see the electron in a useful sense, but our brains are used to being able to construct a visual image or analogy. Our brains work hard to find anything that can form a bridge for us to a visual image of the electron. It is better to accept that this does not really exist. Pretend that a person travels to a new continent a long time ago and they come across a giraffe for the first time. They go back and try and explain what a giraffe is using other animals. The new animal is partially like a horse, but with a lot longer neck. And it has different spots. And the tail is different. But at the end of the day, it's something different from a horse. It's a giraffe. We can explain how an electron is like other things, but at the end of the day it's just an electron. Comparisons will always fall short, and for an electron they will miss by a wider margin.

*****

In order to help explain that strange lack of specific colors, we constructed a description of the electron called a wavefunction. The symbol for the wavefunction is called psi (Ψ) and the mathematics of the wavefunction are beyond the current capabilities of this author. The wavefunction can be used to generate information about the electrons in atoms. We can calculate four features for those electrons that we represent using the letters n, l, m, and s.

The principal quantum number (n) tells us the energy of each electron. These energies can only be positive integers. There is no energy state of 0, instead there is a minimum energy. This energy is a combination of potential and kinetic energy. The potential energy is negative and is based on the distance between the electron and the nucleus and other electrons. The potential energy between a proton and an electron is set to be 0 when they are infinitely far apart. So, as they move closer the potential energy is negative. The kinetic energy is positive, but the sum is negative. If light of a particular energy strikes the electron, the electron and nucleus will separate from each other.

The next two quantum numbers are both descriptions of the angular momentum of the electron. Angular momentum is related to rotational motion. For most of us, the first thought we have about that is what we had been taught about the Bohr model. Electrons moving in fixed circular orbits is not exactly how the angular momentum of electrons works. For s electrons the angular momentum is zero. Linus Pauling describes these electrons as only moving in and out from the nucleus.[8]

There are two angular momentum numbers that can be found from the wavefunction for each electron. The first is represented by the letter l, and this angular quantum number tells us the type of orbital.

Angular Momentum	Orbital
l = 0	s
l = 1	p
l = 2	d
l = 3	f

After l = 3, the next orbital types continue in alphabetical order. For l = 4, the orbitals are g orbitals, then h, then i, and so on. The s, p, d, and f stand for sharp, principal, diffuse, and fundamental if you like trivia. These names originated based on the appearance of spectral lines.

Chapter 8 – Atomic Structure

The second type of angular momentum number is given the symbol m and is called the magnetic quantum number. This quantum number tells you which orbital of the set. For an s orbital, there is only one orbital per energy level. But p orbitals come in a set of three that are all perpendicular (orthogonal) to each other. The m value distinguishes those three orbitals. Note how a magnetic field would impact the three p-orbitals differently (Fig. 8-4).

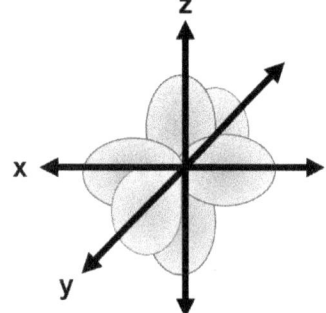

Figure 8-4: a set of p orbitals that includes a p_x, p_y, and p_z orbital.

A set of standing waves can help students to connect how an electron moves while maintaining a wave structure. For orbitals, they struggle to see how the electron moves within the static structure of the orbital. They fill in misconceptions about electrons bouncing around inside the orbital shape or also that the electron is moving in some defined path along that shape.

But with a standing wave, we can show how the motion of the pulse is happening while also presenting a standing wave that has both motion and a somewhat locked in structure. The electron wave mechanics are very similar to this as anyone who is familiar with the particle in a box calculations can attest. The electron is moving, yet in a standing wave like formation. And no particle representation is needed if the students can visualize motion within a standing wave. The orbital description of the electron is similar to this standing wave.

A perk to standing waves is how they mimic quantization. You can have a wavelength of 2L, L, 2L/3, etc. But you can't have a wavelength between 2L and L. It's either one or the other. Likewise, you can have an electron with n = 1, 2, 3, etc. But there is no n = 1.4.

Orbital nodes is the topic I would recommend if you have a student that is excited to learn more about orbitals. A node is where the electron does not exist (ignoring uncertainty). Orbitals have angular nodes and radial nodes. A radial node occurs as you move away from the nucleus if there is a distance where the electron does not exist in all directions. An angular node is when there is an angle where the electron does not exist as you rotate about the nucleus.

As we move from l = 0 to l = 1 to l = 2, we move from 0 to 1 to 2 angular nodes. For the p_z orbital, there is an angular node that is the xy plane (Fig. 8-1). For each d-orbital there are two angular nodes. As we increase in energy levels, the number of radial nodes increases too. The total number of radial nodes can be found using the formula n - l - 1. If you have a 4s orbital, the number of radial nodes is 4 - 0 - 1 = 3. There would be 3 radii where the electron would not exist as you move out from the nucleus. A 4p orbital on the other hand (4 - 1 - 1 = 2) would have 2 radial nodes and one angular node.

Chapter 8 – Atomic Structure

Orbital	Angular nodes	Radial nodes
4s	0	3
4p	1	2
5d	2	2
2s	0	1

Table 8-1: Angular and radial nodes for various orbitals

Angular and radial nodes is the perfect extension topic to suggest for the student that keeps asking you questions that you just don't have a good answer for. These nodes explain patterns and exceptions to trends in reactivity and physical properties. If students struggle with them, have them start with looking at nodes and antinodes for a standing wave and then move up from there.

The last of the four quantum numbers for electrons is spin. Spin does not mean spin. It does not mean that the electron is spinning. When we're using quantum numbers we're describing the electron as a wave, and the wave does not spin. Think of the name spin as you would think of the name strange or charm for a quark (they aren't actually charming). If the electron did spin with the angular momentum it has, the outer edge would be moving faster than the speed of light.[9]

What we do know about spin is that when two electrons that are identical except for their spin, the electrons interact with a magnetic field differently. There are also different rules for particles that have integer spins (bosons) and non-integer spins (fermions). The Pauli Exclusion Principle is only true for fermions (such as electrons), but not bosons. When electrons form Cooper pairs in a superconductor their spins combine which drastically alters their properties.

*****

The set of four quantum numbers leads to a large amount of explanatory power in chemistry. But they are also complicated in what they represent. We compromise what we use to discuss electrons with an organizational method called electron configurations. An electron configuration summarizes the key descriptors from the set of quantum numbers for the electrons within an atom or ion.

The beginning of electron configurations is tough for teachers. The subject becomes uncomfortable for most of us because at some point our understanding falls short. We do not need to introduce all the concepts for students to have a functional model. But we do need to give some rules. And it is difficult to keep student focus if they do not have any evidence to base their learning on. This is why I prefer to start with the light from the spectral emission tubes.

After that lesson, electron configurations are introduced as a description of the different motions of electrons within an atom or ion. Electrons do not necessarily move in the same way we envision them to move or that other things we have seen move. But a 4s electron does move differently than a 3p electron. The 4s electron averages to a further distance from the nucleus and has a different angular momentum and different nodes.

These differences in electronic motion and position have big impacts on the reactivity and properties of chemicals. Much of what we are trying to do is give

Chapter 8 – Atomic Structure

students a solid foundation for models to represent electrons that can later be sufficient when explaining differences, trends, and reactivity.

Showing students how to produce the ground state electron configuration for an element is best done using a periodic table. We start by splitting the periodic table into an s-block, p-block, d-block, and f-block (Fig. 8-5). Each additional square represents one new electron. H has 1 electron. He has 2 electrons. Li has 3 electrons, but 1 of the 3 is moving differently than the others. Two electrons are described by the 1s state while the third is described by the 2s state.

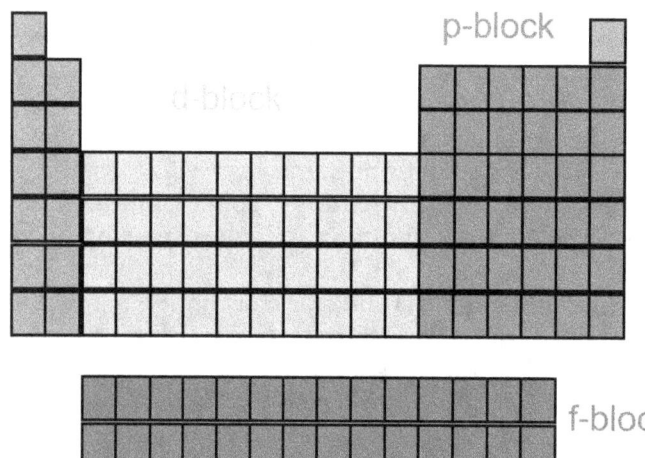

Figure 8-5: The periodic table can be split into blocks to determine electron configurations

$1s^2\ 2s^1$ would be the configuration for Li. The first 1 represents the 1st energy level (which also matches the period). The s represents the orbital type (and matches the block H and He are in) which is determined by the angular momentum of the electrons ($l = 0$). The superscript ^{2} tells us that there are two electrons described by this state (which also matches the number of boxes in this row of the s block). The 2 represents the 2nd energy level (2nd period for Li). The s represents the orbital type (matches the block Li is in). The superscript ^{1} represents the number of electrons described by 2s.

Using the periodic table this way is called dual coding. We are using an image of the periodic table to connect to the written form of electron configurations. This is likely to be easier for students to do the algorithm and remember it later than using a list of orbitals in order of the Aufbau Principle (1s, 2s, 2p, 3s, 3p, 4s, 3d, etc.). But we do not wish for the students to only see the one side of the dual coding. We want them to see beyond the periodic table algorithm. We need them to know that the electron configuration describes the motion and position of electrons in an atom (or ion).

When we reach the d-block, the electrons are now 1 energy level behind. In the 4th period we go from 4s to 3d to 4p. In the 5th period we go from 5s to 4d to 5p. When we reach the f-block, the electrons are now 2 energy levels behind. The 6th period goes 6s to 4f to 5d to 6p. The 7th period goes from 7s to 5f to 6d to 7p (with many exceptions).

This will never not be controversial to students or teachers. *Why is 3d out of order?* It is important for teachers to know that for a H atom with only one electron, 3d and 3s are identical in energy. But as electrons are added, those energies shift, and

Chapter 8 – Atomic Structure

they do not shift equally. There is different repulsion from the core electrons to the 3d and 3s electrons. These discrepancies arise because of the types of nodes discussed earlier. The s orbitals have more radial nodes which places some of the electron density closer to the nucleus. The different amounts of repulsions cause some fascinating (or irritating) inconsistencies.

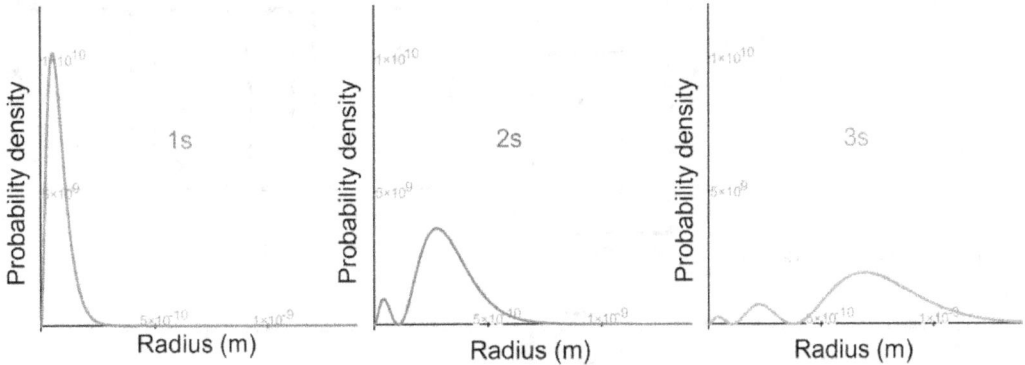

Figure 8-6: Electron density for 1s, 2s, and 3s orbitals. The radial node pattern leaves some of the electron density very close to the nucleus even as n increases.[10]

Chromium and copper are exceptions to the periodic table pattern. For Cr the $4s^1 3d^5$ configuration is slightly lower in energy than the $4s^2 3d^4$. For Cu the $4s^1 3d^{10}$ configuration is slightly lower in energy than the $4s^2 3d^9$ configuration. Most (W does not) of the other elements in chromium's and copper's groups follow the same pattern (Ag is $5s^1 4d^{10}$). Many teachers and students will make the jump that having a half-filled or completely filled shell is stable. This is a bad jump to make.

The word stable implies that you are in a local minimum of energy. The electron configuration exceptions are lower in energy than the alternative configurations expected from the periodic table pattern. But they are not much lower. The exceptions sometimes are so similar in energy that changing the number of protons can flip which is the lowest energy state.

There are a lot of factors for why Cr is $4s^1 3d^5$ instead of $4s^2 3d^4$. One is that the second electron in the 4s state has to pair which causes an increase in repulsion. Sometimes that repulsion is not enough to overcome the energy difference and sometimes it is.

The half-filled or filled subshells are not always the lowest energy configuration. There is a pattern where those commonly show up for exceptions or for a lack of noble gas reactivity. But this is a correlation and not a causation. The alternative to a filled shell is sometimes that an electron must go to the next set of orbitals which are higher in energy.

Mo, Tc, Ru, Rh, Pd ($5s^0 4d^{10}$), Ag, Os, Ir, Pt, and Au are all elements that do not follow the anticipated orbital sequence. Ru for example has its ground state as [Kr] $5s^1 4d^7$. If Ru had [Kr] $5s^2 4d^6$, it would have had a full 4s shell. Sc^+ has an electron configuration of [Ar] $4s^1 3d^1$ even though it is isoelectronic with Ca that is [Ar] $4s^2$.[11] The orbitals are impacted by the amount of nuclear attraction and electron repulsion from other electrons. Those impacts are not felt evenly by different orbitals causing inconsistencies when orbital energies are similar (e.g., 4s vs. 3d).

Chapter 8 – Atomic Structure

This suggests that electrons will enter the 3d before the 4s. But if the 4s is vacant, the electrons will promote to the 4s state. This also explains how the 4s electrons are removed before the 3d when ions form even though K and Ca have the 4s occupied while 3d is vacant.

Once students are capable of writing out electron configurations it is time to connect them to something more visual. Orbital diagrams are the link you want to make things a bit more concrete.

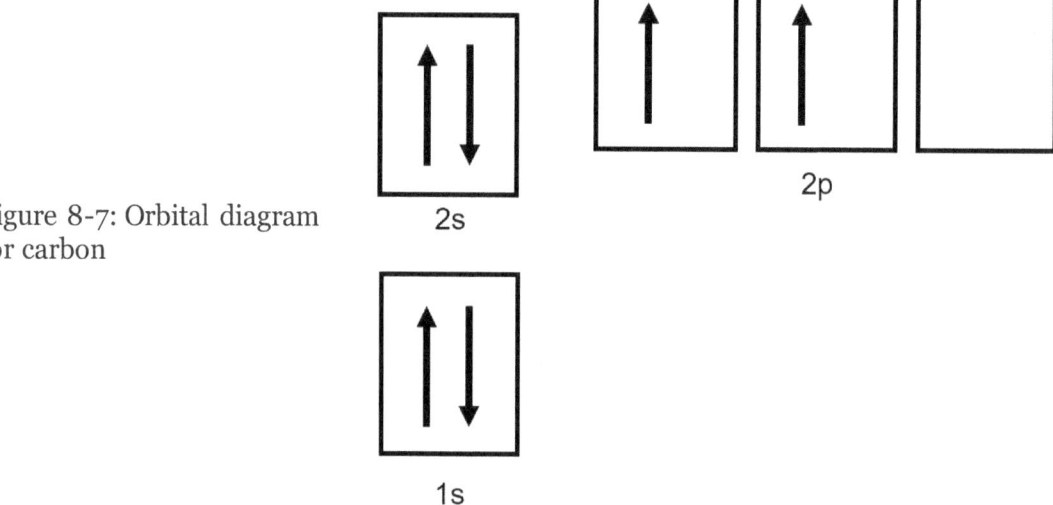

Figure 8-7: Orbital diagram for carbon

We need to move slowly with orbital diagrams. Take a moment to think of some questions that students might not know.
- What is a box (or line)?
- What is an arrow?
- Why do the arrows point up or down?
- Why are some boxes higher?

When I draw orbital diagrams, I put the p orbitals further to the right. A student might wonder what significance being further right or left is. Students will wonder why there is 1 line for s orbitals, 3 lines for p-orbitals, and so forth.

Each line represents a particular type of motion. The height represents the relative amount of total energy (PE + KE) for an electron that is described by that state of motion. The height is influenced by the combination of all electron-electron repulsions and the electron-proton attractions. The arrow is an electron with that particular motion (e.g., 4s), and the direction of the arrow differentiates spin. Some of these electrons have the exact same energy, and we call that being degenerate.

We want students to connect an electron configuration with the orbital diagram. From these the students will be able to have better models for learning about ionization energy, bonding, and electronegativity. Electron configurations are a challenge for teachers since many students can easily determine the electron configuration but understanding what an electron configuration is has a much steeper learning curve. It can be easy to abandon the understanding component but finding patterns later on can be enhanced greatly by giving students a concrete picture.

Chapter 8 – Atomic Structure

There are **three rules** for electrons that guide how orbital diagrams should be filled out to accurately connect with the evidence that we see.

The **Aufbau principle** says that electrons will occupy the lowest available energy states first. This rule is violated whenever an electron is promoted to an excited state. This rule is also tricky because sometimes (Cr, Cu, Pd, etc.) the ground state is not what we would expect it to be. Another interesting example for teachers to know about is that La has a ground state of [Xe] $6s^2\ 5d^1$ and not [Xe] $6s^2\ 4f^1$.

Hund's rule is most easily remembered by describing it as the "urinal rule." If there is an unoccupied space available, you take that rather than pairing. For electrons, this translates to all degenerate (equal energy) orbitals will be half-filled before any become full. There is energy required to pair electrons since they are negatively charged and repel each other. The spins will also all be aligned for unpaired electrons.

The **Pauli exclusion principle** is the final rule. In physics this would state that no two electrons can be in the same place with the same set of quantum numbers. In chemistry we reduce this down to electrons cannot be in the same orbital unless they have opposing spins.

These are pretty easy for students to pick up on. The best way to go about them is to have students construct orbital diagrams that would violate them in various manners. Which rules are followed and violated based on the diagram below?

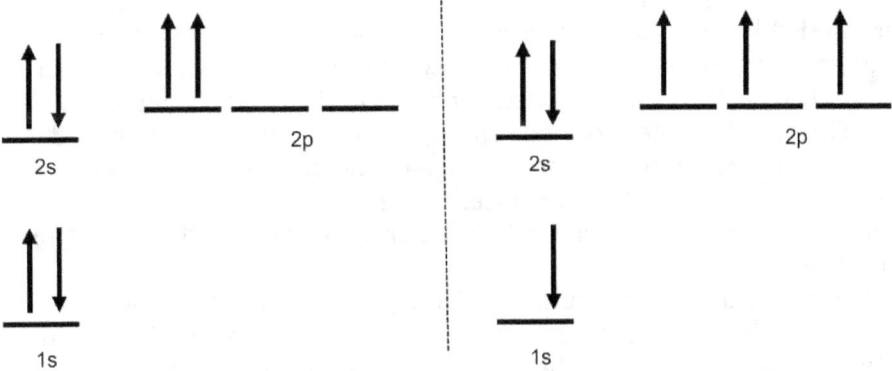

Figure 8-8: Two atoms with various violations of the Aufbau principle, Pauli's exclusion principle, and Hund's rule

Atoms have electron configurations, but molecules are best described by molecular orbital diagrams. Sometimes the electron configurations of atoms forming a bond correlate with the molecular orbital diagram. Others do not. Teachers can fall into the trap of using electron configurations to explain bonding, but this will lead to inconsistencies. If you can address this now your students will benefit when they get to spectroscopy and bonding.

For example, consider an H_2 molecule forming. Each H atom has an electron configuration of $1s^1$. You may then think that the covalent bond completes the 1s shell for both atoms. A superior model is two atomic orbitals are combined into two molecular orbitals. For more information on molecular orbitals, see chapter 10 on bonding.

Chapter 8 – Atomic Structure

Molecules have different abilities with respect to absorbing and emitting light. Molecules can vibrate, rotate, and experience electronic transitions. This means that when we talk about absorbing and emitting light, we need to be cautious about what materials we view.

If you mix concentrated ammonia with aqueous copper (II) ions, you get a stunning blue complex ion. If you mix it just right, you can even get three layers. The bottom is the original copper (II) solution, the top is the brilliant royal blue copper (II) tetraamine complex, and in between is light blue copper (II) hydroxide precipitate.

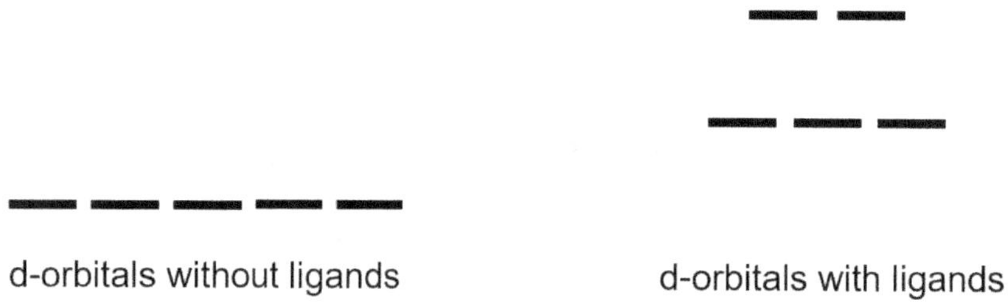

d-orbitals without ligands   d-orbitals with ligands

Figure 8-9: d-orbital splitting in the presence of ligands

What makes the copper (II) complex ion blue is that the ion absorbs orange light. The transition metals have partially filled d-orbital shells and when these electrons are surrounded by ligands (in this case the ammonia molecules), some of those d-orbitals are closer than others to those ligands. Of the 5 d-orbitals in the 3d set, three of these 3d orbitals are along the axes, and two of those 3d orbitals are between the axes. The energy of these orbitals changes when the ligands are added and how they change depends on what shape the ligands take. If the ligands are in a square planar or octahedral shape, they are closer to the 3 orbitals along the axes. If they take a tetrahedral shape, they are closer to the 2 orbitals between the axes. The result is a 2-3 or 3-2 split.

When light hits one of those d-orbital electrons, the electron is promoted to the upper level of the d-orbitals. And here is where things get interesting. When this happens to an atom, the atom absorbs light, and the electron goes from a ground state orbital to an excited state. Then the electron returns to the ground state emitting the same light it absorbed. Sodium emits a lot of orange and absorbs a lot of orange.

But a complex ion or molecule does not have to fit those parameters because there are other particles nearby as well as vibrational and rotational states available. When a complex ion absorbs orange light, it does not re-emit that orange light. Instead, the excited electron will emit multiple light waves as the electron goes back to the ground state in stages. Perhaps instead of an orange light wave, the excited ion emits a radio wave, two microwaves and an infrared wave. The reason why molecules and complex ions can do this is because of their surroundings. Having other particles bonded means that there are vibrational and rotational energy states available. An excited molecule could collide with another and cause a transfer of some of that energy to the other molecule.

Chapter 8 – Atomic Structure

An atom does not have vibrational states or rotational states. An atom can likely only transfer the entire quantum of energy or nothing. The way that atoms function is very different than how molecules work. Glow in the dark stars that you might place on the ceiling of your room absorb blue and violet light. But the stars are not made of atoms, they are made of molecules. Those excited molecules[‡] emit low energy light until they reach a meta-stable excited state. This state gives off green light as the electrons return to the ground state. But that green light does not get emitted in the typical timeframes due to the increase in stability. Thus, the stars glow in the dark for a short period of time. The term for this is phosphorescence, but the important concept to understand is that molecules have abilities for light emission that atoms do not.

Think about the greenhouse effect with glass and infrared light. Visible light gets into your greenhouse (if you don't have a greenhouse currently, think of your car on a sunny day). The visible light hits substances that are composed of molecules. Those molecules absorb the visible light. But they do not re-emit that visible light. Some of it might reflect off, but some is also emitted in stages. The infrared light emitted is not able to make it through the glass. The amount of energy absorbed by the greenhouse is larger than the amount of energy reflected and emitted. This causes the thermal energy and temperature to rise inside. This entire process is contingent upon the type of incident light differing than the emitted light.

Many textbooks guide teachers to teach light by focusing on calculations and the electromagnetic spectrum. You will want to reframe that because this vastly undersells the content. Light is how we learn about the largest and smallest things in the universe. If human brains are designed for survival and not learning, light is one of those topics that can prompt curiosity to learn.

This doesn't work for all cell phones, but if you have a cell phone on you, stop reading for a second and turn on the camera. Find any remote control in your house and see how the glass bulb in the remote looks through your phone camera when you push buttons. The selfie mode camera tends to work more frequently. Many cameras will detect infrared light and will convert that to visible light on the display. The remote control emits infrared light as a signal to change the channel or volume, but nothing happens when the infrared light hits your retina.

Think about that for a moment. That's wild! There are entirely different worlds going on that you don't see at all. Think about how much radio, microwave, and infrared light is traveling around you right now that you aren't detecting.

Here's another great demonstration for your class. Take a strip of glow in the dark material and glue it to a piece of black construction paper. Cut out 6 holes on one side of the paper and put red, orange, yellow, green, blue, and violet filters on the 6 holes. When you fold the filters over, they will let through those 6 colors of light to hit the glow in the dark strip when you shine white light on the paper.

[‡] Note that excited electron and excited atom or molecule is used interchangeably. This is valid because the electron moving to a higher energy level is really the electron and the rest of the atom separating from each other. Both are in the higher energy state. We tend to focus on the electron because it is smaller and more mobile, but both are in the excited state. Just as when we lift a heavy weight, both the weight and the earth have their gravitational potential energy increased.

Chapter 8 – Atomic Structure

Interestingly, when you unfold the paper, you will only see two spots. The two spots are where violet and blue light were coming through (Fig. 8-10). This implies that when white light hits the glow in the dark strip that the red, orange, yellow, and green light in white light do nothing. It is only the blue and violet components that cause the electrons in the glow in the dark strip to change.

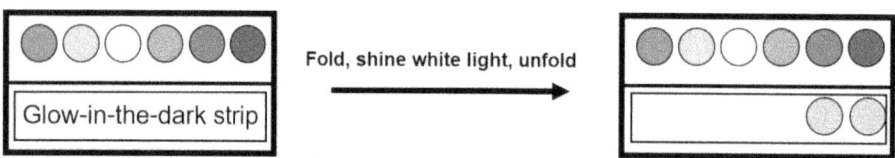

Figure 8-10: Demonstration showing energy of visible light

This demonstration is similar to the photoelectric effect that Einstein experimented with during 1905 and won the Nobel Prize for in 1921. It highlights that different types of light have varying amounts of energy. If you shake an electron harder, you'll get a different type of light. That light can then cause a bigger change when it meets the next electron. You can extend the demonstration by shining ultraviolet light on the glow in the dark strip and having students predict what will happen. You can also spray sunscreen onto a clear object and place that between the glow and the dark strip and UV light. You will get shadows on the glow in the dark strip from the sunscreen.

When we differentiate types of light using the electromagnetic spectrum, we want to emphasize the components that are shared and those that differ. Calculations with wavelength, frequency, and energy can help with this, but only if we are intentional. It is more than possible for these calculations to only be seen as arbitrary numbers.

Students have little knowledge of waves. I recommend getting one of those long metal slinkies and demonstrating standing waves as well as wave pulses. The wave pulse is just a quick shift of the hand on one side of the slinky that causes the disturbance to travel down the slinky and back. The students are surprised by the inversion when the pulse reflects, but I think much of this surprise actually stems from how fast the pulse travels. Most have never seen anything like it because we over emphasize standing waves in textbooks.

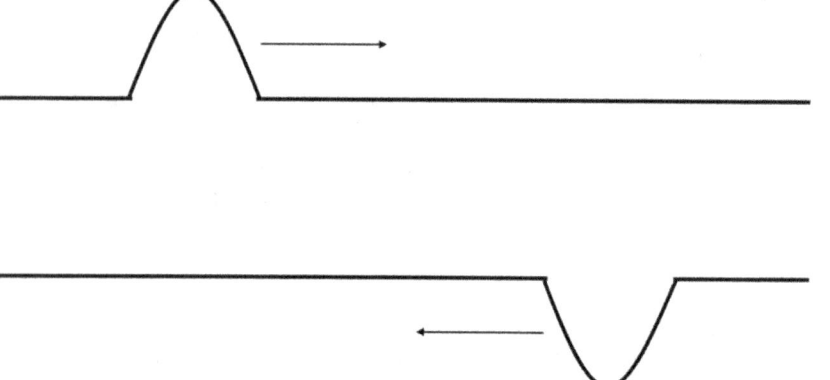

Figure 8-11: A wave pulse is sent down a slinky and the pulse reflects back

Chapter 8 – Atomic Structure

Historically the use of spectral lines was critical for discovering new elements. Helium was discovered via the spectral lines coming from the sun. Rubidium was discovered because of its red spectral line which prompted the name rubidium after the ruby red spectral emission. Cesium was discovered from its blue spectral lines and was therefore named after caesius or blue in Latin. Prior to when Henry Moseley discovered that X-rays could be used to determine the nuclear charge spectral lines were a key method to verifying a new element. Henry Cavendish did an experiment where he removed nitrogen and oxygen from air, but there was always a bubble remaining. When Lord Rayleigh and William Ramsay repeated the experiment and tested the gas they found new spectral lines from the argon gas.[12]

Ernest Rutherford suspected that the alpha particles emitted from radium were actually helium nuclei. He set up a thin glass bulb filled with radium that had a second thick glass bulb surrounding it. When the radium decomposed into radon and helium the helium could pass through the thin glass but not the thick glass. After a few days he hooked up a battery to it to create a "neon" light but with helium.[13]

Calculations for light involve three variables and two constants. The three variables are wavelength (lambda λ), frequency (nu ν), and energy. The two constants are the speed of light ($c = 3 \times 10^8$ m/s) and Planck's constant ($h = 6.63 \times 10^{-34}$ J*s). All five of those can trip students up.

Wavelength, frequency, and speed all work together. When I used to coach track, we would talk about that there are only two ways to run faster. You can take longer strides, and you can take strides more frequently. All our training had to connect to one of those two things. We do plyometrics to get our knees higher during the stride, so we cover more ground. We build endurance so that our bodies can maintain a high turnover rate.

Light works similarly except the speed is the same thing no matter what. It can be helpful to think of an electric field line that travels away from an electron. If you shake the electron there will be a disturbance to that electric field line that will propagate away. If you shake the electron faster, the disturbance will result in a smaller wave packet. Teachers tend to define frequency in the context of the wave pulse. For example, frequency is the number of wave cycles that pass a given point in a second. But it is much more concrete to define frequency by the source of the wave pulse. Even if you end at the same definition, students can connect better to shaking an electron up and down.

Chapter 8 – Atomic Structure

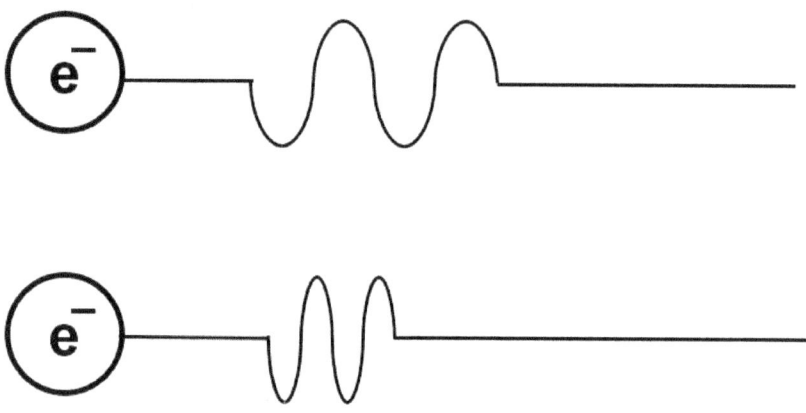

Figure 8-12: The top electron is shaken twice, the bottom electron is shaken faster

What is an electric field? Does the electric field move up and down or does the magnitude of the electric field get bigger and smaller? How do we know there are electric fields? Can we see them? How is an electric field different from a magnetic field?

Because the pulse moves away from the source at the same speed, changing the frequency will also change the length of the wave produced. Students can observe this by shaking a slinky quickly and slowly. They just need to keep the tension in the slinky somewhat steady to avoid changing the wave speed.

Another great demonstration of this is to have students act out various colors of light. Each student represents a color red through violet. They can put colorful paper on their shoes to make this more vivid. Then they have to walk at the same speed but where violet takes more frequent steps. In order for the students to remain at a constant pace the red will take longer and less frequent steps. The violet will take shorter and more frequent steps.

$$c = \lambda \nu \qquad E = h\nu \qquad E = \frac{hc}{\lambda}$$

We want to establish the relationships between the variables, but we also want to use calculations to give some numerical context. Visible light is between 400 nm (violet) and 700 nm (red). A wavelength of 700 nm can also be written as 0.0000007 m or 0.00007 cm. 1 centimeter is pretty small as it is. A single strand of hair is at least 20 times as large as the wavelength of red light. The wavelengths of visible light are incredibly tiny.

The frequency of red light is somewhere around 4.3×10^{14} Hz. 10^{14} can be deceptive for high school students so let's write out the full frequency. 430,000,000,000,000 Hz means the electron is being shaken 430,000,000,000,000 times per second. The light is not being produced for a full second, so the actual number of vibrations will be far less but see if you can shake your hand back and forth more than 10 times in a second let alone 430 trillion.

Chapter 8 – Atomic Structure

The speed of light is 300,000,000 meters per second. Go find a track and think about doing a loop in one second. In that one second light can make it around the entire Earth over 7 times.

The wavelengths of visible light are as small as just about anything we can imagine. The frequencies are incomprehensibly fast. And the speed of light is just a shade below ludicrous speed. Then we get to energy and Planck's constant. Tell the students Planck's constant is named after the famous scientist Dr. Constant. It's funny and you can reuse that joke all year long. Planck's constant is 10^{-34} and my p-chem professors assured me that if it were a normal-sized number that life would be exceptionally weird. When we multiply by the frequency of red light, we end up with an energy of 2.8×10^{-19} J. That is a very tiny amount of energy, but it's only for one red light wave. A single photon if you will. A dim red light will involve a substantial number of atoms that can emit red light. If a mole of red light is emitted that would be 1.7×10^5 J or 170 kJ of energy.

Even though the scientific notation is going to lead to student calculator mayhem, try to keep some focus on the relationships and scales of light. It might be more important for a student to try and consider the scale of a nanometer instead of being able to convert from nanometers to meters.

Light is a unifying topic in all sciences. We use light to gather information from the depths of space to tiny subatomic particles. Light needs to be connected to our understanding of electron configurations in order to get students to move past the abstract symbols used for them. We want to take light and electron configurations and work with students to understand how light teaches us about chemicals as well as how electron configurations give us a useful model to explain electron motion. The more phenomena students can observe the more they will be able to understand. But these concepts are difficult, and this is a great place for students to write a reflection that highlights some experimental evidence along with what the students envision from this evidence.

Electron configurations and light should work together as the cornerstone of the unit. Both have challenging new concepts, and limits of current student mental models.

For light you want to end with students understanding the composition of light, how light originates/terminates, and the similarities/differences of different types of light. For electronic motion you want students to understand electronic motion is quantized, we can't see electrons move, instead we make mathematical descriptions of electrons (electron configurations and quantum numbers) and use light to see if they work.

A great way to assess these ideas is by presenting students with evidence of paramagnetism and diamagnetism. $MnSO_4$ will exhibit paramagnetism that can easily be detected with a strong magnet. $ZnSO_4$ will not interact with a magnet due to being diamagnetic (it's technically going to repel slightly, but I've never noticed the minimal effect). These can be explained using electron configurations and orbital diagrams. The Mn^{2+} ions have 5 unpaired electrons. But if we were to use a Bohr model of the atom, we wouldn't have the same ease to explain the phenomenon we're observing. Gilbert Lewis struggled mightily with trying to rationalize paramagnetism along with pairing of electrons in the early 1900s. Your students can use the magnetism of unpaired electrons to confirm the scientific models in your chemistry course.

Chapter 8 – Atomic Structure

Atomic Structure

The scale of an atom is beyond the brain of a human. This is true even for someone as intelligent as yourself. Atoms are so small that it is not possible for us to have any reasonably understanding of how small they are. Are there more atoms in a grain of sand or grains of sand on the earth? It doesn't matter. Your brain can't grasp how many grains of sand are on the earth. Try and count them the next time you're at the beach. These comparisons aren't helpful even if we find out that there are slightly more atoms in a grain of sand.

The subatomic particles within the atom are that much smaller than an atom. What do we do with such incomprehensibly small scales? A great introduction to atomic structure is to give students some modeling clay. Have them make a small sphere. Tell them this represents a proton that has been scaled up in size. How big would a neutron be? How big would an electron be? Have them make them. I did this for years until I finally realized I was making the spheres based on mass and not volume. How can you even know the volume of an electron? It does not have a definite end, especially when the electron is bound in an atom. But mass is more definitive. By mass a neutron is ever so slightly larger than a proton. Both are larger than an electron by a factor of about 1760.

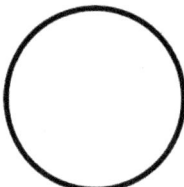

Figure 8-13: A scaled up version of a proton. How big would a neutron be? An electron?

But we can't use many significant figures here because the masses of protons and neutrons change depending on what nucleus they are a part of. But before we look at that, let's imagine our atom made of clay subatomic particles. Using the size of clay to represent mass, we have a proton, a neutron, and an electron. The electron is much smaller. Somehow the neutron and proton attract each other and stick to form a nucleus. How far apart should the electron be from the nucleus? Obviously, the distance varies, but on average for a deuterium atom, how far apart is the nucleus and the electron?

If we use the atomic radius of hydrogen, we would say the distance is somewhere about half of an angstrom (10^{-10} m or 10^{-8} cm). But the radius of the nucleus is much smaller than that. A back of the envelope calculation says that if the size of the clay proton is 1 or 2 cm, the electron should be 1 km away. Picture that clay proton, a tiny clay electron and you're going to put half a mile of space in between them.

What is in the space between them? Nothing. If you want to be technical there are some electric and magnetic fields. If you ask that question to students dramatically you might catch someone off guard who will respond by saying "air." But air is made out of atoms and molecules so there surely is not air in between the electrons and the nucleus of an atom.

The atoms that comprise this book are mostly empty space. The atoms your fingers are made of are mostly empty space. The scale of atoms is so unusual. One analogy that I find a little helpful is an image of a galaxy where the stars look like a big cluster of light. But we know that as you approach and zoom in that the distance

Chapter 8 – Atomic Structure

between the closest of those stars is unimaginably large. I'm probably just drawn to this analogy because both examples are so far removed from my ability to comprehend the scale.

Let's accept our inability to fully comprehend the size of subatomic particles and move forward. Counting those subatomic particles is very simple. An element is defined by the number of protons. All carbon atoms have 6 protons. All helium atoms have 2 protons. All Milamanium atoms have 122 protons.

A mass spectrometer produces simple data that students can easily use to learn more. Mass spectrometers are machines that measure the mass to charge ratio of the atoms after they have been charged somehow. Some types of mass spectrometers measure the mass of the nuclei and the relative frequency of those nuclei. If we put in a sample of pure carbon, we will learn how many isotopes of carbon there are, and how frequently each isotope occurs in that sample. From the relative abundance we can calculate what the average mass of one atom of a given element is.

Students tend to confuse this average atomic mass with mass number and molar mass. And who could blame them? Mass number refers to a single atom, but multiple atoms will have the same mass number. Average atomic mass is the average of every single atom of an element. Some mass numbers are very similar to average atomic masses, and some are not.

The molar mass is the same number as the average atomic mass, but they are different things with different units. The molar mass is the number of grams per one mole. The number is the same because 1 mole is such a large amount that you will get a consistent distribution of isotopes. Make sense? Think of how many different perspectives a student needs to distinguish between for the three concepts of molar mass, average atomic mass, and mass #. Some texts even add in terms like gram atomic mass and formula mass. Don't use those!

In my experience, some students will have had teachers tell them to use the periodic table to determine how many neutrons there are for an atom. It is important not to do this. Sometimes we take shortcuts, but here we are linking mass number and average atomic mass. You're using a system that describes all the atoms that exist with a system to describe one single atom and we instead should be aiming for students to differentiate the two ideas.

Now we get to calculating those average atomic masses. Start with what students know. They know that to find an average you should add up all the values and divide by how many there were. The average of 2, 6, 8, and 12 is 7. But now show them a different way to view that add up and divide method. There are four values we wish to average.

$$2 \quad 6 \quad 8 \quad 12$$

Each value occurs 25% of the time. A different way to look at this problem would be to apply the 25% to each value.

$$0.25*2 + 0.25*6 + 0.25*8 + 0.25*12 = 0.5 + 1.5 + 2 + 3 = 7$$

Note how that also works out to:
$$¼*2 + ¼*6 + ¼*8 + ¼*12 = 7$$

Chapter 8 – Atomic Structure

Here's the thing. You can't add up every atom's mass and divide by the number of atoms. There are bajillions of them. But we can easily measure how frequently each one occurs and find percentages. So even though the two methods look the same for 2, 6, 8, and 12; we can only use the percentage method when working with atoms.

Let's assume that in an alternate universe, 18.4% of Ar atoms have a mass of 38.000 amu and 81.6% have a mass of 40.000 amu. The average atomic mass of argon in this universe would be:

0.184*38 + 0.816*40 = 39.632 amu

A good follow up question to students would be, "Which percentage would have to increase in order to get the same atomic mass as what we have on Earth?"

Part of atomic structure should focus on helping students understand how mass number, atomic number, and charge help us to communicate how many protons, neutrons, and electrons an atom or ion has. Immediately some students are curious why we don't just provide the number of subatomic particles. It seems unusual to them to give an atomic number and a mass number instead of just a proton number and a neutron number. The mass number is more important in nuclear chemistry and the number of neutrons is rarely relevant on its own accord.

When we introduce the term atomic number, we also want to help students see that the difference between each element is fundamentally how many protons they each have. This can be enhanced by talking about alchemy and the work to change lead into gold. By removing three protons we would have success. But this cannot be done through chemical means, and to do so with bombardment requires equipment that has a cost far in excess of the value of the gold produced.

This is also a good time to reinforce a big idea from naming. Elements can exist in a neutral state or a charged state. Neutral iron has 26 electrons. Iron (II) has 24 and iron (III) has 23 electrons.

Nuclear Chemistry

Atoms are incomprehensibly small. A nucleus is about 100,000 times smaller than that. That 100,000 refers to just one dimension too. The volume of a nucleus is about 10^{15} or 1,000,000,000,000,000,000 times smaller than the volume of an atom. And to make things interesting, the non-hydrogen nuclei have multiple positively charged protons in that tiny space.

Our learning about the scale of the nucleus starts with one of my favorite experiments. Ernest Rutherford constructed a beam of alpha particles that struck a thin piece of gold foil and saw some of them bounce backwards. He compared it to firing a gun at a piece of paper and having the bullet ricochet back at you. There were two conclusions to the experiment. The first is that most of the atom is empty space. The second is that there is a massive, dense, and positively charged region of the atom.

Let's break down some relevant details from the experiment. Why did they use gold? Gold can be made into extremely thin sheets. This allows the foil to be thin enough that the alpha particles are unlikely to be absorbed. Gold also has 79 protons. Rutherford didn't know that protons existed at the time. But if something such as aluminum foil had been used the thickness would have interfered with the experiment and the deflections would have been much smaller due to the smaller amounts of positive charge.

Chapter 8 – Atomic Structure

An alpha particle is a helium nucleus. So how did Rutherford use them to effectively learn that atoms have a nucleus. What gives here? Prior to the gold foil experiment, Rutherford had done experiments with alpha particles. They were able to deflect them in a magnetic field and from the deflection determine their mass to charge ratio.

How did they make a beam of alpha particles? Rutherford put a small sample of radioactive americium in a container with a tiny hole in it. The alpha particles fly off in all directions, but in nearly all of those directions they hit the container and are absorbed. Only the alpha particles moving in a straight line towards the gold foil escape through a hole.

How did they detect the alpha particles? When the alpha particle hits the chemical ZnS a flash of light is emitted. Rutherford had two graduate students who would sit in a dark room and count the flashes in one section of the ZnS detector. One of those students, Hans Geiger, went on to construct the Geiger counter that does that work for him.

The design of the gold foil experiment is beautiful.[14] The conclusions were huge developments in chemistry. And the experimental procedure to aim a beam of particles at a surface had so many subsequent follow up experiments. Our techniques for making new elements is contingent upon such a method.

If the scale of an atom is beyond our brains, the scale of nuclei is nearly hopeless. But students should learn that hopelessness regardless. These students are not only going to need to make decisions about nuclear power at some point, they will also likely have the opportunity to educate others. Most are wildly unprepared for this. To verify this, have them construct a list of places that would be good options for storing nuclear waste. You'll make several important observations.

Most of their options are smart aleck responses. The last time I asked a class to do this they selected my mom's house and under my mom's pillow. A few of their ideas are terrible and some are reasonable. But they don't know which are which. They don't know how to evaluate the pros and cons of blasting a rocket full of nuclear waste at the sun or burying nuclear waste in a mountain. They fail to consider transport for many of their destinations, which should quickly rule space out. If even one of the rockets were to fail the results would be catastrophic.

But the most interesting observation is that they do not ask any questions. How much nuclear waste is there? What is nuclear waste? How long will it take to be reasonably safe? Could we use it for something productive? Is it dangerous? Can it explode?

The amount of nuclear waste is small. If you search for a nuclear waste cube, there are cutouts that students can fold into the amount of nuclear waste they would generate from their electricity use over a twenty-year period. They are about 4 cubic inches big. If you had one cube for everyone in the United States, you could stack them 12 feet high and fit all of them into a single football field. If that seems like a lot to you, consider that the average amount of carbon dioxide produced is about 8000 liters of pure CO_2 gas per person per year. That's about 500,000 cubic inches for every one person for just one year (instead of 0.2 in^3 of nuclear waste).

In fact, nuclear waste is nearly the best-case scenario for byproduct from producing electricity. There's very little generated and it is easily contained. When you burn coal, the products are immense and are nearly impossible to contain. Most people don't realize that the radiation from coal pollutants causes severe health

Chapter 8 – Atomic Structure

problems for anyone living nearby a coal power plant. We use coal to adsorb chemicals in water treatment. So, when we burn the coal, all that stuff is adsorbed to it goes into the atmosphere. Many of those contaminants are radioactive, harmful to humans, and harmful to the environment.

That scale applies to all levels. Nuclear fuel for generating electricity requires less transport and less mining. Solar and wind energy have a lot of perks and we should use more of them. But the amount of silica mining for solar, the transport of that material, and the placement of solar panels all must be weighed against the lack of waste. Students are passionate environmental advocates, but we want them to use environmental science as part of that advocacy.

How does nuclear fuel work in the first place? Uranium has two common isotopes. 99.3% of uranium has 92 protons and 146 neutrons. This Uranium-238 has a half-life of 4.5 billion years. 0.7% of uranium has 92 protons and 143 neutrons. Uranium-235 has a half-life of 700 million years. Both isotopes undergo alpha decay where the nucleus splits into a thorium nucleus and an alpha particle.

$$^{238}_{92}U \rightarrow {}^{234}_{90}Th + {}^{4}_{2}He$$

The first question students should have is how does a nucleus stay together in the first place? How can you put 92 positive charges into such a tiny volume without them flying apart? There is a very strong attractive force between protons and neutrons when they get really close together. This really strong force is called the "strong force" or "strong nuclear force." The strong force only works over a very short distance so even within the nucleus there are protons and neutrons that are far enough that they don't experience this attractive force.

If you are interested, I'll do a quick diversion on the origin of the strong force. One model states that forces occur by an exchange of particles. Electric forces occur between electrons by the creation and consumption of a photon particle. But creating a particle out of nothing violates the conservation of mass/energy. This isn't a huge deal for the massless photon, but it is for the meson created in the strong force. This can only occur via the Heisenberg uncertainty principle. Normally in chemistry we look at the uncertainty relationship between position and momentum. But a relationship also exists between energy and time. Thus, the meson particle can be created, but only for a brief time. That timeframe is just long enough for the meson to travel about the diameter of a single nucleon. Therefore, the strong force only occurs at that scale.

The strong force only works over short distances which means that as nuclei get larger the electrostatic repulsion increases faster than the strong force attraction. This is why the end of the periodic table is filled with synthetic atoms that last for less than a second. The protons and neutrons tend to form small clusters in the nucleus, and they move around. At some point a cluster of two protons and neutrons can get far enough apart that the electrostatic repulsion causes the alpha particle to fire off away from the nucleus.

The splitting of a nucleus into smaller fragments is called fission. As the pieces separate their masses can change and their relative speeds increase. The changes in mass are proportional to the total energy changes. $E = mc^2$ can be used to determine the total energy changes, where c is the speed of light. A uranium nucleus undergoes fission, but with a half-life of hundreds of millions of years isn't going to power a city.

Chapter 8 – Atomic Structure

What makes the uranium-235 fission faster is to hit the atoms with a beam of neutrons. When uranium-235 absorbs a neutron, the resulting uranium-236 rapidly fissions and a product of that fission is 3 more neutrons which can cause more fission to occur. Neutrons give us control over the uranium. If we want more decay to occur, we use a "moderator" to slow down neutrons. This causes more neutrons to be absorbed and for the fission to happen more frequently. If we want the fission to slow down, we add a "control rod" to absorb the neutrons. Boron absorbs neutrons well and can be used as a control rod to shut down the uranium decay. Graphite and water can be used as moderators. The single proton of hydrogen in water can cause neutrons to slow down during a collision.

The history of fission is a fascinating tale. Many of the scientists who first figured out what was happening now have elements named after them. William Roentgen (roentgenium, 111) discovered X-rays when he noticed that a uranium sample was illuminating film. He took an X-ray of his wife's hand that showed her ring. Lise Meitner (meitnerium, 109) was the first to conclude that atoms can change via fission. She isolated barium and krypton from uranium decay. Otto Hahn was awarded the Nobel Prize for their work and her conclusion. Marie and Pierre Curie (curium, 96) used radioactivity to find and isolate two new elements (polonium and radium). They tested to see if temperature affected the radioactivity and concluded that temperature had no effect.[15]

Glenn Seaborg (seaborgium, 106) and Yuri Oganessian (oganesson, 118) discovered a large number of new elements. Seaborg was born in Michigan and his team discovered ten elements. Oganessian had at least a role in the discovery of all 13 elements from 106 to 118. These two were the only scientists to ever have an element named for them while they were living.[16]

A general trend is that as nuclei get bigger they become less stable. Heavier nuclei tend to have more neutrons than protons since a neutron will add more strong force interactions without increasing the electrostatic repulsion. But there are also some combinations of nucleons that pack together well and are therefore more stable.

Others are inherently unstable. Element 43 technetium was difficult to discover. There were 7 failed attempts to name element 43. Masurium may have actually been a credible discovery by Ida Noddack and her husband Walter. When students learn about alkali metals and halogens they want to know about francium and astatine. Neither exists in any meaningful quantity, and the atoms that do exist are decay products randomly mixed in with uranium somewhere that soon disappear.[17]

To be clear, when we say disappear, we mean change into a different nucleus (and different element). The nuclei that are unstable will decay when the nucleons arrange themselves in a way where an alpha particle gets too far away. The probability of this decay is constant, and we can therefore use the amount of decay as a means to track time.

Let's say you have a jar with a few fruit flies in it. Every ten minutes the number of fruit flies doubles. After two hours the jar is full. When was it half full?

Your intuition is quick to scream one hour. But that is your system 1 talking. For some, their system 2 notices the flaw in reasoning and corrects to the answer of one hour and fifty minutes. Since the number of fruit flies doubles every ten minutes, the full jar at 2 hours would have been half-full ten minutes prior.

Chapter 8 – Atomic Structure

Our brains are designed to look at patterns. Cutting something in half is one pattern, and the linear progression of time is a different pattern. Students struggle to operate in both at the same time. When you do activities with them, they will be drawn to the cutting in half portion of the activity. I personally am a fan of rolling 100 pennies and removing just the tails each time. But a critical piece to highlight is that each roll represents a consistent amount of time passing. The other fact that is not obvious is that the penny that ends round 6 has been flipped 6 times and has been heads for every single one of those flips.

In order for students to learn about half-lives more effectively we want to highlight the linear changes in time. If you roll pennies for example, have them all roll at the same time at regular time intervals. Have them write down the times and not just the number of pennies remaining.

Teaching half-life tends to focus on the mathematics, but we should also highlight some of the chemistry involved. One key feature to this is making sure that the students understand that the radioactive substance doesn't disappear. It turns into something else. But in some instances, like carbon-14, the product is nitrogen which does leave into the atmosphere. A group from Grand Valley State University designed a set of Target Inquiry Labs. Their nuclear chemistry lesson called "How Old is my Dinosaur Bone?" takes students through a typical set of half-life determinations. But then the lesson quite brilliantly has students determine the age of a "dinosaur bone" by comparing the amounts of carbon-13 and carbon-14. The carbon-13 remains static over time, so as the carbon-14 beta decays into nitrogen and leaves, the ratio of C-14 to C-13 decreases.[18]

When organic material is alive the carbon-14 gets replenished from carbon entering the system through food. We measure the age based on the final amount of carbon-14 as the organic material stops taking in new food. We can even use carbon dating to measure inorganic substances if they have slight amounts of organic material in them. A piece of pottery is not testable, but if it has a tiny amount of food residue from using used to store food, we are able to date it.

Fusion is a topic that is not typically taught much in introductory chemistry. There is some fascinating information though that may be useful for the teacher to be aware of. Students may be curious about the future of electricity using fusion. The biggest draw towards fusion is that the fuel is available, the products tend to be environmentally friendly, and the amount of energy generated is even larger than what we get from fission. The biggest problem is that fusion is really tough to get to happen.

In order for two nuclei to fuse they have to get close enough for the strong nuclear force to work. This happens at very high temperatures and pressures where the nuclei travel fast enough for the nuclei to collide so hard that they can overcome the electrostatic repulsion. Here's the problem with that. You have to heat the mixture to get these high temperatures. From there the mixture will cool down. In order to maintain a high temperature and limit that cooling you need the size of the nuclear reactor to be larger. This is because volume and surface area do not increase at the same rate. As the reactor core gets larger the surface area gets bigger, but the volume gets bigger faster. This makes the temperature more stable, but the size of reactor that would be sustainable is so large that it would not be functional. You would simply produce too much energy.

Chapter 8 – Atomic Structure

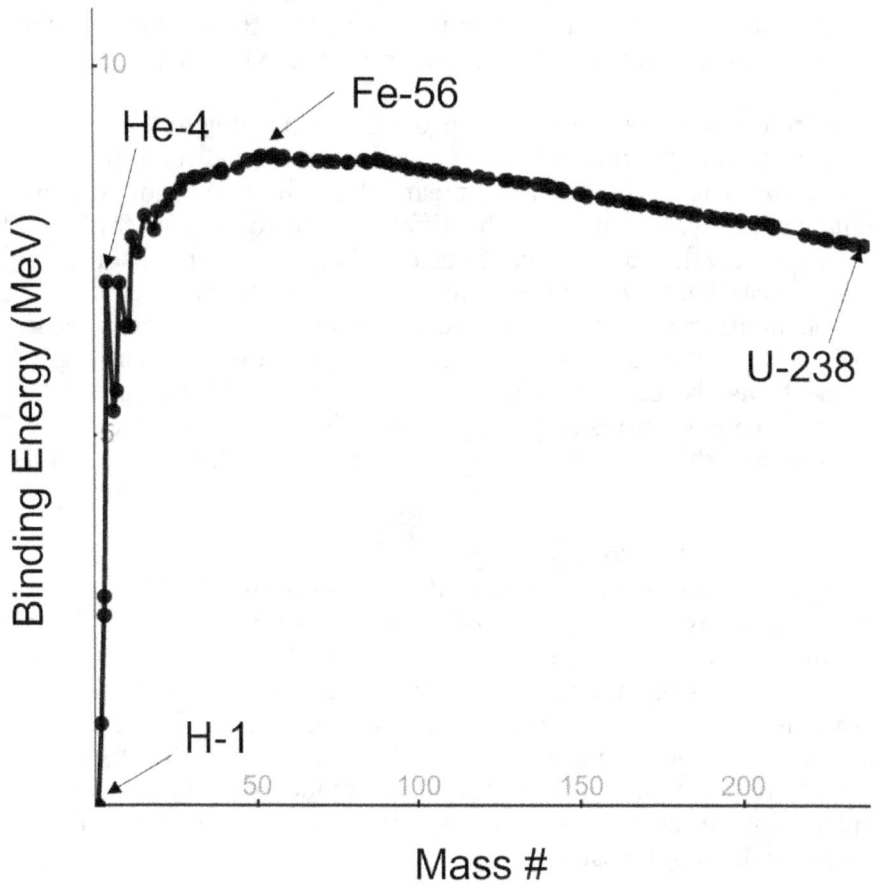

Figure 8-14: Binding energy of different nuclei[19,20]

The most stable nuclei are iron-56. When there are fewer nucleons then fusion releases energy. Elements like H, He, Li, C, etc. all release energy when they fuse. Elements like U, Th, Ra, Rd, Pu, etc. release energy when they fission. This raises an interesting dilemma. How did we end up with anything above iron on the periodic table? If we pretty much started with only hydrogen, how would we ever progress beyond iron?

For elements beyond iron to form, there has to be a substantial amount of energy put in. Stars evolve from fusing hydrogen, to fusing helium, etc. until they reach iron. At the point of iron two results can happen. Either the star stops fusion and becomes a giant dense ball of iron spinning in space. Or if the star is big enough the iron collapsing causes a massive explosion called a supernova.

Supernovae are the source of elements beyond iron (neutron star collisions also supply some of these elements). I think that's awesome. When you're in your chemistry class think of all the objects that have supernova remnants. My gold wedding band has atoms that were in an exploding star. Even a penny made of copper and zinc is the result of a giant space explosion.

Nuclear chemistry is a great opportunity to do some spaced practice from atomic structure. Remember that high school students will wonder why we use atomic number, mass number, and charge instead of just having the number of

Chapter 8 – Atomic Structure

protons, neutrons, and electrons. A nuclear chemist cares about the total number of particles in the nucleus which makes the mass number a more expedient communication tool.

We also want to connect the nucleus to our original topic of light and electrons. The more protons in the nucleus, the larger the force is between the electrons and the nucleus. The increase in force means that it becomes more difficult for electrons to change their type of motion. The different light transitions for H, He^+, Li^{2+}, etc. can be used to show the connection of nuclear charge and light absorbed and emitted. Having this discussion now helps explain how the 1s electrons in gold, mercury and other elements travel near relativistic speeds. These high speeds cause gold's conduction band to shift so that its color differs from most metals. The high speeds of the core electrons also cause mercury to have a lower melting point. The relativistic speeds of the core electrons cause the valence electrons to contract in towards the Hg nucleus and this limits the metallic bonding interactions between Hg atoms.

Student struggles

1. "I have no idea what I'm doing." Most students are successful with the majority of questions involving some type of process or algorithm. They can write out electron configurations, construct element symbols for atoms based on mass # and atomic #, and they can identify violations of Pauli/Aufbau/Hund. But they will struggle with explaining what they are doing, and they will have a lot of questions during instruction. As the teacher you want to scaffold the difficulty by having observations that students can relate back to. Maybe our model of how light/matter interact is incomplete, but we can agree that it was unusual that only certain colors of light were observed in hydrogen emission.

2. Light calculations - students will struggle with calculations of frequency, wavelength, and energy. There are two issues here that you can help with. The first is that many students are not strong with scientific notation. Some of the calculations task students to divide by a number in scientific notation. If they don't use parentheses, the power of ten will move to the numerator in their calculator. For $3x10^8/(2x10^{-7})$ they'll get $1.5x10^1$ instead of $1.5x10^{14}$).

Besides scientific notation, students also struggle with the scale of the numbers. They don't have a strong sense of what $4.3x10^{14}$ Hz means. This makes it difficult for them to evaluate their answers. When you go through solutions with them, talk about what these numbers are. 10^{14} is 100 trillion. That means that an electron is vibrating 430 trillion times in one second when it emits this type of light (but the electron does not vibrate for a full second so it's actually much fewer cycles). If red light has a wavelength of 0.0000007 m, how does that compare to a strand of hair that's 0.00002 m? The hair is about 30 times larger. Light with 700 nm wavelength has an energy of $2.8x10^{-19}$ J. That's a tiny amount of energy. But a mole of red light would have 170,000 J of energy.

3. "How can the electron get from one side of an atom to another if it's never in the middle?" Students love to ask some form of this question. How can it get from ring 2 to ring 3 without being in-between the two? What you want to stress

Chapter 8 – Atomic Structure

is that the electron is undergoing some initial state of motion and changes to a new state of motion.

We treat the electron as a wave. This means that there can be points (or planes) where the wave does not have any amplitude. The electron can simultaneously be on both sides of that plane without having to cross it. Electrons could tunnel through a space, but that isn't necessary. Think of a standing wave on a slinky as a representation of the electron. Is there a spot on the slinky that doesn't move? Does the slinky move on both sides of that point? Yes! Does the electron have a spot where it doesn't exist? Does the electron move on both sides of that point? Yes!

4. "How do we know what electrons do if we can't see them?" We get information about electrons from emitted and absorbed light. We've spent a century making predictions about what electrons do and testing those from the available evidence. We know quite a bit about electrons without being able to track them visually.

5. "Why does 3d come after 4s?" The 4s orbitals have 3 radial nodes. This means there are three radii where the electron does not exist as you move away from the nucleus. This also means that there is a part of the electron very close to the nucleus before that first node. As more electrons are added, the components of the 4s orbitals do not experience as much repulsion as the 3d electrons that only have angular nodes. Because of the differences in nodes, the 3d and 4s electrons are very similar in energy and will fluctuate which is lowest depending on the exact number of protons and electrons.

6. "Why do chromium and copper have different electron configurations than what the periodic table pattern says?" Cr, Cu, and many other elements have situations where the electron configurations differ from the expected patterns because of either energy required to pair electrons or because of orbital energies changing. The 4s/3d, 5s/4d, 6s/4f/5d, and 7s/5f/6d shells will all vary in order from time to time. La is [Xe] $6s^2\ 5d^1$ and not [Xe] $6s^2\ 4f^1$. That doesn't mean that La can't have either configuration, it just means that the ground state or lowest configuration is [Xe] $6s^2\ 5d^1$. The differences occur from a combination of electron pair repulsion, shielding, and effective nuclear charge. They do not occur due to a magical stability of full or ½ full shells.

7. "Why don't the electrons crash into the nucleus?" Electrons do exist in the nucleus. But this question is really getting at why the electrons don't just go to the nucleus since the nucleus and electrons attract each other? A couple of ideas could be prompting this question. One potential issue is that the electrons (and nucleus a little bit) are moving. If you had a static electron and proton, we'd expect them to move towards each other, but as you increase the motion that path changes. The Bohr model just stipulated that the electrons do not fall to the nucleus because that's what we observe. In classical physics, we'd expect the orbiting electron to emit light (it's an accelerating charge when an electron moves in a circle). As the light gets emitted the electron should quickly lose energy until it stops moving in the nucleus. We can now add to these ideas the fact that electrons being in a single location presents some

Chapter 8 – Atomic Structure

challenges with Heisenberg's uncertainty principle as well as the Pauli exclusion principle.

8. "What does the electricity do to cause light emission (or the flame in flame tests)?" The short answer to this is that the electricity or flame causes the electrons to move into higher energy states more frequently. For electricity this happens via collisions of the particles with each other, and also due to the potential difference that can attract and repel electrons to higher states. The flame has a chemical reaction occurring that causes the particles in the flame to be moving faster as new bonds form. These fast-moving particles collide, and those collisions lead to electrons moving into higher energy states.

9. p_x, p_z, p_y Students struggle to understand the interchangeability of the p-orbitals. There is no universal x-direction. We assign an x-direction. In a set of p-orbitals students will often assume that the p_x orbital is closer to the p_y than the p_z because we draw the set of three orbitals in the orbital diagram with p_y in the middle. The reality is that the three orbitals are equidistant. There is no difference between $p_x^1 \, p_y^0 \, p_z^1$ and $p_x^1 \, p_y^1 \, p_z^0$ unless there is a set reference frame (e.g., magnetic field in xy plane).

Phenomena

1. Light Spectra - It is critical that students see light spectra in this unit. This is the piece of evidence that gets students to understand how the physics for electrons is quantized instead of continuous. Without this you are likely to have student only understand electron configurations and other quantum chemistry concepts as abstract associations. I recommend focusing on hydrogen to enrich discussion of the Bohr model, and because the visible spectral lines are easy to identify and link with the abstract orbital representations.

2. Colorful Spheres Demonstration - Buy a pack of colorful bouncy balls. Have the students throw them to you (clear out any fragile items before you start this!) one at a time. When the red one reaches you do not absorb the "photon." Either dodge it at the last second or rudely ignore it if possible. But when the orange "photon" is thrown, absorb the photon and move onto a chair. Once atop the chair, hide the orange "photon" since it has been absorbed. Upon return to the ground, emit the orange sphere in a random direction.

Repeat this with other spheres. When higher energy "photons" are absorbed, move to an even higher level such as a desk, or a desk with a textbook. Don't forget that energy levels become closer in energy as you get higher.

3. UV light - Obtain a black light (ultraviolet light). Show the students fluorescence caused by the ultraviolet light. Black lights will cause fluorescence in paper currency (try a variety of foreign currency), tonic water (due to the quinine), turmeric (dissolve in ethanol), fluorescein, brown spots on a banana peel, Vaseline glass, scorpions, glow in the dark substances, used glow sticks, and more.

4. Types of light - Demonstrate non-visible light using a black light (#3), having students watch remote controls using their cell phone cameras (infrared light, selfie mode works more frequently), put a bar of Ivory Soap (so pure it floats!) into a microwave, heat a large chocolate bar in the microwave (remove the spinner first), or fly on an airplane (higher levels of X-ray radiation).

Chapter 8 – Atomic Structure

5. Polarizing filters - show the class how polarizing filters block out light based on the polarization of the electric field variance. For an added bonus, put a filter above and below a beaker of corn syrup. The corn syrup is optically active and will cause a variety of colors as you rotate one of the polarizing filters. The corn syrup circularly polarizes light, but it does not impact each color of light equally.

6. Old television - Don't do this with a new television but try if you have an old TV that is ok to break. Take a strong magnet and place it near the screen of a large CRT TV. This works best if you can get the channel that shows just a blue screen. The cathode ray tubes inside the TV will experience an additional deflection of the electron beam that is currently being targeted to the screen. The electron beam is targeted towards blue, green, or red pixels that combine to give you a picture. This is very similar to J.J. Thompson's experiments with cathode ray tubes when he was investigating the electron.

7. Playdough - Have the students construct a "proton" using playdough. Then have them make a neutron and electron to go with it based on what they envision to size of each to be. Use these to question students about how far apart the nucleus and electrons should be for an average sized atom. Either have them work out the scale for themselves or come up with an analogy. A common one is that if the nucleus were the size of a marble, the atom would be the size of a large football stadium. That tiny electron is flying around the concourse with the marble sized nucleus hovering over the 50-yard line. Then draw their attention to a solid material like a desk and ask how a solid can exist with so much empty space in each atom. Be cautious that this representation uses a particle model of the electron which is not ideal when teaching orbitals.

8. Candium - Buy 3 types of candy to use as isotopes for a new element called candium. Have the students determine the average mass of a single piece of candium. Then walk them through how to determine the % abundance, relative abundance, and relative mass for each "isotope." The sum of the relative masses should be equivalent to the average mass they found by dividing the total mass by the total number of pieces. If you teach mass spectrometry, ask them to determine how these isotopes would separate in a mass spectrometer.

9. Flame tests - Light a set of Bunsen burners. Have the students put salt solutions into the flame. This can be done by soaking wooden splints in solutions of the salts, or by adding the solutions to small spray bottles. Do not use methanol for this! Every year an unexpected flash ignition of methanol causes harm to chemistry students somewhere.

10. Paramagnetism and diamagnetism - Fill a test tube ⅓ of the way with $MnSO_4$ and another test tube ⅓ of the way with $ZnSO_4$. Suspend each test tube from a string and wait for both to become still. Bring a strong magnet (Nd works well) near the $ZnSO_4$ to show that nothing happens. Then bring the strong magnet by the $MnSO_4$ to show that the Mn is magnetic due to the unpaired 3d electrons.

Chapter 8 – Atomic Structure

Flashcards

$^{56}_{26}Fe^{2+}$

How many protons, neutrons, and electrons?

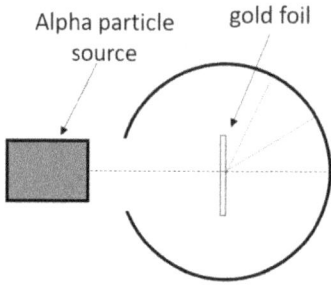

Alpha particle source

gold foil

What would change if the gold foil were thinner?
What would stay the same?

$^{A}_{Z}X$

13 protons
14 neutrons
10 electrons

What should the symbol be?

Chapter 8 – Atomic Structure

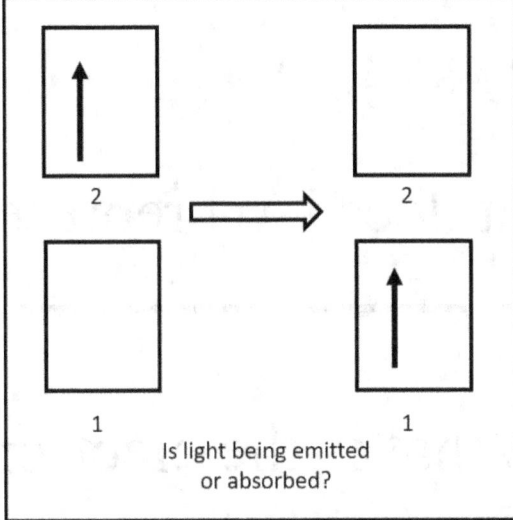

Is light being emitted or absorbed?

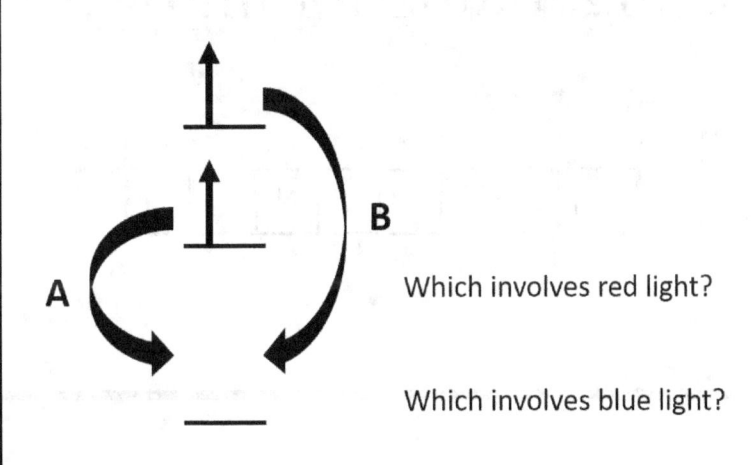

Which involves red light?

Which involves blue light?

What does the "3" tell us?

Chapter 8 – Atomic Structure

[Ar] 4s^1

What does [Ar] represent?

What is the electron configuration for nitrogen?

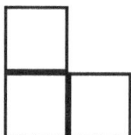 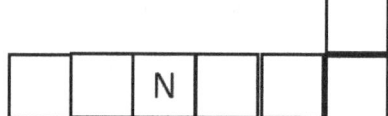

Why is there no orange light for the H-emission spectrum?

Part II

Advanced Chemistry

"No, we are in a jungle and find our way by trial and error, building our road behind us as we proceed. We do not find signposts at crossroads, but our own scouts erect them to help the rest."

- Max Born

Chapter
Periodic Trends

History

The periodic table of elements is a critical feature of chemistry. It represents the history of scientists discovering elements while serving as a roadmap for why different chemical reactions happen. Teachers should aim to share its history and value with students. There are two major milestones of periodic table history that all students should know about.

The first periodic table is credited to Dmitri Mendeleev even though he did not make the first periodic table. If you go back too far before Mendeleev we did not know of as many elements. There is the famous 4 element theory from the Greeks (air, fire, water, earth). Boyle pushed for a superior definition of "element" in the Skeptical Chymist.[1] Berzelius constructed symbols for the elements. And Cannizzaro brought atomic weights for 62 elements to the Karlsruhe conference in 1860. As we discover more elements, a few people began to organize them. Mendeleev did not make the first organization of the elements, but he did make the best ones relative to others during that time period.

Johann Döbereiner communicated in 1817 that he had found that calcium, strontium and barium had a pattern. Strontium had an atomic mass that was the average of the other two.[2] Later in 1826 Jacob Berzelius published more atomic weights that led to a total of 4 such groups, but also many elements that did not fit into any of these triads.

The event that kicked off the first major periodic tables was between 1858 and 1860, Stanislao Cannizzaro measured and published atomic weights for the known elements at the time. The first periodic table to follow that demonstrated periodicity was Alexandre-Émile Béguyer De Chancourtois in 1862. This periodic table was a cylinder with patterns every 16 elements. De Chancourtois was not well connected since he was a geologist and not a chemist or physicist.[3] Very few people knew about his arrangement.

Next John Newland made an arrangement in 1863 of 37 elements in 10 groups. Two years later he made a revised version with 8 groups for 62 elements. This was the one where he compared the elements to octaves in that every 7 elements saw a return to a similar element. At the time there were no known noble gases and many of the groups involved transition metals. Newland is generally considered to be the first to discover periodic law. De Chancourtois did first but had several large flaws in his arrangements (for example he used a giant carbon in place of a missing element) and received little recognition for his publication.

Then came the two best periodic tables. One was made by Dmitri Mendeleev and the other by Julius Lothar Meyer.

Some key components for Mendeleev's periodic table were that he predicted nine elements that had not been discovered yet. Four of those predictions came to fruition with their properties mostly matching what Mendeleev had come up with. Gallium, or eka-aluminum as Mendeleev called it, was discovered in 1875. The close match between Mendeleev's predictions and the properties of gallium sparked a rapidly growing interest in the periodic table. In 1886 germanium (eka-silicon) was

Chapter 9 – Periodic Trends

discovered and again had nearly the same properties as Mendeleev had predicted. Mendeleev pushed back on Paul-Émile Lecoq de Boisbaudran claiming that his measured density of gallium must be incorrect and that he should remeasure. Lecoq did in fact remeasure and this was a prime example of theory correcting the experimental result.[4] The density was updated from 4.7 to 5.9 g/cm^3.

Mendeleev also correctly swapped the positions of tellurium and iodine. At Mendeleev's time they only knew the relative masses of the elements. Because tellurium has more neutrons on average, tellurium has a higher relative mass in spite of one fewer proton than iodine. Mendeleev correctly grouped iodine based on its properties and trends with chlorine and bromine. Meyer also made this correction and so had William Odling. Mendeleev incorrectly reasoned that tellurium's atomic mass had been incorrectly determined. But Mendeleev had made the same claim about boron and later the mass of boron was found in alignment with Mendeleev's prediction.

Meyer published a periodic table before Mendeleev, and he did so twice. His first was in a textbook he wrote in 1862 and published in 1864. This periodic table had 28 elements, included periodicity, switched Te and I based on chemical properties, and predicted a couple of missing elements. In 1868 he made an expanded version that included 52 elements. This was published prior to Mendeleev's 1869 periodic table that has given Mendeleev the status of being the first.

Should Mendeleev be credited as the founder of the periodic table? I would argue yes, but it's complicated. He was not the first, but he made the best of the first ones with a substantial number of bold predictions. Many of those predictions came to fruition, sometimes for the wrong reasons. One thing that Mendeleev did better than many others is that he continued to work and revise his periodic table. Over his career he published about thirty different versions and he worked to include the noble gases when they were discovered much later than his initial periodic table. For his efforts he has element #101 named after him (mendelevium). Not too shabby for the youngest of 17 children.

The development of the periodic table is a wonderful opportunity to teach students about the progression of science. Here we see a variety of scientists each making contributions that are flawed but help us progress. The scientists make predictions, use feedback and evidence to revise those predictions, and the credit is excessively focused to a few individuals.

The second critical milestone for the periodic table was when Henry Moseley (1913) discovered an answer key to the periodic table. Prior to Moseley, element discoveries were uncertain and controversial. But Moseley found that you could determine the nuclear charge of an atom using X-rays. Effectively this tells us the number of protons. The periodic table was organized as we now have it where elements are aligned by their number of protons or nuclear charge. When the dust settled there were seven missing elements between #1 hydrogen and #92 uranium.

Scientists began searching for those seven elements. Protactinium was discovered first (1913) by Kasimir Fajans and Otto Göhring but later a more stable isotope was isolated (1918) by Lise Meitner. Meitner is credited with the discovery even though Fajans and Göhring discovered and isolated the element first. They had named the element brevium in 1913 due to its very brief half-life.[5] The question of what counts for discovering an element is an interesting one that only grows more complicated as we search for elements with shorter and shorter half-lives. The isotope

Chapter 9 – Periodic Trends

for brevium had a half-life of 1.7 minutes while the isotope that Meitner isolated has a half-life of 32,500 years.

As I write this book, the periodic table is complete. There are 118 elements, and all have been verified and named. Yuri Oganessian had a hand in discovering thirteen of them and he is currently the only living person to have an element named after them. Oganessian was hired to be the chief engineer for Georgy Flerov who now has element 114 named after the lab that is named after him (flerovium).[6]

The next element to be discovered will likely be #119, #120, or something close. Its placement will be controversial. Will we create a new row? Do we just add a small list of elements 119 and beyond as they are confirmed? The periodic law becomes meaningless with these atoms as the nuclei barely last long enough for electrons to form around them before they decay. Oganesson is theorized to not have electrons in shells.[7] Instead of a typical electron configuration, the electrons would be delocalized over the atom. These superheavy elements have so many protons that electrons would move at relativistic speeds causing highly unusual deviations from their groups if they were able to last long enough to be studied.

*****

Periodic aliens (Fig. 9-1) is a great introductory activity for the history of the periodic table. Students are given a set of cards that have one alien per card. The aliens have varying numbers of fingers, foot length, hair, and body shapes. But when they are correctly organized there are consistent patterns for each down a group and across a period. But two aliens are missing. Students can predict which ones are missing and their properties much like Mendeleev and other scientists did when constructing the first periodic tables.

Figure 9-1: Periodic aliens activity

Teaching the periodic table is also an opportunity to share stories about elements with students. Share with them the alkali metals, the halogens and the noble gases. When we fill double pane windows with argon instead of air our energy losses decrease. This is because argon has a lower specific heat capacity, so it takes less energy each trip from the one pane to the other. Or when they ask if astatine is a halogen, they can learn that there are so few atoms of astatine in the earth because how unstable the nucleus is that its chemistry is irrelevant. Let them know that some lanthanides are going to be highly impactful as our needs for rare earth metals increases as we use more technology. But those metals are not evenly distributed, and the politics and economics of their mining is about to become very interesting.[8] Or let them know how much total gold is dissolve in all the oceans. There might be less than a billionth of a gram of gold for every liter of ocean water, but there are many liters in the ocean(s).[9]

Chapter 9 – Periodic Trends

Nuclear charge, Energy Levels, and Shielding

Periodic trends is a fundamental concept to chemistry. If you want a student to do well in organic chemistry, this is the critical starting point. At some level, the majority of chemistry can be reduced to opposite charges (+/-) attract while like charges (+/+ or -/-) repel.

Atoms get smaller moving across a period of the periodic table. That is weird. You are adding stuff, yet the atom gets smaller. Students should find this odd. Teachers should present this as a discrepant event. It can be worthwhile for students to propose reasons why lithium ($3p^+$, $4n^0$, $3e^-$) has a radius that is over twice as big as neon ($10p^+$, $10n^0$, $10e^-$).

Unpacking this trend reveals an interesting observation. Lithium atoms have 3 electrons, and yet have an atomic radius more than double the radius of the fluorine atom (Fig. 9-2). Part of the reason is that there are more protons in the nucleus of the fluorine atom. Those additional protons cause a greater force inward on the electrons. But there are also way more electrons in the fluorine atom. Why do they not add a substantial amount of repulsion?

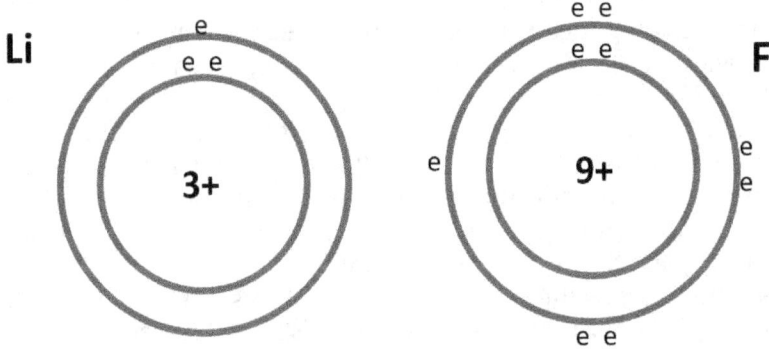

Figure 9-2: Diagrams showing the number of protons and electrons in Li and F atoms

Bohr model representations are misleading (Fig. 9-2). For example, there are two electrons in the first energy level (closest to the protons). In the Li atom, they are drawn right next to the valence electron. But even if we resort to using a particle model to describe the electrons, the electrons aren't static. They move. So sometime later those two electrons might be in a completely different spot (maybe off of the page in the 3rd dimension). Electrons move really fast, and the actual size of those drawings has been scaled up tremendously. The organization, the 2-dimensions, and the lack of motion all lead to misleading concepts about the differences in repulsions between electrons. We'll look at wave representations later.

When we make the claim that electrons in the same energy level exhibit little repulsion towards each other, it's important to know that we're working from a foundation that has some issues. Our ability to reconcile this claim could be challenged by a wide variety of misconceptions.

The big question we want to address is why is there less repulsion between electrons in the same energy level than between electrons in different energy levels? I know the picture has a lot of flaws, but if we look at an electron pair in one level, how can they barely repel each other? They are so close. If electrons are close, they should repel a lot!

Chapter 9 – Periodic Trends

It turns out that they do repel each other a lot. But there is some interesting math behind that repulsion that we see in physics. For example, if you tunnel toward the center of the earth, the gravity force you would experience would decrease. The same amount of stuff is pulling on you, but now some of the earth is pulling you one way and the rest is pulling you the other. In fact, the total amount of force acting on you is the same, but the net force has decreased since you are being pulled in many directions. Electrons in the same energy level have a similar tendency. They repel the other electrons in that level, and they repel a lot, but they repel in different directions resulting in a diminished net force.

Take the lone valence electron from fluorine. Look at how the core electrons (Fig. 9-2) repel that electron outwards away from the nucleus. But the valence electrons repel the electron in directions that are more left and right, rather than outward. Even if the electrons move around, you see that this direction of force is not based on a unique position. Core electrons repel valence electrons away from the nucleus. Valence electrons repel each other perpendicularly to the nucleus.

We have a term for that repulsion, and I wish that we didn't. The word shielding is both misleading and unnecessary. The term shielding to a student implies a shield or a barrier. What shielding means is repulsion between electrons. Students have too little experience with electrical forces to use terminology such as shielding. Expect misconceptions.

When you tell students that electrons in the same energy level do not repel each other much, they become intrigued and curious. But when you tell students that electrons in the same energy levels do not shield each other much, they construct ideas about electrons in the core blocking forces from the nucleus. This wrong idea still allows for appropriate explanation of trends for what it's worth.

Students will sometimes show these misconceptions in responses. They write things such as, "Fluoride (F-) is larger than F because there is an extra electron. The additional electron means that there is less force from each proton on each electron because there is a different ratio." The additional electron does not change the forces between the protons. It only adds repulsion to the other electrons. Students often hint that each proton pulls on 1 electron and that if you shift away from the ratio of 1 proton for every 1 electron, problems result where no proton is pulling on one of the electrons.

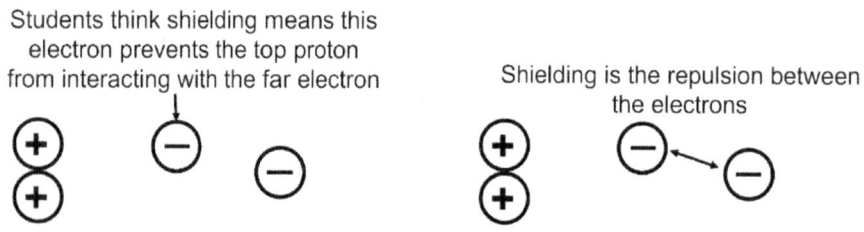

Figure 9-3: What students perceive shielding to mean vs. what shielding means.

Two charges have an attractive force between them (Fig. 9-4). When a 3rd charge is placed in between them, that force is still exactly the same (Fig. 9-5). There is no "shielding" where the 3rd charge blocks the interaction. Instead, there is now a new repulsive force between the 3rd charge and the positive charge. There is also a

Chapter 9 – Periodic Trends

new attractive force between the 3rd charge and the negative charge. But the original force is unchanged, it just has new forces that may affect the total net force.

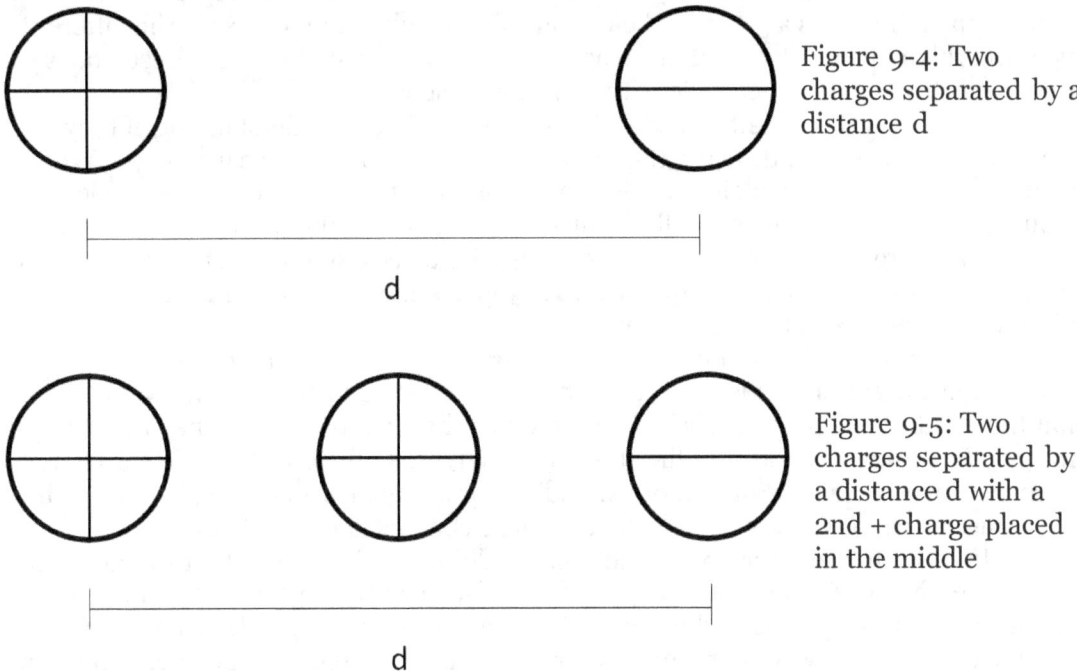

Figure 9-4: Two charges separated by a distance d

Figure 9-5: Two charges separated by a distance d with a 2nd + charge placed in the middle

There are three atomic variables that explain most periodic trends.
1. The number of occupied energy levels has implications for how far the valence electrons are from the nucleus.
2. The shielding (repulsion) experienced by electrons causes the valence and other electrons to experience a force away from the nucleus.
3. The nuclear charge or number of protons determines how much force toward the nucleus.

These three variables overlap at times. If a large number of energy levels is occupied, then the shielding that valence electrons will feel is going to be higher. The atom will also likely have a larger amount of nuclear charge. The nuclear charge and shielding can be combined into a single term called effective nuclear charge. Effective nuclear charge is determined on an electron-by-electron basis. In the same atom, different electrons will experience difference effective nuclear charges.

Effective nuclear charge (Z_{eff}) is the difference between the pull in from the nucleus and the repulsion away from the nucleus by other electrons. Formally we write this as $Z_{eff} = Z - S$ where Z is the number of protons and S is the amount of shielding. Early on the amount of shielding can be estimated as the number of electrons in lower energy levels than the electron that is being analyzed. This has limitations and those limitations will show up to cause some exceptions in your trends data later. Electrons in the same energy level only experience about 15% of the repulsion relative to the attraction they feel from 1 proton in the nucleus. The electrons in the energy level one beneath the electron of study show about 85% repulsion. The exact amounts vary depending on orbital, nuclear charge, and energy

Chapter 9 – Periodic Trends

level. From there the repulsion from an electron 2 or more levels beneath is nearly equivalent to the attraction of a proton.

But these amounts are not consistent. A 3s electron will not repel a 4s electron by the same amount as a 3p or a 3d electron. If you look at the nodes for the different types of orbitals, you will find that s electrons have more radial nodes. These nodes cause s electrons to have more density near the nucleus.

Students will get better feedback and develop better understanding if they consistently explain trends using the three tools (shielding, n, nuclear charge). We want to avoid students explaining the trends using the trend. For example, students shouldn't say that fluorine is smaller than chlorine because the trend is that as you move down a group, atoms get larger. If we use the three criteria, teachers will be able to discern confusions that students have. They give students an organized set of criteria to evaluate and explain trends.

One problem that students have is using the trend to influence their explanation for group trends. If we ask why Na is larger than Li, students will explain that the number of energy levels has increased so the electrons are further from the nucleus. But some will add that the shielding is larger, so the effective nuclear charge is smaller for Na. Na does have more shielding for its valence electrons, but it also has more nuclear charge and so the effective nuclear charges are equivalent (+1).

In order to see where your students are doing this it is important to have them explain trends down groups, across periods, for isoelectronic species, for ionic species, and for exceptions to the trends. Use warmups, exit slips, flashcards, and low-stakes quizzes to give them multiple opportunities for retrieval practice. This is a good topic to use spaced practice, and to teach students about interleaving.

Fe atoms are larger than Zn atoms. Identify whether the following criteria are equal, larger, or smaller for the valence electrons of iron:
<u>Shielding</u>   <u>nuclear charge</u>   <u>effective nuclear charge</u>   <u>energy levels</u>

When students understand shielding, nuclear charge, and energy levels; the explanations of trends start to be redundant. Students must practice explaining how the nuclear charge, the electron-electron repulsions, and the number of occupied energy levels combine to influence trends in electronegativity, ionization energy, and electron affinity. Give them the definition of the next trend you want them to learn about, give them some data and have them construct explanations as to what is happening. If they struggle, have them draw out diagrams with the energy levels as we did earlier.

The key framework that all these trends devolve into is that some atoms and ions pull on electrons more than others. The relative amount of pull can be explained using Zeff and n while the pull on electrons connects a wide umbrella of topics. We can explain the periodic trends, but we can also explain much of chemistry by the fact that some atoms pull on electrons more than others. We want students to recall these ideas even as we proceed to these other topics, so we want to make sure that we link the ideas in this section with the images used for others (e.g., Lewis structures).

Periodic Trends

Which atoms are bigger than others? Are they a lot bigger? The smallest atom is helium with a radius of 38 pm while francium is the largest (so far) at 348 pm. Why

Chapter 9 – Periodic Trends

don't they just get progressively larger as the number of protons, neutrons, and electrons goes up?

Before we look at why the trends for atomic radii are what they are, we really should start by defining what we mean by radius. Remember that an atom is made of a nucleus and electrons. Those electrons are not static points. The cutoff of where an atom is not obvious. One method of determining the radius of an atom is to take a diatomic molecule of the element and use half of the distance between the nuclei as the radius. But a problem with that is that the atoms overlap slightly and so this underestimates the radius (Fig. 9-6). The point here is not to be nitpicky, but to clearly illustrate to the student that there is not a definitive endpoint to the atoms. So, any claim about a radius should be taken with some uncertainty in mind.

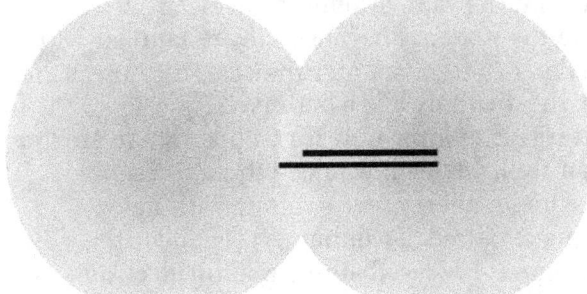

Figure 9-6: Two atoms overlapping where the top line shows halfway between the nuclei and the lower line shows the radius of one of the atoms.

Give students atomic radii data. Have them make observations, come up with group and period trends, and find exceptions to those trends. This is superior to telling them that atoms get bigger as we move down or left. When they look at numbers, they can notice how the alkali metals have a bigger size difference with the alkaline than the halogens and noble gases do. They'll get to see how there are inconsistencies in the f-block. They'll have a visual that will help them to dual code this information.

Next have them construct an explanation for the trends down a group and across a period. They'll want to use energy levels, shielding, and nuclear charge in their explanations. Which atoms are the biggest? Why? Which atoms are the smallest? Why?

Across the period the occupied energy levels are constant, the shielding is either constant or nearly constant, and the nuclear charge increases. Thus, the size of the atoms decrease across a period because of the increase in proton pull with minimal change in shielding. Decide ahead of time whether you want to treat the shielding as constant or nearly constant.

Assuming that all core electrons show the same repulsion as a proton attracts would result in effective nuclear charges of +1 for lithium and +7 for fluorine. But using Slater's rules (a more precise shielding model) the effective nuclear charges would be +1.279 for lithium and +5.10 for fluorine.

Down a group the number of occupied energy levels increases, the nuclear charge increases, and the shielding increases. The net effect of the nuclear charge and shielding increases mostly balance so we often focus on the energy level changes as the primary factor. If you use Z_{eff}, you will find some students will make errors on how shielding and nuclear charge vary down a group even if they determine and explain Z_{eff} correctly.

*****

Chapter 9 – Periodic Trends

    First ionization energy is the minimum energy required to remove an electron from a gaseous atom in the ground state. The reason why the term gaseous is included is that if the atoms are in contact with something, the energy required will shift. For example, a sodium atom in a metal lattice has a conduction band that is easier to remove an electron from than a single sodium atom. The reason why the word first is included is that more than one electron can be removed.

    For this trend I start helping students make connections for future topics. Redox chemistry focuses on the movement of electrons, and we want to prime that now by making it clear that halogens are excellent at pulling on electrons while alkali metals exhibit much weaker forces on electrons. While we're learning about first ionization energy trends, I want that point to come out. Many chemical reactions can be explained by how good the different chemicals were at pulling on electrons. The order of the activity series is a good bridge for students because they're familiar with that and this will help them connect reactions to trends. Later they can reconnect the two when we look for how to make the highest voltage for a battery.

    Students sometimes draw toward rote memorization for trends. There are too many similar variables for this to function well. Instead, we need them to use the three core principles (shielding, nuclear charge, energy levels) along with an understanding of the different trends being explored. To enhance this, focus the different trends along one theme. That theme is how well atoms pull on electrons. We want them to understand that francium is the worst at pulling on electrons (or Cs if you want something more common), and that the corner by F, He, and Ne is the best at pulling on electrons. Students have prior knowledge and experiences with similar phenomena (e.g., tug of war) that will help them chunk information better with this theme.

    The real fun for ionization energy starts when we begin taking away more than 1 electron. This is an opportunity to provide evidence of energy levels and show how shielding varies depending on the electron being discussed.

$$Na\ (g) \rightarrow Na^+ + e^- \quad\quad 1st\ IE = 496\ kJ/mol$$
$$Na^+ \rightarrow Na^{2+} + e^- \quad\quad 2nd\ IE = 4560\ kJ/mol$$
$$Na^{2+} \rightarrow Na^{3+} + e^- \quad\quad 3rd\ IE = 6910\ kJ/mol$$

What do you notice for Na? There is a substantial increase between the 1st and 2nd ionization energies. The 3rd is largest but does not show the same relative increase.

$$Mg\ (g) \rightarrow Mg^+ + e^- \quad\quad 1st\ IE = 738\ kJ/mol$$
$$Mg^+ \rightarrow Mg^{2+} + e^- \quad\quad 2nd\ IE = 1450\ kJ/mol$$
$$Mg^{2+} \rightarrow Mg^{3+} + e^- \quad\quad 3rd\ IE = 7730\ kJ/mol$$

What do you notice for Mg? How do these compare to the ionization energies for Na?

$$Al\ (g) \rightarrow Al^+ + e^- \quad\quad 1st\ IE = 578\ kJ/mol$$
$$Al^+ \rightarrow Al^{2+} + e^- \quad\quad 2nd\ IE = 1820\ kJ/mol$$
$$Al^{2+} \rightarrow Al^{3+} + e^- \quad\quad 3rd\ IE = 2740\ kJ/mol$$

    In each set the ionization energies always increase. As an electron is removed there is less repulsion on the remaining electrons. This means that the next electron

Chapter 9 – Periodic Trends

will always be more difficult to remove since there is less pushing it away. Note that this is contingent upon some repulsion from electrons within the same energy level. If you teach shielding within an energy level to be 0 this will conflict with your instruction.

Next, we want to make sure that students notice the large increases. Sodium has the ionization energy increase by a factor of 9 between the 1st and 2nd. Magnesium has an increase by over a factor of 5 between the 2nd and 3rd. Aluminum does now show such an increase, but from 3rd to 4th would exhibit an increase by a factor of 4. We want to explicitly connect this to the charges that these elements form. Students likely learned this in the formula writing and naming unit. Now we want them to explain why Na forms a 1+ charge and not a 2+. We want them to explain why Mg forms a 2+ charge and not a 1+ or 3+ charge. We can also remind students that elements are also sometimes neutral when nothing is available to take these electrons.

For magnesium atoms the first two electrons removed are from the 3s sublevel. These electrons have 10 core electrons repelling them and 12 protons attracting them towards the nucleus. But the third electron removed is from a 2p subshell. This electron only has 2 core electrons repelling it but still has the 12 protons attracting it towards the nucleus. This and the lower energy level being closer to the nucleus make it substantially more difficult to remove the electron. Including the removal of the 3rd electron helps minimize the misconception about half-filled and filled electron shells having a magic stability.

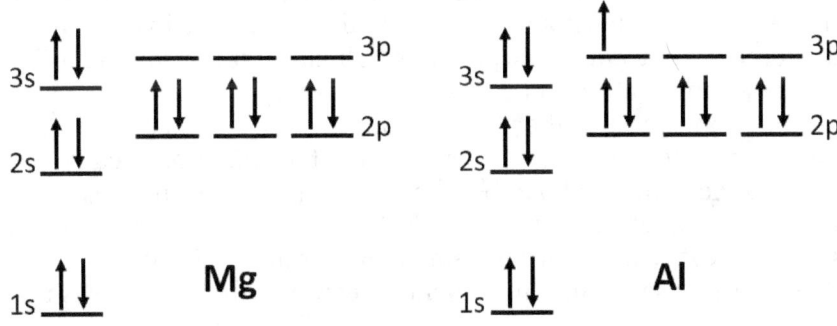

Figure 9-7: Orbital diagram for magnesium and aluminum

Aluminum has three valence electrons. Those electrons experience a much smaller effective nuclear charge than a fourth electron in the core energy levels. But we should also notice something new about aluminum's first ionization energy.

While moving through ionization energy trends, there are discontinuities. From sodium to argon there is a trend that the first ionization energies increase. But for some reason that is not the case for aluminum. Aluminum's first electron removed is from the 3p subshell. This electron has 13 protons attracting it towards the nucleus and yet is somehow easier to remove than magnesium's first electron removed from the 3s subshell. This would imply that the 3p subshell experiences a larger amount of electron repulsion than the 3s subshell.

Chapter 9 – Periodic Trends

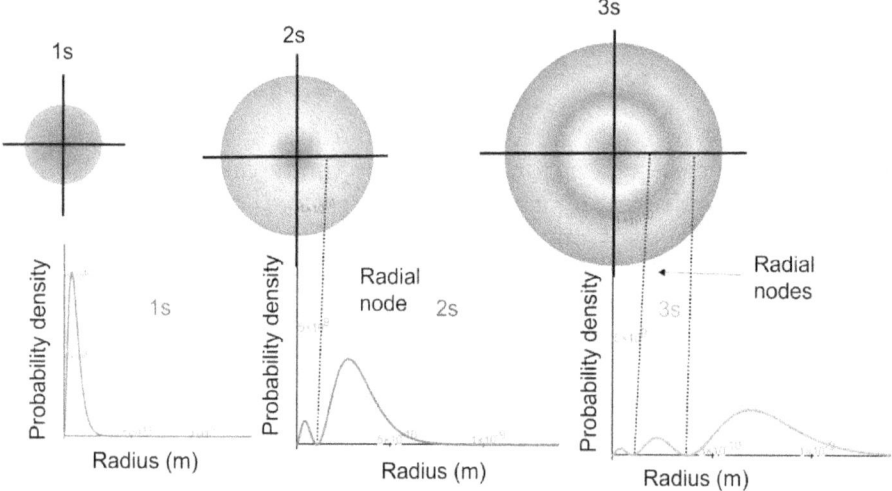

Figure 9-8: A 3s orbital has a substantial amount of electron density near the nucleus

Remember that s-orbitals have one more radial node than p-orbitals. This means that there is more electron density closer to the nucleus (Fig. 9-8). That electron density near the nucleus experiences far less shielding. In the last chapter I had mentioned that nodes were a great subject for curious students to explore more. The connections in the periodic trend data help clarify these abstract ideas. Note that the trend of ionization energies being smaller for p-orbitals is countered with the trend to increase the number of protons. Silicon shows a higher 1st ionization energy than Al or Mg, but Si has a combination of the two factors at play.

Nitrogen (1400 kJ/mol) has a higher first ionization energy than oxygen (1310 kJ/mol). This does not occur because energy levels that are half-filled or filled have a magic stability. This has to do with repulsion between electrons paired in the same orbitals. There are quantum limitations for paired electrons with opposing spins that cause the electrons to have restricted positions relative to one another. These restrictions cause them to be closer than other valence electrons which increases their repulsion.[10]

*****

Electronegativity was the key for me doing well in organic chemistry. When I would work with other students in the class, my anecdotal impressions were that they struggled to visualize which portions of the molecule would be positively and negatively charged. Not only do you want students to know the trends for electronegativity, but you also want them to connect electronegativity to Lewis structures. This is especially important for elements common to organic chemistry. When you draw a Lewis structure, they should visualize which portions have positive and negative charges.

Electronegativity correlates with first ionization energy unless the element does not participate in covalent bonding. I find it best to avoid the noble gases. Some do bond, but when they do the bond formation involves some anti-bonding molecular orbitals that make trends overly complicated. Xenon has an electronegativity of 2.6 which is very similar to iodine (2.7). But if we travel down that road, we're adding in too many new features at once and the return is minimal.

Chapter 9 – Periodic Trends

To teach about noble gas electronegativity you must start with how we determine the values in the first place. Linus Pauling first derived a method where he could compare the average bond enthalpy for two diatomic molecules with the bond enthalpy of the different elements.

$$H_2 \text{ (g)} \rightarrow 2H \text{ (g)} \quad \Delta H = +436 \text{ kJ/mol}$$
$$F_2 \text{ (g)} \rightarrow 2F \text{ (g)} \quad \Delta H = +159 \text{ kJ/mol}$$
$$HF \text{ (g)} \rightarrow H \text{ (g)} + F \text{ (g)} \quad \Delta H = +567 \text{ kJ/mol}$$

The average bond enthalpies for diatomic fluorine and hydrogen is 298 kJ/mol while the bond enthalpy for hydrofluoric acid is 567 kJ/mol. Pauling postulated that the larger the difference in bond enthalpy corresponds to a larger difference in electronegativity. Every difference of 1.0 in electronegativity would lead to a difference of approximately 100 kJ/mol in the bond enthalpy comparisons.[11]

This can be rationalized by starting with two elements that have the same electronegativity. If these two elements are used to form a molecule, they should have the same strength of bond since both pull on the shared electrons with equivalent force. But as the electronegativity changes the electrons are pulled closed to the more electronegative element and this causes the bond to be more stable.

This method becomes more complicated when any of the molecules involve double or triple bonds, but hopefully this gives you an idea. Now if we apply that to noble gases you can see why we have trouble. Forming a diatomic molecule of Xenon is not going to happen. There are other methods that have since been developed that incorporate the combination of electron affinity and ionization energy to get a sense of how much an atom can attract a new electron (EA) along with how much energy is required to remove an electron. The results show that noble gases do have decently large electronegativities and much of that stems from their high ionization energies and strong attractions to their current electrons. But the bonding is not as simple as adding a $6s^1$ electron to a Xe. Instead, we are looking at a molecular orbital arrangement.

Electronegativity has far reaching implications for acid and base chemistry as well as organic chemistry. Carbon and hydrogen both have middling electronegativities that provide a flexibility to shift electron density towards or away from these elements. We can think of electronegativity as a tug of war on the bonded electrons where Z_{eff} and n combine to determine the relative strength of each pull.

That tug of war can be expanded into redox chemistry where we study how electron density shifts. Oxidation states are a means for us to track where electron density is high and low. The basis for oxidation state assignments is placing electrons with the element that has the higher electronegativity. When oxidation and reduction occur, we are moving electrons from one chemical species to another and the primary mechanisms for this correlate with our electronegativity trends.

The energy tracking during these changes also connects electronegativity to thermodynamics. Electronegativity is a foundational concept. When students understand a foundational concept well, it gives them a starting point to draw from that allows them to make future connections in more difficult content.

*****

Electron affinity has the most erratic trends on the periodic table. This is really a topic you want to consider your objectives before you start teaching. My

Chapter 9 – Periodic Trends

principal goals for electron affinity are that students can define the term using a reaction, and that they know the halogens have large electron affinities.

$$Cl\ (g) + e^- \rightarrow Cl^-\ (g) \qquad EA = -349\ kJ/mol$$

Chlorine atoms exert a large force on an additional electron. This correlates to large electron affinities, but with more inconsistencies than other trends. Some of the best elements at pulling on electrons have small electron affinities. Having an excess of charge in a small volume is a large constraint and so we find exceptions such as fluorine having a smaller electron affinity than chlorine.

Nitrogen has a slightly positive electron affinity meaning it requires energy to add an electron. This is due to the repulsion experienced when the new electron is paired with another electron in a 2p orbital. Magnesium has a nearly 0 electron affinity as the new electron is going to an empty 3p subshell where the new electron experiences nearly even nuclear pull inwards and electron repulsion outwards. In this particular case I fear we are shifting too far from chemistry. Magnesium and many other metals are unlikely to form negative charges and we really aren't setting up anything by exploring these electron affinities.

I do think there are two valuable lessons within electron affinity trends. Many of the elements that you would consider to be good at adding electrons have largely negative electron affinities. The halogens and other nonmetals typically have the biggest electron affinities. The other lesson is that we want students to know about electron affinities for the purposes of understanding ion formation. This will be foundational for Born-Haber cycles, redox chemistry, and electrostatics in physics.

Oxygen is an interesting example in that the first electron affinity is largely negative, but the second electron affinity is positive. It takes energy to move an electron closer to a negatively charged object (O^-).

$$O\ (g) + e^- \rightarrow O^-\ (g) \qquad EA = -142\ kJ/mol$$
$$O^-\ (g) + e^- \rightarrow O^{2-}\ (g) \qquad EA = +844\ kJ/mol$$

This means that in order for oxygen to form a 2- charged ion that something else somewhere had to input some of that energy.

Even more confusing to me than electron affinity trends are those of metallic character. I don't see any reason why more specificity is better than just separating the periodic table into nonmetals, metalloids, and metals. The argument that copper is more metallic than zinc doesn't strike me as serving much purpose in education.

There is an underlying trend that is sometimes grouped with the metallic and nonmetal character. This trend is that as you move down and left towards cesium that metals become more reactive in losing electrons to other species. As you move up and right towards fluorine that nonmetals become more reactive in taking electrons from other species. To me this is important yet redundant from trends in ionization energy, electronegativity, and electron affinity.

There are a couple more trends that I feel are better connected when discussed in bonding rather than in this section of periodic trends. The melting points of alkali metals and halogens is one. The melting points of the alkali metals is related to their sizes but is also contingent upon understanding a metallic bond. The melting points of halogens is based on the strength of dispersion forces. Both of these topics are better to address in bonding because they are too disconnected from our original three criteria of nuclear charge, occupied energy levels, and electron shielding.

Chapter 9 – Periodic Trends

There is also a trend of metal oxides to nonmetal oxides as we move across the periodic table about how these oxides react with water. The metal oxides form basic solutions, and the nonmetal oxides form acidic solutions when aqueous. This trend is intriguing because you move from basic (metal oxides), to amphoteric (metal oxides near the metalloids), to acidic (nonmetal oxides). Additionally, there are trends from P to S to Cl that show increasing acid strength.

But there are also inconsistent trends with the nonmetal oxides because the number of oxygens attached change things. H_3PO_4 is weaker than H_2SO_4 which is weaker than $HClO_4$. But that is not the case for $HClO$ or $HClO_2$. Students potentially do not know much about acids and bases by this unit. But we could use a simple method of comparing acids such as the pH of equimolar solutions. There are pros and cons to including this trend here and which is the better option depends on what the goals for the class and teacher are.

If students can tie this back to earlier units, they will retain information better. This will lead to better application in future units. You want them to know that if calcium bromide is $CaBr_2$, that Magnesium chloride will have the same ratio of ions ($MgCl_2$). If they swap out an element for another from the same group that there will be patterns that can be used to make predictions. These substitutions don't always work (e.g., there is SF_6 but not OF_6), but students will have observed these substitutions frequently during chemistry and it is worth drawing their attention to them now.

At the conclusion of periodicity, we want students to connect the models we've built for trends with the models we use in other subjects. They want to connect a picture of an atom that has a large effective nuclear charge for its valence electrons with the symbol F in the formula CH_3F. We want them to look at a Lewis structure for HCl and to see the shift in electron density because of the electronegativity differences. We want them to begin to visualize the implications for what that does to the bond itself as well as the charge surrounding the H and Cl atoms in the molecule.

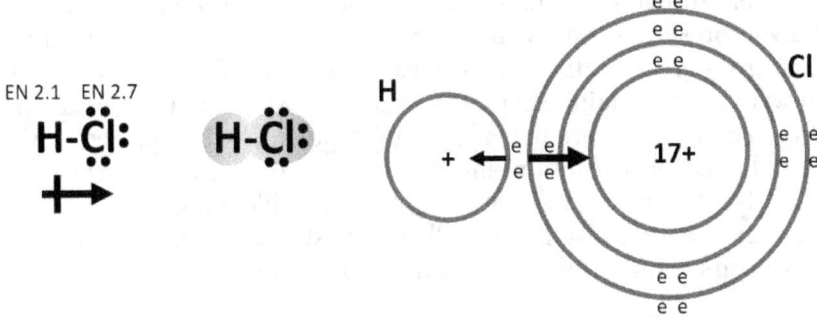

Figure 9-9: Various representations showing electronegativity differences for HCl

This will help them have a conceptual framework for why the redox potentials are what they are in batteries. This will help them with reaction mechanisms in organic chemistry. This will help them understand why some reactions are exothermic and others are endothermic when we learn about enthalpy.

A common complaint I have towards students is that they respond to difficult questions in chemistry by saying, "I have no idea where to start." *Periodic trends is*

Chapter 9 – Periodic Trends

a great place to start. What do you know about the elements involved? How well do they attract electrons? What are they similar to? What properties do these types of elements have? Key properties that distinguish metals and nonmetals stem from their abilities to pull on electrons.

My anecdotal experience in organic chemistry was that many students found the course to be overwhelming. But a small minority of students found it to be somewhat simple. Many of those students were also in future chemistry courses with me. In my judgement, we saw more when looking at structures than the students who struggled. We saw electron deficiencies and excesses while struggling students saw each new thing as completely independent from the last. We noticed more patterns whereas they had to memorize everything as if it were brand new.

Student Struggles

1. *"I have no idea what I'm doing!"* Students who struggle to determine trends are best served to circle back to the trends in atomic size. This is the simplest concept for a trend. Students can easily visualize atoms getting bigger down a group or smaller across a period. Many visual representations of data exist that can be shown, but these are superior when observing atomic size because the trend does not include an additional step of rationalizing ionization energy or electronegativity.

From here, students require an understanding of how amount and position of charged particles combine to determine what these trends will be. Electrons repel other electrons. The nucleus attracts electrons. Electrons in higher energy states are further away from the nucleus. Electrons in the same energy level do not repel each other much. Students must spend time applying these principles to a variety of examples. Explain why Sb is larger than As. Explain why Ne is smaller than O. Explain why O is larger than Ne.

2. *"Oxygen has a smaller ionization energy than nitrogen because nitrogen has a stable half-filled shell, right?"* The idea that a full or half-full valence shell is stable is very difficult to undermine. Be wary of your language in presentation. Anytime you present atoms with anthropomorphism you risk pushing this misconception. Electrons do what they do because of their motion, position, and forces acting on them. When we explain exceptions to trends, we need to be clear that one of the factors is that electrons experience additional repulsion when paired. This additional repulsion can lower the ionization energy required for one of those electrons to be removed. But there are multiple examples of stable ions, and ionization energy exceptions that do not fit the "full shell is stable" theory. Many of these exceptions and patterns extend into bonding in our next unit.

3. *"What do you want us to put for the trend for electron affinity?"* I still am not sure what the appropriate answer is to this. As we move toward nonmetals there is an inconsistent increase in electron affinity. But there are so many discontinuities that it really does not add much value to assume a trend exists. Moving down a group has an inconsistent trend of electron affinity decreasing. I tend to provide these trends but remind students that the large number of exceptions makes this trend less meaningful than the others.

Chapter 9 – Periodic Trends

4. "Who did make the first periodic table then?" I'm not sure it's important. A large number of people constructed periodic tables with varying components that made theirs valuable or in error. Boyle, Lavoisier, Dalton, Berzelius, and others had written down collections of elements but not arranged to demonstrate periodicity. Mendeleev's was probably the best of the early ones in many ways. I think it is more important that students understand that multiple scientists were working toward an organization of the known elements, that patterns were helpful to determine chemical information, and that the current periodic table allows for a lot of accurate predictions. With that said, the periodic table is a central feature of chemistry, and its history is relevant. But I'm not convinced that history lies with a single person.

5. "Ar is larger than Ne because Ar has more energy levels occupied, equal shielding, and larger nuclear charge." Often students will make mistakes like this where they explain a trend correctly but include one piece of information that is wrong. In this case, Ne and Ar do not have equivalent shielding for their valence electrons. Ar has more core electrons repelling its valence electrons, while also having more nuclear charge attracting them. The effective nuclear charges would be similar, but the shielding is not. It can be worthwhile to have students just compare shielding for a variety of atoms and ions (Cu^{2+} vs. Cu^+, a valence electron in Ca vs. a 3s electron, Mg vs. Ca, etc.)

6. "I keep getting the trends backwards." If a student is mixing up the directionality of trends frequently then it is likely that they have tried to memorize the directions of them using rote memory. They are just trying to remember that ionization energy decreases down a group and increases across a period. Instead, they should combine what ionization energy means with the core properties that affect the trend, Teachers sometimes inadvertently encourage this by presenting an image of a periodic table with arrows showing the directionality of all of the trends simultaneously. Students cannot adequately remember this many trends if they all remain abstract. You'll see students in this group have greater success with size given its visual correlation. But students need to practice using effective nuclear charge, shielding, and energy levels by applying them to the trends in order to have meaningful memory of the trends.

Phenomena

1. Periodic data - At some point during this unit you absolutely must have students search for patterns and exceptions with data for a periodic trend. This can be for the first trend you teach as a way for students to search for prior knowledge, or it can be used as an assessment of using effective nuclear charge, and energy levels to explain trends.

2. Alkali metals reacting with water. Showing the advancement of reaction rate from lithium to sodium to potassium is a vivid demonstration that connects well with periodic trends. This also provides a concrete example that students can retain a strong memory of. Be wary of a few things though. The explosion is not the trend. In fact, sodium often gives the largest explosion because it has the most time for hydrogen gas to build up prior to ignition. You also need to consider that part of what impacts the reaction rate is the bonding within the metal. Lithium metal has stronger

Chapter 9 – Periodic Trends

bonds, and this impacts the reaction rate obviously. Additionally, you might have coatings on the metals if they have been stored for a while. Sodium and potassium will develop an oxide coating that can further evolve into more dangerous peroxide or superoxide coatings.

3. Halogen displacement. Showing how $Cl_2/Br_2/I_2$ react with $Cl^-/Br^-/I^-$ is a fantastic way to delve deeper into the key concepts in this unit (f you have the means to generate liquid bromine and chlorine gas). Periodic trends should bridge with redox chemistry since we are working to explain the combination of forces on electrons. Redox of course studies how electrons shift between elements.

4. Naming and chemical reactions. By now the students have learned quite a bit that relates to periodic trends. One thing they can do is generate lists of patterns that they have observed from formulas of compounds as well as patterns of chemical reactions that align with periodic table placement. For example, students have seen that combining a group 2 element with a halogen results in a 1:2 ratio for the compound ($MgBr_2$).

5. Window insulation. Double pane windows are often filled with argon gas. Argon has a low heat capacity and is relatively cheap due to the amount of air being argon. But other noble gases would function even better. Kr has a similar heat capacity on a molar basis. But Kr atoms are more massive, so at a given temperature they would take fewer trips from one side of the window to the other. Each trip allows the Kr to pick up energy on one side and transmit that energy to the other. Xe would be even better than Kr.

6. New element discovery. At the time of writing this book there are 118 elements with names for the periodic table. What did chemists do to discover those elements (in particular the superheavy ones)? What are chemists doing now to move beyond element 118? How does the naming process work? Elements have been discovered/synthesized in labs in Livermore, Darmstadt, Berkeley, Oak Ridge, Dubna, and Wako City.

7. Lewis structures. If you have not yet taught Lewis structures, make sure that you connect these trends to them when you do teach them. We want students to be able to visualize the movement of charge within a Lewis structure so that they can appreciate why organic mechanisms function the way that they do. Students who can describe how electronegativity influences the electron distribution in a Lewis structure are going to be able to identify intermolecular forces, polarity, and predict reaction mechanics with greater success.

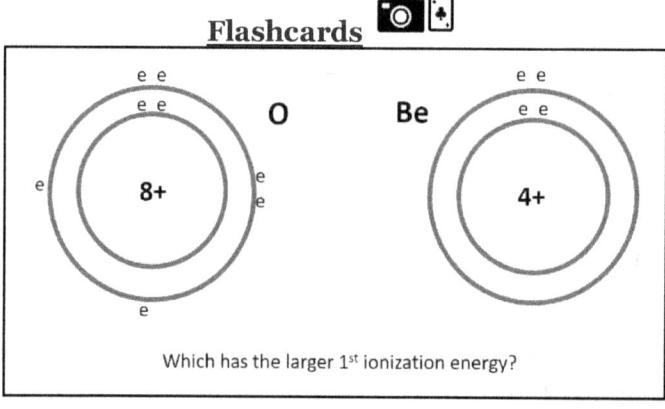

Chapter 9 – Periodic Trends

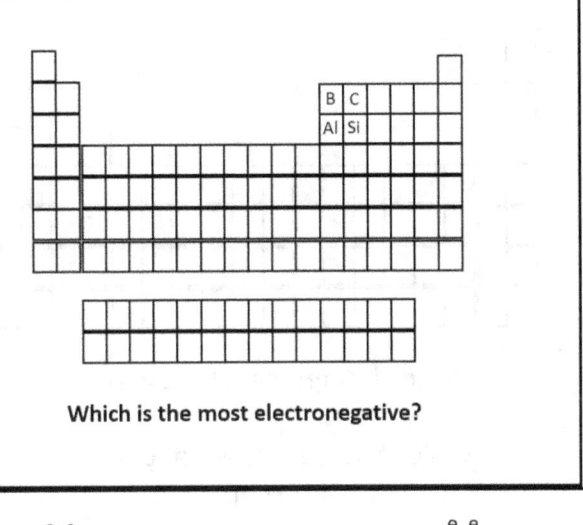

Which is the most electronegative?

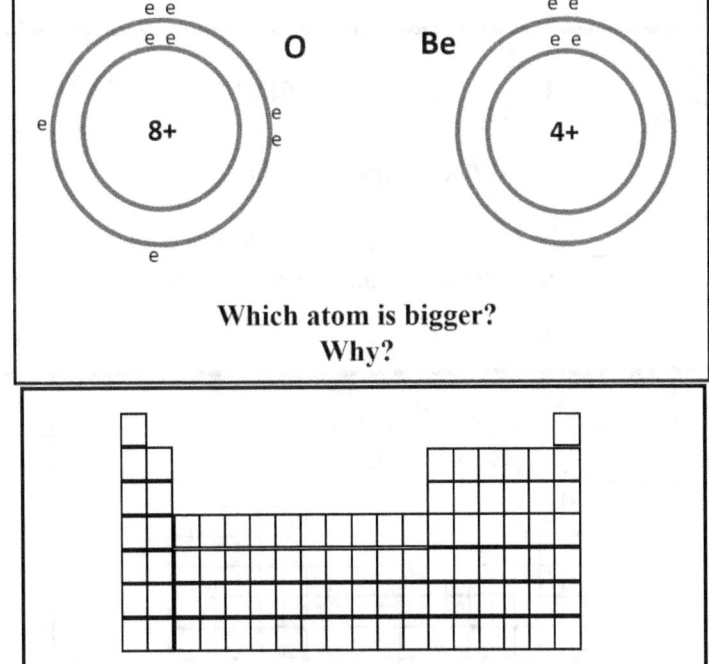

Which atom is bigger?
Why?

How did Mendeleev's periodic table differ from this one?
How were they similar?

Chapter 9 – Periodic Trends

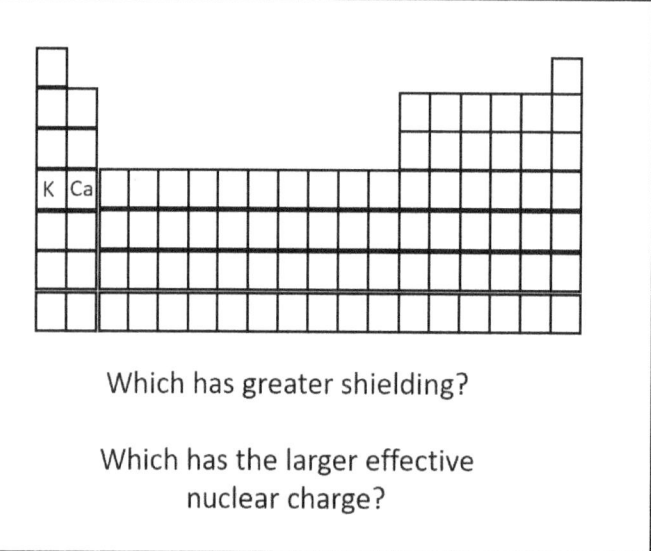

Which has greater shielding?

Which has the larger effective nuclear charge?

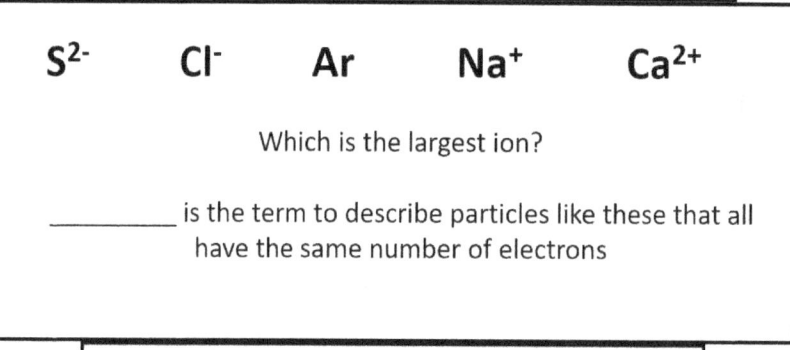

S^{2-} Cl^- Ar Na^+ Ca^{2+}

Which is the largest ion?

_____ is the term to describe particles like these that all have the same number of electrons

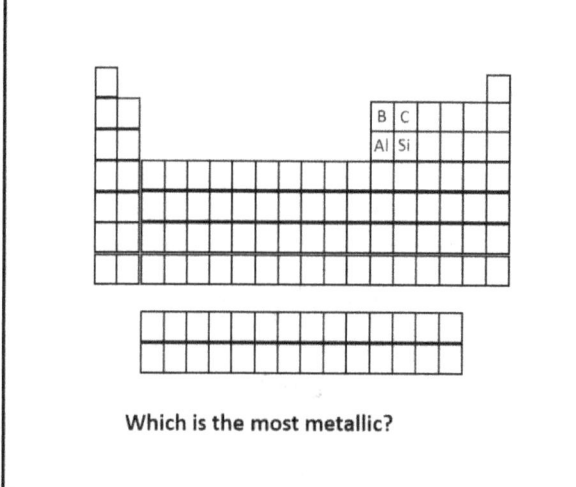

Which is the most metallic?

Chapter 9 – Periodic Trends

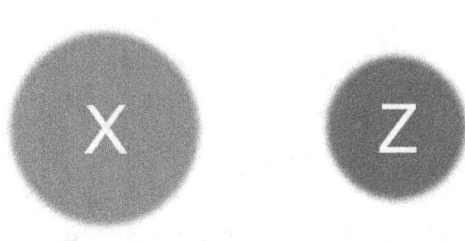

These two elements are both in period 3.

Which has the larger effective nuclear charge?

Provide 2 reasons why Li is larger than Li⁺

$$\text{Li vs. Li}^+$$

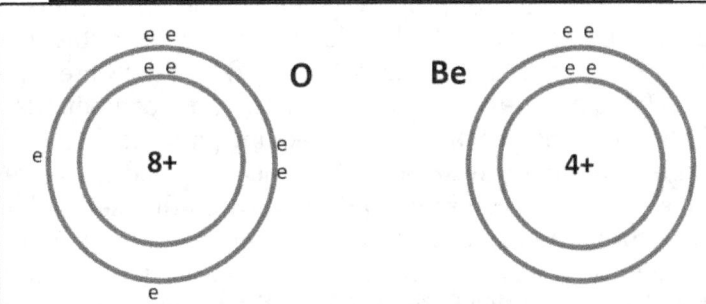

How do Zeff, n, and shielding compare for the valence electrons?

If both atoms are neutral, which will be larger?

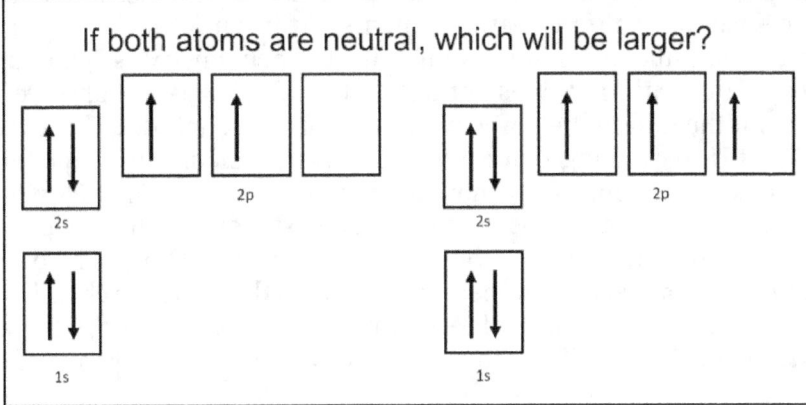

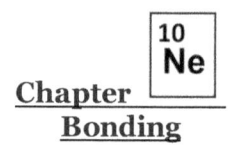

Chapter
Bonding

History

Knowing the details of bonding would have solved a lot of chemistry problems. It's quite impressive what we were able to decipher about chemistry prior to knowledge of bonding. But how do particles stick together? Many details were clarified in the 1910s by Irving Langmuir and Gilbert Lewis. The octet rule (guideline) was proposed in 1916 by Lewis, Lewis's cube model was proposed in 1903, and in 1919 Langmuir published "The Arrangement of Electrons in Atoms and Molecules."

Jons Jacob Berzelius's dualism model (1803) was what we used prior to valence bond theory. Berzelius used electrolysis to assign elements to either be + or - by which electrode of the battery they were formed.[1,2] This worked well for ionic compounds.[3] It led to confusion with molecular compounds.

In the 1830s the model was advanced by radical theory where organic chemists proposed that there were radicals in molecules that had consistent composition (this would be functional groups today). Radical theory was instituted when chloroacetic acid (CCl_3COOH) was reduced to acetic acid (CH_3COOH) and this conflicted with dualism since positive H should not be able to replace negative Cl particles.[4] Type theory was another proposed model of bonding where there were types of compounds. The water type had O along with two constituents. H-O-H, K-O-H, ethyl-O-H were all the water type. Gerhardt (1853) had the water type along with 3 others (NH_3, HCl, H_2) to account for classifying all organic compounds.

In the 1850s and 1860s chemists were seeking patterns of chemical formulas as the periodic tables were being designed. Mendeleev studied the patterns of hydrides for the elements. The combining power showed that some elements tended to combine with one other atom, while others combined with 2, 3, or 4. We would say that carbon has a valence of 4 and nitrogen a valence of 3. August Kekulé build sausage structures that are similar to the LEGO activity seen later in this chapter.[5]

Inspiration for the need for constitutional structures stems from the discovery of isomers. Liebig and Wöhler discovered silver fulminate and silver cyanate have identical composition (1827). They assumed two different chemicals that are made of identical pieces can have different arrangements of the particles.

Isomers continue to inspire advances in stereochemistry as Jean Baptiste Biot (1832) discovers that tartaric acid is optically active, Louis Pasteur observes that the crystals differ for the different enantiomers of tartaric acid (1848). This eventually builds into Van't Hoff claiming that carbon must be tetrahedral in shape (1874) so the symmetry results in the appropriate number of stereoisomers. Van't Hoff likely was building on a claim from one of his instructors August Kekulé who made a similar claim (1862). Initially scientists paid little attention to Van't Hoff, so he became more aggressive and build models of tetrahedral carbons to distribute to chemists. Joseph Le Bel made an independent claim of carbon being tetrahedral (1874), but more credit is assigned to Van't Hoff due to his success in growing acceptance of the claim.[6]

Chapter 10 – Bonding

Lewis worked out a cube model for atoms (1902, published in 1916) where electrons resided in the corners of atoms. Cubes with two gaps could add two electrons. This model allowed for differences between ions where electrons are transferred from one cube to another, and molecular compounds where electrons are shared between cubes. A double bond was explained by having two faces of a cube share 2 electrons from each cube. The triple bond could not be explained. This model was refined as quantum theory starts to emerge. In 1916 Lewis proposed the octet rule along with Lewis structures to represent electrons and chemical bonds.[7]

Irving Langmuir (1919) read Lewis's work and published work adding on to Lewis's ideas. The two started off friendly toward each other, but there are rumors of bitterness. Lewis died within a day of meeting with Langmuir (Langmuir won a Nobel prize while Lewis never did). It is not clear how Lewis died, but he had access to cyanide and could have died of suicide.[8]

Pauling (1932) published a hybridization theory that explained how carbon could form four equivalent bonds (it doesn't). He also developed the first set of electronegativity values (1932) and hydrogen bonding (1939). Pauling (like Marie Curie) was awarded two Nobel prizes in two fields (Chemistry and the Nobel Peace Prize). Valence bond theory (1927) and molecular orbital (MO) theory (1929) both evolve quickly at the end of the 1920s. Hund, Mulliken, and many others advanced MO theory while Pauling popularized valence bond theory (VBT).

Am I Breaking Bonds?

My first year of teaching fell apart when I got to bonding. I was the only chemistry teacher. When I started the bonding unit, I realized that I had no idea what to teach. For stoichiometry I could show them how to do a stoichiometry problem. For acids and bases we could calculate pH or classify acids as Arrhenius, Bronsted-Lowry, or Lewis acids. But for bonding I couldn't think of algorithms for students to follow. The idea of teaching a concept was unknown to me. After we completed Lewis structures I was stuck. Somehow, I ended up deciding to teach a lesson on how to determine the number of nonbonding lone pairs in a Lewis structure. I remember dedicating a lecture, a worksheet and a quiz to this topic. It was brutal.

It was a crossroads in my teaching career. I had loved learning about chemistry and physics throughout high school and college. But so much of what I had learned about was abstract. As a new teacher I struggled immensely with teaching bonding because I wasn't used to starting with an idea or with experimental evidence. I still believe that bonding is the most difficult unit to teach in all of chemistry. The start of bonding should not be with seeking algorithms to follow like drawing Lewis structures. The beginning should be with the idea that chemicals stick together.

Why do atoms stick together? Why do molecules stick to other molecules? Why do ions stick together? The answer that students want to respond with is one word, and one word only. Bonds. To a student, a bond is circularly defined. A bond is what makes chemicals stick together. Chemicals stick together because of bonds.

A great method to disrupt their circular definition of "bonds" is an activity I call *"Am I breaking bonds?"* In this activity I perform ten to twenty simple changes. In each one the question to the students is, "Am I breaking bonds?" They must choose yes or no. There is a minimum of at least one yes answer and at least one no answer.

Chapter 10 – Bonding

    Am I breaking bonds?
1. Ripping a piece of paper in half
2. Breaking a paper clip
3. Writing with chalk
4. Burning a piece of paper
5. Breaking a sugar cube into pieces with a hammer
6. Breaking a chunk of copper (II) sulfate into pieces with a hammer
7. Dissolving a sugar cube in water
8. Dissolving copper (II) sulfate in water
9. Playing with a slinky
10. Acetone evaporates from a table

    If you'd like to, write down what you think the answers would be. There are many other variations that you can add. I usually do a mild explosion at the end such as the whoosh bottle or lighting a piece of flash cotton.

    Something interesting develops during this introduction. The students have an intuition that ripping a piece of paper does not break bonds. But they don't know why they think that. They will choose that no bonds are broken pretty consistently. But if you ask them how it is possible that you aren't breaking bonds, they do not have many answers. The paper particles were stuck together, right? What did I break if I didn't break bonds?

    This discrepancy is the distinction between intermolecular forces and bonds. They are probably identifying the piece of paper ripping as a physical change, and they assume physical changes do not break bonds (this is not always true, but that's what they are thinking). When they later see the piece of paper lit on fire, they identify that as a chemical reaction and assume that bonds are broken there.

    This shows that students have an initial model where all bonds are covalent. They see bonds as those sticks from model kits and have only seen covalent bonds as lines in Lewis structures. We don't build model kits for ionic compounds or metals. Because changes to metals and salts are all physical changes in the examples 1-10, they think that there must not be changes to bonds. Numbers 5-8 are particularly interesting. Some students will choose "yes" for 5-8 while others choose "no." And they tend to put the same answer for all 4 scenarios.

    Breaking a sugar cube with a hammer does not break bonds. Breaking a salt crystal with a hammer does break bonds. Dissolving a sugar cube does not break bonds. Dissolving a salt crystal in water does break bonds.

    Metallic bonding is too confusing for students to understand it well prior to high school chemistry. Therefore, it is not surprising that they select breaking a paper clip to not break bonds. But bending a paper clip would break bonds. The bonding in metals is based on an attraction between free electrons and the lattice of cations. As that lattice changes positions you break bonds. Even the slinky moving involves bonds breaking. The students on the other hand will not select the paper clip breaking as they associate it with the paper ripping. I have never found a chemistry teacher who agrees with me about the slinky, but I don't see a way that there are not atoms in the slinky moving far enough apart from each other that bonds don't break. Even if they return to the same positioning after, I still think that would be a case of bonds breaking and forming. If someone can provide me with evidence that confirms or counters my current position, I would love to see it.

Chapter 10 – Bonding

My answers for 1-10 are:
1. No
2. Yes
3. Yes
4. Yes
5. No
6. Yes
7. No
8. Yes
9. Yes
10. No

A lesson from this activity is that we have different mechanisms of how atoms can stick together. There are covalent bonds, metallic bonds, ionic bonds, and intermolecular forces. Many teachers expect students to walk in with a strong understanding of those distinctions because of how we defined molecules, atoms, and formula units earlier in the year. But students need work developing those ideas. We need to remember that most students have limited physics knowledge and that bonding requires applying physics knowledge to an abstract concept. It requires experimental evidence and model development to have a thorough understanding of bonding.

When we define bonds, we should make clear that we are talking about the attractive force between two chemicals. An ionic bond is not a transfer of electrons. An ionic bond is the force that exists between anions and cations arranged in a lattice. A covalent bond is not a sharing of electrons. It is the set of forces between a pair of electrons and two nuclei. A metallic bond is the attractive forces that exist between a lattice of cations and a sea of delocalized electrons.

Remember that the students walk in with a lot of strange ideas about what a bond is. To many of them it really is a stick. They've never taken the time to think about what the stick is made of until they realize that a bond in an atom can't be made of a stick that would consist of atoms. They might say that a bond is made of energy. When they do that, they are communicating that they do not know, so instead they reply with a single scientific vocabulary term. Some students may even picture the bonds exploding like they were sticks of TNT when chemicals react. It is better for students to wonder how atoms stick than to state that "bonds" hold atoms together.

If students have a strong representation of the forces involved in a covalent bond, they will have an easier time with distinguishing intermolecular forces. Using a potential energy diagram with particle representations for different points accomplishes this.

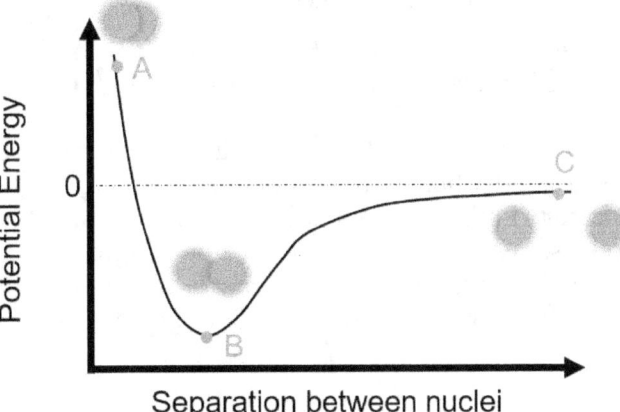

Figure 10-1: Diagram showing how potential energy varies based on internuclear distance for a diatomic molecule

Chapter 10 – Bonding

We assign the energy to be zero when atom 1 is a large distance from atom 2. There is minimal attraction between the electrons from atom 1 and the nucleus of atom 2. There is minimal repulsion between the nucleus from atom 1 with the nucleus of atom 2. The tiny amounts of attraction and repulsion that do exist are balanced so the net force is basically zero.

When we move atom 1 and atom 2 closer the repulsive and attractive forces both increase. But the attraction between the electrons of atom 1 and nucleus of atom 2 (and the electrons of atom 2 and nucleus of atom 1) increase more than the repulsion between the two nuclei. If the electrons are positioned between the two nuclei without experiencing too much repulsion a stable bond forms. The electrons can be positioned closer to the nuclei than the nuclei can be positioned to each other.

But at some point, the nuclei get too close. As the space between the nuclei shrinks, the electrons must get closer together. The increased repulsion between the electrons and the increased repulsion between the nuclei starts to overpower the increased attractive forces between the nuclei and the electrons.

The result of this is amazing. There is some distance where you can maximize the net force between two atoms. If the atoms get closer, they repel. If they move further from that point they still attract. So, the two atoms vibrate around this point. If you want students to connect with this diagram, have them draw the particle representations. Then after some checking in, have them add in the forces, net forces, and future direction of motion for those three spots.

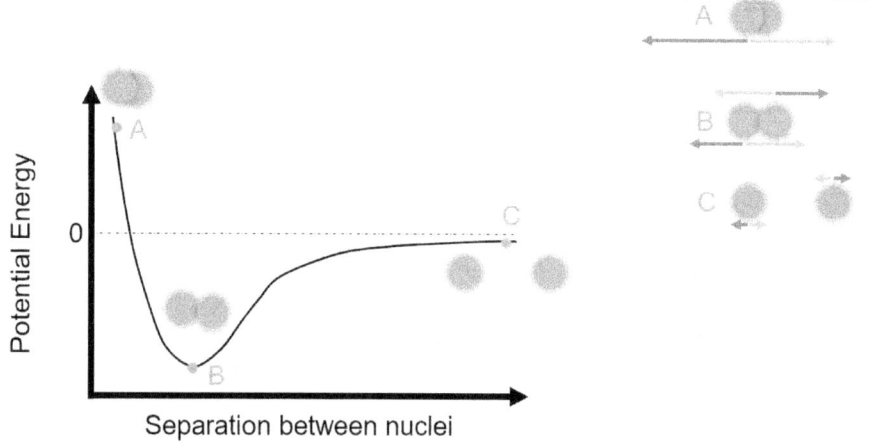

Figure 10-2: Forces of attraction and repulsion labeled for three bonded positions. At A there is more repulsion than attraction, at C there is more attraction than repulsion

Now is the time to deal with a pressing misconception. Students do not understand what breaking a bond or making a bond entails. And it isn't a biology teacher's fault! Students lack a conceptual model for bonding changes, and the phrase "releases energy" gives them the opportunity to avoid developing better understanding. The potential energy diagrams should be linked with the discussion from Chapter 7 (Fig. 7-8, 7-9) about how particle speed varies as bonds break and form.

Chapter 10 – Bonding

Lewis Structures and LEGO

After establishing why atoms stick together, we want to keep those ideas present as we start to make Lewis structures. Many teachers use the octet rule to introduce Lewis structures. The octet rule has some issues. The octet rule has many exceptions and to me works best as a correlation rather than a causation. How you present the octet rule has the potential to lock in some misconceptions with students. The octet rule is a pattern with many exceptions. It should never be used as an explanation as to why something happens in chemistry.

When sodium reacts with chlorine both atoms change the number of electrons until both have a valence of 8. One interpretation of this result is that 8 is a stable number of valence electrons. The chlorine with 7 valence electrons and the sodium with 1 valence electron were less stable. Alternatively, the chlorine has a high effective nuclear charge that causes the atom to have a large pull on electrons. The chlorine also has the capacity to add one more electron to the current energy level. The sodium has a small effective nuclear charge. When the two particles are in contact, the chlorine is capable of pulling the electron from the sodium to itself.

Does this mean that the chlorine with 7 valence electrons is unstable? That's a difficult question to provide a simple yes or no to. Here are some data pieces to consider. The electron affinity of chlorine is -349 kJ/mol while the ionization energy of sodium is +496 kJ/mol. That would imply that in the gaseous state that Na (g) + Cl (g) would be more stable than Na^+ (g) + Cl^- (g). It is the lattice energy that causes the reaction to be a net exothermic process.

The octet rule has even more issues when we delve into molecular orbitals. When we construct a molecular orbital diagram for Cl_2 (Fig. 10-3), the final four electrons go into anti-bonding orbitals. It's important to keep in mind that the removal of 1 electron can shift the energies for all the other orbitals, but the bond length of Cl_2^+ is shorter than Cl_2. *The octet rule would not explain this.*

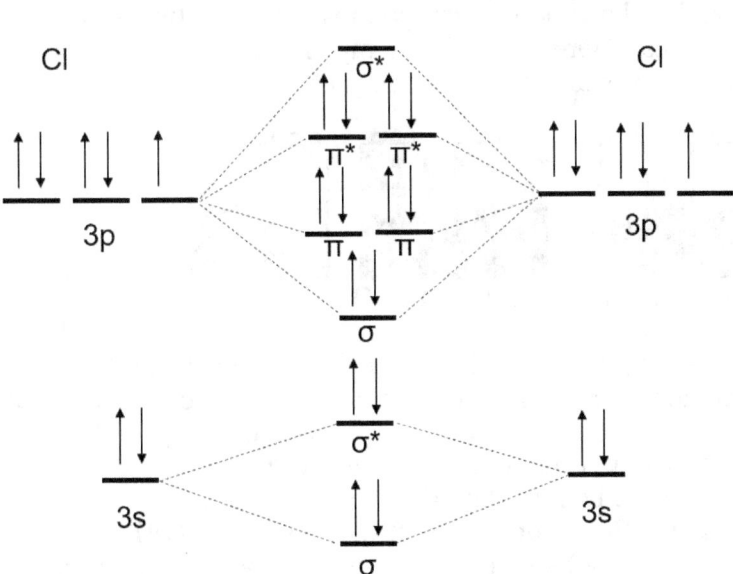

Figure 10-3: MO diagram of Cl_2

Chapter 10 – Bonding

Chlorine can also have more than eight electrons. In Chlorate (ClO_3^-) the chlorine has 12 valence electrons and in perchlorate (ClO_4^-) the chlorine has 14 valence electrons. We must approach the octet rule carefully. It is meant to be an initial model that helps ease into constructing Lewis structures. But that model is quickly improved upon and discarded. Even in a high school chemistry course the exceptions of too few, too many, and having an odd number of electrons are introduced within the unit.

We don't want students to justify their models by giving the atoms human traits. A chlorine atom doesn't want to add an electron. Atoms don't have feelings and emotions. The octet rule is going to be heard by students in this manner. They are going to hear you say, "atoms want 8 electrons." Even if you explicitly say otherwise. We must be vigilant as teachers against this anthropomorphism because it precludes future thinking from the students. We want them to wonder why nitrogen often forms three bonds and oxygen usually forms two (but not always). They'll notice patterns in the periodic groupings better as well as the exceptions.

An alternative model to the octet rule is to use LEGO bricks. Provide a list of stable compounds (CH_4, NH_3, C_2H_4, H_2O etc.) and compounds that do not exist (CH_3, NH_2, C_2H_5, etc.). The best combination of LEGO bricks for this introduction is to have the four types of bricks (Fig. 10-4). Ideally each element is one color such as carbon-black, nitrogen-blue, oxygen-red, and hydrogen-white.

Figure 10-4: LEGO bricks for an initial model prior to Lewis structures

For a single set of LEGO, I use 2 carbons, 2 oxygens, 1 nitrogen, and 9 hydrogens so that students can build a variety of compounds. The objective is to create two rows using all the pieces. There should be no gaps in either row including at the ends. You can see four examples of these in Fig. 10-5 below.

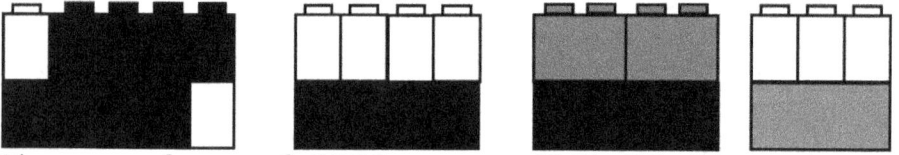

Figure 10-5: four sample LEGO structures (C_2H_2, CH_4, CO_2, NH_3)

There are advantages to using the LEGO bricks in place of the octet rule. The octet rule uses the same model for the rule and the exceptions. The Lego model shows us an initial model that is more easily discarded when we get to more advanced structures. The LEGO model draws attention to the number of connections and simplifies the use of double and triple bonds. The LEGO bricks are based on observations of stable compounds. We can observe that C_2H_2 exists but C_2H_3 is not a stable compound. From these observations we can then propose that carbon forms 4 bonds, nitrogen 3 bonds, oxygen 2 bonds and hydrogen one bond as an initial model.

Chapter 10 – Bonding

The octet rule can also do this, but you carry more risk of students disconnecting from the observations because of the word rule.

The LEGO structures are not oriented in a believable shape. When we take the methane (CH_4, Fig. 10-5), students are likely to understand that the actual molecule is shaped differently. But the Lewis structure is not as easy to move away from. The Lewis structure looks like a plausible shape for the molecule. The LEGO thus forces students to work out which parts are connected to which. This can set students up for success when they have to evaluate hydrogen environments for NMR spectroscopy.

Using isomers can really stretch the LEGO model for a few different compounds. Have the students see if they can construct two different arrangements for C_2H_6O (dimethyl ether and ethanol). Benzene (C_6H_6) could be made in theory, but in practice the double bonds prevent the shape from being able to connect back around. But the benzene structure allows for some great connections with isomers. Dichlorobenzene ($C_6H_4Cl_2$) has 3 isomers. If we assign each carbon a number 1-6, have them determine what the 3 isomers would be. If they use carbon-1 and carbon-3, why wouldn't carbon-1 and carbon-5 be an isomer of that?

The isomers were helpful historically because we could find evidence of the formulas and properties. When two compounds had identical formulas but different properties, that led us to know that there were differences in structural positioning. The fact that dichloromethane does not have isomers helped us to know that tetravalent carbon atoms must be symmetrical with respect to each attachment. This insinuates that CH_2Cl_2 must be tetrahedral in shape instead of square planar.

When we construct Lewis structures, there are multiple approaches and none of them work all the time. Before we go through the pros and cons here are some underlying hints to help students.

It is helpful for students to count the valence electrons for each component as well as the total for all atoms/ions. Generally, if there is one of an element it goes in the center. It is helpful for students to have some Lewis structures memorized. Students should be aware that elements from the same group usually work the same. If I memorize the structure of CO_2, then when I get asked for CS_2 I can easily replace each oxygen with sulfur. Lastly, it is helpful for students to know the common variations for each element (Fig. 10-6). Formal charge can be helpful for students at this juncture.

	# valence e⁻	- (1/2*bonded e⁻)	- nonbonding e⁻	= formal charge
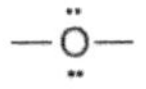 —Ö—	6	2	4	0
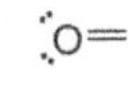 :Ö=	6	2	4	0
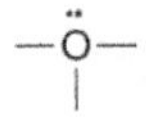 —Ö— (with +)	6	3	2	1+
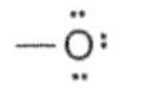 —Ö:	6	1	6	1-
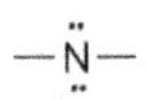 —N̈—	5	2	4	1-
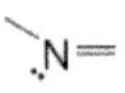 :N=	5	3	2	0
 —N̈—	5	3	2	0
 —N—	5	4	0	1+

Figure 10-6: Various formal charges and bonding arrangements for N and O

239

Chapter 10 – Bonding

There is a mathematical way to build Lewis structures, but the formula only works when the octet rule is followed.

# of bonds = (# of atoms * 8 - # of valence electrons)/2

For CO_2 we have 3 atoms that have 16 (4+6+6) valence electrons. Our equation then becomes (24-16)/2 = 4. So, we form four bonds. That would put carbon at the center with double bonds to each oxygen. After the bonds are set, you evaluate how many electrons you have remaining. For CO_2 we had 16 electrons to work with. We've used 8 for the double bonds so we put 4 more on each oxygen atom.

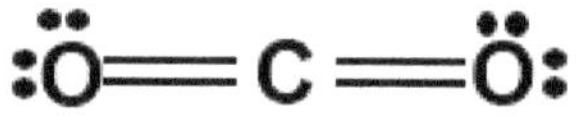

Figure 10-7: Lewis structure of CO_2 :Ö=C=Ö:

The second method of determining a Lewis structure is to start with each atom having their initial valence electrons and just keep sharing them until we get to 8. Here carbon would start with 4 and both oxygens with 6. We put carbon in the center and form single bonds for both. Single bonds leave C and O electron deficient, so we form double bonds and find the Lewis structure to now be satisfactory.

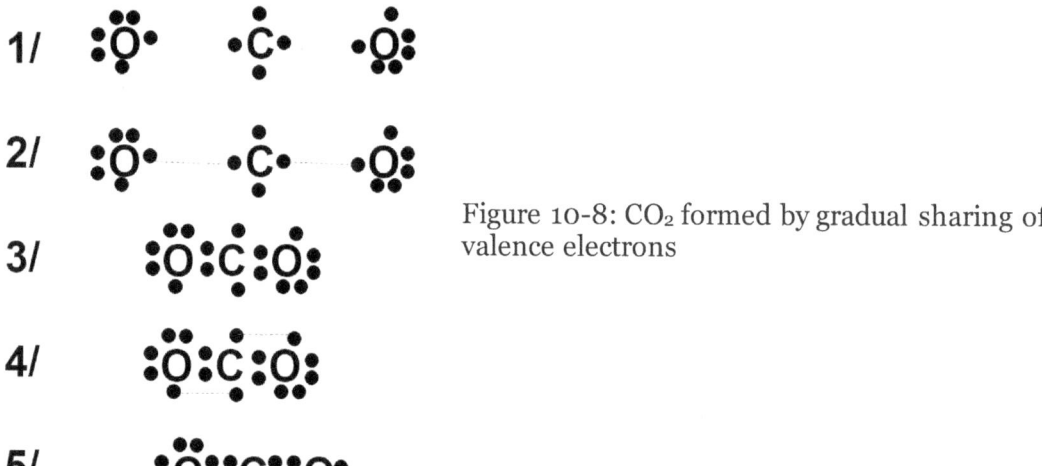

Figure 10-8: CO_2 formed by gradual sharing of valence electrons

The third method of determining Lewis structures is my personal favorite. It is the most confusing at first but works the best long-term. In this one students look at the common patterns for elements in Lewis structures based on groups. Each formula is a puzzle, and we determine the Lewis structure by guess and check. Carbon can form 4 single bonds, 2 single bonds and 1 double bond, 2 double bonds, or 1 triple bond and 1 single bond. Oxygen can from one single bond with 3 lone pairs, 2 single bonds with 2 lone pairs, 1 double bond with 2 lone pairs, 3 single bonds with 1 lone pair, or 1 double bond and 1 single bond and 1 lone pair. Given that carbon is in the middle of two things the best arrangement is to use two sets of double bonds.

The third method is the least prescriptive. I find that it forces students to look at the molecules the same way that experts do. Experts substitute elements from the same group easily. Expects notice how many bonds and lone pairs are common within a group. It is more challenging at first, but this encompasses the most

Chapter 10 – Bonding

exceptions (non-zero formal charges, octet expansion, coordinate covalent bonds). This processing can be enhanced by teaching your students about formal charge (next section).

Students with competency in constructing Lewis structures must realize there are limitations to the images being produced. Students will commonly ask about placement of electrons and that is a great opportunity to reinforce the problems with these structures. A student asks, "Does it matter if I put the electron pair on top or to the side of the oxygen?" That is your chance to talk about limitations such as 2-dimensional representations of 3-dimensional objects, the lack of time (the particles are moving yet we represent them as stationary), and the size of the symbols can mislead us from the scale of the nucleus and surrounding electrons.

One common student error that you should anticipate is that students will put no lone pairs or way too many lone pairs on any atom that is not central. For CH_4 you might see 8 electrons on each H because the student has learned to focus on the central atom and is overapplying an example for the outer atoms. For CCl_4 you might see no lone pairs on the chlorines for the same reason or just because the student was overly focused on the central carbon.

Some students will struggle with the number of lone pairs on a central atom. This stems from not counting the total number of valence electrons. The best element to test them on this is sulfur. If a student can figure out SO_3 and SO_3^{2-} they are usually on a good path. Regardless of resonance, the SO_3 has no lone pairs on the sulfur while SO_3^{2-} has one pair. SF_2 and SF_4 are also good for testing what they do with the central sulfur. SF_2 should have 2 lone pairs on the central sulfur and SF_4 should have 1 lone pair. The students will often recognize that the fluorine atoms should have single bonds with 3 lone pairs but are unsure how to figure out the sulfur. If students get these correct and want additional practice; the noble gas compounds are a great final check. XeF_4 and XeO_2 will test students' abilities to determine electron placement.

Some students will be a mess at the beginning. They might make a long chain such as C-H-H-H-H instead of a central carbon with four hydrogens. Some will read the formulas literally and try to make a C-H-C connection. If a student is really struggling it can be a big help for them just to memorize a few simple structures. This will help train their brains to identify patterns that work and do not. It gives them something to start with. Whatever you have 1 of tends to go in the center with everything else attached. That doesn't hold for carbon chains in organic chemistry, but it will work for many of the simple Lewis structures that we start with. Hydrogen does not form two bonds and so it is never central.

Tips for students learning about Lewis structures:
1. Count all the valence electrons first
2. If you have a polyatomic ion, remember to adjust for the charge
3. Whichever element you have 1 of usually goes in the center
4. You'll need at least one bond connecting each atom
5. Hydrogen can't have more than one bond
6. If you don't know, guess something and evaluate if it works

Chapter 10 – Bonding

Formal Charge

Formal Charge adds a lot of value to Lewis structures. The purpose of formal charge is to evaluate Lewis structures by tracking where electrons are. The three goals of formal charge are to keep formal charges to a minimum, prevent positive and negative formal charges from being adjacent, and aim for placement of formal charge based on electronegativity. If we compare Lewis structures of NCO^-, the Lewis structure with the negative formal charge on the oxygen is preferred because oxygen is more electronegative than nitrogen.

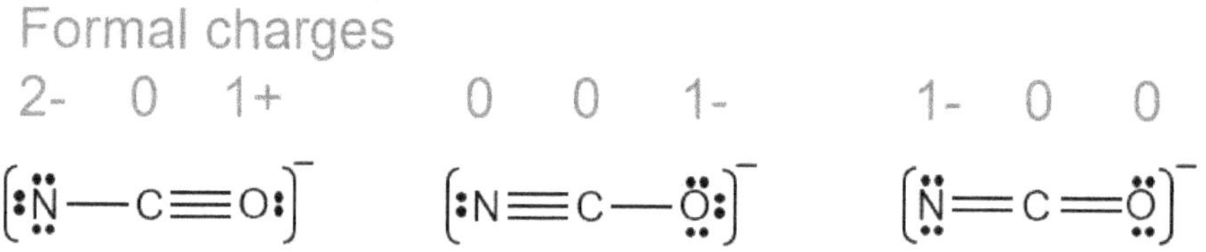

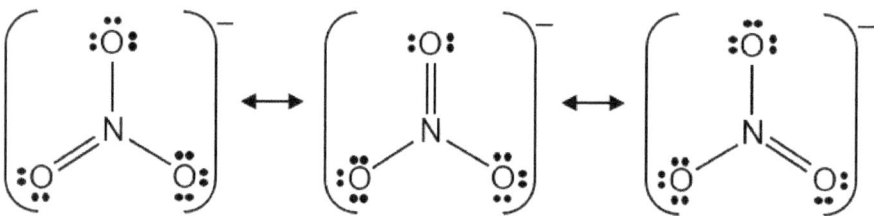

Figure 10-9: Formal charges dictate that the 2nd/3rd Lewis structures are preferred for NCO^-

Don't start formal charge by giving students the equation. Instead, you're trying to assign every electron in the structure to one atom. The equations and algorithms detract from the process when the evaluation comes naturally to students. When you draw the Lewis structure of NO_3^- students will sometimes ask how nitrogen can only have 4 valence electrons in it.

Figure 10-10: Set of resonance structures for the nitrate anion

You want to use that vision and maintain it. Using an algorithmic formula pulls students away from their intuition. The only thing we need to distinguish is that for formal charge we're counting electrons on each atom, but they only get 1 of the 2 bonded electrons. In Lewis structures we are counting both shared electrons for both atoms. For the purposes of electrons in the Lewis structure, the nitrogen has 4 bonds and thus has 8 valence electrons. But for formal charge we only count 1 electron from each bond as assigned to the nitrogen. The nitrogen would have 4 valence electrons by this assignment and therefore would have a 1+ formal charge since nitrogen originally had 5 valence electrons.

We want students to be get to a point where they recognize electron deficiencies and excesses by just seeing the elements. As a chemist, when I see an oxygen with 3 bonds and a lone pair I know that it has a 1+ formal charge. As an experienced chemist, I do not plug into an equation to recognize this. When an oxygen has 1 bond and 3 lone pairs it has a 1- formal charge.

Chapter 10 – Bonding

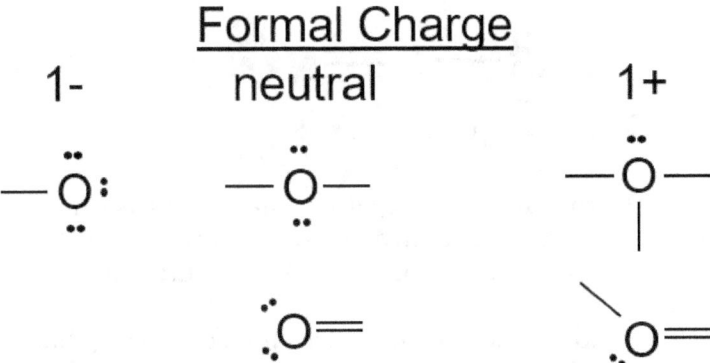

Figure 10-11: Formal charge possibilities for oxygen

You could start formal charge by showing a variety of compounds and ions with oxygen and asking students to group them by which oxygens are electron deficient and which ones have an excess of electron density. This would give students an opportunity to think about the difference between a lone pair of electrons and a bonded pair of electrons and how that might impact the charge of an atom within a molecule (Fig. 10-12).

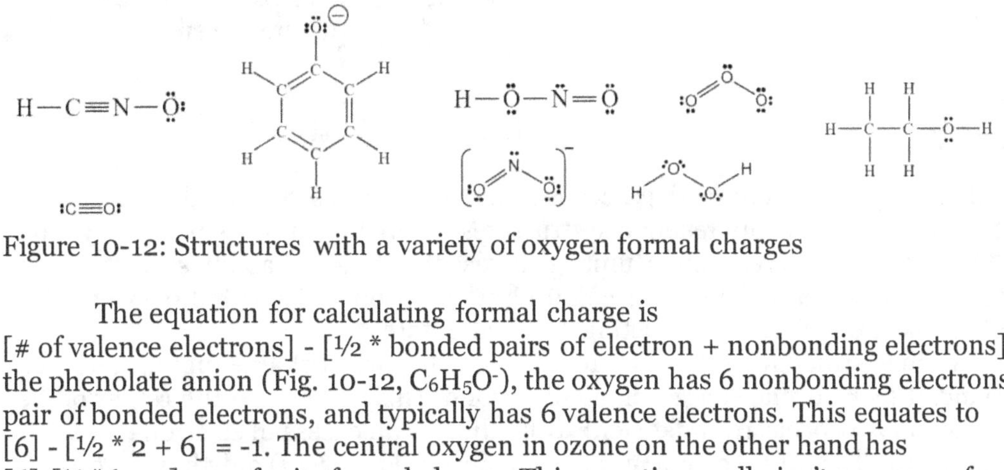

Figure 10-12: Structures with a variety of oxygen formal charges

The equation for calculating formal charge is
[# of valence electrons] - [½ * bonded pairs of electron + nonbonding electrons]. For the phenolate anion (Fig. 10-12, $C_6H_5O^-$), the oxygen has 6 nonbonding electrons, one pair of bonded electrons, and typically has 6 valence electrons. This equates to [6] - [½ * 2 + 6] = -1. The central oxygen in ozone on the other hand has [6]-[½*6 + 2] = +1 for its formal charge. This equation really isn't necessary for students. At best it should be used to reinforce the visual cues they take from the Lewis structure.

A final check for formal charge is that the sum of all formal charges should add up to the total charge of the molecule or ion. This is because you are assigning all the electrons to individual atoms. Every electron must get assigned somewhere. If a diatomic molecule has 16 protons and 16 electrons, no matter how you assign the electrons the overall charge is still zero.

As a teacher you want to focus the students on the concept of counting and assigning electrons first. Next you want to use formal charge as an evaluation tool. How do two different potential Lewis structures compare? Thiocyanate (SCN^-) and sulfate (SO_4^{2-}) are both helpful examples to help students accomplish this.

Chapter 10 – Bonding

$:\ddot{S}=C=\ddot{N}:^{-}$ $:\ddot{S}-C\equiv N:^{-}$

Figure 10-13: two possible Lewis structures for thiocyanate

The preferred Lewis structure from an electronegativity standpoint is where the negative formal charge is on the more electronegative nitrogen. However, the sulfur is much larger and so it is reasonable to expect both Lewis structures to contribute to the actual molecule.

The bond length of the C-N bond is experimentally determined to be 115 pm and the C-S bond length is 169 pm. The C-S single bond is 182 pm and the C=S double bond is 156 pm. The C=N double bond is typically 130 pm, and the triple bond is 116 pm. This somewhat shows the C-S bond length of thiocyanate being partially due to a single bond and partially to a double bond. The carbon-nitrogen bond length is the typical length of a triple bond. Perhaps the excess negative charge causes a small amount of lengthening. How would OCN⁻ differ from SCN⁻?

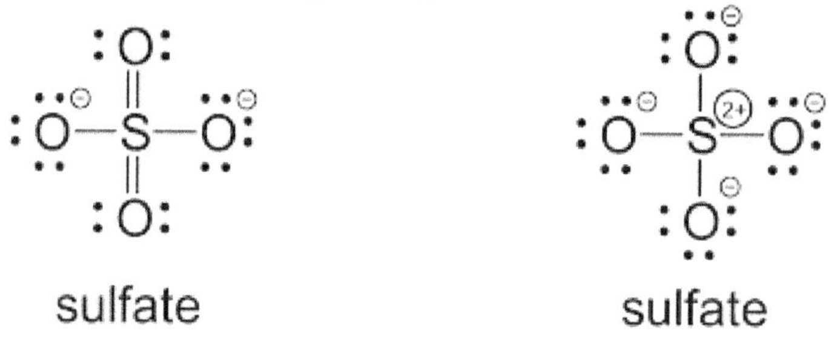

Figure 10-14: Two Lewis structures of sulfate

sulfate   sulfate

Sulfate (SO_4^{2-}) is interesting because there is a Lewis structure that satisfies the octet rule, but also many resonance structures where the sulfur expands beyond the octet. This makes a great opportunity to assess the students' abilities to distinguish between formal charge accounting for electrons relative to how we account for electrons when constructing Lewis structures.

The octet structure (right, Fig. 10-14) has a 2+ formal charge on the sulfur and is surrounded by 1- formal charges in all directions. This means that the center of the polyatomic ion is very electron deficient while there is an excess of electrons everywhere else. The other resonance structures (left, Fig. 10-14) only have two 1- formal charges on oxygens and those are spread over the four oxygen atoms through delocalization.

The bond lengths in sulfate are 149 pm while the S-O single bond is 161 and the S=O bond is 153 pm.

After Lewis structures there is a lengthy list of objectives that follow. The order is not critical by any stretch. Shapes and bond angles are fun and help give students a bit more context to Lewis structures. For shapes you are trying to help students visualize what the molecules look like in 3-D. Whatever you use to accomplish this, it should be in 3-D. I know many use toothpicks and marshmallows to build the shapes, but I prefer spice drops instead of marshmallows. Most students hate the taste of them, and for some reason that brings me joy.

Chapter 10 - Bonding

When you build the shapes, start with the philosophy. We want to get the bonds as far away from each other as possible. The bigger the bond angle the better. This is enhanced when they can manipulate the bond angle, hence the toothpicks. They will be able to determine the linear shape (AX_2)[§] and the trigonal planar shape (AX_3) on their own. Planar is a fancy way of saying flat in chemistry while pyramidal means not flat. The tetrahedral shape (AX_4) is where students might incorrectly form 90-degree bond angles to form a square. If you start with a square, you can bend a bond out of the plane to increase the angle above 90 degrees in all directions. Showing the changes from square planar to tetrahedral is a powerful move to help a student distinguish between the Lewis structure shape in 2-D with the actual shape in 3-D.

Trigonal pyramidal (AX_3E) introduces a new concept. What do lone pairs on the central atom do? A lone pair is two electrons, and a bond is two electrons. Are they going to act the same? It turns out that lone pairs have two differences from bonded electrons. The first is that bonded electrons have another atom attached. The bond therefore has more positive charge relative to the lone pair. But even more important is that bonded electrons shift in position away from the central atom and towards the midpoint between the two nuclei. This means that lone pairs are closer to the central atom than bonded pairs. Both result in larger repulsion between lone pairs and bonded pairs, than between two sets of bonded pairs. That additional repulsion causes bond angles to shrink slightly. The bond angle changes differently depending on what elements are involved, but they often change by a couple of degrees.

This can be seen for the bond angles for CH_4, NH_3, and H_2O. CH_4 has the typical 109.5° bond angle for a tetrahedron. For NH_3 the bonds are slightly closer at 107°. For H_2O the bonds experience a little more repulsion and have a bond angle of 104.5°. This might imply a linear relationship since both change by 2.5°. But there are numerous examples that do not fit into a linear relationship. For example, H_2Se has a 91° angle which suggests the bonds are using p-orbitals due to the large size of the sulfur atoms.[9]

The bond angles are simpler for the students who accept that lone pairs repel more than bonds. There are some students who push a little deeper and try and connect this to the limits of our models. What does a pair of electrons actually look like? When we use a space filling model to represent a trigonal pyramidal molecule like NH_3 each atom is a sphere. We have a space where the lone pair exists, but we don't fill in an atom there. Afterall, the electrons are very tiny. But an atom consists of mostly empty space as well. Why does the atom get the sphere and the lone pair does not?

This is not a simple thought experiment because at the very end of the road is our inability to see what the molecule actually looks like in real life. It is not always simple to remark on the various ways we can differentiate a lone pair from a bonded atom in a representation of a molecule.

Something I didn't notice until midway through my career is that every bond angle is either 90, 120, or 180 if the shape is not tetrahedral or derived from a tetrahedron (trigonal pyramidal or bent triatomic). A student will probably ask why

---

[§] *A is the number of central atoms, X is the number of atoms attached to the central atom, E is the number of lone pairs on the central atom*

Chapter 10 – Bonding

tetrahedron has four things attached but octahedron only has six. The "hedron" component means faces. A tetrahedron has four faces, it is not because there are four attachments. When you have six attachments, eight faces are formed. Hence the name octahedron.

Polarization

As we shift into polarity, we want to keep connecting back to Lewis structures and shapes. Not only do we want to use spaced practice for Lewis structures, but we want students to connect subsequent ideas in future content such as curly arrow mechanisms in organic chemistry.

Polarization is a concept that has a lot of explaining power in chemistry. Before we get into specifics of bond polarity and molecular polarity, we want to establish what polarity is. Take a ruler and balance it on an upside-down watch glass. Rub a balloon on your head to charge it (or borrow some equipment from a physics class). The ruler is neutral. But as you bring the charged balloon close to it there is an attraction between the ruler and the balloon. The ruler will start to spin. You can even reverse the spin of the ruler by holding the balloon on the other side as it spins. Throughout this the ruler is neutral. If you pick it up, you can show the students that it indeed does not electrocute you. There should be a balance of attractive and repulsive forces.

But the charged particles in the ruler are not locked into a position. When the negative charges shift (negative charges are much easier to move than positive ones) to one side we end up with more attraction than repulsion. Recall that distance influences electric force. If the balloon is negatively charged and the ruler is polarized (Fig. 10-15), we will get more attraction between the balloon and the positive charges, than repulsion from the balloon and the negative charges.

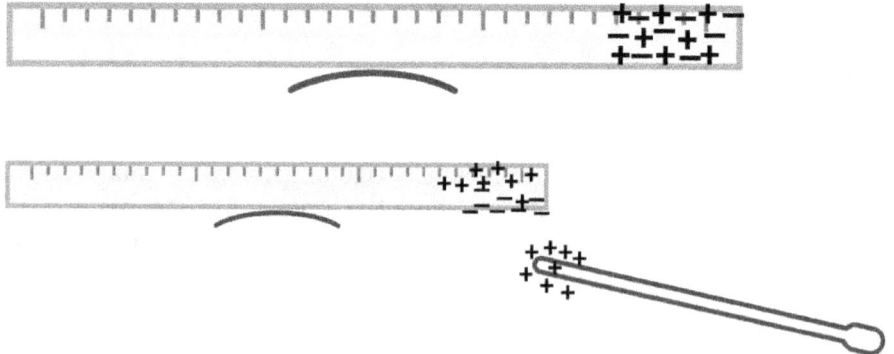

Figure 10-15: A neutral ruler on a watch glass and the same ruler when it is polarized by a charged object

- What would this look like in an atom?
- What would polarization look like for a molecule?
- What would be the results of having a polarized molecule?
- Would it change the reactivity?
- Would it matter for how molecules can stick together?
- Have we mentioned polarization when discussing bonds?

Chapter 10 – Bonding

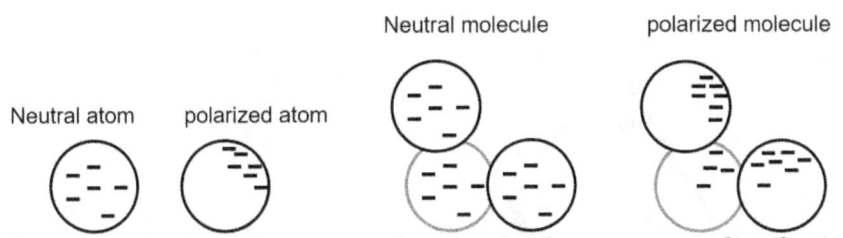

Figure 10-16: Neutral atoms and molecules have an even distribution of charge, polarized molecules have a balance of charge, but the charge is distributed unevenly.

If you want to stretch this further, you can do the same experiment on a thin strip of aluminum foil and a thin strip of paper (I use chromatography paper). The aluminum foil will have a much larger attractive force since the electrons are delocalized. This allows them to move to one side of the foil. The paper has an attraction, but it is noticeably smaller because the paper is an insulator. This means the charges can move within the molecules, but not from molecule to molecule.

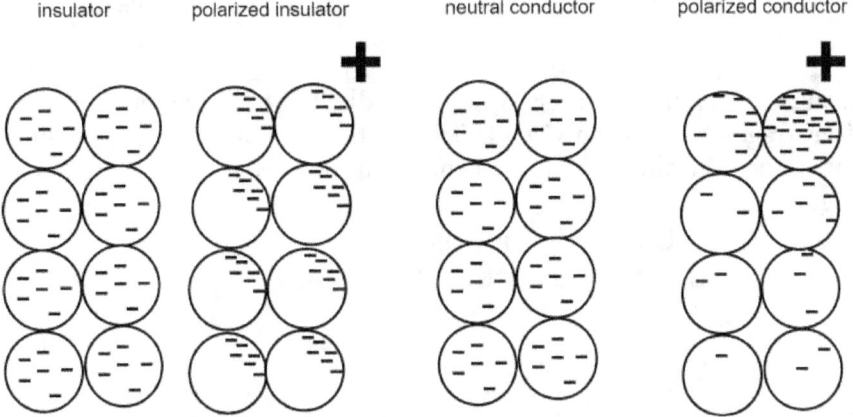

Figure 10-17: Polarization of an insulator vs. conductor

With this initial model we now want students to link three similar concepts. *Our aim is for them to understand that bonds are typically polar (to some degree), molecules can be polar or nonpolar, and that molecular polarity influences how molecules stick together.*

Bond polarity depends on electronegativity in two ways. The bigger the difference in electronegativity the more polar the bond will be. The average electronegativity also has some influence on how bonds function. A Van-Arkel Ketelaar triangle plots these two variances to help differentiate nonpolar, polar, metallic, and ionic bonds.

Chapter 10 – Bonding

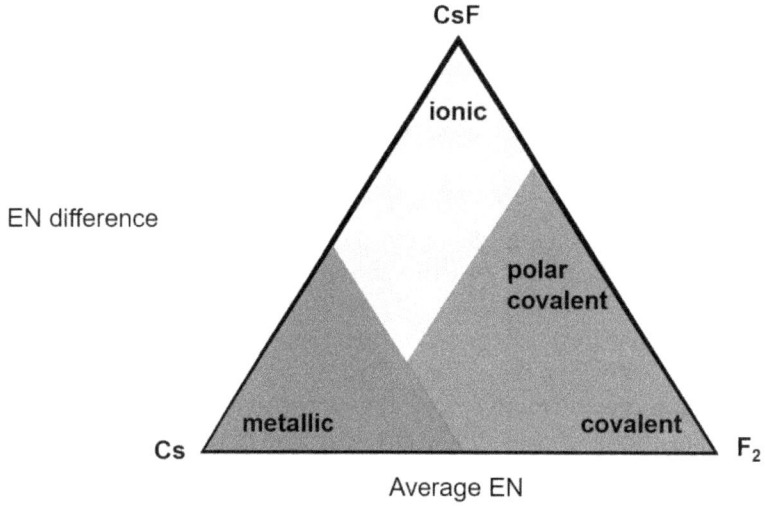

Figure 10-18: A Van-Arkel Ketelaar triangle plots electronegativity difference and average electronegativity

In most introductory chemistry classes, we just talk about electronegativity difference and omit average electronegativity. This leads to some inconsistencies, but rarely are they problematic. We should note that there will be exceptions to scales such as the following:

> 0.0 to 0.4 nonpolar
> 0.4 to 1.0 moderately polar
> 1.0 to 1.7 polar
> Above 1.7 ionic

Before students can even start asking you about what happens if it's 0.4 exactly you want to start clarifying that these are not incremental. As the electronegativity difference goes up, the polarity increases. 0.3 is more polar than 0.1. You can save yourself some headaches by telling students that either answer is fine when you are on a boundary and that anywhere from 0.4 to 1.7 should just be considered polar. When you get to molecular polarity there really is no reason to differentiate the two anyways as long as we are clear that the bond becomes more polar as the electronegativity (EN) difference becomes larger.

Because much of our organic and biochemistry is based on the idea that C and H form nonpolar bonds, we list 0.4 as nonpolar. But a much simpler evaluation system would be to all bonds between two different elements are polar. Some are so similar that the bonds are negligibly polar. Bonds on the other extreme tend to be ionic with some rare exceptions (Al and I can form a molecular compound).

With molecular polarity you will have students struggling to translate the 2-dimensional Lewis structure into a 3-dimensional image. They have to determine the symmetry while evaluating the various bonds to be polar or nonpolar. Before we get into an algorithm, I'd like to point out the introduction to polarity. If students understand polarity to mean neutral overall, but charges are separated they are more likely to be successful in envisioning a polarized molecule.

Chapter 10 – Bonding

Determining molecular polarity starts with the symmetry of the molecule. Students struggle with symmetry (in particular for tetrahedral shapes). A linear shape is easy for them to ascertain as symmetrical. Their intuition is that trigonal planar is also symmetrical. It is helpful to push back on that briefly. Why is trigonal planar symmetrical if there is nothing directly across from any of the bonds?

One of the fundamental tenets of VSEPR (Valence Shell Electron Pair Repulsion) theory is that bonds will maximize the bond angle to get as far from each other as possible. A perk of VSEPR is that symmetry will be favored. A molecule will not be symmetrical under two conditions. When there are bonds to different atoms, and when there are lone pairs on the central atom.

There are some rare exceptions to those two rules. You can have lone pairs on the central atom for square planar (AX_4E_2) and the linear derivative of trigonal bipyramidal (AX_2E_3) and still be symmetrical. You can also have different bonds for trigonal bipyramidal if the axial and equatorial positions are the same. The octahedral shape can also have different atoms bonded as long as each atom is the same as the one opposing it (180 degrees). Note that many of these examples would be extremely rare, especially in an introductory course.

Otherwise, these two criteria hold up. HCN is asymmetrical since it has a hydrogen on one side and nitrogen on the other. NCl_3 is asymmetrical because the lone pair on the nitrogen distorts the shape of the molecule causing there to be a nitrogen side and a chlorine side.

Dipole moments might help students. A dipole moment is a product of how much charge is separated and the distance of that separation. As charge increases and distance increases the dipole moment gets bigger. Dipole moments are vectors so they can help students visualize the symmetry of the molecule. If the sum of the dipole moments is zero, then the molecule is not polar. Some teachers explain this as a cancelling of dipole moments, but it works better to understand that there is no + and - ends (dipoles).

Your flow chart for molecular polarity should determine if the bonds are polar and if there is an asymmetry to those polar bonds.

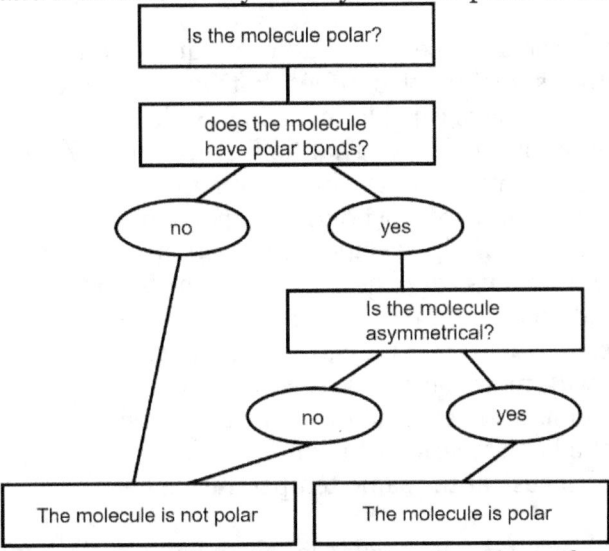

Figure 10-19: Flow chart to determine if a molecule is polar or nonpolar

Chapter 10 – Bonding

*****

At this point in the unit things begin to drag. All these concepts are related, but they lack a major landing point up until now. Who cares if a bond is polar, or a molecule is polar, or if I can draw a Lewis structure? What's the point?

The first major landing point is that molecules stick together to form liquids and solids. But different molecules stick together differently. The degree of molecular polarity changes how sticky molecules are. We want to emphasize stickiness before we introduce terms such as dispersion forces or dipole-dipole interactions. This helps students connect the concept of intermolecular forces between particles directly to the Lewis structure (Fig. 10-20).

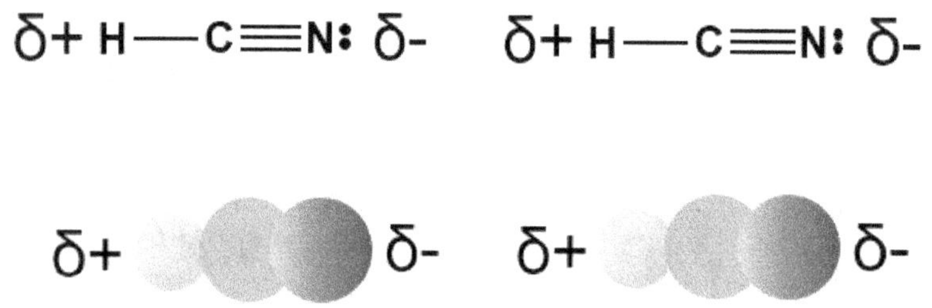

$\delta+$ H—C≡N: $\delta-$ $\delta+$ H—C≡N: $\delta-$

$\delta+$ $\delta-$ $\delta+$ $\delta-$

Figure 10-20: Two polar molecules of HCN shown with Lewis structure and particle diagrams

If a molecule has a large dipole moment, then that molecule is likely to be sticky. Stickiness influences a wide range of properties from melting point to reactivity. While the stickiness is because of an electrostatic attraction, the strength is much weaker than a typical covalent bond. So, we use the term intermolecular force in place of bond. When a molecule is polar the intermolecular forces are called dipole-dipole interactions.

But there are also examples of nonpolar molecules that form liquids and solids. They must be able to stick together somehow. How does that work? Electrons aren't static in a molecule. A nonpolar molecule is capable of having its electron density shift. These shifts result in a temporary polarity. That temporary polarity can cause other temporary shifts in nearby molecules. We call the resulting attraction from these temporary polarizations dispersion forces or London dispersion forces.

Dispersion forces tend to be weaker than dipole-dipole interactions, but that tendency is inconsistent. Just as dipole moments vary based on electronegativity differences, dispersion forces vary based on the number of electrons. Some texts will erroneously state that dispersion forces vary based on mass, but mass changes are not the cause. Mass just usually correlates with more electrons. If we had two samples and one used heavier isotopes, the dispersion forces would be equivalent because of the commonality of electrons despite the larger mass for the one sample. The melting points would not be identical however because at the same temperature a heavier isotope moves slower.

An example that I struggled with when I was a high school student was carbon dioxide. I was told that there are polar bonds but that the dipole moments cancel and

Chapter 10 – Bonding

so the molecule is nonpolar. But in my head, I was wondering why the carbon could not be aligned with an oxygen on a different molecule so that positive and negative charges are aligned. The problem with such an arrangement is that it only works for a small number of molecules. When we expand to 3-dimensions and trillions of molecules these arrangements don't function. This is evidenced by the fact that carbon dioxide is a gas at room temperature because the intermolecular forces are weak. Your students may find similar sticking points if you always present 1 or 2 molecules. Part of chemistry is that we tend to work with quadrillions of molecules at a time.

Hydrogen bonding is a big deal in chemistry and biology. But hydrogen bonding is just an extreme form of dipole-dipole interaction. When we track properties, we see an unusual blip when hydrogen is bonded to a small electronegative element such as N, O, or F (Fig. 10-21).

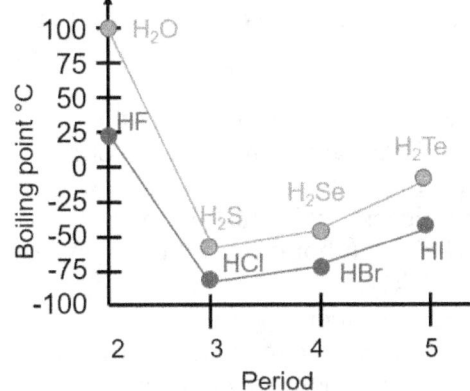

Figure 10-21: Boiling points trends for HF, HCl, HBr, and HI show that HF is an anomaly.

When an atom is bonded to fluorine, the fluorine is going to shift electron density away from that atom. But a hydrogen atom is particular because there are no other core electrons. This exposes the proton of the hydrogen as opposed to a nucleus with some core electrons (cation). The exposed proton is what causes hydrogen bonding to be unusually strong.

The released 2013 AP Chemistry exam asked a fantastic question. It stipulated that formaldehyde can from hydrogen bonds with water and asked students to identify whether the oxygen or the hydrogen of the formaldehyde is the part capable of hydrogen bonding. In most hydrogen bonding cases, the oxygen (or F or N) that causes the exposed proton is also involved in the intermolecular force. The oxygen on water molecule 1 attracts to the exposed proton of hydrogen on water molecule 2 (Fig. 10-22). But really the needs for hydrogen bonding are the exposed proton of a hydrogen and anything that is negatively charged. So, the formaldehyde (methanal) can hydrogen bond via its oxygen with the exposed proton of water. The hydrogen on formaldehyde has carbon adjacent and therefore cannot fulfill the role of having the exposed proton of a hydrogen (due to the lack of polarization). The dipole-dipole label is questionable as well as the carbon-hydrogen bond is mostly nonpolar.

Chapter 10 – Bonding

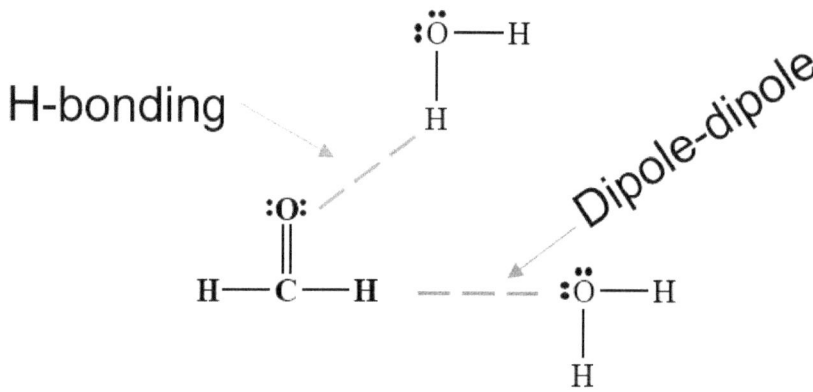

Figure 10-22: Hydrogen bonding between the oxygen of methanal and the hydrogen of water differs from the dipole-dipole interaction between the hydrogen of methanal and the oxygen of water.

    What should students know about hydrogen bonding? We need them to understand that the name is frustratingly not ideal. Hydrogen bonds are not bonds even though their name has the word bonds in it. But the attractive forces in bonds and intermolecular forces are all electrostatic. The biggest difference between intermolecular forces and bonds is the strength. Hydrogen bonds vary from about 4-13 kJ/mol while the H-H covalent bond is about 440 kJ/mol.

    Intermolecular forces impact how well molecules stick together. This has implications for melting points, boiling points, reactivity, protein structures, material science engineering, solubility, capillary action, and more. Soap is a fantastic discussion point because soap has a polar end that sticks to water well, and a nonpolar end that sticks to fat, oil, and other nonpolar substances. Emulsifiers such as soy lecithin do the same thing, but they help make chocolate delicious (soap does not).

    Surface tension is an interesting phenomenon that teachers often bring up. The molecules at the surface of a polar liquid have stronger attractions inward towards other liquid particles than they do outward towards the surrounding air. This causes a tension to form at the surface. One way to demonstrate this is to add small objects to a full glass of water. The water will not spill for some time and the surface of the water will curve upwards to a noticeable height above the top of the cup before it does.

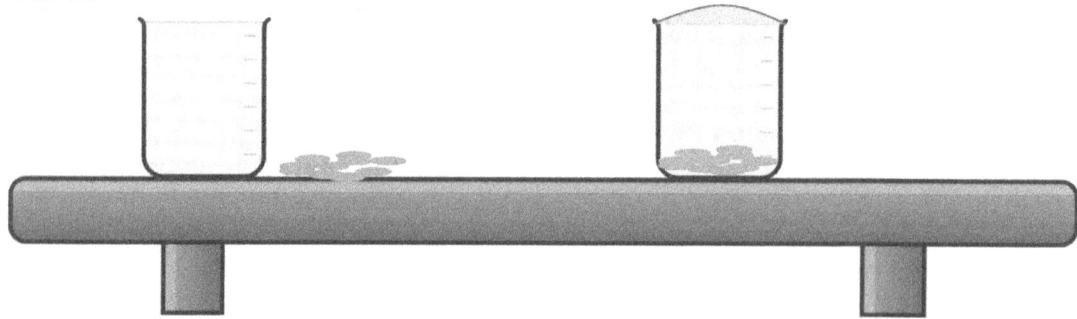

Figure 10-23: Adding pennies to a full container of water will not cause a spill for some time

Chapter 10 – Bonding

You can add a paper clip to this surface and the paper clip will be supported by the surface tension. If you add soap to the water the paper clip will quickly fall and the water above the top of the glass spills. If you do add soap, make sure you rinse everything thoroughly before the next class or the soap residue that lingers will prevent the curvature from forming. A similar demonstration is to count how many drops of water can be added to the top of a penny.

Another demonstration I enjoy is to drop marbles through a homologous series of organic compounds to compare how long the marble takes to fall. A marble will fall faster through methanol than propan-1-ol. This begs two questions. Why is the propan-1-ol stickier, and what has to happen for the marble to fall?

The propan-1-ol has more electrons, therefore, propan-1-ol has stronger dispersion forces than methanol. Both have hydrogen bonding. Molecules that are in contact must separate in order for the marble to descend.

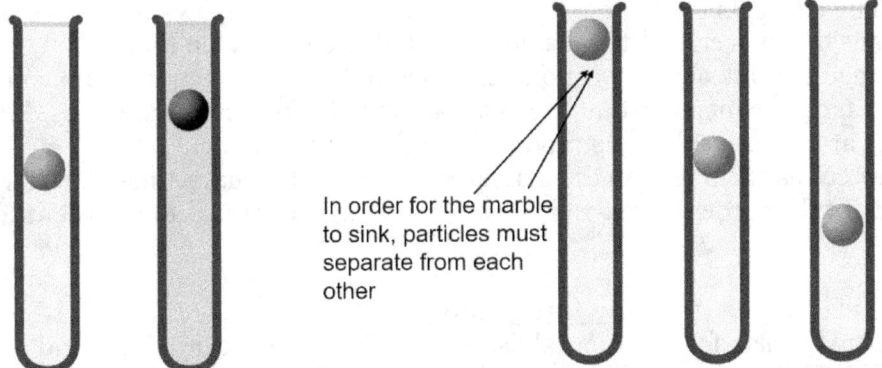

In order for the marble to sink, particles must separate from each other

Figure 10-24: A marble will sink faster in methanol than propan-1-ol due to weaker intermolecular forces for the methanol

When I know the students in the class are going to be learning organic chemistry, I want to make sure they start to analyze organic structures with regards to intermolecular forces. One simple experiment involves dipping a Q-tip into an organic liquid. A line of the liquid is traced onto a dark table and the students must construct a series of trials to determine a ranking for the stickiest and least sticky liquids by how quickly they evaporate. How would you predict heptane, acetone, ethanol, propan-1-ol, and butan-1-ol compare?

After the students rank them (I have them put stickers onto a chart, high school students love stickers), I put the structures on the board and ask them to analyze why the order is the way that it is. Why are acetone and heptane so weak? Why is acetone similar to heptane even though heptane only has dispersion forces and acetone has dipole-dipole interactions?

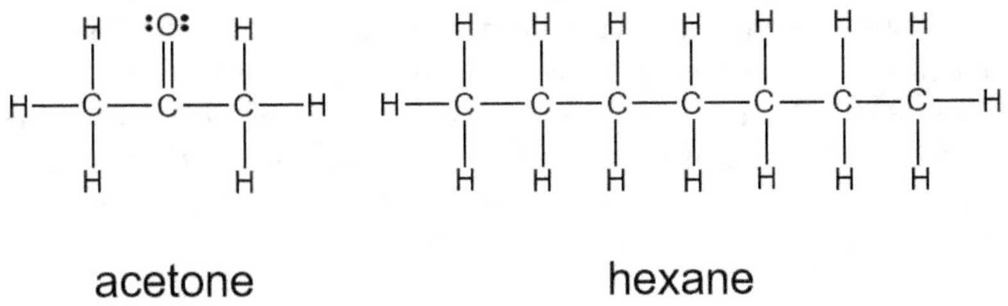

acetone   hexane

Figure 10-25: Lewis structures of acetone and hexane

Chapter 10 – Bonding

Before concluding intermolecular forces, take a moment to think of all the knowledge that takes you from the formula NH_3 to the type of intermolecular force. You have to be able to draw the Lewis structure, determine the shape, visualize the shape, find the bond polarity using electronegativity differences, determine the symmetry of the dipole moments, and remember that hydrogen bonded to N, O, or F leads to hydrogen bonding. You have to understand the concepts of polarization, electronegativity (nuclear charge, shielding, principal quantum number), charge, and more. You have to be able to distinguish between intramolecular and intermolecular.

Then consider how many implications there are for the intermolecular forces. The boiling point, meniscus in volume measurements, viscosity, reactivity, solubility, and so much more are all impacted by the type and strength of intermolecular forces. There can be implications for spectroscopy, reaction mechanisms, catalysis, Le Chatelier's principle, and many other complex topics that are a step or two removed from these bonding concepts.

This means that it is critical to take time for students to develop these concepts. They need to think about questions that inspire these connections. And you need to develop as many concrete connections as possible. Make sure the students do an experiment or at least see a demonstration where they must explain intermolecular forces as stickiness on the macroscopic level. Then have students draw these phenomena at the particle level, and again using Lewis structures to represent the particles.

How things stick

After intermolecular forces are established, it is time to move into ionic and metallic bonding. Many teachers do these first, but I prefer the focus to be on covalent bonding. Moving ionic and metallic bonding to the middle gives me more time to space out practice with Lewis structures and that is the most important concept from the unit that students will need for future content.

One method of introducing ionic bonds is to start by using Lewis dot structures of the elements as they change from neutral to ions.

Figure 10-26: Lewis dot structures for sodium and chloride ion formation

The sodium and chlorine do not have to transfer electrons to each other in order to form ionic bonds as sodium chloride. The elemental potassium must lose its electron to something, but the bond itself is just based on the attractive forces between the ions. For example, the Na metal could react with H_2O to form NaOH, the NaOH could be neutralized by HCl, then the water and excess HCl evaporated to form crystals of sodium chloride. The Na^+ ions are bonded to Cl^- ions, but there was not a transfer of electrons from Na to Cl.

Chapter 10 – Bonding

When using these dot structures, it is critical to connect back to the distinction between charged forms and neutral forms of elements that we discussed in the naming and formula writing chapter. The sodium with a dot is neutral, it is a metal, and it is highly reactive. The Na with the + sign is positively charged, it is a part of an ionic compound (probably), and its reactivity has changed because of the absence of the valence electron.

Another consideration we want to make is that the reaction (Fig. 10-26) shows only one atom of Na and one atom of Cl, but that an ionic compound would be a 3-dimensional crystal lattice with gajillions of both ions. We don't want students to think of KCl as a molecule, but as a formula unit. That means we need to clearly distinguish the two at the particle level and so we need to clarify representations that muddy that distinction.

While learning about ionic bonds, I find that the melting point trends are great for students to connect how charge and size impact bonding. NaF, Na_2O, and MgO have melting points of 993 °C, 1132 °C, and 2852 °C. The melting points increase as the charges of the ions increase. NaF, NaCl, and NaBr have melting points of 993 °C, 801 °C, and 747 °C. Here the trend is that as the anion gets larger the larger separation between the ions causes a weaker attractive force. There are two ways to influence an electrostatic force. More charge results in a larger force. A larger separation results in a smaller force.

Melting ionic compounds involves the breaking of ionic bonds. Since the ionic bond is the electrostatic attraction between a cation and an anion, as the ions change positions that interaction is being reduced. When we talk about ions being able to move, we set up students to be able to understand what happens in a salt bridge and solutions of a galvanic cell for redox chemistry. Students do not have a concept of electrical current as a result of moving ions and we can start that framework here.

The strength of ionic bonds relates to solubility. For an ionic compound to dissolve in water the bonds must be broken which means the water must have stronger interactions with the ions than the other ion does. Most insoluble ionic compounds have larger charges or have large ions that do not interact as strongly with the smaller water particles.

Soluble
All alkali metal ions are soluble (Na^+, K^+, etc.)
All ammonium compounds are soluble (NH_4+)
All nitrates are soluble (NO_3^-)
All chlorates are soluble (ClO_3^-)
Most chlorides are soluble (Cl^-)
Most sulfates are soluble (SO_4^{2-})

Generally Insoluble
Carbonates, hydroxides, phosphates, sulfides (CO_3^{2-}, OH^-, PO_4^{3-}, S^{2-})

Insoluble
$BaSO_4$
AgCl, AgBr, AgI
PbI_2, $PbBr_2$, $PbCl_2$
$HgCl_2$, $HgBr_2$, HgI_2

Chapter 10 – Bonding

Metallic bonding is a fascinating topic with so many different directions to go in. How much time do we have left in the unit? I have seen two representations of models for metallic bonds that I like. The best one for an introductory class is where the metal atoms are treated as cations with the valence electrons as weakly held. These electrons can move about the entire lattice of cations with minimal restriction ("delocalized"). The other representation uses conduction bands in lieu of individual orbitals which is likely not productive for this level.

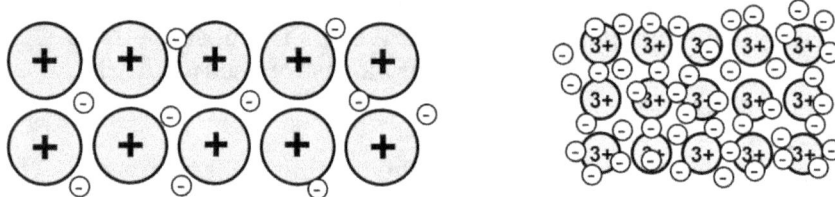

Figure 10-27: Particle model of metallic bonds for Na and Al. Which is stronger?

This image is functional for the introduction because it explains why metals with higher charged cations have stronger bonds. Aluminum has such a valuable combination of strength, durability, and light weight because of its small 3+ charged cations. Sodium is soft and can be easily cut with a butter knife because of its large 1+ charged cations.

The melting point trends provide insight into metallic bonding. The alkali metals (Li, Na, K) have melting points of 180 °C, 98 °C, and 64 °C. These all have 1+ charged cations, but the increase in size causes a larger distance between the electrons and cations and so a weaker bond is formed with potassium being the weakest. Cesium melts at 28 °C so it would melt in your hand if it didn't burst into flame in contact with the air.

Metallic bonding is an opportunity to talk about alloys. Alloys are frequently present in your daily life. Chromium is added to steel for your silverware to prevent rusting. The metals mixed in with your gold jewelry make the gold more rigid. The students who were just in band class were just surrounded by brass. In World War I, Germany discovered that adding molybdenum to steel allowed the steel to be used for firing larger explosives. Germany didn't have a molybdenum source but were able to take over a mine in Colorado that they used to make the Big Berthas that could hit Paris from a 75-mile range.[10]

If you really want to have some fun, then I recommend going briefly into superconductors. When metals are cooled to extremely cold temperatures (even colder than winter in Michigan), the electrical resistance drops to zero. Mercury does this below 4K (-269 °C, -452 °F). As the electron travels through the lattice of cations, the cations shift in towards the electron. This causes a second electron to attract to the first electron because of the proximity of the cations. Any disturbance or change in energy of the first electron is transferred to the second electron and so they combine to experience no resistance.

The best result of superconductors is that the zero electrical resistance causes the Meissner effect where magnetic fields cannot penetrate the superconductor. This allows for some fascinating magnetic levitation.

Before we move into technical portions of bonding, I'd like to revisit the demonstration we started with. After instruction do students now understand why

Chapter 10 – Bonding

hitting a sugar cube with a hammer does not break bonds, but hitting a copper (II) sulfate crystal does? Do they understand why breaking a paper clip breaks bonds, but ripping a piece of paper does not? And can they give a sufficient explanation discriminating between the intermolecular forces of the paper and the metallic bonds of the paper clip?

A powerful lesson to tie together some of these concepts is for students to work at explaining the particle level of metals being malleable and ionic compounds being brittle. I like to exaggerate these properties by hitting solids with an actual hammer. If you hit something with a hammer, does it shatter into a million pieces (brittle) or does it deform (malleable)? Metals deform.

But what happens at the particle level? Do the cations slide past one another? Yes. But as they do they retain attractions to the sea of delocalized electrons (they are still bonded). When you hit a chunk of salt with a hammer the ions slide because of the force but this forces them out of alignment. Now you have cations by cations and anions by anions and so the ionic bonding fails causing a shattering into lots of pieces.

You may teach a lesson about the types of solids. Some teach four (ionic, molecular, network covalent, metallic), but I recommend adding in plastics as a fifth. One of the key ideas being stressed with types of solids is that the molecular solids have intermolecular forces holding the object together but covalent bonds holding the molecules together. When we heat ice, it is the hydrogen bonding that is disrupted during melting or sublimation. The covalent bonds remain intact and so the molecular structure remains the same.

There are a variety of plastics, but thermoplastics are what we usually think of. These can be molded when heated because the polymer strands can slide past one another as the intermolecular forces are disrupted at higher temperatures. But the polymer chain remains intact because the chain is held together by covalent bonds. If you heat an empty gallon milk jug the plastic will become clear and colorless in the spot you heat, and you can form a bubble by blowing into the jug (wash it first!). The plastic organizes into a more structured crystalline arrangement at higher temperatures causing the interactions with light to change.

You can also purchase pop bottle preforms. These are the 2-liter or 20 oz. bottles before they are molded. They are made of polyethylene terephthalate (PET). Pop is stored with heavy pressure of carbon dioxide so that the pop does not go flat until well after being opened. Therefore, the plastic used for pop bottles needs to be more durable than for bottled water. The pop bottle preforms are what is shipped to the factories where they make the pop itself, so they don't waste money by shipping large empty containers. Instead, they can just mold them at the factory.

Any container that has the recycling code 2 or 4 is made of polyethylene. The 2 code is for high-density polyethylene (HDPE) and 4 is for low density polyethylene (LDPE). The monomer ethylene (or ethene) is C_2H_4. To make polyethylene, multiple monomers of ethene are linked together. But as they link, branches are formed. LDPE has more branching and thus more empty space due to the inefficiency of packing chains together. HDPE has less branching and so the solid can be packed more efficiently. LDPE tends to be more flexible and transparent while HDPE is opaque and more rigid.

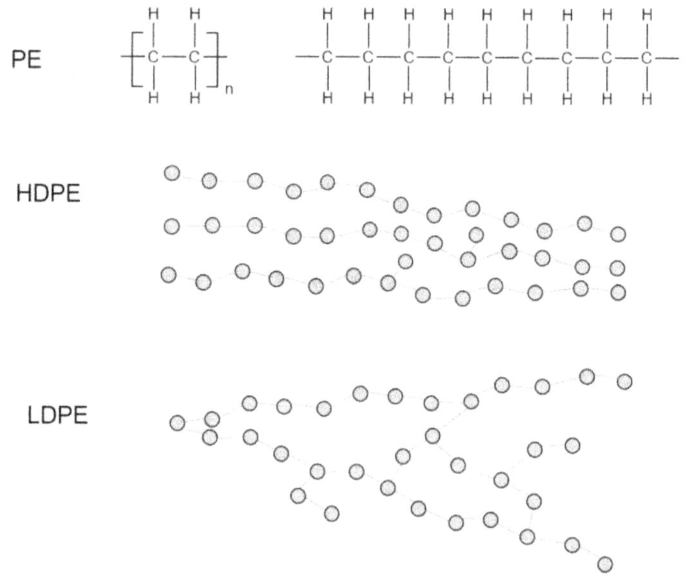

Chapter 10 – Bonding

Figure 10-28: HDPE and LDPE branching differences

PE

HDPE

LDPE

There are so many interesting applications and stories about plastics. The accidental discovery of Teflon when Roy Plunkett was working with tetrafluoroethene, and the container stopped releasing the gaseous monomers because they polymerized inside the container. Kevlar is the polymer used for making bullet resistant vests and the synthesis must be done in concentrated sulfuric acid in order to disrupt the strong intermolecular forces between chains. Or if you really want to talk about some biochemistry you can talk about proteins or DNA. The sheer length of a DNA chain makes its storage and replication an impressive feat (how does it not get tangled?).

The final type of solids are network covalent compounds. There aren't as many of these as some of the other solid types, but they are interesting and really help students differentiate between covalent bonds and intermolecular forces. Network covalent compounds are held together in a web of covalent bonds. In diamond the carbon atoms are all connected to four other carbon atoms. Graphite has interconnected hexagonal arrangements of carbon atoms similar to a sheet of benzene rings. One individual sheet would be called graphene and a graphene sheet can form into a cylinder called a carbon nanotube. Graphite has strong intermolecular forces between the sheets, but these can be broken much easier than the covalent bonds within each sheet. When you write with a pencil the layers are sliding past each other with some graphite remaining on your paper.

Quartz (SiO_2) and boron nitride (BN) are two other network covalent compounds. The melting points of network covalent compounds are very high (Quartz is around 1700 °C). Many do not melt. They either sublimate or decompose at extreme temperatures before they can melt. The strong covalent bonds are reflected by the properties of network covalent compounds such as high melting points and generally being insoluble. But there is variation in conductivity among the group. Graphite is able to conduct electricity due to the delocalization of the electrons in the sheet.

Obviously, there is variability within a group (e.g., ionic solids). But there are key characteristics (melting point, solubility, conductivity, etc.) that allow us to make predictions for what type of bonding an unknown solid has. These predictions reinforce many of the core bonding principles that distinguish covalent, ionic, and metallic bonding.

Chapter 10 – Bonding

Resonance

Resonance structures are difficult for students who have just learned Lewis structures. Delaying them until later gives them a chance to have some spaced practice and it gives you a chance to introduce some more concepts. But resonance structures are a valuable tool to demonstrating some of the limitations of Lewis structures. Ozone and benzene are good initial examples. Students are familiar with ozone from earth science lessons. Benzene is good to use because the phenyl group is common in organic chemistry.

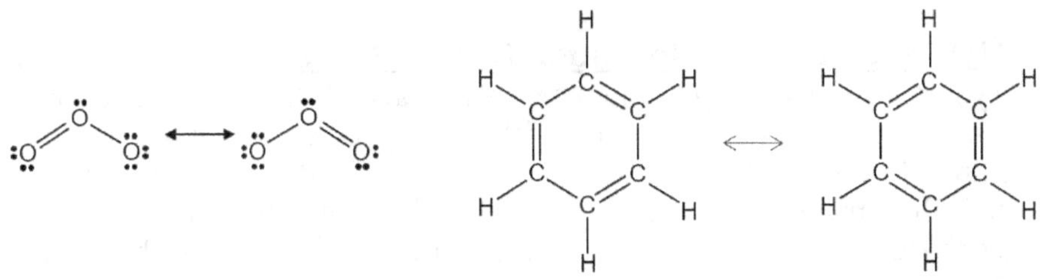

Figure 10-29: Resonance structures of ozone (O_3) and benzene (C_6H_6)

The bond length of an oxygen-oxygen double bond is 121 pm, and the oxygen-oxygen single bond length is 147 pm. What are the bond lengths in ozone? They are both the same length, 128 pm. Note that the bond length is smaller than a single bond and longer than a double bond. Now take a second as a teacher to think about what changes are needed to move from the one resonance structure to the other. We would need some electrons to move, which is possible without the nuclei of the oxygen moving much. What we have are oxygen atoms locked in place (besides their vibrations and rotations), but some of the electrons are free to move about the molecule. The term for this is delocalization. Instead of the electrons being localized in a single spot, they are delocalized and move over a wide range of the molecule.

The 1st idea we want to stress to students is that the nuclei remain in place. The 2nd idea is that some of the electrons are not localized on one atom or between two atoms. The 3rd idea is that a single resonance structure is insufficient to describe the molecule. We arrive at these conclusions from the observation of the bond lengths and bond energies.

The carbon-carbon double bond and single bond lengths are 134 pm and 147 pm. What do we expect the bond length to be in benzene? All six bonds between carbons are 138 pm. Benzene was trouble for chemists until in 1865 August Kekulé theorized that benzene was a cyclic ring. Kekulé got this idea from Archibald Couper in 1858. Prior to this, chemists knew the formula C_6H_6 from combustion analysis, but had trouble envisioning a structure that would accommodate the units of unsaturation as well as the number of isomers of benzene that have side groups. The ortho, meta, and para versions of dibromobenzene make sense with the cyclical structure.

There was some lingering pushback for the next couple of years, however, as the resonance of the double bonds was not part of the proposal. If the double bonds in benzene were locked in place, this should mean that there would be two different isomers of 1,2-dibromobenzene. One with a single bond and another with a double bond between the 1st and 2nd carbons. No such isomers exist because the bond lengths are identical at 138 pm each.

Organic chemistry extends resonance into a wider range of rules and exceptions. Which elements can violate the octet and under which circumstances plays a huge role in explaining reactivity and stability of organic molecules? For further practice in inorganic chemistry, polyatomic ions make for excellent practice with drawing resonance structures. Students can even start to draw how electrons shift from one resonance structure to another as a precursor to curly arrow mechanisms.

Hybridization vs. Molecular Orbital Theory

I was many years into teaching chemistry when I learned that hybridization was outdated. The theory in hybridization starts with discrepancies between atomic orbitals and molecules. If we look at a methane molecule, the four bonds seem to be the same even though the carbon atom is using electrons from different atomic orbitals to construct these bonds. How can an electron from a 2s orbital produce the same bond as an electron from a 2p orbital?

The theory of hybridization states that the four valence electrons of carbon go from a set of 2s and 2p orbitals to having four identical hybrid sp^3 orbitals. These sp^3 orbitals align with the observed shape of a tetrahedron. For an ethene molecule, the carbon uses three atomic orbitals to form a set of 3 sp^2 hybrid orbitals. The remaining p-orbital is used for the double bond.

Thus, students can use two methods to algorithmically determine the hybridization of an atom (it is always the atom that forms hybrid orbitals, not the molecule). They can identify the shape and connect the shape to the hybridization. Or they can count the number of atoms attached (and lone pairs) to an atom. The number of atoms + lone pairs will be the number of atomic orbitals that hybridize. CH_4 has four atoms attached to the carbon so the carbon is sp^3 hybridized. NH_3 has three atoms and 1 lone pair so the nitrogen is also sp^3 hybridized. A carbon in ethyne (C_2H_2) has two atoms attached (a triple bond still just counts as one atom attached) and no lone pairs so the carbon is sp hybridized.

It turns out that the bonds are not equivalent. Molecular orbital theory gives much better agreement than hybrid models. But we still use the nomenclature of hybridization in organic chemistry. Perhaps for that reason the subject continues being taught.

Whenever you have options, molecular orbital (MO) theory is better than hybridization. In hybridization the atomic orbitals of an atom change to hybrid orbitals. In molecular orbital theory we make new orbitals for the molecule. They sound similar, but there is evidence that supports MO theory that is not easily explained under hybridization theory. For a hybridization model the hydrogen in methane (CH_4) are just the regular atomic orbitals (1s) and the carbon forms hybrid orbitals (sp^3). But with molecular orbitals we would construct a new set of orbitals for the entire molecule.

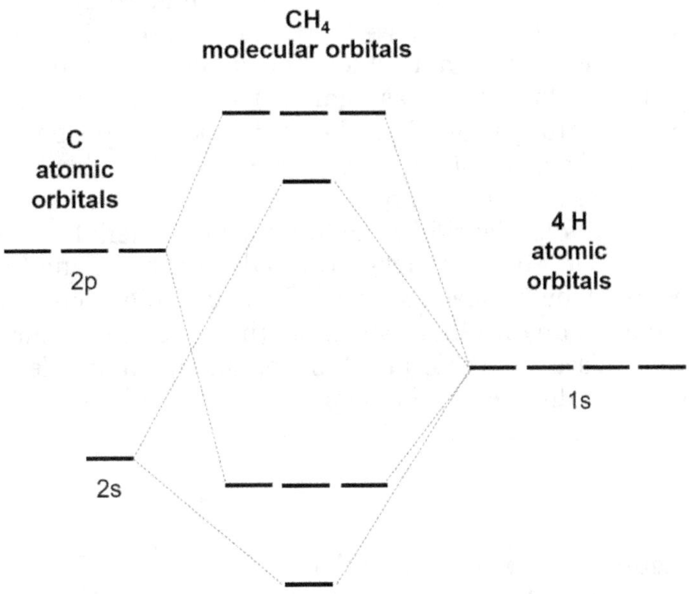

Chapter 10 – Bonding

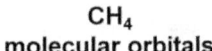

Figure 10-30: Molecular orbital diagram for methane (CH_4)

If hybridization theory were correct, then the four bonds in methane would be equivalent as they would all involve a hybrid orbital from carbon and a 1s orbital from hydrogen. Molecular orbital theory on the other hand predicts a molecular orbital from the overlap of the 2s carbon orbital and 1s orbital of hydrogen. There will be 3 equivalent molecular orbitals from the 2p orbitals of carbon and the 1s orbitals of hydrogen. And there will be unoccupied antibonding orbitals for those as well. We can confirm the molecular orbital theory by striking the methane molecule with light that has the same energy as the gap between the highest occupied molecular orbital (HOMO) and the lowest unoccupied molecular orbital (LUMO). The absorbance of that light provides evidence that molecular orbital theory is a superior model. The different bond energies in methane also provide additional evidence.

In molecular orbital theory the focus is on constructing sigma and pi orbitals. A sigma orbital lies between the two nuclei while a pi orbital is above and below the plane. The other big concept is that there are bonding orbitals, nonbonding orbitals and anti-bonding orbitals. When an atomic orbital from atom 1 combines with an atomic orbital from atom 2 they form a bonding and antibonding set of orbitals for the molecule.

The antibonding orbital will have electron density away from the midpoint of the two nuclei and therefore has higher energy than the atomic orbitals had. The bonding orbital will be between the two nuclei and therefore is lower in energy. The antibonding always increases in energy by more than the bonding orbital decreases. If both the bonding and antibonding orbitals are filled the molecule is not stable.[11]

The evidence supporting molecular orbital theory is fascinating. MO theory tells us why oxygen molecules are paramagnetic. MO theory helps explain why the C in carbon monoxide ligands is the better source of electron density even though oxygen is more electronegative. The color from beta carotene comes from the excitation from the highest occupied molecular orbital (HOMO) to the lowest unoccupied molecular orbital (LUMO) for the long chain of alternating double bonds.

But MO theory also helps explain basic chemistry such as molecular stability, which portions of molecules are the most reactive, and molecular polarity. The diversity of applications for MO theory makes it a great topic to put into an

Chapter 10 – Bonding

introductory chemistry course even with its complexity. Using the HOMO and LUMO to explain Lewis acid-base theory can demonstrate the versatility of the model.

Even though most teachers do not teach MO theory, they do touch on components of it such as sigma (σ) and pi (π) bonding. They may teach that a sigma bond is stronger and that when a double or triple bond form, the first interaction is a sigma bond, and all the rest are pi bonds. A double bond has 1 σ and 1 π bond. A triple bond has 1σ and 2 π bonds.

Molecular orbital theory is also a benefit to the teacher even if you choose not to teach it. Many teachers incorrectly equate Cl⁻ and Cl_2 as having the same octet. The chloride ion has had an electron added to the atomic orbital. The chlorine atom in the molecule on the other hand is sharing an electron pair in a molecular orbital. Some view this as both chlorine atoms obtaining an 8th electron but that doesn't explain the clear discrepancy in the properties of molecular chlorine and chloride ions.

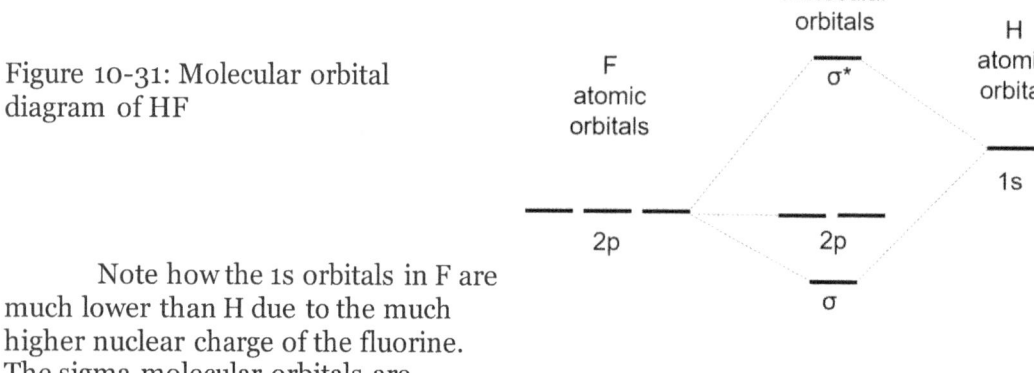

Figure 10-31: Molecular orbital diagram of HF

Note how the 1s orbitals in F are much lower than H due to the much higher nuclear charge of the fluorine. The sigma molecular orbitals are formed between a 2p-orbital of fluorine and the 1s-orbital of hydrogen. One molecular orbital is the bonding orbital, the unoccupied σ* orbital is the antibonding orbital. By hitting an HF molecule with light, an electron could be promoted from a ground state molecular orbital to the unoccupied σ* orbital. This could destabilize the bond, but also would provide evidence for the MO theory that would not be explained using hybridization theory.

Bonding is about things that stick together. There are a large number of concepts that converge in this chapter. But you will also have a lot of opportunities in thermodynamics, kinetics, equilibrium, redox, and organic chemistry to reinforce those concepts. Therefore, your two keys for the unit are to get students thinking about how atoms, ions, and molecules can stick together. Also, you are trying to make students proficient in drawing and interpreting Lewis structures. The amount of information within a Lewis structure is substantial. Students will be stressed in organic chemistry and kinetics when we start to really stretch the rules of Lewis structures in the middle of high-speed collisions during a reaction. That's why it is important they have developed some tools for the basics now.

Chapter 10 – Bonding

Student Struggles

1. "I have no idea what I'm doing!" The two most likely places where this would occur would be in determining intermolecular forces or drawing Lewis structures. Intermolecular forces require the student to be able to draw a Lewis structure correctly, determine bond polarity, and apply the bond polarity and 3D symmetry to determine molecular symmetry. There's a lot going on and we should expect issues. One helpful means is to start the students with a set of particles that represent polar or nonpolar molecules. Walk the students through the final image without the Lewis structures (Fig. 10-20 particle diagrams). They can better fill in the details needed to get there later.

For students that are stuck on Lewis structures, have them put more work into the process. Have them identify the number of valence electrons that they are working with and the core electrons that are being omitted from the structure. Then have them try different ways to put the pieces together. Most students who feel stuck just sit there until you show them the next step. Instead ask them to put together some wrong guesses. This helps them decipher what they are and are not supposed to be doing.

2. "I keep getting resonance structures incorrect." When students struggle to produce a second resonance structure after the first, they have been looking at the structures without envisioning how to change one into the other. You'll want to do a bit of curly arrow action with them. This helps them understand the conservation of charge that we're adhering to. Students may also struggle with examples that involve unstable resonance structures. If I draw a carbocation as part of a set of intermediate resonance structures, I'm confusing students who are not able to place this "unstable" structure into their current format of guidelines. When starting resonance structures, it's best to use stable chemicals such as ozone, phenyl groups, and polyatomic ions.

3. "I don't understand what resonance structures are. Do the molecules change from one to another? Are they a blend of them?" The key idea for this student is that the nuclei are much more stationary than the electrons. The nuclei aren't moving (relative to one another, ignoring vibration). When we draw a double bond in one resonance structure that becomes a single bond in another, we're indicating that two of those electrons are not localized (stuck in place) between those two atoms. That doesn't mean that the electrons are moving back and forth though. The electrons are distributed over multiple spots of the molecule all of the time. The student is likely thinking of the electrons as particles, but waves are better. In fact, another way to describe these delocalized electrons is by using molecular orbital (MO) theory.

4. "How can sulfur have more than 8 electrons?" One thing many teachers do when teaching bonding is take atomic orbitals and combine them to make a molecule with those same atomic orbitals. See how hydrogen has a $1s^1$ configuration? It can add another electron! But when we put together an H_2 molecule, we're not making a $1s^2$ orbital. We now have a set of molecular orbitals in place of the two atomic ones previously. Now we have bonding and antibonding orbitals. When we limit our descriptions to combinations of atomic orbitals, we end up limiting

Chapter 10 – Bonding

students' abilities to view more complex structures. I say that knowing that we can construct molecular orbitals by combining atomic orbitals, but when combined they differ from their initial representations.

5. "I don't understand how nonpolar things can attract each other."
Polarization is the key concept at play here. A neutral object can polarize, where you split charges so that one side is + charged and the other is - charged. Yet the object remains neutral overall. If electrons shift more to one side than the other, one side will have a surplus of - charge while the other has a deficit of - charge. That might satisfy most students. But for the remaining skeptics, we also need to note that having a polarized (yet neutral) object means that there will be a net attraction or repulsion to something else charged because of that separate of charge. If we place a + charge by the polarized object, the object can turn so that the - side is closer. This proximity makes the attractive forces larger than the repulsive forces from the further + side. Distance plays a role in Coulomb's law. A student might also confuse neutral with zero charge.

6. "Why is the hydrogen bond so much stronger than dipole-dipole interactions?"
This conversation can get heated with teachers. I have yet to see any explanation of a hydrogen bond that shows that the hydrogen bond is anything other than an unusually strong dipole. The bond is polarized, and the attraction is from that polarization. What is key for hydrogen bonding is that hydrogen only has the one electron (or two for a bond). When that electron shifts away, you're left with a proton. If another element polarizes, you have a core set of electrons. But hydrogen exposes a bare proton to whatever negative charge the hydrogen is attracted to via hydrogen bonding. The lack of core electrons and the proximity of the exposed hydrogen nucleus explain why hydrogen bonding is a uniquely strong interaction.

7. "I'm not sure how to start calculating formal charge."
Formal charge is difficult for students who are new to Lewis structures. In Lewis structures we count any electron an atom has whether it is shared or a lone pair. With formal charge we count half of the bonded electrons and all the lone pairs for each atom. This makes it difficult for students to distinguish. It can help to teach formal charge when introducing Lewis structures so students can be asked questions about both and use formal charge as a tool to construct the Lewis structures.

When students are stuck, they should start by counting. How many bonded and nonbonded electrons are there? How many valence electrons does this element typically have? It helps students to show elements in isolation (Fig. 10-11) and taking time to digest why the formal charges change the way that they do.

8. "The Lewis structure for CO doesn't make sense to me."
Coordinate covalent bonds throw students for a loop. Their brains naturally see them as deviating from other structures. Carbon monoxide is a great opportunity to let students see how the formation of a coordinate covalent bond fits in with what they had previously learned. You'll want to show the structure forming bit by bit. MO for carbon monoxide (CO) shows the (HOMO) lone pair exists primarily on C even though O is more electronegative which explains the phenomena where the carbon is the better nucleophile.

Chapter 10 – Bonding

9. "How did you know how many lone pairs go on the central atom?" If you ask the students to construct the Lewis structure of XeF_2, many will know that fluorine will form a single bond. But they might get stuck determining how many lone pairs should go on the xenon center. One method they could find a solution would be by counting how many total valence electrons they have to distribute. Another would be to use the fact that both fluorine atoms have a formal charge of 0, so Xe must also have a formal charge of 0 for the neutral molecule (really this is similar to counting electrons). But more important than method is that teachers recognize that students need practice doing this with elements like S, Se, P, As, N, Xe, I, Br, etc. that can have a variety of lone pairs of electrons.

10. "I thought H_2S would be linear!" Students will see Lewis structures and their brains will automatically identify the 2-dimensional shape. It is important that you draw bonds in different directions when possible. A water molecule can be drawn with the three elements in a line, with the oxygen as a corner, or in a bent shape that clearly shows the angle is not 90°. Expect students to require multiple instances of this before they can detach from the 2-dimensional shape. They haven't had organic chemistry yet!

Phenomena

1. Water on a penny - The big theme in bonding is that things stick together. When water droplets are placed carefully onto a penny, the water can fill to an appreciable extent before it loses its shape and runs over the edge. The water particles stick together well. You can try adding other liquids such as ethanol, or acetone. But they won't exhibit the same cohesive forces and the drop that forms will be minimal.

2. Ethanol vs. heptane - Did I mention what the theme is in bonding yet? Take 2 Q-tips and draw a thin line of ethanol next to a thin line of heptane. The heptane evaporates much faster than the ethanol. This is because ethanol sticks together well. Heptane does not. If you have a safe space to do so, you can repeat the two lines, but this time ignite them (make sure the sources of the flammable liquids has been moved out of the room before ignition!). The flame for the heptane will be larger than the flame for the ethanol. This provides a visual representation of the heptane particles vaporizing faster, therefore occupying a larger volume as a vapor. That is in spite of the fact that the ethanol molecules move much faster at a given temperature since their molar mass is lower (46.1 g/mol vs. 100.2 g/mol)

3. Crumbling salt - show students that ionic compounds are brittle. You can strike a larger crystal with a blunt object, or crush salt using a mortar and pestle. Either way you'll want students to interpret what happens at the particle level that causes ionic compounds to be brittle even when subjected to minimal force. Dissolving the salt into water is also useful to show that the water particles exert comparable forces on the ions as the other ions in the crystal.

4. Molecular model kits - It is imperative that students use molecular model kits. They need to build structures. The processing is different when students use these than when they draw structures. They should see how double bonds restrict the rotation about a bond. They should experience the 3-dimensional shape of molecules

Chapter 10 – Bonding

that Lewis structures obscure. And you can use multiple models to give students a better sense of intermolecular forces.

5. LEGO - The LEGO (Fig. 10-4, 10-5) brick activity is helpful for students. You aren't using element symbols, so the cognitive load can be applied towards making observations about connections. But as an introduction into connectivity of molecules, this activity works well. It can be referenced in more complex topics such as structural isomers, formal charge, and organic chemistry. The activity also immediately highlights that these models are not going to look like the actual molecules. Students will need to expand the connections later. For example, oxygen doesn't always form 2 bonds.

6. Chromatography - Separating a mixture is a fantastic way to start a discussion on how a set of particles stick together. Simple experiments can be done using coffee filter, markers, and water. There are also some apparatus that can be purchased that separate food dyes. I have used grape Kool-Aid in the past to resolve the blue and red food dyes. You can extend this conversation into TLC plates or liquid chromatography or even gas chromatography with a mass spectrometer (GC-MS). There are some GC-MS apparatus that are affordable for high school lab budgets (some budgets anyway).

7. Burning Mg - Again be careful with the light. Have students point their faces away from the flame to prevent damage to their eyes. Burning Mg is an excellent demonstration because it shows all three types of bonding. The Mg particles stick together through metallic bonds. The oxygen molecules stick together through covalent bonds. And the final product (MgO) has ionic bonding. Students will be surprised to learn that the white smoke particles are tiny pieces of solid MgO product.

8. Melting sulfur - Elemental sulfur is not discrete particles. Sulfur forms rings with 8 sulfur atoms. When heated, these rings of sulfur melt. At a slightly higher temperature, the bonds within the ring begin to break apart and long polymer chains are formed. Students will observe a color change from a straw/yellow color for the initial liquid to a red that darkens into brown. The initial melting involves breaking apart intermolecular forces, but further heating causes a change in the covalent bonds. Don't expect this to smell good.

9. Capillary action - Have a single horizontal tube that has tubes of varying diameter extending up. As the tubes are filled with water, the thinner tubes will show higher water levels than thicker tubing. The adhesive forces between the glass and water increase as the ratio between area of contact to volume of water increases. Capillary action is how trees pull water up without using air pressure. You can also create a simple version of this by putting two square glass plates together. Put a paper clip in between them on one end so that they touch on one side and have a gap on the other. When you dip this into water, the water level will rise much higher as the plates approach each other. Using food dye can help see the rising water better.

10. Toothpicks and spice drops - When teaching students shapes (tetrahedral, trigonal pyramidal, bent, etc.) have them make them using toothpicks and candy. The candy must be able to hold the toothpick in place, but also have toothpicks inserted and removed. I use spice drops because I find them to be delicious and most students don't care for them. Marshmallows, play dough, and many other items would substitute well.

Chapter 10 – Bonding

Flashcards

How many valence electrons do X and Y have?

What groups are they from?

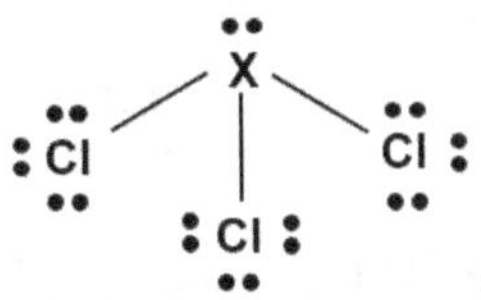

XCl₃   YCl₃

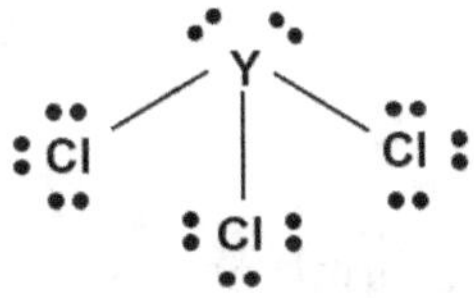

Which structure has oxygen with a formal charge of -1?

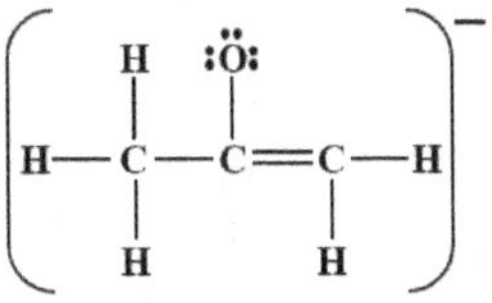

Structure #1 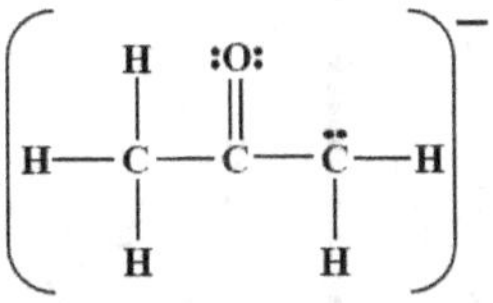   Structure #2

What formal charge does oxygen have in the other structure?

:C=Ö: → :C≡O:

The green bond is a _____ bond

Chapter 10 – Bonding

```
   H   H   H   :O:
   |   |   |   ||
H—C—C—C—C—H
   |   |   |
   H   H   H
```

The bond angle of this carbon is _____°
and the hybridization is _____.

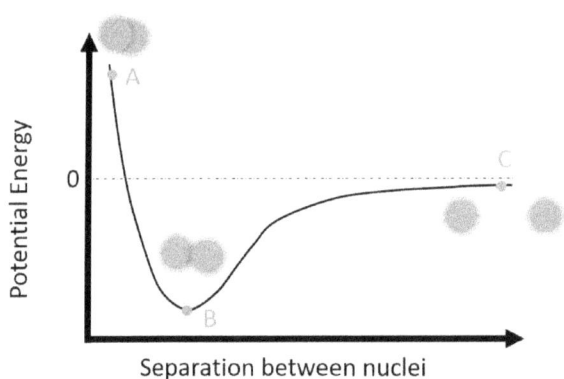

Separation between nuclei

Do the particles speed up or slow down when going from C to B?

A. :Ö—N≡N:

B. Ö=N=N̈:

_____ is the better Lewis structure

Chapter 10 – Bonding

Fe   NaCl

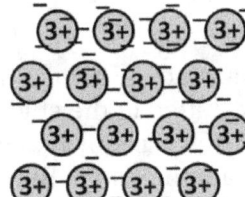 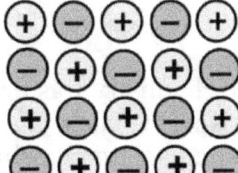

Which is brittle?
Why?

:O:
‖
H–C–H

What is the hybridization of the C atom???

A. Hg (l)
B. $NaC_2H_3O_2$ (s)
C. C_2H_5OH (aq)

Which conducts electricity???

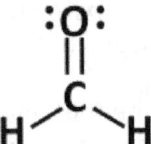

Bond angle???

$CH_3-\ddot{O}-CH_2CH_3$

Type(s) of intermolecular forces???

Chapter
Organic Chemistry

History

Organic chemistry was originally the study of chemistry for living organisms (vitalism theory). This shifted to the study of carbon-based compounds. Organic chemistry is now a major branch of chemistry with applications in medicine, material-science, and industry. But organic chemistry is also interwoven with much of chemical history. We find exceptions to rules here that prompted further experimentation. We also have a wide variety of hydrocarbons that required more precision to differentiate than other elemental combinations.

Gay Lussac (1815) set a wedge between organic chemistry and dualism theory when we showed how a cyanide radical persisted in multiple organic molecules.[1] The radical theory is what we would describe as functional groups today. Not the radicals like triphenyl methane that Moses Gomberg discovered in 1900 at the University of Michigan (Go Blue!).[2] But this deviation was important in constructing explanation that went beyond the limits of a positive fragment and a negative fragment for organic molecules.

The problem with the cyanide radical, was that a chemical cyanogen chloride (ClCN) existed as well as hydrogen cyanide (HCN). According to dualism theory, Cl was a negative element and H was a positive element. How could the cyanide radical form a stable compound with both?

Jean-Baptiste Dumas (1833) showed that chlorine could also substitute for hydrogen. The following year he found that oxygen and hydrogen could be displaced by chlorine. Then Auguste Laurent showed that nitro groups (NO_2) could replace hydrogen. Dumas produced trichloroacetic acid (1839) which has similar properties to acetic acid. Then Louis Melsens (1842) converted trichloroacetic acid back into acetic acid.[3] The electron (1897), the Bohr model (1913), and Lewis structures (1916) would not rescue organic chemistry from these conflicts with dualism theory for a long time.

After radicals (functional groups) were established, the next big step forward came in 1824 when Wöhler found the formula for silver cyanate to be identical to the formula (AgCNO) that Liebig had found for silver fulminate (1823).[4] How could two different chemicals be composed of identical parts? Thus, we begin our journey of isomers (same parts) with structural isomers. Liebig challenged Wöhler's claim, but the two quickly resolved that both were correct and established a brilliant friendship.[5]

At the time (1820s) formulas were a mess in part due to inconsistent atomic weights being used. Liebig and Dumas used 6 as the atomic weight for C. Berzelius used 12 for C. Dalton assumed diatomics violated the dualism bonding model. Other scientists noted how nitrogen was denser than ammonia and therefore nitrogen must be diatomic.

In Familiar Letters on Chemistry (1851) Liebig explains how the formula for sugar was determined to be $C_{12}H_{12}O_{12}$ ($C_6H_{12}O_6$) because when sugar ferments it forms ethanol $C_4H_6O_2$ (C_2H_6O) and when the sugar dehydrates it becomes $C_{12}H_9O_9$ ($C_6H_6O_3$). Any smaller formula and the results wouldn't make sense! His relative amounts of C, H, and O were in a 6:1:8 mass ratio. But Liebig assumes that water is

Chapter 11 – Organic Chemistry

HO, and thus treats H_2 as H. This results in many of his relative masses being off by a factor of 2.[6] These incorrect formulas could function with the faulty assumptions, but confusion reigned when different scientists were using other faulty (and correct) assumptions.

Even with flawed formulas, chemists continue advancing our models of matter. Kekulé proposed sausage structures that established a "valence" for each element. Carbon had a valence of 4, nitrogen was 3, and oxygen was 2. He used sausage structures to show the connectivity of compounds. Crum Brown (1861) and August Hofmann (1865) extended these into arrangements that were similar to Lewis structures, but without electrons. The colors used in the models Hofmann built are the same as what we use in model kits today.

Stereoisomers have the same structural arrangement, but different in spatial arrangement. Yet evidence of stereoisomers came about much earlier in organic chemistry than we'd expect based on the limits of bonding theory at the time.

Stereochemistry started with Auguste Laurent (1837). Laurent proposed a nucleus theory that molecules would take on a particular shape in 3-dimensions. If one of the atoms was replaced by a different one, but the shape maintained, then the properties would be similar. But if a substitution caused a change in the structure, then the properties would differ. Louis Pasteur (1848) studied Laurent's ideas as they applied toward crystal structures and optical activity. He discovered a tartrate that had both an optically active form and an optically inactive racemate.[7]

Pasteur noticed that the sodium ammonium tartrate crystals differed in orientation. He separated the crystals by hand to show that the clockwise and counterclockwise versions rotated plane-polarized light in opposite directions. Since the solutions were also optically active, he concluded that the molecules themselves must somehow be mirror images of each other.[8]

Two scientists took this a step further (1874). Jacobus Van't Hoff and Joseph Le Bel independently proposed a tetrahedral shape for carbon atoms within molecules. Neither scientist had the outreach for this claim to take hold. But Van't Hoff persisted. He wrote a book and included models of these tetrahedral carbons that he sent to famous chemists. Organic chemists were able to build a considerable lead in determining the geometric structure of molecules, and optical activity was the primary component of their advantage.

Without tetrahedral carbon arrangements, how many isomers would exist for a compound such as CH_2Cl_2? If the compound were square planar for example, we'd expect there to be an arrangement with the 2 chlorine atoms at 90° or 180° bond angles. These would be different stereoisomers with different properties. Without evidence of their existence, we can rule out the square planar arrangement as a possible shape.

It was 1919 when Langmuir popularized Lewis's method of using electronic structures to represent molecules. Organic chemistry had a wide opportunity for application of Lewis structures. Robert Robinson (1932) first used the curly arrow convention. Christopher Ingold (1934) deduced two competing mechanisms for substitution (S_N1 and S_N2) where he used curly arrow mechanisms as we use them today.[9] Ingold (1937) showed the curly arrow mechanism for the S_N2 reaction between methyl bromide and iodide that included the inversion. Ingold's book "Structure and Mechanism in Organic Chemistry" (1953) was widely used for quite some time.[10]

Chapter 11 – Organic Chemistry

NMR spectroscopy unfolded in stages during the 1940s with some of the scientific methodology coming just before. IR spectroscopy began in 1881 but was significantly advanced by William Coblenz (1905). The history of mass spectrometry extends over a range of time. J.J. Thompson (1913) separated neon into two isotopes using a beam of Ne ions. Thompson's student (Francis Aston, 1919) constructed the first mass spectrometer.[11] He used this to study more elements in search of isotopes.

If you are lucky enough to teach organic chemistry, make this unit awesome. Some teachers use this unit to intimidate students, but instead we should be pushing them to challenge themselves. Organic chemistry is the perfect opportunity to do just that by applying the chemistry they know.

You're going to want to structure this unit so that students get multiple opportunities for learning. To me there are three sections that need to be present in an introduction to organic chemistry: nomenclature, reactions, and stereochemistry.

Basics and Nomenclature

We often start organic chemistry with naming. But nomenclature includes more than just learning the names of the chemicals. You are showing students functional groups, patterns with the common elements, and the variety of structures that students will encounter. You aren't going to show them every organic compound, so plan the end of the unit before you start making decisions about the start. Are you teaching them cyclical compounds? Do they need to know what a nitrile group is? Will we be looking at polymerization? These decisions have impact on what nomenclature needs we will have.

If you have ever played Minecraft, you know that using a better pickaxe pays for itself. The time you save by mining faster leads to more ore being found. Likewise, even if you are not going to test students using skeletal structures, the time that you will save by being able to use them is worth it. In a skeletal structure, every dot represents a carbon atom and the lines connecting them are bonds.

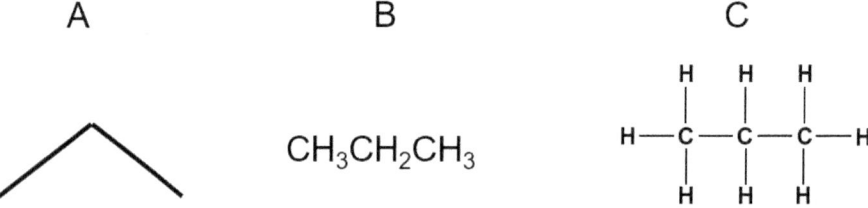

Figure 11-1: Skeletal structure, condensed structure, and structural formula for C_3H_8

The time saved by drawing the skeletal structure in A isn't much. But when we get to larger molecules and reaction mechanisms skeletal structures become vastly more efficient. We need our students to connect the skeletal representations all the way back to the Lewis structures that they used during the bonding unit.

Start by having students identify how many hydrogens are present on each carbon. Start with structure A where each dot is a carbon. The first dot is connected to one other, so that first carbon has capacity for 3 more bonds. Hence the CH_3 in structure B and C. The middle dot has two carbons attached and therefore has 2 hydrogens attached. Students should connect this to structures B and C.

Chapter 11 – Organic Chemistry

They should also note the zig-zag nature of the skeletal structures. We don't always draw Lewis structures in the same pattern (although we can). We want students to recall that the shape in C is somewhat deceptive since our bond angles are 109.5° and not 90°. You could introduce the wedge and dashed lines used to create 3-D structures at this point.

We want students to notice patterns in the number of hydrogens attached to a carbon. Why is the formula for an alkane C_nH_{2n+2}? Each carbon in a chain has two hydrogens attached and two bonds to carbons in the chain. The two carbons at the end each have an additional hydrogen. If a branch extends from the chain, we remove a hydrogen to accommodate the branch, but now we have a new ending. Branches do not change the relative numbers of carbon and hydrogen.

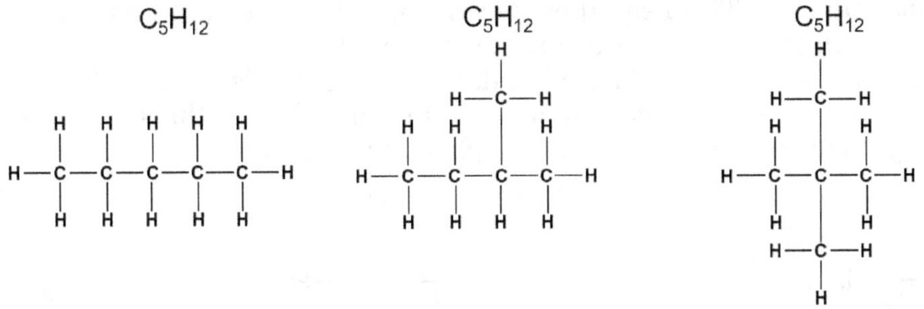

Figure 11-2: The number of branches for an alkane does not change the formula for a given number of carbons

Students can begin differentiating between primary, secondary and tertiary carbons. The number of carbon atoms a carbon is bonded to influences the electron density of that carbon. This has implications for reactivity, optical activity, and for NMR spectroscopy. The distinctions students make in NMR can have similar implications within reactivity.

When we add a double bond between two carbons, we remove a hydrogen from each. Therefore, alkenes have the formula C_nH_{2n} (if there is only one double bond). A triple bond results in a reduction of four hydrogens. A fun way to test for unsaturation is to mix orange bromine (or bromine water) with oils. For oils with double and triple bonds the bromine reacts at those sites causing the orange color to fade. Often the oil will also solidify as the double bonds cause the shapes of the oils to distort and increase intermolecular forces. Students have likely heard of saturated and unsaturated fats and those double and triple bonds are the difference.

The number of possible combinations of atoms in organic molecules is outrageous. This makes systematic naming a challenge, but also critically important. Before we begin to learn that system, we need students to also be aware that some organic molecules existed and were named prior to IUPAC (International Union of Pure and Applied Chemistry) constructing systematic rules for naming organic molecules. This leads to inconsistencies that are mostly trivial. However, it is helpful for the teacher to determine their acceptability in your classroom. Is it 2-propanol, propan-2-ol, isopropanol, isopropyl alcohol, or 2-propyl alcohol? *Will you accept multiple names, all those names, or just the IUPAC?* I see 2-propanol more than the IUPAC name propan-2-ol, so there is some flexibility for your decision. You also

Chapter 11 – Organic Chemistry

want students to be prepared for variations that they should expect to encounter outside of your classroom.

Naming organic compounds is quite fun for students. There is minimal transfer from concrete to abstract needed and so the content is approachable. There are some predictable struggles. When students are presented with chains, they tend to read them like they would a novel, from left to right and top to bottom. You'll want to give them some chains that are intentionally drawn to mislead them.

For initial naming you want to give students examples that show multiple possible answers. Don't draw 2-methylbutane where the branch is on the left side. Draw it on the right so that students have to grapple with the fact that 3-methylbutane does not exist. This helps them understand that compounds do not have a singular position in space, and we must have names that are absolute with respect to orientation. This can be enhanced by doing some examples with multiple modes of presentation. Have students build a "3-methylbutane" and others build a 2-methylbutane. Mix them up and ask them to decipher which one is which.

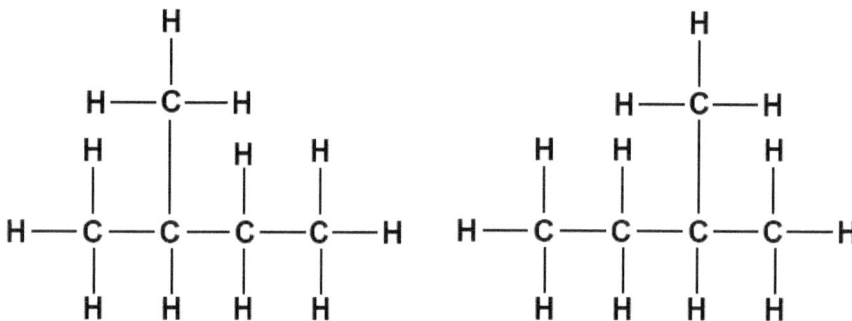

Figure 11-3: 2-methylbutane structures

You'll want to emphasize chain numbering as a means to undermine the student instinct to read a Lewis structure from left to right. Halogenoalkanes are a strong practice tool for numbering chains and following priority in naming. Priority becomes more difficult when we expand to other functional groups.

Students can overcome their initial struggles assigning priority with a little bit of practice and feedback. For instance, when there is a side branch on a carboxylic acid students might not assign the carboxyl group as carbon #1 in the chain. Alkenes must include the double bond in the main chain even if the chain is not the longest possible.

Do both 2-methylbutanoic acid and 3-methylbutanoic exist? This is different than before because now we have the carboxyl group locking in the 1 position carbon. These structures can easily be differentiated even if you put them in a shoebox and mixed them thoroughly. So now we're giving students the concept of priority by using concrete examples along with some time for students to think critically about the concept. This process of using concrete examples as phenomena will help students view structures as an expert would.

Students will commonly add or remove letters from the names of the main chains. The inclusion of "an", "ane" is a spot that lingers for longer. Ethan-2-ol will be

Chapter 11 – Organic Chemistry

incorrectly written as eth-2-ol or ethane-2-ol. As they learn more organic chemistry these issues become less frequent.

Naming ethers and esters is challenging. One must remember which side is which. Is $CH_3OCH_2CH_3$ methoxyethane or ethoxymethane? Is $CH_3COOCH_2CH_2CH_3$ ethyl propanoate or propyl ethanoate? One key concept for naming organic compounds that continues into functional groups is the idea of the longest chain. The ether is methoxyethane and this can be rationalized by having the two-carbon chain where the methoxy group is attached to that longest chain. That doesn't work with the ester because we need to differentiate the two sides. There is no such thing as ethoxymethane, but ethyl propanoate and propyl ethanoate both exist. The best course of action is to include the ester group in the "-oate" portion of the name. The ester $CH_3COOCH_2CH_2CH_3$ is propyl ethanoate.

If students just practice naming followed by feedback; the students will learn many rules, but they will make fewer observations about the structures. It is important to give them some questions that give them time to process and think. Does 3-methylbutane exist or not? Otherwise, students will struggle with curly arrow mechanisms because they lack the ability to draw information from organic structures. Work on providing some examples where students get that quick feedback, and others where students can do some deeper thinking and analysis.

*****

Are you going to include stereochemistry into naming, teach that nomenclature later, or omit it completely? It might be possible to leave out R and S designations in a high school class. But it is probable that you'll be teaching about cis and trans (E and Z) for alkenes or cyclic compounds. Does that lesson start now or later? The perk of starting these earlier is that students get more time to grapple with them along with some spaced practice. The tradeoff is that you're likely to have a wide range of initial abilities that might put some students into cognitive overload and frustration early for minimal gain.

It is worthwhile to consider purpose. The purpose of students learning more about resonance, stereochemistry, acid-base chemistry, and other introductory organic chemistry is to help them understand what happens in the middle of a reaction. We are building towards understanding reaction mechanisms, drawing curly arrows, and verifying/falsifying potential mechanisms. How does the inversion of an enantiomer provide us insight into whether a nucleophilic substitution reaction was bimolecular (S_N2)?

Resonance is critical because we need to be able to understand electron delocalization for some mechanisms. The ability of a carbonyl double bond to form a single bond allow nucleophilic attack to produce a tetrahedral intermediate. One thing to keep in mind for resonance is that we're trying to show that the electrons are not static within the molecule and may be delocalized. Students could potentially struggle to differentiate resonance with conformations (or something else) due to the overlap of concepts.

Consider the addition of a halogen to an alkenyl group. Liquid bromine is added to propene. When the bromine cation adds to the double bond, there are multiple possible arrangements. The bromine can bond to both carbons (+ formal charge on bromine), or the bromine can bond to just one of the carbons leaving the other as a carbocation. Think of how unusual these arrangements are to a novice student! They're seeing a bromine with a positive charge, a bromine with two bonds,

Chapter 11 – Organic Chemistry

and carbon atoms with three bonds and a positive charge. All of these contradict what that student has worked to build schema about. If we expect students to later work on the curly arrow mechanism for this, we want to begin addressing this now in the introductory section.

I can't stress enough that most students learn these differences in context of curly arrow mechanisms but never stop to think about the fact that curly arrow mechanisms highlight unstable intermediates. It is a struggle to get students to differentiate why it is suddenly acceptable for a carbocation to exist. Teachers should be explicit that the timeframe these exist is very small and in the midst of large collisions. A baseball doesn't just hover in the air, but under the right conditions (being thrown, being hit by a bat, etc.) the ball can be in the air unsupported for a short time. That short time period is sufficiently long enough for the next collision to happen in many instances. There are many concrete examples of elementary steps from kinetics in organic chemistry.

Acids and bases impact the amount of positive and negative charges. This is why acids and bases are frequently involved in organic reactions (often as catalysts). Students may encounter a new base in organic chemistry, sodium hydride. The hydride ion H^- can bond with a proton H^+ to form hydrogen gas which easily separates to the surroundings. This can be used to avoid producing water or other products that might interfere with the reaction or remain with the solvent.

Acids and bases are common catalysts. We want students to understand how a proton addition or removal can lead to a reaction intermediate that helps a reaction proceed faster. Sulfuric acid is a common catalyst because the acid is strong, but also because the anion is unlikely to participate in the reaction. The chloride in HCl is more likely to act as a nucleophile, especially when dealing with alkenes or alkynes.

Acids can be used to improve a leaving group. When a hydroxyl group has a proton added, the water is much more likely to leave since the positive charge weakens the bond between the oxygen and the carbon.

Be intentional about your choices in the introductory sections of organic. If you're going to work with alkenes, then now is the time to teach about cis and trans (Z and E) forms for alkenes. If you're going to do a lot of work with enantiomers, you might want to focus on naming compounds with halogens to provide easy examples with four different groups. If you're going to be teaching $S_{N}1$ and $S_{N}2$ mechanisms, you might want to show examples where a carbocation is resonance stabilized. Organic chemistry is like drinking from a fire hose, so we need to build knowledge now that will be emphasized later.

Stereochemistry

Stereochemistry is a tough subject. It is difficult for students to visualize chemicals in 3-dimensions and their success will depend on them being able to do so. Before we start stereochemistry, it is important to think about what we are assuming that students already know. We are assuming they can translate a 2-dimensional Lewis structure into a 3-dimensional image. We are assuming that students can evaluate the symmetry of those 3-dimensional images. We are assuming students can use those images to visualize collisions happening during a reaction.

The best way to start gradually with stereochemistry is to begin with the concept of isomers supplementing with model kits frequently. The ability to manipulate the model kits and rotate their view is helpful for the students to be able

Chapter 11 – Organic Chemistry

to visualize the same things happening in their heads when they look at Lewis structures. Our brains work well with spatial objects. It is the translation from 2-D to 3-D that we aren't good at. You'll want students to build the isomers while asking review questions from bonding. There are some very quality simulations that also present space filling models that students can manipulate, but those work best in conjunction with the student holding a model kit in their hands.

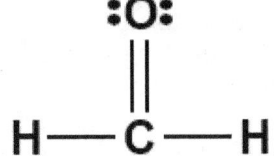

Figure 11-4: Lewis structure of methanal

When you see formaldehyde or methanal, what do you notice about the Lewis structure? Do you know that the molecule is planar? Do you know which parts are + charged and - charged? A student might be processing a lot of other information prior to thinking about those things. They might be looking at the formal charges. They might be counting electrons. They might be distracted by the lone pairs on the oxygen and wondering if there is a 120° angle between them. They might be wondering if the oxygen can be directed down instead of up. In terms of cognitive load this means that they might need help or experience in seeing the relevant features for the topic. The model kits help with this. The brain is well-equipped to visualize actual objects.

The term isomer has the prefix "iso" which means same or equal, and the suffix "mer" which means parts. An isomer is made of the same parts but must have something different about it. "Structural", "optical", and "stereo" isomers all have the same parts but differ in some way.

Structural isomers (e.g., C_2H_5OH vs. CH_3OCH_3) are when two molecules are made of the same pieces but are connected differently. Stereoisomers are arranged differently in 3-D space (cis-but-2-ene vs. trans-but-2-ene). Optical isomers are a specific type of stereoisomer that impact plane-polarized light differently depending on the 3-D arrangement. Instruction needs to emphasize that there will be commonalities and differences in both structure and properties of isomers.

One of the earliest structural isomers discovered was silver fulminate and silver cyanate. Justus Liebig was working for Gay-Lussac, and he measured the composition of silver fulminate to be equivalent to what Friedrich Wöhler found for the composition of silver cyanate. Gay-Lussac concluded that the structures must be different, and Jacob Berzelius coined the term isomers.[12] This was in the 1820s when structural arrangements were assumed to be hidden from our abilities to deduce. It was 50 years later when Van't Hoff advocated for the tetrahedral arrangement of carbon atoms.[13]

Silver cyanate

Ag^+ $:\ddot{O}^- - C \equiv N:$

Silver fulminate

$:\ddot{O}^- - N \equiv C:^- Ag^+$

Figure 11-5: Structures of the isomers silver cyanate and silver fulminate

Chapter 11 – Organic Chemistry

Students could construct ethanol and methoxymethane (dimethyl ether). When working with any type of isomers, the first two questions to ask are what is the same and what is different? For ethanol and methoxymethane, the formulas are the same, the composition is the same, and the molar masses are the same. But because the structural arrangement or connectivity is different, the physical and chemical properties will differ. Both chemicals would burn, but the methoxymethane is much more volatile without the hydrogen bonding. The ethanol could be oxidized using permanganate or dichromate to form ethanoic acid or ethanal. The melting points (-141 °C vs. -114 °C) and boiling points (-24 °C vs. 78 °C) of methoxymethane are lower than ethanol's.

These two molecules can be set up with the LEGO activity in the bonding unit earlier in the year. They can be built again with model kits now. *The key for students should be to build one from the other.* This emphasizes the key component of structural isomers. The same parts with different structural arrangements. In order to convert between structural isomers bonds would need to be broken.

This is a good opportunity to assess whether students have mastered carbon chain assignments by asking them how many structural isomers they can make from 1-chloropropane. If they end up with 3 isomers, they are likely making an incorrect 3-chloropropane that is identical to 1-chloropropane. Some might even try and make one of the CH_3 groups into a branch before realizing that it's still part of the main chain. If they have the model kit built, they can rotate the structure over to show that the position can be flipped but the molecule is still identical. It is helpful to make that point explicitly, that a molecule should not be considered to be changed just because we change our orientation in space. An upside-down water molecule is still water. An upside down 1-chloropropane is still 1-chloropropane and not 3-chloropropane.

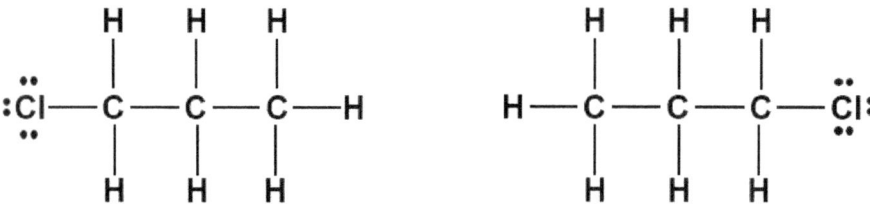

Figure 11-6: 1-chloropropane and 1-chloropropane

Enantiomers are non-superimposable mirror images. This requires the same connectivity, but results in a different arrangement in 3-dimensional space. There is a lot of terminology in those requirements so we must help break this down for students. Hands are a good start. Ignoring rings or skin imperfections, your hands are enantiomers. They are mirror images and students can see this by holding their hands, so their palms face each other. But the non-superimposable description should be broken down slower. To superimpose means to place on top of each other. Non-superimposable means that when you put your right hand over your left that they are not the same. Not just because your right hand is on the other side of your body, but your right hand is different than your left. If you were to chop off both hands (DON'T ACTUALLY DO THIS!!!!) and mix them up in a box, you would easily

Chapter 11 – Organic Chemistry

be able to tell which is which afterwards. They are different. So, with our isomer starter questions, what is the same about your hands and what is different?

Now when are chemicals like your hands? Well, when a carbon atom has four different things bonded to it, the carbon is chiral and the molecule will usually have an enantiomer where the structure is identical, but two of the groups attached to the carbon have been switched in 3-dimensional space.

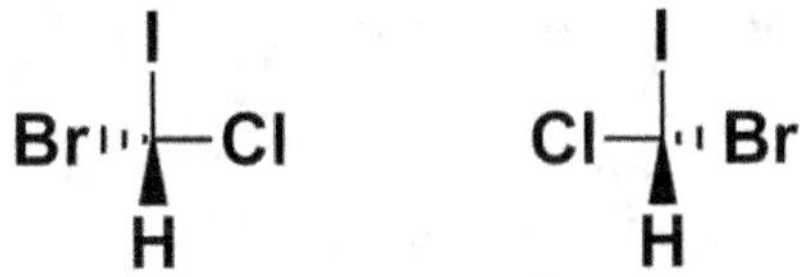

Figure 11-7: enantiomers are non-superimposable mirror images

Before we go anywhere, give the students a chance to make the enantiomers using different colored model kits. Have them build them as mirror images, and then let them try and superimpose one on the other. What they will find is that if they align two colors, the other two will be switched. If blue and yellow are superimposed, then green and red will be switched. If green and blue are superimposed, then yellow and red will be switched. If they switch two bonds, the enantiomer will change to the other (e.g., R to S).

What do they have in common and what is different? This question is much more difficult than it was for the structural isomers. The formulas are the same, the Lewis structures would be the same (for the 2-dimensional format), and the connections are the same (blue, green, red, and yellow are all connected to the same carbon). Would the melting points be different? Would they react differently? This isn't clear just by looking at them.

Before we clarify the answers to those questions, I like to shift into a new question. If you put the two enantiomers in a box and mixed them up, would you be able to tell them apart afterwards? Are they like your right and left hands? How can we communicate which one is which?

We are able to distinguish enantiomers. We assign them R or S designations. In order to do this, we require a means to always align the enantiomers in space so that we don't confuse them because of our current viewpoints. This is done by assigning priorities to each group attached. Remember that all four groups attached to the carbon are different. The highest priority is whichever attachment has the highest atomic number. If there are multiple atoms in the group, we start with the first atom attached and move out from there until we have a tiebreaker.

Once priority is assigned, we orient the molecule so that the path from our eye goes to the carbon and the 4th priority group is directly behind the carbon. From this view the 1st, 2nd, and 3rd priority groups either rotate clockwise (R) or counterclockwise (S).

Rather than just having the students follow the algorithm, it is much more effective for them to understand why that algorithm works at distinguishing between the two enantiomers. If you can somehow distinguish your model kits (e.g., springs for R vs. wooden bonds for S) you can have them assign R and S to each and then see how they can rotate the molecule in space but always get the same assignment. As chemists we need that permanence to be consistent in how we label.

Chapter 11 – Organic Chemistry

Most properties are the same for enantiomers. They have two key differences. Enantiomers rotate the plane of polarized light in opposite directions. The different arrangements of the four groups cause a different impact on polarized light. The rotation can be measured. In addition to R and S labels, the direction of polarized light rotation provides a second set of labels (D and L). The D and L labels are experimentally determined and do not align with R and S consistently. Enantiomers also interact differently with other enantiomers. This happens in biological systems. An example that is helpful to students is thalidomide. One of the enantiomers works at reducing nausea. The other causes birth defects.[14]

A polarimeter is the machine used to measure the amount of optical rotation (rotation of plane polarized light). Light goes through a polarizing filter, then the sample, then a second polarizing filter before hitting a detector. If the two polarizing filters were perpendicular (90 degrees) to each other, no light would go through. If they were parallel (0 degrees) then 50% of the light would go through. When the sample is placed in between the filter angle with the most and least light changes.

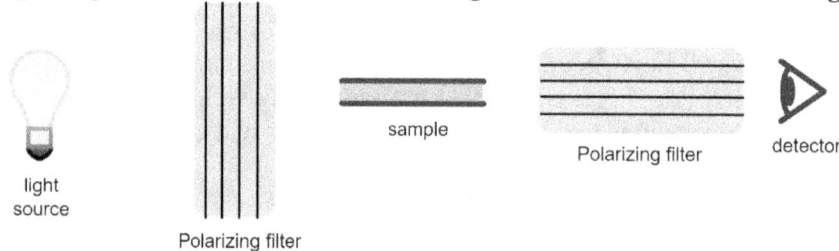

Figure 11-8: Polarimeters have light move through a polarizing filter, the sample, and a 2nd filter before reaching the detector

A polarimeter can be constructed as a demonstration. A unique demonstration is to use corn syrup which will not only show the effect of optical activity, but it will also work differently for different colors. If you have one of those ancient educational devices known as an overhead projector, place a polarizing filter on it, then a beaker with corn syrup, and then a second filter. As you rotate either filter the color of light will change because the optical rotation will differ for different wavelengths of light.

Another interesting consideration is how can you end up with only one enantiomer? Can you separate the R enantiomers from the S enantiomers? It turns out that you can if you are creative. Remember that the enantiomers interact differently with other chiral molecules. There are chromatography separations done where the stationary phase used is a chiral molecule. Depending on the molecule it can also be possible to react the enantiomers to form diastereomers that have different physical properties.

In 1968 William Knowles used a rhodium atom attached to a chiral compound to help synthesize one enantiomer instead of a racemic mixture. The drug is called L-Dopa, and it helps treat Parkinson's disease, but only the one enantiomer. The placement of the rhodium catalyst in a chiral position causes only the one enantiomer to be synthesized.[15]

Make sure to help students with all the very similar vocabulary. Chiral, Chiral center, enantiomers, stereoisomers, diastereomers, optically active, racemic mixture, non-superimposable, R/S, D/L, and mirror images are all very similar terms. As

Chapter 11 – Organic Chemistry

teachers we tend to use some of these interchangeably, but students will struggle with them because of their similarity. The worst thing to do is to start with the vocabulary. They are far too similar for students to be able to differentiate them while they are abstract. You want students to struggle through the similarities and differences between enantiomers conceptually before they start to add in the labels.

Diastereomers are defined as any set of isomers that are arranged differently in 3-dimensional space but are not enantiomers. Alkenes with different groups attached to both carbons of the double bond are diastereomers. You'll want to stress to students that the connections are the same for the two diastereomers. They won't easily see this because the shape differs, so to help them view the molecules as having the same connectivity you'll want to ask what is attached to each carbon. For (E)-2-bromopent-2-ene (Fig. 11-9), what is attached to carbon 2? What is attached to carbon 3? What is attached to C-2 and C-3 for (Z)-2-bromopent-2-ene? This helps them understand what we mean when we say the connectivities are the same.

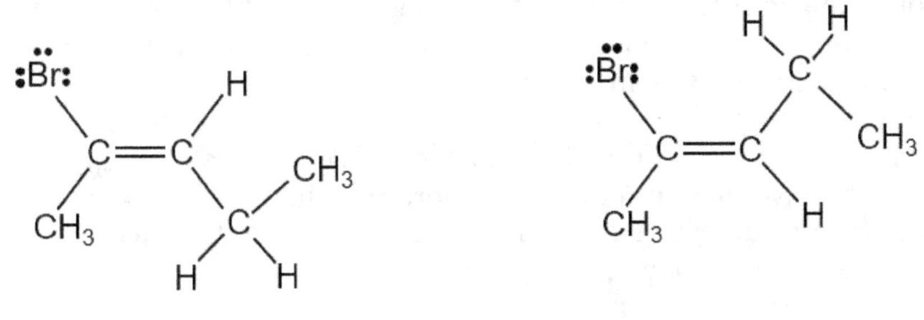

Figure 11-9: (E)-2-bromopent-2-ene and (Z)-2-bromopent-2-ene

For this set of diastereomers what is the same and what is different? The connectivity and the molecular formulas are the same. But the relative positions across the double bond differ. Students will see the four positions around the double bond but will not automatically differentiate which matter for positioning in 3-D.

Figure 11-10: Lewis structures of 2-methylpropene and trans-but-2-ene

But-2-ene has two diastereomers (cis and trans), but 2-methylpropene does not. To a student they will initially see these both having the potential for diastereomers. Remember that we've been trying to make them skeptical about translating between 2-D images and 3-D molecules. It helps to guide them towards the trigonal planar shape of each carbon or the construction of the model kits.

The double bond is the key feature, so we want to highlight that focus by explaining how the double bond prevents the molecule from rotating to convert between the two diastereomers. By adding a strong acid this can temporarily protonate the double bond forming a carbocation, and the molecule can rotate before

Chapter 11 – Organic Chemistry

that proton is removed to reform the pi bond. The rotation of double bonds is a key feature in optics in your eye. When light hits retinal, the double bond breaks, and this changes the shape of the retinal so that it does not fit into a space in the retina the same. This leads to an electrical signal being generated that goes to your brain where your vision is decoded into what color of light was seen.[16,17]

There are other diastereomers that are not alkenes. Molecules with two different chiral centers can end up as stereoisomers that are not enantiomers. If you have an RR molecule with two chiral centers, the RS form will be a diastereomer for example.

Conformational isomers do not really need to be two separate molecules. They can be the same molecule in two conformations at two different times. For these you really want to think ahead to what your goals are. For me, I want students to understand that parts of a single molecule are not static relative to the rest. Atoms are vibrating and sections of the molecule can rotate about a single bond. Some of these conformations are more likely to result in a reaction taking place. But before we get to reactions from conformations, I want to give students an opportunity to visualize 2-dimensional images from different views in 3-dimensions.

I want students to be able to draw an eye somewhere near a 3-D structure and be able to determine what that view would be for that eye. It is unbelievably helpful to do this by using model kits. Have them hold up a 1,2-dibromoethane and draw what they see. Then they should draw an eye on a paper staring down the 3-D structure to show the same view translated to the paper.

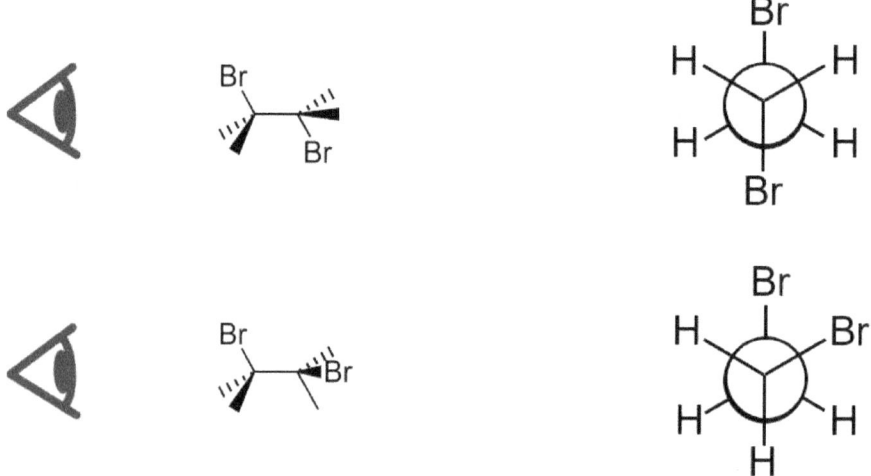

Figure 11-11: An eye placed in position to view 1,2-dibromoethane

Some courses want students to understand how the relative positions of groups inhibit rotation around bonds. If a large group such as a bromine or an iso-propyl group is attached to a carbon, then the steric hindrance will reduce rotation, and this can be seen in an energy diagram (Fig. 11-12). This can be used to set up spectroscopy (how light and chemicals interact) later.

Chapter 11 – Organic Chemistry

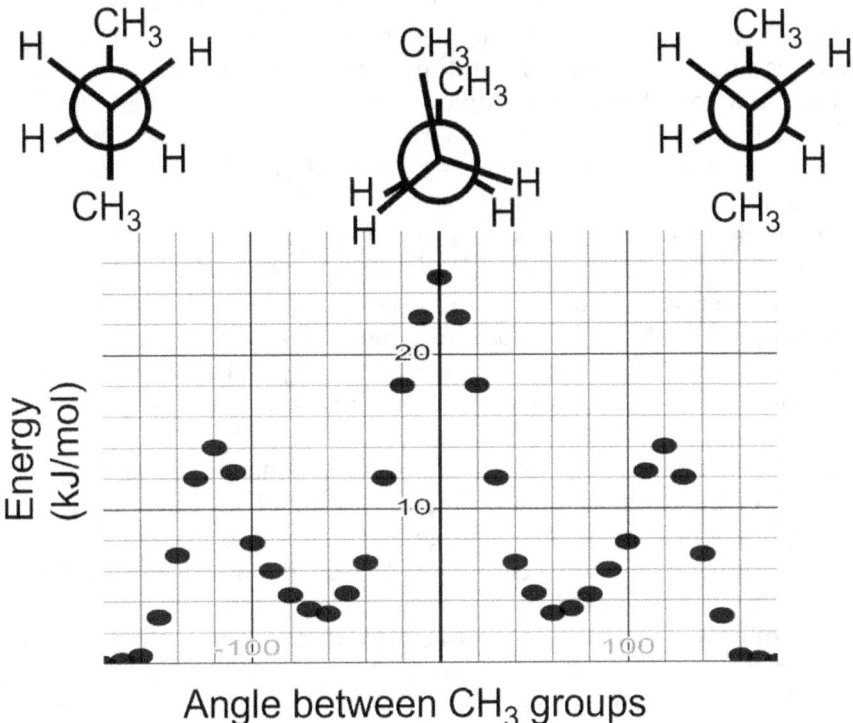

Figure 11-12: energy diagram showing how the steric hindrance limits rotation[18]

The other consideration for conformational isomers is cyclohexane groups being in the chair or boat stage. The conformations for cyclohexanes are an excellent test of a student's ability to translate a non-static molecule into the dependency of a reaction on the various conformations. For a cyclohexane molecule each carbon has two bonds that go outside of the ring. One is vertical and the other is angled. These change based on the current conformation and this changes how groups position themselves relative to one another. Again, this is something that the teacher should be thinking ahead for. Are you going to be working with rings opening and closing? How does the final result that you are aiming for impact what examples you want to work out with your students now?

Curly Arrow Mechanisms

Curly arrow mechanisms are gibberish to many students. One reason for this is that students have learned primarily about compounds that are stable. We have built their intuition of chemistry about stable chemicals that are quasi-permanent. But the fragments that form in the midst of a curly arrow mechanism are sometimes unstable transition states that last for less than a second. It is a completely different set of dynamics. Just like how significant figures was a transfer from a world of numbers to a world of measurements, there is a transition that students need to take here. They need to know that some of the intermediates being formed are a result of a large collision and the intermediates will quickly revert or turn into a new intermediate upon their first opportunity to do so.

Chapter 11 – Organic Chemistry

A second issue is that many students do not recognize that most curly arrows represent a collision. Electrons do not fly through space to bridge two reagents. Rather the molecules collide with proper orientation for the displayed mechanism step to occur. For the $S_{N}1$ arrow that is on a single atom, a collision may be occurring with a solvent molecule that does not otherwise participate in the reaction. The arrows in resonance do not represent collisions as they are describing delocalization of electrons which is a different representation.

A third reason curly arrow mechanisms take time to develop is that they represent a single reaction when the macroscopic view involves a mixture of all these pieces together. We draw a series of transitions occurring that depend on collisions within a mixture of components from all steps and others. All of this prevents students from easily connecting what happens macroscopically in the lab with the symbolic representations we do with Lewis structures. This is why it is critical to introduce reactions that are affected by solvents.

The students who do the best on curly arrow mechanisms see the underlying patterns of positive and negative charges attracting each other. They see that the carbon of a carbonyl group will have positive charge and that if a source of electrons collides in the vicinity, that changes can occur. When these students view a Lewis structure, they can identify the regions that are positively and negatively charged automatically. Beware of introducing the terms nucleophile and electrophile too soon or students will use those abstractly without making the appropriate connections. If a student struggles to identify the regions of a Lewis structure, ask them to consider the charges of the N in NH_3 (-) and NF_3 (+) by the lone pair of electrons on the N.

A student that will be successful in organic chemistry should be able to tell whether a chemical (or portion of a chemical) will act as a nucleophile or electrophile just by looking at it. Because there are so many underlying concepts that are prerequisite to curly arrow mechanisms it can be good to scaffold to reduce cognitive overload. Present the lesson in three steps (Fig. 11-13):[19]

A. Students are given an initial structure with curly arrows drawn and must draw the final product
B. Students are given an initial structure and a final structure and must draw the curly arrows needed for the conversion
C. Students are given an initial set of reactants and must propose some curly arrows that could occur

Chapter 11 – Organic Chemistry

A. Draw the final product

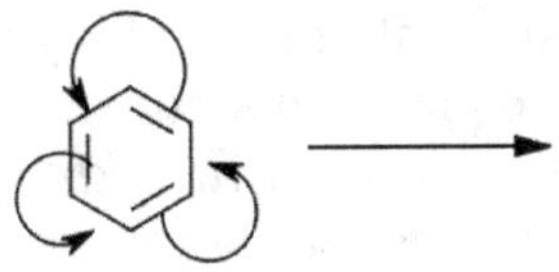

B. Add arrows needed to change initial reagent to final product

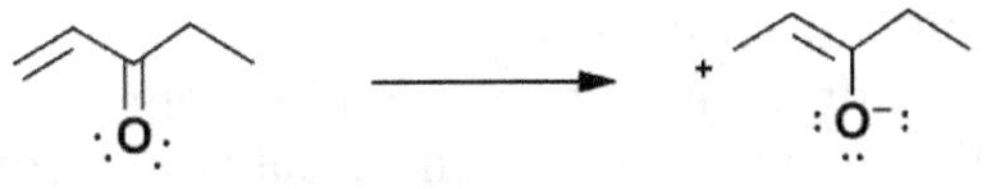

C. propose a set of reasonable curly arrows for the reagents

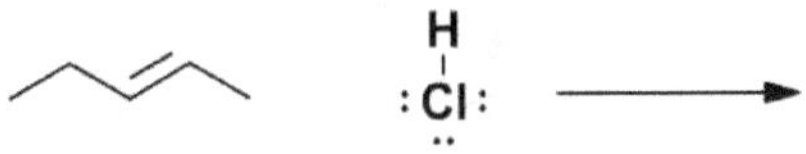

Figure 11-13: Introduction to curly arrow mechanisms

If you show a sample curly arrow mechanism, there will likely be too much going on for some students. Scaffolding this in segments gives students the time to understand and notice patterns. The students will be better prepared to analyze what they know and do not know yet.

The best mechanism to begin with is the addition of a hydrogen halide to an alkene (Fig. 11-14). The first arrow shows a bond being formed between an obvious source of electrons (the double bond), and an obvious electrophile (H$^+$). The H-Br bond breaks as hydrogen cannot form two bonds at once leaving a bromide ion and a carbocation intermediate. The two ions will obviously attract to each other.

Your first curly arrow mechanism should use structural formulas and not skeletal structures. Otherwise, students will have to determine where the hydrogens are and are not for a carbocation and this will tax the short-term memory.

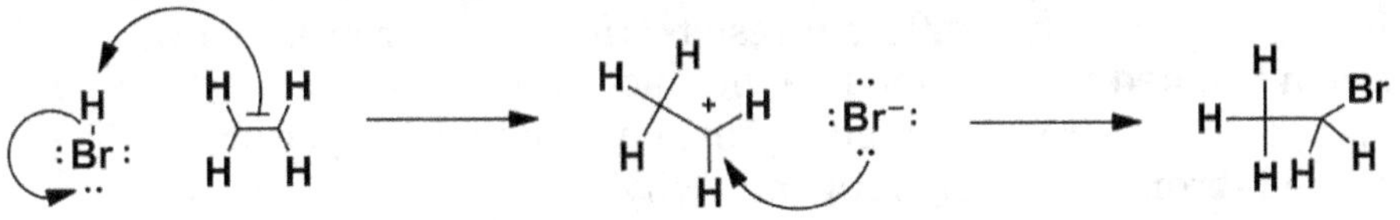

Figure 11-14: Curly arrow mechanism for addition of HBr to ethene

Now let's push their thinking a little bit more.
- What has to happen for that first step to occur? We want students to verbalize that a collision must take place and we want them to visualize what that would look like.
- What would happen if the Br portion of HBr hit the double bond?
- Since ethene is planar would the HBr have to hit from above or below the plane or could a collision happen where the HBr hits within the plane of the hydrogen atoms?

When we draw attention to collisions, we also want to highlight the possibility of a reversible reaction. *Could the $C_2H_5^+$ carbocation form and then lose the H$^+$ back to the Br$^-$? Are both chemicals capable of reacting in the gas phase? If not, what else would be present in terms of solvents?*

Chapter 11 – Organic Chemistry

As a teacher you want your students to be asking and answering questions like these so that their model of this moves beyond static images to picturing moving chemicals that collide. *It should seem like a mini cartoon to them.* Each arrow represents a collision where bonds change. Some collisions have a solvent molecule bump into a molecule and the change only occurs with the reactant molecule. For example, the first step of a S_{N1} mechanism (where the carbocation forms) is initiated by a collision with solvent typically.

Next, we want to begin to make minor changes. What would happen if the HBr was HCl? What would change and what would remain the same? What if instead of ethene we used propene? Propene is interesting because now there is a lack of symmetry. *Would your students realize that before you show them?*

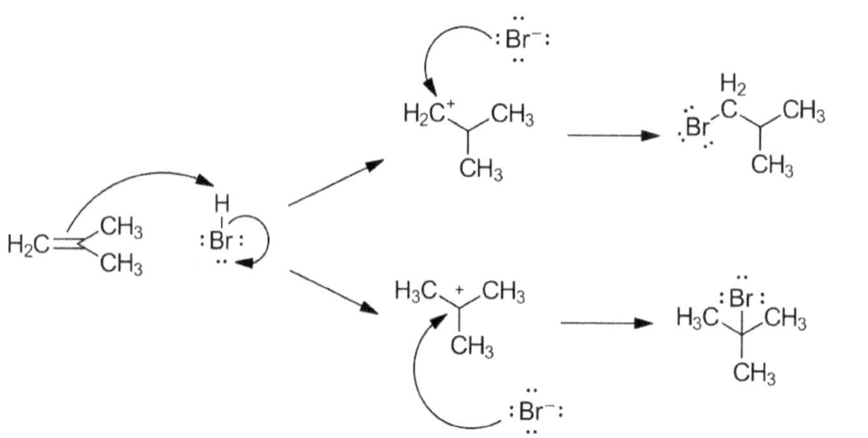

Figure 11-15: Curly arrow mechanism for HBr to 2-methylpropene

If we take 2-methylpropene and add HBr (Fig. 11-15) we get a very similar mechanism as our previous example. But now there are two options for the initial placement of the H+ that give us two different results. In the earlier example with ethene, either carbon resulted in the same final product. But now we can end with 2-bromo-2-methylpropane or 1-bromo-2-methylpropane. When the experiment is run, we end up with more 2-bromo-2-methylpropane. Why?

There are two major steps in this reaction, the formation of the carbocation and the addition of the bromide ion. The carbocation is unstable because there is a large amount of positive charge on a small atom that has a decent electronegativity. When we compare the two possible mechanisms, the top carbocation has that positive charge on a carbon surrounded by three other carbon atoms while the bottom carbocation has that positive charge on a carbon with 1 carbon and 2 hydrogen atoms attached. Since a H atom only has 1 electron and carbon has 6 electrons, the top carbocation is better at sharing electron density with neighboring atoms.

By presenting these two examples first, we are giving students an opportunity to analyze the evidence of which product is obtained with how the two mechanisms to prepare the potential products. This gives students a better opportunity to make observations, and construct questions about the mechanisms.

In order to communicate faster we use the terms primary, secondary, and tertiary to describe carbon atoms. A primary carbon has only one other carbon attached and makes a very unstable carbocation. A secondary carbon has two other

Chapter 11 – Organic Chemistry

carbons directly attached, and a tertiary carbon has three carbons directly attached. All carbocations are unstable but remember that these form in the middle of a collision and then quickly change into something else. For those fleeting moments it is not unreasonable for a carbocation to form. This is especially true for a tertiary carbon. Carbocations are stabilized by electron sources, including neighboring atoms and solvent molecules.

S_N1 and S_N2 are another good introductory curly arrow mechanisms. But they have more variables and contributing factors. For that reason, your best bet for starting curly arrow mechanisms is the addition of a halogen acid (HX) to an asymmetrical alkene. S_N1 and S_N2 reactions open up a wealth of connections. Kinetics and bonding play key roles in addition to some new concepts of organic chemistry. I like to start with S_N1 and S_N2 and return after stereochemistry to look at how the final products work. But if you teach an organic chemistry course (my course only has 2 organic units) you might start stereochemistry earlier since students will have had more prerequisite chemistry.

The type of nucleophile, the leaving group, the solvent, the temperature, and whether the leaving group is on a primary, secondary, or tertiary carbon all impact whether the mechanism will be unimolecular (S_N1) or bimolecular (S_N2). This will draw students into a line of questions where they want to know how to tell which will happen. But it is far more important that they can explain how each impacts the mechanism. The end result is better determined experimentally, but we want students to be able to visualize how the solvent influences the nucleophile, rather than trying to put a quantitative assignment to how much S_N1 mechanism will occur when the reaction occurs in ethanol.

Don't just use qualitative statements such as HS⁻ is a strong nucleophile. Talk about what makes it strong and what we are comparing it to. HS⁻ is strong because it has a sulfur that can share electrons easily due to its large size. It is a better nucleophile than OH⁻ which is better than H_2O. The comparison shows how the size matters and also how the amount of charge matters. It is always helpful to compare fluoride (F⁻) because students will want to assign fluoride as a strong nucleophile since they identify it as having the highest electronegativity. The tiny size of the fluoride causes the electrons to be highly attracted to the nucleus which limits fluoride's ability to interact with electrophiles.

The biggest impact on S_N1 and S_N2 is of course whether the leaving group is primary, secondary, or tertiary. This is due to the carbocation formation in the S_N1 mechanism. Something that most teachers completely omit is how that carbocation forms. A leaving group doesn't just leave on its own! Positively charged and negatively charged pieces don't spontaneously separate. A collision must occur, but the collision is not with the nucleophile. The collision can be with a second molecule of reactant, or it can be with the solvent. The solvent is most likely just because there are so many of them. But tert-butyl chloride does not form a carbocation intermediate without something hitting the molecule in the first place. Because the tertiary carbocation has a much greater stability the tert-butyl chloride will temporarily lose the chloride more frequently than something like chloroethane. And the more frequent formation of the carbocation means that S_N1 is the favored mechanism.

When you add in E1 and E2 reactions you'll want to use as many concrete examples as possible. When you talk about dielectric constants for solvents you are

Chapter 11 – Organic Chemistry

talking about an abstract idea for students without physics background. Discussing how ethanol has hydrogen bonding capabilities as a solvent is much more concrete. The more examples of chemicals you use the better students will be able to compare similarities and differences among them.

I still remember tert-butoxide from when I was in college many years ago because the combination of bulkiness and its ability to act as a base is very apparent from its structure. That example helps highlight how a base can initiate an elimination reaction by removing a proton for an E2 reaction. In the E1 mechanism the initial step is a leaving group leaving to form a carbocation and so a strong bulky base is not necessary. If the base had been smaller, the base could act as a nucleophile and a substitution reaction could take place.

Don't start with a flowchart filled with abstract rules. Start with multiple examples and show how the different leaving groups, solvents, nucleophiles, and steric hindrance impact the reactions and mechanisms that take place. It all comes back to opposite charges attract and like charges repel.

*****

Now that we've gone through a couple of different examples of organic reaction mechanisms let's step back and look at something different. What if you start with just an alkane such as ethane or propane? The bonds are not polar, so we don't have a reactive portion of the molecule. But a lot of the organic material starts from crude oil which is largely made up of alkanes. What do we do with alkanes besides burn them?

To change an alkane into something more reactive we start by adding a halogen such as chlorine or bromine. But the two molecules do not easily react because the alkane is so inert. In order to initiate a reaction, we then have to increase the reactivity and we do this with ultraviolet (UV) light. The UV light causes the halogen to separate into radicals ($Cl\cdot$ or $Br\cdot$). The dot represents an unpaired electron, but it should be noted we are omitting the other valence electron pairs. Radicals are highly reactive and when they react with a molecule a new radical is formed. Only when two radicals meet do they form a molecule without any new radical formation.

Radicals were discovered at the University of Michigan (Go Blue!) by Moses Gomberg.[20] In 1900 he made triphenylmethyl which is a carbon radical with three phenyl groups attached. The large amount of resonance from the phenyl groups helps stabilize the radical so that it can exist as more than just an intermediate.

The initiation of a free radical reaction is when two radicals are formed by means of UV light. After this those two radicals can independently cause a substantial number of subsequent reactions. We label these steps as initiation, propagation, and when two radicals finally meet it is termination.

Cl_2 (g) → $2Cl\cdot$ (g)   Initiation
$Cl\cdot$ (g) + C_2H_6 (g) → $C_2H_5\cdot$ (g) + HCl (g)   Propagation
$C_2H_5\cdot$ (g) + Cl_2 (g) → C_2H_5Cl (g) + $Cl\cdot$ (g)   Propagation
$Cl\cdot$ (g) + $Cl\cdot$ (g) → Cl_2 (g)   Termination

The organic chemistry purpose of these radical mechanisms is to provide students with a means to start with unreactive alkanes that are abundant and convert them into halogenoalkanes that are more reactive. Therefore, we want to highlight reactions such as the substitution via S_N1 and S_N2 mechanisms where these

Chapter 11 – Organic Chemistry

halogenoalkanes can be turned into something else such as an alcohol which could later be reduced to a carboxylic acid, aldehyde, or ketone.

Additionally, we want students to understand that the propagation steps can continue on for quite some time. This means that the products will be a mixture of various halogenoalkanes. The alkyl radical could even form a longer chain under the right conditions.

Students might be too young, but teachers are likely familiar with the CFC bans to protect the ozone layer. Ozone (O_3) was deteriorating because chlorofluorocarbons (CFCs) that were used as refrigerant fluids were mixing in the atmosphere and forming chlorine radicals. A single chlorine radical might be responsible for converting 10,000 ozone molecules into oxygen molecules.

$$Cl\cdot (g) + O_3 (g) \rightarrow O_2 (g) + ClO\cdot (g)$$
$$ClO\cdot (g) + O_3 (g) \rightarrow 2O_2 (g) + Cl\cdot (g)$$

Once students are beginning to master simpler reaction mechanisms, we can move on to longer curly arrow synthesis. I like the Fisher esterification mechanism because it is a highly memorable experiment for students due to the aroma. It's not that other organic chemicals don't have odors, but the synthesis of a recognizable aroma latches to a student's memory. You can make banana smell using isoamyl alcohol and ethanoic acid along with a sulfuric acid catalyst. Typically, students will make the ester in a small test tube, but if they pour the test tube onto a flat surface such as a sink, the increase in surface area will allow them to get an even better whiff.

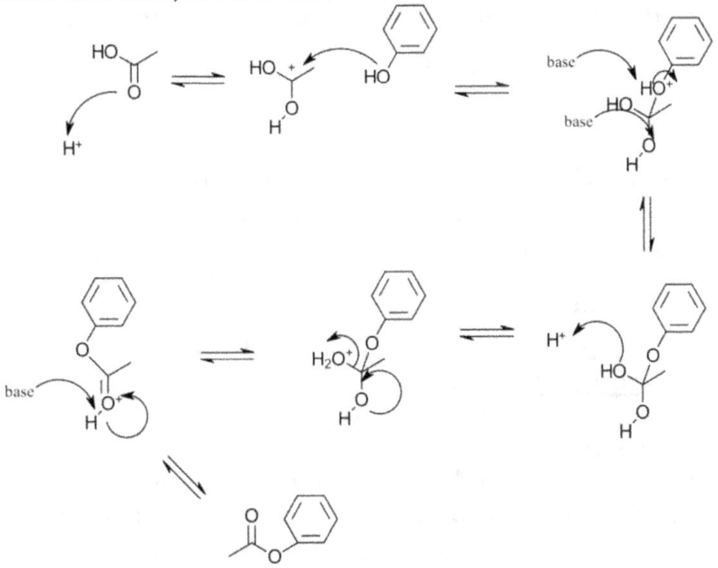

Figure 11-16: Curly arrow mechanism for the esterification between ethanoic acid and phenol.

The steps in this mechanism are all reversible. Each step represents a collision between two particles and the changes from those collisions can be reversed. This helps students make connections to reaction conditions. What combination of temperature, glassware, and timeframe will produce the most product and why? Is there a way to easily separate the final product from the reaction mixture? How can we determine how much product has formed? By connecting to other chemistry

Chapter 11 – Organic Chemistry

topics, we strengthen the mental model for the mechanism itself by providing the student an opportunity to engage in elaborative interrogation.

This is a great reaction to perform while the mechanism is discussed. This allows students to construct a memorable storyline while observing the macroscopic changes that occur. We want to emphasize that the nucleophilic attack of the hydroxyl group on the carbonyl can lead to a new bond, but that the alcohol is likely to remain protonated and be removed just as quickly as it attached. It is only when one of the hydroxyl groups gets protonated that the alcohol remains in place. As more and more ester form, the rate of the reverse reaction will also increase leading to an eventual equilibrium unless the ester is removed as it forms.

Follow up questions can be done with the wintergreen smelling methyl salicylate or acetylsalicylic acid (aspirin). There's a great joke about a chemist in a pharmacy searching for a long time. The pharmacist asks if she needs any help, and the chemist informs her she's looking for some acetylsalicylic acid.

"Do you mean aspirin?" the pharmacist asks.

The chemist responds, "Yes! That's it. Sorry, I can never remember that word."

Acetylsalicylic acid is interesting because the salicylic acid has been used as a pain treatment for some time, but the consumption causes irritation to the throat. But by replacing the carboxyl group with an acetyl group allows the consumption without irritation. Then once the acetylsalicylic acid reaches the acidic stomach the salicylic acid is formed by hydrolysis. If you smell old aspirin, you will note a heavy vinegar smell due to the acetic acid being formed as a byproduct of the acetylsalicylic acid decomposition.

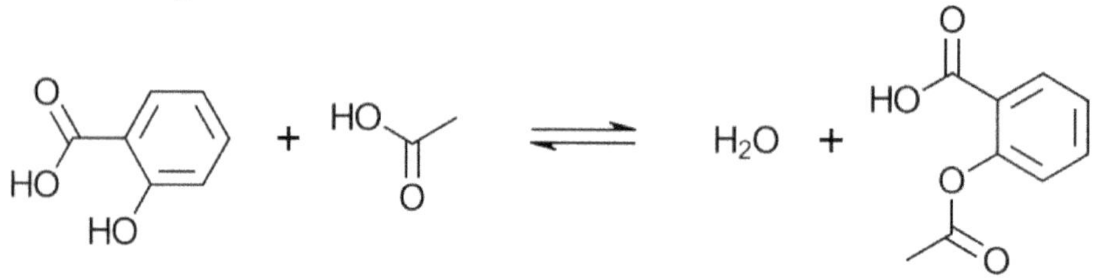

Figure 11-17: Reversible synthesis of aspirin

Spectroscopy

Infrared (IR)

Infrared spectroscopy is as simple as shining infrared light through a sample to find patterns of which frequencies are absorbed. IR is used to identify functional groups present and the fingerprint region can be used to validate what you believe the chemical to be.

As the teacher you'll want to be aware that reduced mass and the bond strength are the two factors that influence where a peak will be. The frequency of infrared light that will be absorbed also depends on the specific type of vibration, since bonds have multiple vibration states that are caused by different frequencies of infrared light. CO_2 has three vibrational states (one does not lead to a dipole shift and thus does not appear on an IR spectrum).[21]

The reduced mass is a combination of the mass of both elements. What you want to know is that hydrogen having a mass of 1 makes the reduced mass very different for bonds with hydrogen as one of the atoms. This is why C-H, O-H, and N-

Chapter 11 – Organic Chemistry

H bonds are all the way on the left side of the spectra with wavenumbers around 3000 cm^{-1} or higher. If you can picture a spring with two masses on each side, you can imagine the spring reacting very differently if it has two large masses (C-C), or one large and one small mass(O-H).

Novice students should begin by searching for three peaks. The C-H peak is around 3000 cm^{-1} but will shift higher for saturated carbons and lower for unsaturated carbons. The C=O peak will be a sharp, narrow peak near 1700 cm^{-1}. The exact type of carbonyl will shift the peak slightly, but the ranges overlap too much for this to be used to differentiate between ketones, aldehydes, esters, etc.

The easiest peak to note is the O-H peak. This is a broad peak meaning that it absorbs a wide range of frequencies. The wide range of absorption is because the hydrogen bonding that occurs between that functional group causes the frequency of excitation to shift. Hydrogen bonding is not a consistent interaction like a bond, meaning that the hydrogen bond between two molecules might be much shorter or longer than two other ones.

After students have some proficiency in these common peaks it can be useful to discuss inconsistencies that arise. You can take an IR spectrum in variety of solvents and even in the gaseous phase. This will alter the size and range of the peaks. For example, O-H has a broad peak because hydrogen bonding causes a range of lengths between atoms and thus a range of infrared frequencies are absorbed. Typically, students aren't going to be using IR by itself since it is mostly used for qualitative determinations and there isn't enough information to obtain to identify a molecule just from the IR. When students do use IR, you want them to have some flexibility for how they view peaks.

NMR

Students do not need to understand how nuclear magnetic resonance (NMR) works in order to be successful at interpreting spectra. But how NMR works is fascinating. The sample is placed in a strong magnetic field that causes nuclear spins to align either with or against the field. When a radio signal passes through the sample some of the spins flip and then gradually return to the other spin state. The return back emits a radio signal, and that signal is affected a tiny amount by the electron density around the hydrogen nuclei. Hydrogen that have a large electron density emit a slightly different signal than hydrogen nuclei with less electron density.

The key here is that students connect that some hydrogens emit the same signal, and some do not. The difference is caused by what they are attached to. That difference is incredibly tiny. The shifts in frequency are a few parts per million of the frequency. But we are able to measure them to a very high degree of precision.

Ethanol is a great spectrum to start with so that students can begin to understand the concept of equivalent hydrogens. There are three peaks in ethanol for the 6 hydrogens present. This shows that some of the hydrogen atoms emit the same signal. The condensed formula CH_3CH_2OH gives us a good initial breakdown of those three groups. But we really want to stress using the Lewis structure how those 3 hydrogens attached to carbon #2 are all equivalent.

Chapter 11 – Organic Chemistry

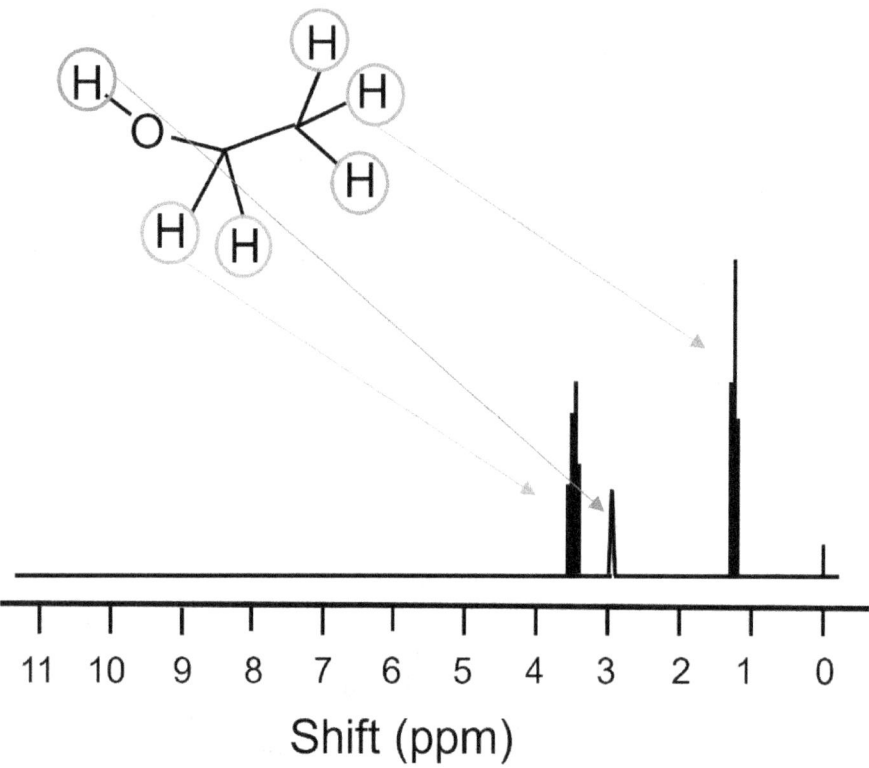

Figure 11-18: NMR spectra of ethanol

All three green hydrogens are connected to the same carbon even though one H points up and another points down. Illustrating the pathway of bonds/connectivity from different green H to the red H can help students appreciate the symmetry of equivalent hydrogens. Again, the principle equivalent property with implications for NMR is the electron density around the hydrogen, but symmetry in position is how we assess that electron density.

A student might see one of the green hydrogens as being further from the rest of the molecule than those that go up and down. It is their intuition to use the 2-dimensional image. If they build actual molecules, it becomes much clearer that the three hydrogens are not only equivalent, but also that the carbon-carbon single bond can rotate causing the three hydrogens to change position relative to the other parts of the molecule.

The biggest thrill in NMR spectroscopy is teaching about splitting. Splitting combines some unique mathematics with complicated physics while providing students additional means to puzzle together connectivity.

Splitting happens when one peak is split into multiple peaks that shift slightly away from each other. For some resolutions the peaks blend into fewer peaks and potentially just one peak. Splitting happens for signals of hydrogen nuclei that have other hydrogen nuclei on adjacent carbons.

The reason splitting happens is that the spin of the nuclei cause a magnetic field that impacts other nuclei. The spins can be spin up or spin down and so the neighboring hydrogens can either cause a shift in two directions. But each of the neighboring hydrogens causes this impact and so we end up with a variety of shifts in different molecules in the sample. Some shifts are more probable than others.

The best way to start is by looking at a single hydrogen that has two neighboring hydrogens. Assume the fragment (Fig. 11-19) has no other hydrogen atoms in the molecule. The signal from hydrogen 1 will be altered slightly by the magnetic fields of hydrogens 2 and 3. If hydrogen 2 is spin up there will be a shift one way, if it is spin down the shift will be the other way. Students will see hydrogen 2 as closer to hydrogen 1 in this drawing, so point out that we are looking at a 2-dimensional representation and that the carbons can rotate about the bond between them.

From this shift upfield or downfield we end up with two possibilities for every hydrogen. If both hydrogen 2 and 3 push the signal upfield we will see a shift two small shifts left. If both hydrogen 2 and 3 push the signal downfield, we will see a shift two small shifts to the right. But if both have opposing spins the peak will remain. Since there are lots of molecules in the sample, all these possibilities are happening, so we see three peaks.

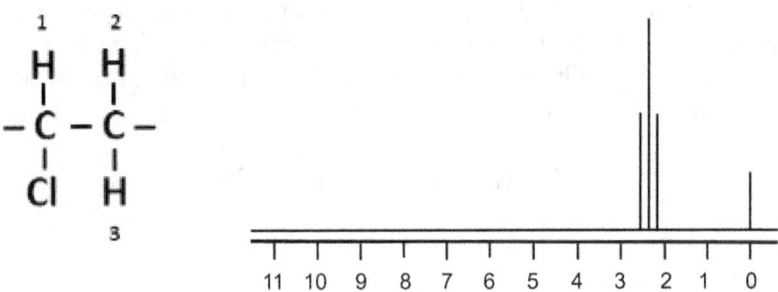

Figure 11-19: Hydrogens 2 + 3 can align with or against the spin of H-1 causing a triplet (signal for H-2 + H-3 is omitted)

But those three peaks are not all equivalent because there are two ways the spins can be opposing. Hydrogen 2 can shift upfield while 3 is downfield, or hydrogen 2 can shift downfield while 3 is upfield. The relative sizes of the split peaks are therefore 1:2:1. Pascal's triangle can be used to determine the relative peak sizes.

When starting students out you'll want to give them small sets of NMR spectra to match to the compounds. For example, when students reach splitting, ethyl butanoate and propyl propanoate are really helpful for students to work out how splitting and downfield shifting can be used to differentiate the two spectra. Ideally you give students comparisons that include the same number of hydrogens, the same number of peaks, but different splitting patterns or different shifts of the splitting patterns.

Chapter 11 – Organic Chemistry

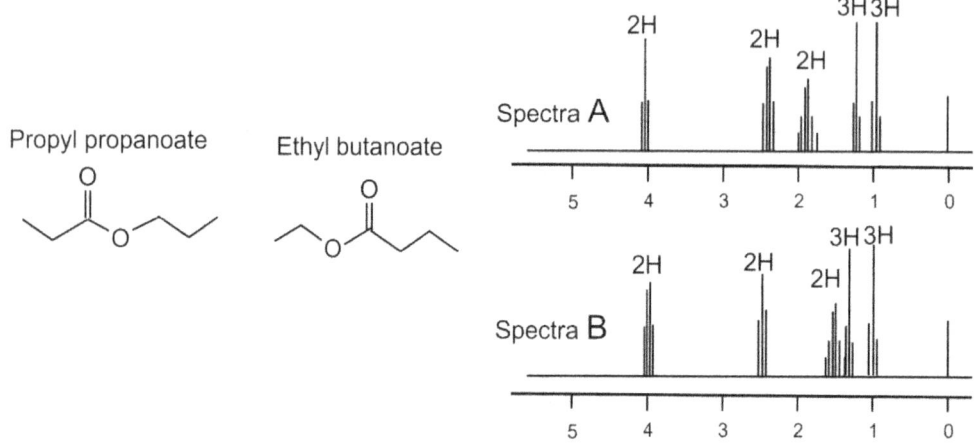

Figure 11-20: NMR spectra for ethyl butanoate and propyl propanoate[22]

Another good thing to test students with is ortho, meta, and para positions for benzene constituents. For the ortho form there will be two equivalent hydrogen groups, meta will have three equivalent groups and para will have only one. The symmetry concepts here will be built upon later for students that continue into advanced inorganic chemistry.

Figure 11-21: Ortho, meta and para dichlorobenzene

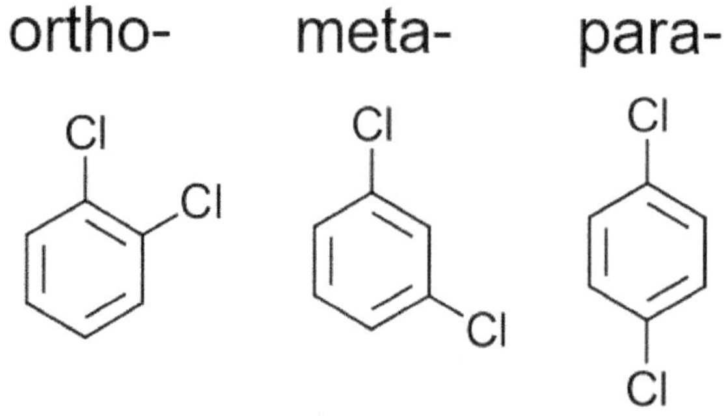

Use spectra together. Don't just teach IR and then NMR both in isolation. Give students an IR spectrum, a mass spectrum, and an NMR spectrum. Have them figure out what they might have and justify their selection. Inorganic chemistry classes frequently have a massive qualitative analysis lab. Spectroscopy is your chance for the same thing in organic chemistry. If you have the lab equipment, give the students an unknown sample where they can take the spectra themselves.

When they write their conclusions, have them look at similar compounds and explain how they know those are incorrect instead of just verifying their selection. Remember to guide them towards falsification philosophically.

Student Struggles

1. "I have no idea what I'm doing!" - These students need to focus on Lewis structures. They aren't visualizing the charges within them properly. Without seeing how charge distribution implicates reaction mechanisms, students are forced into rote memorization that remains too abstract to be functional. Students who complain about having too much material to memorize are likely suffering from this issue.

Chapter 11 – Organic Chemistry

2. "I don't get how you know how many hydrogens there are." Students struggle when they first start converting skeletal structures to condensed to Lewis structures (Fig. 11-1). Their brains often wonder why the number of hydrogens varies from 1-3. The simple answer is that carbons form 4 total bonds, so whatever is missing from that is where a hydrogen goes. But the bigger picture here is that students need time for sensemaking between the various representations of structures. If you teach organic chemistry, you'll find yourself teaching simple introductory components right after the previous cohort had been doing complex mechanisms with stereochemistry. Give the students some opportunities to remind yourself of where they are by listening to them as they rationalize what they see.

3. "I keep getting the numbering wrong on the main chain." Students have been taught to read from left to right their entire lives (some languages differ with direction, but the point remains). Now in chemistry they have to work with chemicals that must end up with the same name regardless of how the structure is oriented on the paper. The words nap and pan are different because of how the letters are sequenced. But 2-methylbutane must remain 2-methylbutane regardless of how you draw it. Students must see examples that test this. From there, the best way to resolve their confusion is to build using model kits to show how you can flip the molecules in any direction. Students must also be directed to try counting multiple pathways to see that there is often symmetry where multiple pathways result in the same chain.

4. "I don't understand why sometimes you can have carbocations and others you can't." Organic chemistry is heavy on mechanisms and chemicals in solutions. Both of these provide opportunities for chemicals that are not stable in isolation but can exist for a fraction of a second when connected to a solvent particle or in the midst of a violent collision. The rules change here, and often we do not address why all of a sudden, a radical can exist. We don't always explain when carbon can have three total bonds, and when that is not plausible. Without that guidance, students will start to doubt other rules. These might be possible during a large collision, but do not produce anything meaningful and thus are not part of instruction.

5. "Why does the solvent matter for these reactions, but not in others?" Solvent molecules tend to outnumber the solute by quite a bit. When a reaction forms something unstable, the solvent is likely to reach the intermediate before another reactant. Other reactions only require one reactant particle in a collision; hence the solvent can be the initiator of the bond breaking.

6. "I don't know how to start assigning hydrogens for an NMR." When starting NMR spectra, it helps to draw out the full Lewis structure including hydrogens. Then circle equivalent hydrogens in a way that you can determine how many environments there are. You can do this by using circles, squares, and other shapes. Or you can use different colors (beware of students who have colorblindness). Then you want lots of repetition. The more problems you try, the better students get at sensing what how much of a shift a functional group will cause. They'll learn how to use splitting as a tool.

Chapter 11 – Organic Chemistry

7. "There are so many different variables for S_{N1}/S_{N2} and E_1/E_2. How do I know what's most important?" A reaction happens when a collision occurs that successfully causes a change in structure. For S_{N1}, any molecule can initiate the reaction. A solvent molecule can bump into the reactant. It's probably not likely, but even an air molecule could. We don't mention that in the flow chart, but some of these components make bigger impacts than others. S_{N1} tends to happen faster than S_{N2} because there are far more collisions that can initiate the reaction. The structure (primary, secondary, tertiary, etc.) of the carbon with the leaving group plays a big role in determining how fast the collision must be in order for the activation energy requirements.

8. "I get the definitions of the terms, but I struggle to look at structures and know whether it is an R or an S enantiomer." Even with the right-hand rule, you need to do some writing. The first thing to do is label the 4 groups by priority. If the 4th priority is into or out of the page, you can usually figure out whether it's R or S just by drawing the path from 1-3. If that's a struggle for a student, have them build structures to help translate the 2-dimensional image into 3D.

9. "My free radical mechanism is a disaster." Students struggle to distinguish an element symbol with a single dot from what they have seen in Lewis structures. Why aren't all the valence electrons shown? What would these atoms look like using a Bohr model, or an orbital diagram? This leads to students constructing a new model for radicals instead of integrating them into existing ideas. This maintains an abstract view of radicals causing students to struggle with manipulating them. Students often do not process what a propagation step involves at the particulate level. The solution to this is to help students integrate what they already know about elements with the new presentation of radicals. They need some time and questioning about how Cl· could be represented differently and why Cl· would react with Cl· (and the answer isn't "stability" or filled shells).

10. "How am I supposed to learn this without knowing the answers? How can I tell if I'm right or wrong?" Organic chemistry is hard for students. This can undermine confidence and cause students to revert toward toxic ideas about learning and grades. We should be prepared to redirect them. When I was in college, our organic chemistry courses had large problem sets from previous exams. But there were no solutions. Students raged about them and how absurd it was not to provide solutions. But science isn't about someone showing you how to do something. It is about building your understanding and confidence to a point where you feel comfortable making (and defending) an assertion or prediction. Students must learn to find a balance between someone showing them everything and obtaining a reasonable and healthy level of feedback as they learn.

Phenomena

1. Corn syrup - Put corn syrup into a beaker. Set the beaker on top of a polarizing filter that's on top of a light source. Place a second polarizing filter on top of the beaker (final order should go light source, filter 1, corn syrup beaker, filter 2). The second filter should now change what color light is transmitted as you rotate it. The

Chapter 11 – Organic Chemistry

corn syrup is optically active, but different colors of light are affected differently by the chiral carbon centers. This causes the colors being absorbed by the filter to change as the path length of the corn syrup and angle of the filters is changed.

2. Carrots and bromine - Take a small piece of carrot and add it to bromine or bromine water. The orange color of the carrot fades to white as the bromine reacts with the double bonds in the beta-carotene. This demonstration functions to show addition of a halogen to an alkene, but also opens up the conversation about dyes and pigments that have long chains with alternating double and single bonds. The HOMO to LUMO transition is often in the visible spectrum for such structures.

3. Tert-butyl chloride and NaOH - The tert-butyl chloride is dissolved in acetone solvent. Dilute NaOH is added to water along with an acid-base indicator. As the t-butyl chloride is mixed with the aqueous solution, the color of the indicator will shift from basic to neutral and possibly slightly acidic. The hydroxide is acting as a nucleophile in an S_N1 reaction. The hydroxide may also be acting as a base to cause some elimination reaction to occur as well. As the reaction proceeds, the hydroxide decreases in amount causing the shift in indicator color.

4. Hexane, bromine, UV - In a test tube, put a small amount of hexane and a small amount of bromine or bromine water. The two liquids will form layers. Then put a UV light next to the test tube. The orange color of the bromine gradually fades as the bromine forms radicals and reacts with the hexane. The UV light will work best if it emits the higher energy UV-C rays.

5. Underwater fireworks - Generate chlorine gas to bubble through a large graduated cylinder (or another similar container). I use a test tube with permanganate and hydrochloric acid to make the chlorine gas and then a one-hole stopper with tubing directs the chlorine to the bottom of the cylinder. While the chlorine gas bubbles through the water, add calcium carbide to the cylinder. As the carbide reacts with water, ethyne (acetylene) gas is produced. When the acetylene and chlorine mix, small fireworks are created under the water. It can be a challenge to get the two gases to mix. I use glass tubing on the end of the rubber tubing to help direct the chlorine stream toward the acetylene source.

6. Model kits for enantiomers - While hands make great examples of non-superimposable mirror images, students need to use model kits as well. The reason is that model kits allow students to see how flipping two of the substituent groups causes a change from R to S or S to R. You should also encourage students to try to overlap the R and S enantiomers. They will quickly find that they can get 2 of the groups to superimpose, but never all 4 groups.

7. Model kits for nomenclature - Model kits help students lock in nomenclature for the typical reasons of being visual and spatial. But they also eliminate issues with symbols such as reading left to right and having the appropriate 3-dimensional shape. When students are told to build 3,4-dichlorohexane they can quickly shape the chain, then add the chlorine groups, and finally fill in the hydrogens. The sequence they use to build the model kits is superior to what they use when they draw.

8. Unsaturation in oils - Prepare some bromine water. Then add the bromine to various oils (vegetable, peanut, olive, etc.). Most liquid oils have carbon-carbon double bonds that the bromine will add to without a catalyst or heating. As the addition reaction proceeds, the brominated oils increase in intermolecular forces. For many oils, they form a white grease as the double bonds decrease in quantity. The double bonds cause the large molecules to have a shape that prohibits strong

Chapter 11 – Organic Chemistry

interactions between molecules. As those double bonds become single bonds, the shapes allow better alignment between molecules.

9. Esterification aspirin and others - Aspirin can be synthesized using salicylic acid and acetic acid. Aspirin tends to have higher yields when acetic anhydride is used in place of acetic acid. The equilibrium amounts can be measured using a complex of iron (III) salicylate. Old aspirin containers will have a heavy odor of vinegar as the acetyl group is hydrolyzed slowly over time. Students can also prepare esters such as wintergreen (methanol and salicylic acid) by combining different carboxylic acids and alcohols with a catalyst.

10. Spectral analysis - Organic chemistry studies a wide range of compounds composed of a few component elements (carbon, hydrogen, oxygen, etc.). Spectroscopy allows students to identify structural differences between very similar compounds. Provide students with opportunities to distinguish between similar compounds using NMR, IR, and other spectroscopy methods. If you have the means, have them gather the spectra themselves. But even if spectra are provided, this is an essential activity for students to partake in.

Flashcards

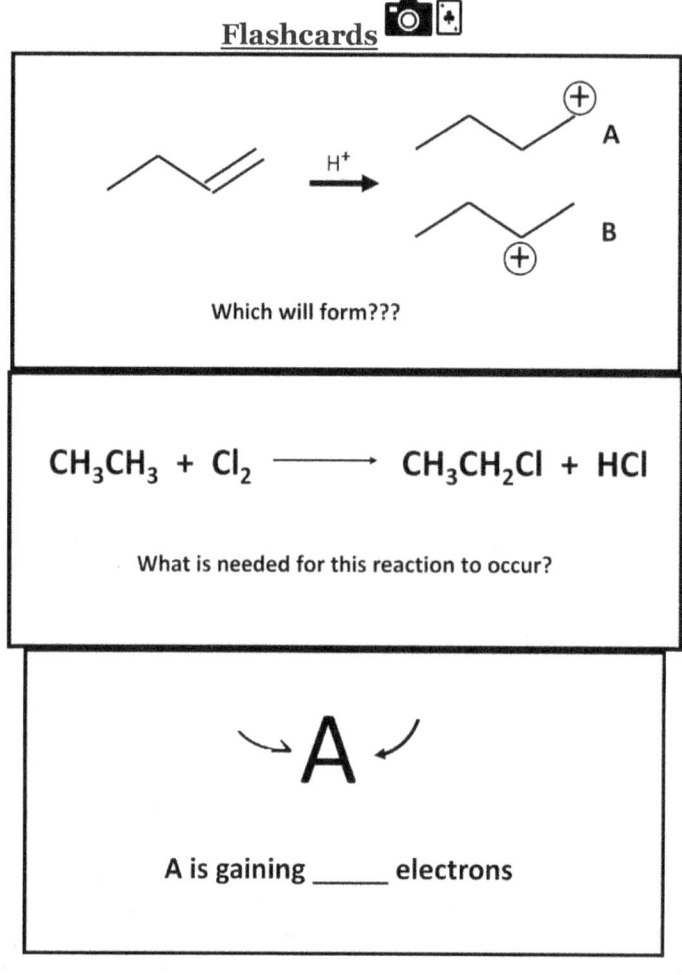

Chapter 11 – Organic Chemistry

These are _____ isomers

HCOOH

Name it!

C_5H_{12}

alkane, alkene, or alkyne ???

•Br + •CH₃ ⟶ CH₃Br

Initiation
Termination
or
Propagation?

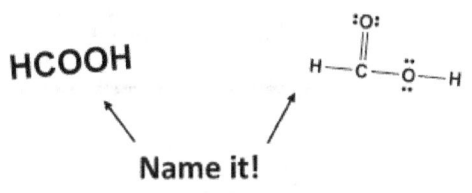

KMnO₄
H₂SO₄

The C was : oxidized or reduced ???

Chapter 11 – Organic Chemistry

Better nucleophile???

OH⁻   or   H₂O

Name a polar protic solvent

Best leaving group???

Cl⁻   Br⁻   I⁻

???

Carbocation intermediate forms….

S_N1 or S_N2???

Carboxylic acid + Alcohol ⟶ ??? + H₂O

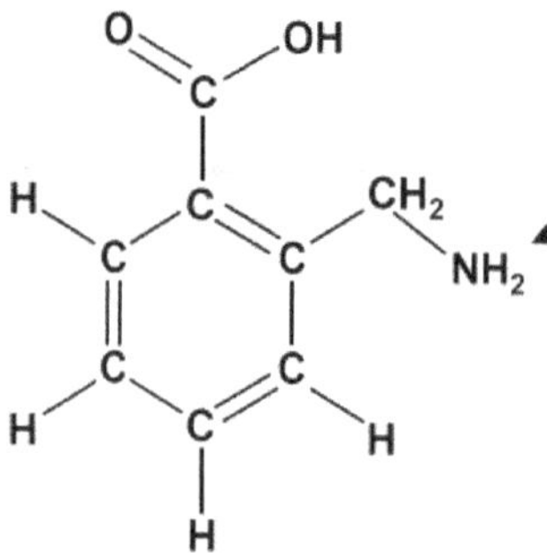

What functional group?

(structure: benzene ring with a –COOH group and a –CH₂NH₂ group on adjacent carbons; arrow points to the CH₂NH₂ group)

Chapter 11 – Organic Chemistry

ketone $\xrightarrow[\text{catalyst}]{???}$ secondary alcohol

_____ have the same physical properties except direction of rotation of polarized light

Inversion of a stereocenter every time.

S_N1 or S_N2???

$CH_3\text{-}\ddot{\underset{..}{O}}\text{-}CH_2CH_3$
name it

Polar protic solvent favors….

S_N1 or S_N2???

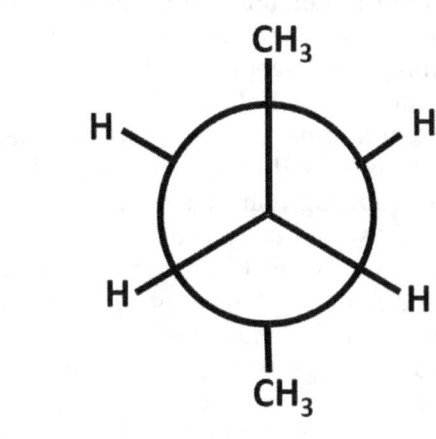

staggered or eclipsed???

Chapter
Redox Chemistry

History

Redox chemistry describes chemical reactions where charge distribution changes around two elements. One of those elements increases in electron density while the other decreases. Before we had a strong concept of charge in chemistry, we used the presence of hydrogen and oxygen as a means to track organic redox reactions. Prior to that redox chemistry was applicable in the conversion of ores to metals.

Our sequence begins with the stone age, then the bronze age, and finally the iron age. These are arranged by higher temperature requirements. Bronze requires a high temperature furnace to convert ore into copper. Producing iron requires even higher temperatures yet. As we increase our ability to produce better kilns, we open up more possibilities of metallic production. Using charcoal during heating lowered the required temperature for both.

Redox plays a big role in combustion reactions. As fuels transition from hydrocarbons to carbon dioxide the particles speed up allowing us to use the increased molecular motion to do work. As charge is reallocated in a redox reaction, particles tend to speed up. This increase in thermal motion can be used to produce usable work.

When we think about the history of redox reactions, it's important to emphasize the challenge of deconstructing reality without advanced knowledge of charge and electricity. Our utility of redox reactions was advanced beyond our knowledge of redox reactions.

Many critical experiments advanced our understanding of redox chemistry. The ability to plug something into an outlet to obtain electricity is something we take for granted today. But without a source of moving charge, it is difficult to perform redox experiments. The first voltaic pile by Alessandro Volta (1800) was an alternating stack of zinc and copper plates with a brine-soaked interphase between them. These allowed Humphry Davy to run electricity through chemicals. He isolated potassium and sodium (1807), and then the following year he isolated barium, calcium, strontium, magnesium, and boron. He would run the electricity through liquid (molten) KOH or NaOH. The metals were produced at high temperatures where they immediately burst into sparks and flames.[1,2] Berzelius assigned elements in his dualism model by which electrode they were drawn toward.

The Daniell cell (1836) is the most common battery shown in introductory textbooks. A copper cathode in copper (II) solution connected via wire and a porous membrane to a zinc anode. These produce direct current just like the voltaic pile. Michael Faraday watched Hans Christian Oersted (1820) demonstrate that electricity could impact a magnet. He worked to show the inverse was true as well (1830). The conversion between electricity and magnetism is how we currently produce much of our electricity.

All of this took place well before JJ Thomson proposing a small particle within atoms called the electron (1897). Thomson had been working with cathode ray tubes and had shown that a ray was produced for a variety of metals and other chemicals.

Chapter 12 – Redox Chemistry

He was able to determine some of the properties of this electron. But he was limited to only knowing the mass to charge ratio. It wasn't until Millikan's oil drop experiment (1913) that the charge of an electron was determined that allowed the calculation of the mass of the electron.

Look at all that was discovered and rationalized prior to the discovery of the electron. Our understanding of charge and redox chemistry could have advanced much faster, but the amount of knowledge we had accrued before knowing of electrons is impressive. Claude Berthollet studied hypochlorite (1789),[3] Frederick Margueritte (1846) first did the permanganate reaction with iron (II) common to balancing redox reactions,[4] and explosives from nitrocellulose (to nitroglycerin ((1846) to trinitrotoluene (1867)[5] were all done before knowledge of the electron existed.

Much of this experimentation started in 1780 when Luigi Galvani showed that connecting two different metal wires together could cause a dead frog to induce muscle contractions.[6] Not only did this inspire Volta, but it also shows how biochemistry today is dependent upon redox reactions. Digestion of food, contraction of muscles, communication between cells, transport of resources in blood, and maintaining homeostasis all involve redox reactions.

Oxidation States

Redox is my favorite topic. Even if you do not teach students this unit, it applies to other chemistry you teach. Redox chemistry begins with very simple ideas about charges attracting and repelling. But redox chemistry becomes more challenging as you delve deeper.

Teachers typically start redox with assigning oxidation states. Much like significant figures, you want to get the concept down first and worry about the rules later. Do not start by showing students the rules and algorithms for assigning oxidation states.

I recommend teaching formal charge and oxidation states together. Both are methods to count electrons and assign them to a single atom in a molecule or ion. The difference between them is that oxidation states award bonded electron pairs to the more electronegative atom while formal charge splits the electron so that each atom gets one. I refer to formal charge as "everybody gets a trophy" and oxidation states as having "winners and losers." If I have a molecule of fake elements X-Z and Z is more electronegative what will the formal charges and oxidation states be? The formal charge will be 0 for both X and Z (assuming there are no coordinate covalent bonds). But the oxidation states will be -1 for Z and +1 for X (assuming 1 single bond).

Before you move on from the concept, ask the students to identify on the periodic table the elements that will tend to have negative oxidation states. Which element will most frequently be negative (Fluorine)? When would oxygen not be negative? What will the alkali metals tend to be? What will single atoms be? What will single ions be?

Then show them the rules. Ask them to figure out some of the exceptions. Why does oxygen have a -1 oxidation state in peroxide? The oxygen has a bond to another oxygen which causes a tie for the oxidation state assignments. A lot of students get told that the sum of the oxidation states should be the same as the overall charge of the molecule or ion. But very few consider why that is the case. Oxidation states are just like formal charge in that they are accounting for electrons

303

Chapter 12 – Redox Chemistry

by assigning them a location. All electrons are assigned somewhere, and thus the total charges should be equivalent to the total of all oxidation states.

After they have had an introduction to the concept and have seen the rules it is time for them to think about this a little bit harder. The middle of the periodic table is filled with things that have variable oxidation states. You want them to practice with some transition metals that have less obvious oxidation states. $KMnO_4$ or $K_2Cr_2O_7$ are good ones for them to figure out (Mn is +7 and Cr is +6). Before we move on to other content though I like to have students figure out an oxidation state where the element has multiple oxidation states within the molecule. Propane is a good choice. For C_3H_8, we would assign each H to be a +1. This means that there is a total of +8 and the three carbons must total to be -8. But that results in -8/3 or -2 ⅔ per carbon.

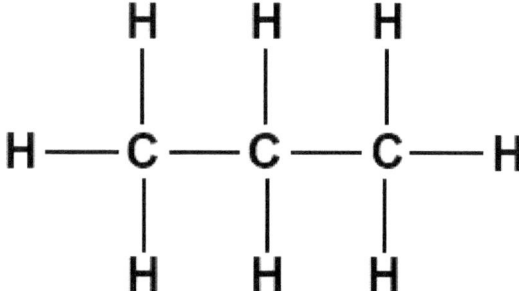

Figure 12-1: Lewis structure of propane (C_3H_8)

The central carbon has two bonds to hydrogens where the carbon is more electronegative and two bonds to other carbons that result in ties. The carbon gets 6 electrons assigned for an oxidation state of -2 (6 valence electrons assigned vs. 4 for a neutral C atom). But the carbons on the ends have three bonds to H and only one to another carbon. These carbons have an oxidation state of -3 (7 valence electrons assigned vs. 4 for a neutral C atom). When we average -2, -3, and -3 we get the -8/3 or -2 ⅔ from above.

This implies that the central carbon and the carbons at the end are not equivalent in electron density. This could tie back to implications for carbocation formation from organic lessons or to spectroscopy when students learn about equivalent hydrogens environments for NMR (nuclear magnetic resonance).

Our end goal for oxidation states is to track when they change so that we can discuss the movement of electron density within a reaction. We want to draw students' attention to the electron density when we are assigning oxidation states. When manganese has a +7 oxidation state that means that the manganese is very electron deficient. In the propane example the central carbon has a bit less electron density than the two carbons at the ends of the molecule. The end carbons have more hydrogens nearby and carbon is more electronegative than those hydrogens.

It is possible that students learned parts of this during Lewis structures in bonding. They might already recognize which portions of molecules have electron deficiencies and excesses. But if they had not mastered that concept, this is a great opportunity for them to learn it again. Use some spaced practice to help students lock in this key idea.

Chapter 12 – Redox Chemistry

Oxidation and reduction are best learned when students have a solid foundation of periodic trends. They need to know that some chemicals will pull on electrons differently than others and be able to use effective nuclear charge and energy levels to explain why. After students learn oxidation states, it is powerful for them to see reactions where oxidation states change. They are already familiar with some redox reactions if they've done single replacement earlier.

When we dip a copper strip of metal into a zinc nitrate solution, nothing happens. When we dip a zinc strip of metal into a copper (II) nitrate solution, the zinc starts to form something on it. Eventually the blue color of the copper (II) ions disappears, and the zinc metal dissolves into the solution as zinc ions. Why did one reaction happen, and the other did not?

$Cu + Zn(NO_3)_2 \rightarrow NR$
$Zn + Cu(NO_3)_2 \rightarrow Cu + Zn(NO_3)_2$

One chemical (Cu^{2+}) can pull on electrons more than the other (Zn^{2+}). When that chemical has the electrons (Cu), the other (Zn^{2+}) cannot pull them away. When the chemical (Cu^{2+}) is missing those electrons, it pulls them back from the other (Zn). Copper (II) ions pull on electrons more than zinc ions do. In a tug of war, the copper (II) will win. When electrons started with Cu atoms, nothing happened. But when the zinc metal was neutral, the copper (II) ions were able to pull those electrons away.

1. $Cu^{2+} + Zn^{2+} + 2e^- \rightarrow Cu + Zn^{2+}$
2. $Cu + Zn^{2+} \rightarrow NR$
3. $Cu^{2+} + Zn \rightarrow Cu + Zn^{2+}$

Reactions 2 and 3 are commonly shown. But think of the value of also showing the first reaction. This allows us to directly compare the two ions and their capabilities of pulling on electrons.

There are two methods to rank how good chemicals are at pulling on electrons. We can construct an activity series, or we can assign them reduction potentials. The thing about reduction potentials is that voltage must compare two different components. We had to arbitrarily choose something to be the zero point and we chose $2H^+ + 2e^- \rightarrow H_2$ (g). This Standard Hydrogen Electrode (SHE) is defined to have a reduction potential of 0. When we say that $Cu^{2+} + 2e^- \rightarrow Cu$ (s) is +0.34 V, that number is relative to the standard hydrogen electrode (SHE).

If we had set the SHE to be 10.00 V, the copper (II) reduction would have been +10.34 V. Either way the same voltage exists between the two cells. We could also have set $Cu^{2+} + 2e^- \rightarrow Cu$ (s) to be 0. This would make the SHE -0.34 V and the $Zn^{2+} + 2e^- \rightarrow Zn$ would be -1.10 V. Notice how we would still end up with the same potential difference for any combination of the three half-cells.

When we teach this to students, how should we define voltage? We want students to understand that the more positive the voltage for a reduction potential, the better that species is at pulling on electrons. The more negative the voltage, the worse that species is at pulling on electrons. If that species manages to take some electrons, the product of that reaction will be excellent at losing these electrons. If you put something with a large and positive reduction potential by the product of

Chapter 12 – Redox Chemistry

something with a large and negative reduction potential, electrons are going to move. They will move from the species with the large negative potential to the species with the large positive potential.

$$Ag^+ + e^- \rightarrow Ag\ (s) \qquad E = +0.80\ V$$
$$Li^+ + e^- \rightarrow Li\ (s) \qquad E = -3.04\ V$$

Silver ions will react with solid lithium as the silver ion is much better at pulling on electrons. The silver ions are strong reducing agents, and the neutral lithium atoms are strong oxidizing agents. Note that this doesn't mean silver ions will pull electrons away from lithium ions as the lithium ion has already lost its valence electron. Li^+ turning into Li^{2+} would be a very different reduction potential.

Which of these describes halogens and which describes alkali metals? Alkali metal ions have large and negative reduction potentials. Halogens have large and positive reduction potentials. This ties back to effective nuclear charge. Halogens have large effective nuclear charges and can pull on electrons strongly. Alkali metals have effective nuclear charges near +1 and have a very weak pull on new electrons as well as their valence electrons.

Something that the teacher should keep in mind here is that there is a reciprocity between reduction and oxidation. A chemical can't lose electrons (oxidation) unless something takes them away (reduction). I would even go as far to make the controversial claim that this codependency is more critical than the LEO proclaiming GER. And I was born in August!

When we want to describe a chemical in isolation, we use the term agent. An oxidizing agent is something that causes something else to oxidize. Since oxidation always occurs alongside reduction, we use the term agent so that we can talk about a single chemical at a time. Oxygen is a good oxidizing agent. Gasoline is a good reducing agent. Solid lithium is a strong reducing agent. Silver ions are a strong oxidizing agent. The agent term is useful for evaluating the storage of a chemical to make sure that strong oxidizing agents are kept apart from strong reducing agents.

The last piece of oxidation and reduction basics is to have students balance redox reactions. You may have some good pieces in place if you've taught organic because the big new idea here is that the solvent participates. In addition to the reducing and oxidizing agents, you also have H^+, H_2O, and e^- available for balancing in acidic solutions. You'll have OH^- instead of H^+ in basic solutions. A simple example to start with is Fe^{2+} reacting with MnO_4^- in acidic solution. The iron (II) oxidizes to iron (III) and the permanganate reduces to Mn^{2+}.

$Fe^{2+}\ (aq) \rightarrow Fe^{3+}\ (aq) + e^-$ **Balanced oxidation ½ reaction**
$MnO_4^-\ (aq) + 8H^+ + 5e^- \rightarrow 4H_2O\ (l) + Mn^{2+}\ (aq)$ **Balanced reduction ½ reaction**

$MnO_4^-\ (aq) + 8H^+\ (aq) + 5Fe^{2+}\ (aq) \rightarrow 4H_2O\ (l) + Mn^{2+}\ (aq) + 5Fe^{3+}\ (aq)$
Overall reaction

There are a couple of things worth mentioning. The number of each element is balanced (they are familiar with this but might not have seen such a large equation), and the total charge is the same on both sides (+17). The total charge balancing is why you need 5 of the oxidation ½ reaction. By splitting into half-reactions and making

Chapter 12 – Redox Chemistry

the electrons the same for both, you'll automatically balance the charge. If you combine ½ reactions without balancing the electrons, you'll balance mass but not charge.

In basic solution we have H_2O, OH^-, and e^- available to help balance the reaction. The simplest way to balance these is to start by pretending the solution is acidic. Then at the very end use hydroxides (OH^-) to neutralize the H^+ ions.

$$Cr_2O_7^{2-} (aq) + 8H^+ (aq) + 6e^- \rightarrow 2Cr(OH)_3 (s) + H_2O (l)$$
$$2Cl^- (aq) \rightarrow Cl_2 (g) + 2e^-$$

This is how the ½ reactions would be balanced in acidic solutions (except for the chromium (III) product). From here we are going to add 8 hydroxide ions to both sides of the reduction half-reaction. Those hydroxides will neutralize the H^+ ions on the reactant side to make 8 water molecules.

$$Cr_2O_7^{2-} (aq) + 8H^+ (aq) + 6e^- + 8OH^- (aq) \rightarrow 2Cr(OH)_3 (s) + H_2O (l) + 8OH^- (aq)$$
$$Cr_2O_7^{2-} (aq) + 8H_2O (aq) + 6e^- \rightarrow 2Cr(OH)_3 (s) + H_2O (l) + 8OH^- (aq)$$

And we can cancel one of those water molecules with the product water.
$$Cr_2O_7^{2-} (aq) + 7H_2O (aq) + 6e^- \rightarrow 2Cr(OH)_3 (s) + 8OH^- (aq)$$

This leads to a final balanced reaction of:
$$Cr_2O_7^{2-} (aq) + 7H_2O (aq) + 6Cl^- (aq) \rightarrow 2Cr(OH)_3 (s) + 8OH^- (aq) + 3Cl_2 (g)$$

Balancing redox reactions in basic solution is difficult. The alternative to the method shown above is to balance oxygen by adding water, and then balance hydrogen by adding H_2O to one side and OH^- to the other. This method also ends up with the possibility of needing to cancel water molecules.

The final piece is that most redox products cannot be predicted by novice students. You'll want to be clear about which ones you want them to memorize or how you'll provide products to them if they aren't expected to be memorized. Concentration can impact which species forms as well. Dilute nitric acid forms NO, while concentrated nitric acid forms NO_2. The final stopping point along the chain of Cl^-, Cl_2, ClO^-, ClO_3^-, ClO_4^- depends on the relative concentrations (and other conditions) as well.

It's also worth noting that in order for students to be successful in this unit they are going to need to be proficient at distinguishing chemical species that differ only by charge. The difference between the neutral and ion form is subtle. If students struggle, be sure to include some particle level representations where they can represent changes in charge. The plum-pudding model works well for this because the electron is represented as a particle.

Batteries

Reactions between alkali metals and halogens are violent and fun to watch because you have electrons shifting rapidly from the alkali metals to the halogens. But is there a way to harness these reactions where electrons move from one chemical to the other? Could we separate the part of the reaction where electrons leave away from

Chapter 12 – Redox Chemistry

the part where electrons go and connect them with a wire? It turns out that with some engineering we can intercept this flow of electrons and use these redox reactions for batteries.

You want students to think of batteries in this context, half of the reaction occurs over here, the electrons flow through a wire, the second half of the reaction occurs on the other side. These three things occur simultaneously, but students will benefit from dividing the process into segments.

For most batteries, using alkali metals or halogens is unrealistic due to their extreme reactivity. Lithium is the exception. But for most batteries we use a combination of smaller reduction potentials and low-cost materials. Silver and gold would give you a good voltage for your battery, but they would not be cheap.

Otherwise, a battery is mostly simple. You put half of the reaction on one side, half of the reaction on the other, and connect them with a wire. But in order for the circuit to be complete you also need to have some way for cations and anions to flow between the two half-cells so that the charge build up does not shut down the reaction.

At this point students have sufficient content knowledge to put together what will happen in a battery on their own. Given the following half reactions:

$Zn\ (s) \rightarrow Zn^{2+}\ (aq) + 2e^-$ $E = +0.76V$
$Cu^{2+}\ (aq) + 2e^- \rightarrow Cu\ (s)$ $E = +0.34\ V$

Students should be able to determine:
What is being oxidized?
What is being reduced?
What is the oxidizing agent?
What is the reducing agent?
Does the concentration of Zn^{2+} increase or decrease as time goes by?
Where do the electrons flow to/from?
Where do cations in the salt bridge go?
Where do anions in the salt bridge go?
Does the mass of Cu increase or decrease?

These are good questions to ask because they will help the student connect the symbolic representations to the macroscopic level of what actually happens in a battery. However, they will struggle with the idea of ions moving. They don't really get it. Part of the problem may be that they have a vision that electrons move down a wire for electricity, but they don't really know what the start and end of electricity looks like. They don't have a visual for the electrons leaving a molecule or ion. They don't have a visual for the electrons moving to a new molecule or ion. When we shift to ions they struggle with what the beginning and the end look like.

A solution to this struggle is to invoke the particle level. Have students represent particles that have two spots for electrons and to make a representation of how those electrons and particles move around in the battery. The reaction (Fig. 12-2)[7] starts at the anode where two electrons leave the metal atom forming the metal cation that has two vacancies. The electrons move from the anode to the wire to the cathode. When the ions reach the cathode, the electrons join to fill the two vacancies for the metal cation on the other side. As this is happening, cations are moving

Chapter 12 – Redox Chemistry

through the solution towards the cathode and anions are moving through the solution towards the anode. A continuous loop of moving charges results.

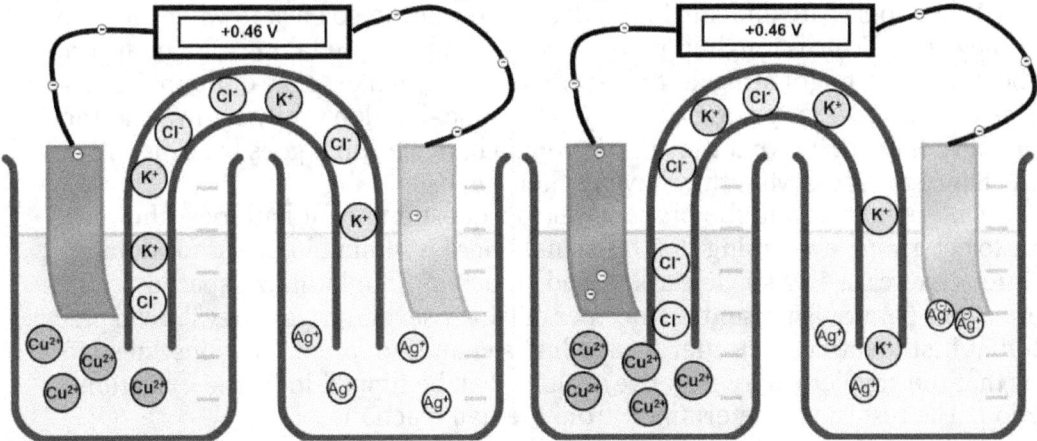

Figure 12-2: Galvanic cell represented at the particle level for copper + silver with a KCl salt bridge

These animations help students connect the uncertainty they have with a definitive start and end for both the ions and the electrons.

At the cathode $Ag^+ + e^- \rightarrow Ag$.
At the anode $Cu \rightarrow Cu^{2+} + 2e^-$.

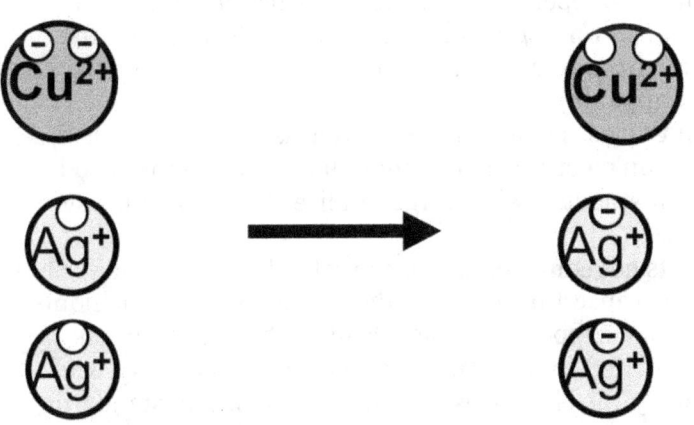

Figure 12-3: Electrons are transferred to vacancies on the silver ion causing vacancies on the copper (II) ion product

From the anode to cathode electrons flow through the wire, and ions flow through the solution. The anions approach the anode, and the cations approach the cathode.

This breakdown gives students a better particle model for what current is. Many students have had minimal physics. They require time to conceptualize the differences between voltage and current. Chemistry can be a great avenue for that to happen, but we should expect the learning to unfold gradually as students compare new concrete examples and observations. We can help by identifying voltage as a predictor of what happens while current is connected to how much of a reaction will

Chapter 12 – Redox Chemistry

happen. We also want to compare current and voltage in galvanic and electrolytic cells.

When topics are difficult for students, we want to go back to what they see and what they know. This is tough in redox because students are taking a lot of abstract symbols and they might struggle to identify how they make sense of them. Try adding time as a variable. If they draw a battery schematic, but don't fully know what they're doing, have them draw what they think would change as time goes by. This will give you a chance to assess what they have in their head.

Once students find the first answer to a question about batteries, they are quick to determine everything that happens. There is minimal benefit for doing numerous exercises in a single class period. Teachers should utilize spaced practice where a new battery is presented once per day over several days. They'll need practice with that first identification after having had a chance to forget. They just need to know that the smaller (more negative) voltage will be flipped to be the oxidation reaction. The rest can be determined from the half reactions.

Some students will walk away with the proper algorithm to flip the smaller reduction potential reaction, but they won't know why they do so. Assume that a battery is made from Ag/Ag^+ (aq) and Cu/Cu^{2+} (aq). The redox potentials would be:

$$Ag^+ (aq) + e^- \rightarrow Ag (s) \quad E = +0.80 \text{ V}$$
$$Cu^{2+} (aq) + 2e^- \rightarrow Cu (s) \quad E = +0.34 \text{ V}$$

Many students will recognize that they should change the copper (II) reduction to oxidation. This has to happen because the construction of a battery is contingent upon losing electrons on one side and gaining those electrons on the other. We need electrons to flow from one end to the other. Therefore, we must see both reduction and oxidation happening.

The reason why we flip the less positive half-reaction is because that is the species that has the weaker pull on electrons and is more likely to lose them. What you will see as the teacher is that students will overgeneralize this algorithm later in electrolysis.

You will also see students get confused about the role of stoichiometry. Should I double the +0.80 V for silver because I need two of them? The role of stoichiometry is confusing in batteries. Every one copper that reacts leads to two silver atoms forming. But the voltage is not related to the amount of reaction happening. The voltage is comparing how strongly the chemicals pull on electrons. The standard conditions include a concentration of 1 M solutions, which sometimes negates the stoichiometry confusion. The voltage is not influenced by the balanced reaction, but stoichiometry does still apply.

Stoichiometry and voltage are difficult to untangle. Stoichiometry is based on the relative amounts reacting. The voltage is based on the pull of electrons due to which chemicals and their concentrations. But as the reaction proceeds the concentrations change based on stoichiometric coefficients. The concentrations are altered, and this impacts the voltage.

At this point, many teachers move into the Nernst equation. I am not a big fan for a couple of reasons. *The Nernst equation isn't critical.* Most batteries have voltages that are very close to the standard voltage for the majority of conditions. Also, the mechanism by which the voltage changes is usually explained using energy

Chapter 12 – Redox Chemistry

and Le Chatelier's Principle. Neither of these allow for students to construct a physical model of what is happening and why.

If you are intent on teaching the Nernst equation, it does allow you to provide challenging calculations for students that incorporate multiple chemistry principles. One thing that I would like to see added is a physical explanation of why concentration influences the voltage.

Let's take the reaction $2Ag^+$ (aq) + Cu (s) → Cu^{2+} (aq) + 2Ag (s). What happens in the reaction is that silver ions pull electrons from the copper atoms. If we increase the concentration of silver ions, that means that we would have more silver cations with more positive charge pulling those electrons (for a given surface area). This would lead to a net increase in force on the electrons and thus the electrons would gain more energy as they move from anode to cathode. Thus, the voltage increases. If we also increase the concentration of copper (II) ions, we would now have more cations pulling those electrons back towards the anode and the electrons would gain less energy. The voltage decreases.

As a general principle of physics, this mostly holds. The more we increase the cation used at the cathode the higher the voltage. The more we increase the cation concentration at the anode, the lower the voltage. The voltage isn't changing due to a stress to re-establish equilibrium, these are non-equilibrium systems. Otherwise, the voltage would be 0. But there are some problems with this model. If the concentrations of silver ions and copper (II) ions are both increased by the same amount, this would seem to favor the silver ions since they pull harder to begin with. Instead, we see the voltage remain at the standard. It is also possible to create a voltage merely with a concentration differential. If 0.10 M Ag+ (aq) is on one end of the battery, and 4.00 M Ag+ (aq) is at the other, a potential difference will be present. I know that fits the general narrative, but to move from the charges pulling on electrons to the Nernst equation mathematically is something I have never seen derived and am not sure it has been attempted since the equilibrium approach is so simple. More on the Nernst equation can be found in the final thermochemistry chapter.

Electrolysis

Electrolysis is a nightmare for students. The fact that the water can participate in the electrolysis and the additional emphasis on voltage causes frustration with students. Expect a steep uphill climb.

A great observation to lead into things goes back to the electrolysis of water. When we first present this reaction in a reaction types unit, we write it as $2H_2O$ (l) → $2H_2$ (g) + O_2 (g). But if you show students a Hoffman apparatus, they've seen that the hydrogen forms on one side and oxygen on the other (Fig. 12-4). Occasionally a student will generate a brilliant question at that time. What happens to the rest of the water? If the water turns into hydrogen gas at one electrode, what happened to the rest of the water molecule on that side? The water is part hydrogen and part oxygen. If a water molecule reacts, how can just hydrogen appear at one side? How can just oxygen appear on the other? Wouldn't that mean the oxygen and hydrogen would have to split and then move to the other side through the solution somehow?

Chapter 12 – Redox Chemistry

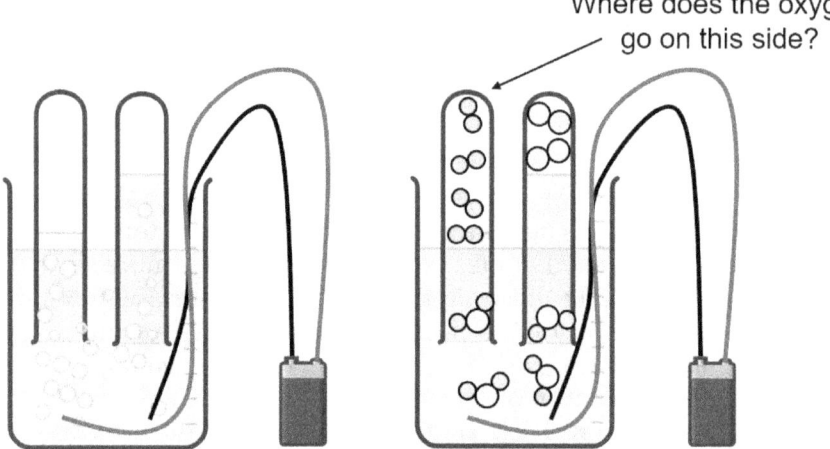

Figure 12-4: particle representations of the electrolysis of water reveal a conflict. If hydrogen gas appears on one side and oxygen on the other, what happens to the other portion of the water particles?

They probably were not ready for this question early in the school year. But now it sets up our quest to understand the complicated reactions that happen when we add electricity to an aqueous solution. The cathode where hydrogen is formed also forms hydroxide ions. The reaction is $2H_2O\ (l)\ +\ e^-\ \rightarrow\ H_2\ (g)\ +\ 2OH^-\ (aq)$. But we don't see the hydroxide ions in solution because hydroxide solutions are clear and colorless when dissolved in water.

At the anode we form oxygen gas, but we also form hydrogen ions. The half-reaction is $2H_2O\ (l)\ \rightarrow\ O_2\ (g)\ +\ 4H^+\ (aq)\ +\ 4e^-$. Again, the hydrogen ions are clear and colorless in the water. But if we add indicator to the water, we will see the indicator change color indicating the pH changes from those ions being generated. Even if we don't have indicator, we could add another solution to form a precipitate with the hydroxide or bubble with the acid to show evidence of their presence.

An alternative starting point for electrolysis is electrolysis of liquid salts. Beware that students will take some time to distinguish between liquid and aqueous. Students frequently mix up aqueous and liquid during the reaction unit. When you think about their life experiences, most of the liquids they are taught about prior to chemistry are water-based solutions. In order to help them differentiate the two you can try melting a salt in class. Zinc chloride melts at a reasonable temperature and you can successfully run a small amount of electrolysis on the molten zinc chloride using graphite electrodes. The salt will solidify around the colder (room temperature) graphite electrodes and shut down the electrolysis, but you can usually get enough zinc to form that you can show evidence of it by putting the cathode into hydrochloric acid afterwards. Bubbles of H_2 gas form as the excess $ZnCl_2$ dissolves allowing the Zn from the electrolysis to react.

If you value your lungs too much to try that, you can find a video online as well.[8] When we run electricity through a liquid salt, there are not multiple options for the electrons. The electrons at the cathode can only add to the zinc ions. The electrons at the anode can only leave the chloride ions. There is no water to compete with.

$ZnCl_2\ (l)\ \rightarrow\ Zn\ (s)\ +\ Cl_2\ (g)$

Chapter 12 – Redox Chemistry

For aqueous zinc chloride, water is present. Now there is competition for which chemical species is easiest to reduce or oxidize. Sometimes it is the water and sometimes it is the dissolved ion. The voltages give us a guide, but the amount or concentration can influence the results when voltages are similar. You will get some chlorine and some oxygen for typical concentrations of chloride ions since the reduction potentials are so close.

$Zn^{2+} + 2e^- \rightarrow Zn$ $\qquad$ E = -0.76 V
$2H_2O\ (l) + 2e^- \rightarrow 2OH^-\ (aq) + H_2\ (g)$ $\qquad$ E = -0.83 V

Both of these two reduction ½ reactions can occur at the cathode during electrolysis. At the anode we have the following two oxidation ½ reactions possible.

$2Cl^- \rightarrow Cl_2 + 2e^-$ $\qquad$ E = -1.36 V
$2H_2O\ (l) \rightarrow 4H^+\ (aq) + O_2\ (g) + 4e^-$ $\qquad$ E = -1.23 V

The end task for students will be determining what products form when you run electricity through an aqueous solution. Along the way they will encounter liquid salts, aqueous salts, and observations of the products formed (splint tests, color changes, electrode coatings, etc.). In order to be successful, students must identify the two possible half-reactions at the anode, and the two possible half-reactions at the anode. Then they will determine which is more likely for each electrode using the voltages. The students should inquire about why that algorithm works and contrast the process with batteries.

I start by using consistent half-reactions for the water. In acidic solutions (e.g., if Al^{3+} or Zn^{2+} ions are present), the half reaction for the cathode could also be $2H^+ + 2e^- \rightarrow H_2\ (g)$. But for the majority of reactions, you are going to use the following oxidation and reduction half-reactions for water:

Cathode $\qquad$ $2H_2O\ (l) + 2e^- \rightarrow 2OH^-\ (aq) + H_2\ (g)$ $\qquad$ E = -0.83 V
Anode $\qquad$ $2H_2O\ (l) \rightarrow 4H^+\ (aq) + O_2\ (g) + 4e^-$ $\qquad$ E = -1.23 V

If we run electrolysis through an aqueous solution of NaF, we have the two half-reactions of water competing with the following two:

Cathode $\qquad$ $Na^+\ (aq) + e^- \rightarrow Na(s)$ $\qquad$ E = -2.71 V
Anode $\qquad$ $2F^-\ (aq) \rightarrow F_2\ (g) + 2e^-$ $\qquad$ E = -2.87 V

At the cathode, the water will react because it has a much more positive voltage than the sodium half-reaction. At the anode the water will react as well because it has a much more positive voltage than the fluoride half-reaction. Remember that the voltage is related to how well the chemical pulls on or releases electrons. The largely negative voltage for reduction means that the chemical is bad at pulling on electrons and the largely negative voltage for oxidation means that the chemical is good at pulling on electrons. Since the aim is the opposite of that we do not see these half-reactions occurring. Of course, sodium ions aren't going to pull on electrons well and of course fluoride is not going to easily give up electrons.

Students will commonly confuse the products and reactants for the aqueous species. You have to stress that they can't start with a reactant that isn't present. They will try to flip the half-reactions to make them more positive because of their algorithm for starting a battery problem. They might write $Na(s) \rightarrow Na^+\ (aq) + e^-$ as

Chapter 12 – Redox Chemistry

the reaction that occurs at the anode because that results in a +2.71 V half-reaction. And if we had sodium atoms, that reaction would be occurring. But we don't have solid Na (especially in water!). In a battery we start with 4 chemical species and so we have options about what will or will not react. But in electrolysis we start with 3 species. Water, the cation, and the anion.

Now let's look at the electrolysis of copper (II) iodide. We begin with H_2O, Cu^{2+}, and I^-. There is no Cu metal nor any iodine that can be a reactant. We have four potential half-reactions.

Anode: $2I^-$ (aq) → I_2 (s) + $2e^-$ E = -0.54 V
$2H_2O$ (l) → $4H^+$ (aq) + O_2 (g) + $4e^-$ E = -1.23 V

Cathode: $2H_2O$ (l) → $2OH^-$ (aq) + H_2 (g) E = -0.83 V
Cu^{2+} (aq) + $2e^-$ → Cu (s) E = +0.34 V

At the anode the elemental iodine is going to form because it is easier to remove the electron from iodide based on the relative voltage to water. At the cathode the copper metal will form because the copper (II) ions are better at pulling on electrons than water as evidenced by the more positive voltage.

If we had run electricity through a solution of CuF_2 on the other hand, the copper would form at the cathode and oxygen gas would form at the anode. In order to produce fluorine gas, you would need to run electricity through a molten fluoride salt. That is not recommended regardless of what safety equipment you have. Multiple chemists died or nearly died trying to do this.

Coulometry

After students can determine what chemicals will form, you want to move into current and how much chemical will form. The units for current are amperes, or amps, or A. But 1 amp can also be written as $1 C*s^{-1}$. Coulomb (C) is the unit for quantity of charge. So current effectively represents how much charge passes by a point in a unit of time. More current means more electrons (or other charged particles) flowing. In electrolysis all the current leads to reaction. If 20 electrons pass a point in 1 s, that means that 20 electrons must have been involved in the reaction taking place as well. If 20 billion electrons pass a point in 1 s, that means that 20 billion electrons were used in the reaction.

Current can be used to determine how much product will form. Before we get into the calculations, there are four key relationships students should know about.
1. Increasing the current (2A to 6A) will increase the amount of product formed.
2. Running the current for a longer time (2 minutes to 6 minutes) will increase the amount of product formed.
3. Increasing the charge of the ions (Sn^{2+} to Sn^{4+}) will <u>reduce</u> the amount of product formed.

Let's look at a specific half-reaction where tin (II) ions are reduced to tin metal.
Sn^{2+}(aq) + $2e^-$ → Sn(s)

Chapter 12 – Redox Chemistry

If the current is increased or if the current runs for a longer period of time, the amount of electrons supplied to the reaction will increase. These changes are proportional. If we double the current, the amount of tin will double. If we triple the time, the amount of tin made will triple.

But what if we used tin (IV) instead of tin (II)?

$Sn^{4+}(aq) + 4e^- \rightarrow Sn(s)$

Now we require twice as many electrons to produce the same quantity of tin metal. If we run the same current for the same time, we'll end up producing half as much tin. If we go back to our stoichiometry example from chapter 7, this would be making a sandwich with four pieces of salami instead of two per sandwich.

The last piece of information before we work out mathematics of coulometry is that many students struggle with the inclusion of electrons in the reaction. The representation (Fig. 12-2) where students track electrons and ions around a circuit can help make this more concrete for students. But in general, they will not quickly associate how the electrons function at the particle level from the symbols in the reaction. This is a great opportunity to reinforce the difference between neutral and charged forms for elements like in the naming unit. They also struggle because we do not always make it abundantly clear that we don't write the electrons in the overall reactions even though we include them in the half-reactions.

$Ag^+ (aq) + e^- \rightarrow Ag (s)$
$Cu (s) \rightarrow Cu^{2+} (aq) + 2e^-$
$2Ag^+ (aq + Cu (s) \rightarrow 2Ag (s) + Cu^{2+} (aq)$

This is very similar to naming and formula writing. Ionic compounds are composed of ions, but we do not show the charges. Many students then assume molecular compounds must also be composed of charged ions. The inconsistencies give them pause and uncertainty.

There are three helpful tips for students to do calculations with coulometry.
1. Students need to write amps as C per 1s. 4 A is 4 C per 1s.
2. Students need to understand what Faraday's constant is.
3. If you are given the length of time the current runs, start with that first.

Coulometry calculations involve a lot of abstract ideas for students. The amount of information given in a problem can easily cause cognitive overload. We need to guide the start and simplify the middle to help students learn the algorithm and make connections between the algorithm and the concepts.

Amps are not going to connect to any other unit. If the student fails to replace amps with coulombs per 1 second, they will get stuck. If they do change them, they can see the connection to the length of time which is how they should start the problem if possible.

The Faraday constant involves many new concepts. The Faraday constant ($\mathcal{F}$) is 96,500 Coulombs. But more importantly for our purposes, it is the amount of charge when we have 1 mole of electrons. It is similar to the molar mass of an element, except charge instead of mass and for electrons instead of elements. If we have 96,500 C of charge flow past a given point, we will have had 1 mole of electrons

Chapter 12 – Redox Chemistry

react in our electrolysis set up. The best way to fully get the point across of what to do with Faraday's constant is to do some questions where we just change between charge in C, moles of electrons, and moles of the metal.

$$Zn^{2+} (aq) + 2e^- \rightarrow Zn (s)$$

If 48,250 C of charge is passed through the solution of zinc nitrate, how many moles of Zn are produced?

We can see that 48,250 C is half of the Faraday constant. That means we have 0.500 moles of electrons. In our reaction, for every 2 moles of electrons we get 1 mole of zinc metal. Since 0.500 moles is ¼ of that recipe, we should end up with 0.250 mol of Zn. Students may not easily see this relationship. Dimensional analysis can help them to get the final result, but a worked-out BCA table might help them see the relationship even better.

*****

An experimental set up runs 4.24 A of current through a copper (II) bromide solution for 18 minutes. Determine the mass of copper metal produced.

What are all the things that a student needs to identify from that problem stem? They need to be able to write out the half-reaction for the copper. They need to know that 4.24 A = 4.24 C per 1 s. They need to know to use the Faraday constant to convert coulombs into moles of electrons. They need to know where to begin the problem. Two of those four "needs" are not explicit in the problem itself.

If this is too much for students, start them with just some basic proportionality to get them to have a better grasp on the components. If you run 2.0 A through a solution of $CuCl_2$ for 1 hour you get 2.4 g of Cu metal. How much Cu metal will you get if you increase the current to 6.0 A? How much Cu would you get if you ran 4.0 A for 2.5 hours?

You can also give them a set of data that shows the amount of Cu formed after various currents and times and ask them to figure out what patterns they notice.

Trial	Current (A)	Time (s)	Initial charge of Cu	Mass of Cu formed
1	2.00	3600	Cu^{2+}	2.4 g
2	4.00	3600	Cu^{2+}	4.8 g
3	2.00	7200	Cu^{2+}	4.8 g
4	4.00	7200	Cu^{2+}	9.6 g
5	2.00	3600	Cu^+	4.8 g
6	4.00	3600	Cu^+	9.6 g
7	2.00	7200	Cu^+	?
8	???	1800	Cu^+	9.6 g

Table 12-1: Data for 8 trials of electrolysis of copper (I) or copper (II) chloride

Chapter 12 – Redox Chemistry

After working through the patterns, students might see the symbols in $Cu^+ (aq) + e^- \rightarrow Cu (s)$ and $Cu^{2+} (aq) + 2e^- \rightarrow Cu (s)$ with greater clarity. They would still require some intervention about what current is and how to use the Faraday's constant.

Another technique is giving students some starting information and have them figure out everything they know. 12.0 A is run for 8.0 minutes through $ZnBr_2$. What do you know? This gives students an opportunity to reflect about what each given represents, and to connect the chemistry concepts with the abstract algorithms.

After an initial introduction to all these concepts you must try interleaving them. Students that can follow a sample coulometry calculation will often struggle when presented with questions that incorporate concepts from galvanic and electrolytic cells. Students will struggle to differentiate voltage and current. Many will try and create a positive voltage for electrolysis reactions.

It can be helpful to now try and create questions that make connections between oxidation states with galvanic and electrolytic cells. Organic chemistry, redox titrations, and reactions with inert electrodes are good prompts to push student thinking.

Keys for redox chemistry:
1. Delay the rules for oxidation states until after the concept is clear. Use the concept to develop why the rules are what they are.
2. Expect students to mix up neutral and charged forms of elements. Particle diagrams can help.
3. Expect students to struggle with what current is. If 2.0 A of electricity is flowing, how many cations move to the cathode every one second?
4. Use electrolysis to help students understand how voltage is used to predict what will happen. A smaller negative voltage means that electrons are easier to push in that direction. It's easier to push electrons onto Zn^{2+} (E = -0.76 V) than onto water (E = -0.83 V).
5. Voltage tells us what will react. Current tells us how much will react.

Student Struggles

1. "I have no idea what I'm doing!" There are many spots in redox chemistry where students can fall apart. But one common initial idea that students struggle with is splitting up a redox reaction into half-reactions. Why do we use half-reactions? Why don't we use them in other content areas? How can the splitting of the reaction into two components be demonstrated at the macroscopic and particle levels?

All of these are wonderful questions. I would bet that many students would struggle with most of them before we even begin to address spectator ions. The purpose of the half-reaction is we are taking reactions that have electrons transfer from one substance to another, and we want to separate the loss of the electrons from the gain of electrons. By separating them physically in real life, we allow electricity to be intercepted between the two half-reactions. To establish the particle level picture, start with a particle level representation of a single replacement reaction where the particles change in charge.

2. "Once I start figuring out the battery details, I'm good, but I don't know where to start." When students get a reduction potential list, they might not know what the reactions and values mean. The reaction shows the addition of electrons to an ion or substance. The corresponding voltage tells us how much energy change occurs for that reaction relative to the standard hydrogen electrode. This should be simplified into a comparison of how strong of a pull on electrons these chemicals exert. Since $Au^{3+} + 3e^- \rightarrow Au$ has a standard potential of +1.52 V, we can infer that Au^{3+} will pull on electrons a lot. This means that Au^{3+} will be very reactive in gaining electrons and that Au will be very unreactive in resisting the loss of electrons.

Since a battery is going to run (one way or the other), the reduction potentials are used to determine which way the electrons will move. The half-reaction with the more positive voltage will remain as the reduction half-reaction. The other half-reaction will run backwards as an oxidation reaction. Once students establish the half-reaction to reverse, most of the other questions fall into place.

3. "I keep getting the products mixed up for electrolysis." In a battery, students typically have 2 sets of reactants and 2 sets of products and the water. They have options for what will be the reactants and products. In electrolysis we don't have the same arrangement of split-up half-reactions. Electrolysis reactions only have one chemical present along with the water (unless it's a molten salt). Right after the student learns to ignore Na^+ as a spectator or that Na^+ won't easily reduce, they now have no option but for the Na^+ to react. If we run current through molten NaCl, there's nothing else that can occur (at the cathode). Students need to have a strong macroscopic picture that shows that the galvanic cells are split into half-cells with a salt bridge while electrolytic cells are a single container with one reactant (two ions).

4. "I don't know how to figure out the oxidation number for the elements that aren't on the list of rules." For $KMnO_4$ a student might know that O is -2 and K is +1, but is unsure how to determine the charge of Mn. The key idea here is that oxidation states account for all the electrons in the substance. If each oxygen is assigned a surplus of 2 electrons (8 total), and the potassium has a deficit of 1 electron, the manganese must have a deficit of 7 electrons. All the electrons must be accounted for. By knowing all the elements but one, the one can be determined using the total charge of the chemical (often neutral).

5. "My redox reaction seems balanced, but it got marked wrong."
Balancing redox reactions involves all the traditional balancing components, plus charges. Students miss the charge piece of it. They do this frequently when the teacher does not connect the balancing of the electrons in half-reactions with the purpose of balancing charge for the overall reaction. We should expect students to miss the idea of charge balancing, even if we state it explicitly. This is because students are being put into cognitive overload when starting to balance redox reactions. The half-reaction idea is new and abstract. The inclusion of charges is new. Spectator ions is either new or developing. And these are all easy steps for the teacher. It's easy to progress too quickly or fail to give students time to think through why some of these components are used.

6. "Am I supposed to just memorize the products for redox reactions?"
There are some redox reactions that students can predict. They should know that chlorine oxidizes to become chloride. But they are not able to determine that the

products of permanganate being reduced are either MnO_2 (basic conditions) or Mn^{2+} (acidic conditions). Students will also wonder how they would know what forms when there are multiple options. If chlorine is oxidized, will it form hypochlorite, chlorate, or perchlorate? The answer depends on the concentrations used and the reaction conditions. A Frost diagram (improved version of Latimer diagrams), and Pourbaix diagrams would be something an advanced student could read about to learn more.

7. "I don't know how to start coulometry problems." Starting a coulometry problem involves a lot of knowledge. It is likely that students have a shaky understanding of electrolysis in general. They might not easily identify what is being made, they might not differentiate reactants and products consistently, they might not have a strong particle model for electrolysis, and students might not understand what current is. They might also not know that amps are equivalent to charge relative to a time interval. They might not know that coulometry is an application of stoichiometry and chemical recipes. To help initiate some of these, start the students with an initial picture and ask what would change if the current were increased, or the charge of the tin being reduced were increased, or the length of time were decreased. Ask them to imagine how the final product would differ and how that difference could be represented symbolically.

8. "Where do the electrons in the salt bridge come from?" Teachers sometimes have unfair expectations for students on what happens in a salt bridge. You can tell them directly that ions conduct the current, but most students don't have much to build from with that. One helpful tip is to pretend there is no salt bridge. Show students what would happen at the particle level if you had a net influx of charge to both half cells. They'll see that electrons would not be able to continue to flow away from a positive charge and toward a negative charge. This might help set them up with a framework to fill in what is taking place.

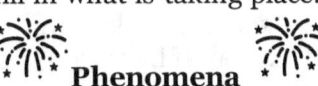

Phenomena

1. Zinc metal reacts with copper (II) solution - Copper metal does not react with zinc ions in solution. This simple demonstration of placing a piece of Zn metal into $CuCl_2$ and a piece of Cu metal into $ZnCl_2$ can be used for a lot of concepts in redox chemistry. We are providing students an opportunity to understand that one chemical is better at pulling on electrons than the other. We can emphasize charge even though the symbols obscure it. We can describe the changes using oxidation numbers. We can talk about balance with respect to the chemicals and the charge. We are even setting up some of the thermodynamics ideas such as Gibbs free energy. And all of that starts with an engaging discrepant visual. One set of chemicals reacts. The other similar set of chemicals does not.

2. Electrolysis of liquid salt - Students do not know many liquids that are not water (or aqueous). They are unlikely to easily differentiate an aqueous solution of an ionic compound with the molten version. Using the term liquid instead of molten can help press them. Zinc chloride happens to melt at a reasonably low temperature. You can run electricity through zinc chloride using graphite electrodes. On the cathode a thin layer of zinc metal forms along with some solid zinc chloride (the electrodes cool

Chapter 12 – Redox Chemistry

the hot liquid salt). Evidence of the zinc metal forming can be demonstrated by dipping the electrode into hydrochloric acid to show hydrogen gas being evolved.

3. Colorful salt bridge - Current is a new idea in chemistry class. Students have some familiarity with electricity, but they likely have never heard of current being conducted without electrons in a meaningful manner. Many students insist that electrons move through the salt bridge. If you construct a salt bridge with copper (II) chromate, the students will see the blue color moving toward the cathode and the yellow chromate ions moving toward the anode. This motion will be gradual but evident.

4. Combustion reactions - Fire is such an engaging tool. It would be a shame to teach redox and not ever discuss how electrons shift from carbon toward oxygen in combustion. The carbon fuels start with lots of hydrogen. As carbon dioxide forms the electrons shift from carbon to the oxygen due to electronegativity differences as the bond changes occur. Teachers can use many demonstrations as long as you are properly trained and knowledgeable about the safety hazards and risks. The whoosh bottle, heptane vapor ramp, flash cotton, or even just burning a candle can have a rich discussion of redox principles.

5. Silver nanoparticles - Using distilled water, the addition of a reducing agent such as sodium borohydride to dilute silver nitrate can lead to silver nanoparticles forming. Silver nanoparticles are small collections of silver atoms somewhere around the order of 10,000 or 100,000 silver atoms per nanoparticle. These collections of atoms behave differently from monatomic silver and from bulk solid silver. Metal nanoparticles are often utilized for color in glass. Yellow stained glass can be made by mixing silver in with the glass and red stained glass can be made using gold.

6. Gold and aqua regia - Gold is resistant to oxidation. But with the extremely reactive combination of hydrochloric acid and nitric acid gold can be dissolved. Gold leaf is an affordable form of gold that can be purchased for food decorating. This method was used to hide two Nobel prizes from Nazis by George Hevesy.

7. Electrolysis of water - An observant student will question what happens when water is split into hydrogen and oxygen. If hydrogen is produced at the cathode, what happens to the oxygen in the water at the cathode? If oxygen is produced at the anode, what happens to the hydrogen in the water at the anode? Obviously, the particles don't separate and then move to the other electrode. The reaction produces other components at each electrode. At the cathode, water turns into hydrogen gas and hydroxide ions. At the anode, water turns into oxygen gas and hydrogen ions. This can be observed by running electrolysis in water with universal indicator added. The cathode will turn violet while the anode turns red. Mixing the solution will return the solution to the neutral green.

8. Tin (II) electrolysis - Aqueous $SnCl_2$ will produce thin segments of Sn metal when reduced. The fractal-like patterns can be reversed by changing the battery hook ups. Filtering the aqueous $SnCl_2$ solution can help this function more effectively.

9. Blue bottle (pumpkin version) - Dissolving methylene blue with dextrose and hydroxide allows for an interesting reaction. When the mixture is shaken, more oxygen is suspended in the solution. The oxygen binds to the methylene blue causing the color to change to blue from colorless. A fun twist is to draw a pumpkin face on the glassware and add orange (red + yellow) food coloring. This starts with orange and colorless, but when shaken becomes blue + orange which appears black. As the oxygen exits the solution, the color fades back to orange.

Chapter 12 – Redox Chemistry

10. Dichromate breathalyzer - Don't actually do this one live. Find a video instead. Chromium (VI) is highly carcinogenic, and its use/disposal should be minimized. But the change from orange chromium (VI) to green chromium (III) is a good visual. The reaction is a quality problem for students to balance, and the application of the reaction for use in breathalyzers can be used for discussion in the classroom. Once I had a student use a similar reaction (with permanganate) to test whether more alcohol was removed from ignition during a flambe dish, or if the hot dish caused the alcohol to evaporate.

Flashcards

$$Cl^- \rightarrow Cl_2 \rightarrow ClO^- \rightarrow ClO_3^-$$

Oxidation

or

Reduction?

$Li^+ + e^- \rightarrow Li$ $E = -3.05$ V

$Sn^{2+} + 2e^- \rightarrow Sn$ $E = -0.136$ V

Which is the better reducing agent?

$A^{2+} + 2e^- \rightarrow A$ $E° = +0.6$ V

$B^+ + 2e^- \rightarrow B$ $E° = +1.4$ V

$E°_{cell} = ???$

Chapter 12 – Redox Chemistry

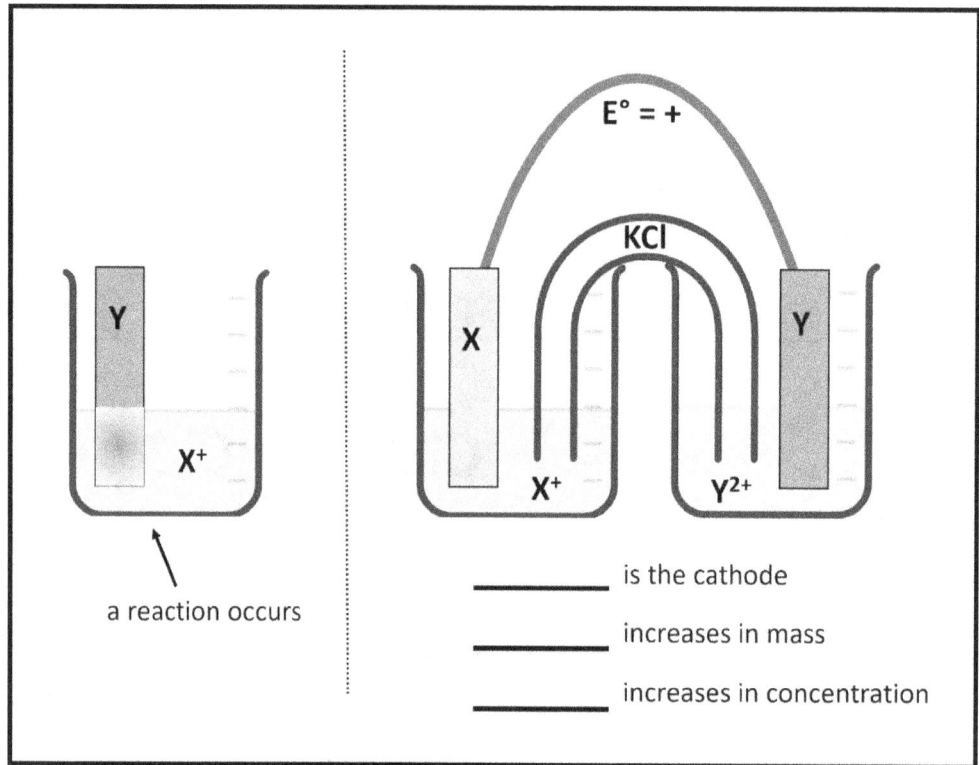

a reaction occurs

_____ is the cathode

_____ increases in mass

_____ increases in concentration

$ZnCl_2$ vs. $CrCl_3$

If equal current is applied for equal time,
_____ will produce more moles of metal

$$2Al + 3Cu^{2+} \rightarrow 2Al^{3+} + 3Cu$$

$$n = ???$$

Chapter 12 – Redox Chemistry

$K_2Cr_2O_7$

↑

Oxidation # = ???

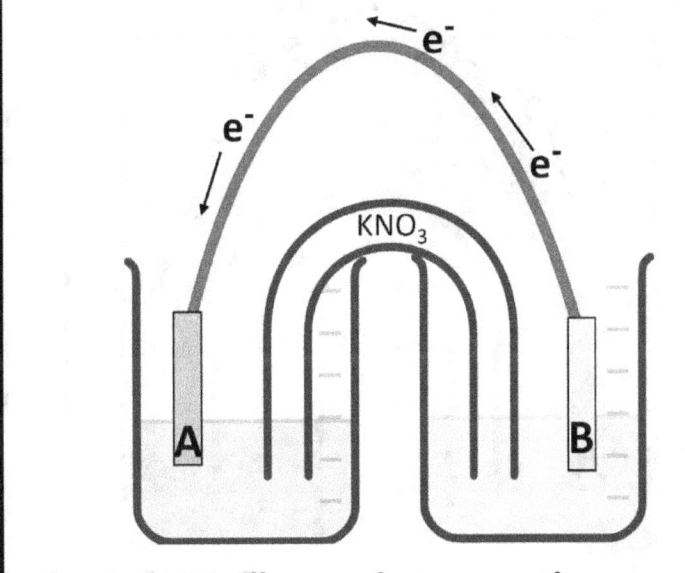

Electrode A will _____ in mass as time moves forward unless it is an _____ electrode

$$1 \text{ Amp} = 1 \frac{?}{s}$$

Chapter 12 – Redox Chemistry

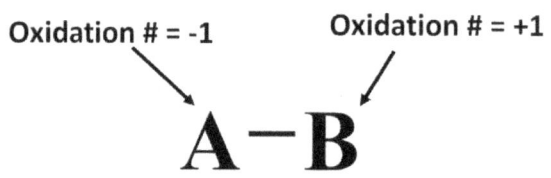

Oxidation # = -1   Oxidation # = +1

A−B

_____ is more electronegative

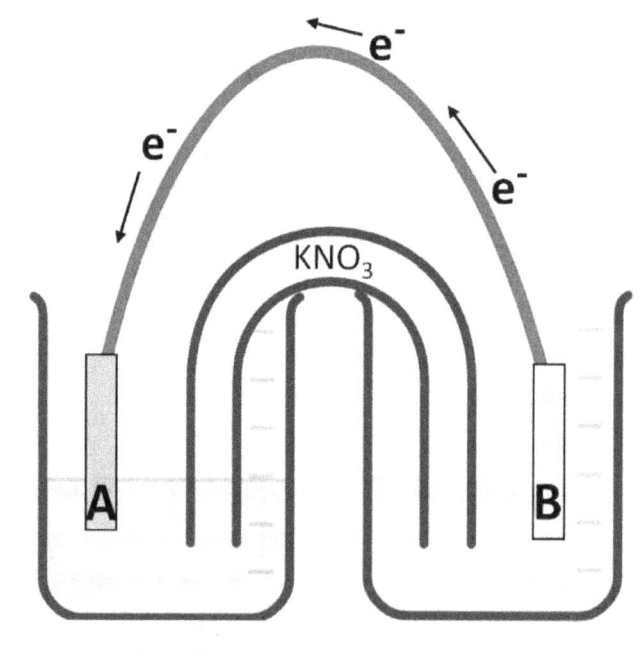

Which is the cathode???

1 mol e⁻ = _____ C

Faraday's constant

Chapter
Kinetics

History

Kinetics took a while to get attention from chemists. Many of the components that we use in education today weren't formalized until later. Without means to measure and communicate concentration, it is challenging to report how quickly concentration changes. Chemistry teachers are of course aware of the challenges in finding a chemical reaction where the macroscopic change is easy to track visually.

The first kinetics experiment published was by Ludwig Wilhelmy (1850) who studied the inversion of sucrose using polarimetry. He was able to determine that the concentration of sucrose and of acid both influenced the reaction rate.[1]

William Esson and Augustus Harcourt worked collaboratively on kinetics. Esson derived the mathematical relationships for 1st-order and 2nd-order kinetics. Harcourt worked on making precise measurements to confirm the theoretical relationships. Harcourt's first reaction studied (1865) was quite complicated:

$$K_2Mn_2O_8 + 3H_2SO_4 + 5H_2C_2O_4 \rightarrow K_2SO_4 + 2MnSO_4 + 10CO_2 + 8H_2O$$

This reaction was productive since it could be stopped by the addition of hydroiodic acid (HI) and then the amount of iodine produced titrated to determine the reaction's progress.

Next, he studied the reaction between hydrogen peroxide and hydroiodic acid:

$$H_2O_2 + 2HI \rightarrow 2H_2O + I_2$$

This reaction was tracked by adding sodium thiosulfate to react with the iodine produced. When the thiosulfate was completely consumed starch was used as an indicator. This reaction is familiar to most teachers as a version of the iodine-clock reaction.[2,3]

Their work was used by Van't Hoff in his book Studies in Chemical Dynamics (1884).[4] Van't Hoff's book details how temperature, pressure, and concentration influence the reaction dynamics for a variety of concrete examples. This book proposed multiple equations for the relationship between temperature and reaction rate. Svante Arrhenius selected $k = Ae^{\frac{-E_a}{RT}}$ from the ones proposed by Van't Hoff. In this book Van't Hoff also noted how there were few examples of 3rd-order rate laws. He posited that this would indicate that the reactions that were would likely occur in multiple steps so that it was not required that 3 particles collide simultaneously for a reaction to proceed.

Catalysts have been used throughout history. The term was proposed by Jacob Berzelius (1835) to distinguish substances that sped up reactions without being consumed as a reactant. This was extended to biological catalysts by Willy Kühne (1878) by naming these catalysts to be enzymes.[5]

A large hurdle that required clearing was that the rate of a reaction only depending on the chemicals involved. The law of mass action dictated that both the amount or concentration was just as relevant as the nature of chemicals reacting. This was not limited to kinetics, however, as we see in equilibrium next.

Chapter 13 – Kinetics

Rate

Kinetics can appear simple for students when they remain in a purely abstract model. The patterns and relationships are frequently simple ratios that students can easily follow without depth of understanding. A lack of understanding of rate can persist in spite of their ability to do calculations. Rate is how fast a reaction occurs. There are many representations of rate that are important if a student is going to show proficient comprehension.

People learn by adding context. Teaching generic information about rate is not going to automatically transfer when we start discussing rate in the context of a chemical reaction. But at the same time rate is a critical feature of kinetics. We should start by giving students opportunities to critically examine what they know and don't know about rate.

- What is rate when we're talking about non-chemistry topics?
- What is a fast rate of driving?
- What does a high interest rate mean?
- Why is the word rate similar to ratio?

A rate is a comparison of two changing things, usually one of them is time. The speed of a sprinter is a good example of a rate. In track sprinters typically compete in a 100 m, 200 m, or 400 m event. In the 1996 Olympics Michael Johnson set a new world record for the 200 m sprint in 19.32 seconds. The first 100 m took him 10.12 seconds and the final 100 m he ran in 9.20 seconds. Usain Bolt currently holds both the 100 m (9.58 s) and 200 m (19.19 s) world records. How was Michael Johnson able to run 100 m faster than Usain Bolt? Should he have the world record for the 100 m?

The rate of speed of the sprinters is not uniform throughout the race. At the start the runners have a velocity of zero. They quickly accelerate, but it takes time for them to reach top speed. Michael Johnson averaged 10.35 m/s while Usain Bolt averaged 10.42 m/s. But during the final 100 m of the sprint, Michael Johnson averaged 10.87 m/s. The reason for the increase in pace could have been explained by not running in a curve, not having to accelerate for the start, or just being able to increase his leg turnover rate. These ideas should allow students to differentiate between average rate and instantaneous rate. The instantaneous rate during a sprint varies.

- What would this look like for a chemical reaction?
- What would a fast or slow chemical reaction look like?
- What would happen at the particle level?
- How could we measure and report out the rate of a reaction?

Figure 13-1: Concentration vs. time graph

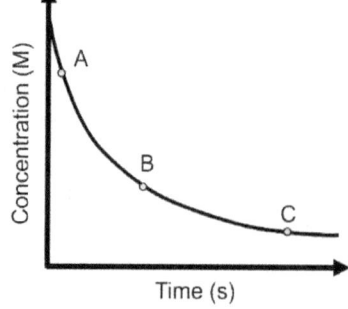

Chapter 13 – Kinetics

A fast reaction has the amount of chemical change a lot during a given time interval. A slow reaction experiences a small change in amount of chemical for a given time interval. Often, we communicate the change in amounts using concentration. The perk of using concentration is that we standardize the volume of solutions we use. A larger solution volume can have more chemicals reacting but isn't necessarily faster in terms of change than a small volume with larger concentration.

Just like with sprinting, rate is often not constant during a chemical reaction. The reason a reaction happens in the first place is because of particles colliding that cause changes in bonding. As the concentrations change, those collisions are going to occur at different frequencies which will likely impact the rate of reaction. The reciprocity of concentration and rate are critical for students to explore especially with equilibrium looming next.

If the concentrations of reactants are large, we would expect a faster rate of reaction. But a faster rate of reaction means that the concentrations of reactants will decrease rapidly. This leads to a change in reaction rate as the frequency of collisions decreases. We see this in Fig. 13-1 where the slope at A is steep. As we move to B and C the slope approaches zero indicating the reaction is slowing down as the concentration of reactants decreases.

It's worth noting that the term rate is not descriptive in isolation. Rate describes a change of something. We can improve that term to reaction rate to describe the change in reaction and we can make that even better by stating rate of change in concentration. To a teacher all those developments might be intuitive, but to a student rate might not land the same.

A student should understand the units of rate, relative rates, graphical representations of rate, macroscopic observations of rate, particle diagrams showing concentration changes, how rate and concentration affect each other, how bonding influences rate, and how temperature influences rate. That is a lot of stuff. Fortunately, we have rate constants, activation energy, rate laws, and various graphical relationships to use as tools in developing that understanding.

For students to develop a strong particle model of rate, they also should understand that rate is influenced more by the likelihood of a collision leading to a reaction than how quickly collisions occur. Let's assume that particles of A have a collision frequency of about 50 collisions per second. Particles of C on the other hand have a collision frequency of about 200 per second. But each collision between A particles has a 14% chance of converting into a particle of B. Each collision between C particles only has a 1% chance of converting into a particle of D. *Which reaction will happen faster?*

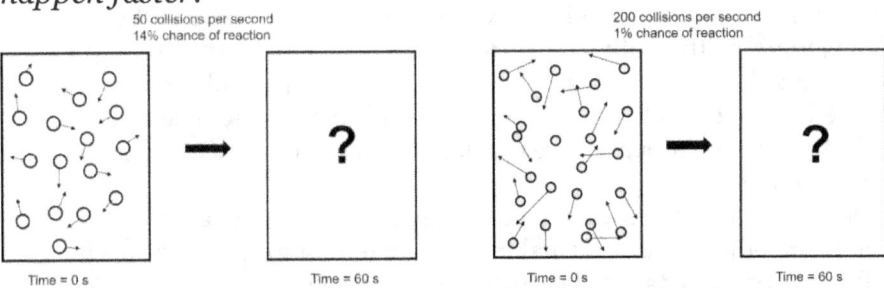

Figure 13-2: Which matters more, collision frequency or probability of a collision resulting in a reaction?

Chapter 13 – Kinetics

About 7 particles of A react every second initially. About 2 particles of C react each second initially. A → B is the faster reaction due to the fact that a collision is more likely to succeed in changing A particles to B. That is in spite of C particles colliding 4 times as often. A key question that sets up from this thought experiment is why would A be more likely to react?

Collision speed could be a factor, but if C has the higher collision frequency it's unlikely to explain what we're seeing. Steric hindrance could play a role. Perhaps C particles have a bulky component that interferes with the ability of the C particles to collide in the proper orientation. The most likely scenario is that the bonding in the C particles is stronger. This means that only very high-speed collisions are capable of breaking the bonds that are needed to convert a C particle to a D particle. We often shortcut that thinking by introducing the term activation energy prematurely. This can be particularly dangerous because students do not have a strong grasp on what breaking a bond entails.

*****

There are potential roadblocks to understanding rate. The chemicals formed in the middle of a collision do not fit with the rules or norms that students know from earlier units in chemistry. Because the timeframe of a collision is very brief, we will see things that are otherwise too unstable to exist for a longer period of time. We tell students that oxygen is always diatomic in the elemental state (ignoring ozone), but it is plausible to form an oxygen radical (a single atom of oxygen) during a collision.

The units of rate are $M*s^{-1}$ or another unit of time in place of seconds. Regardless of whether you're using seconds, minutes or hours the units don't fit into the same schema as other units in chemistry do. When we say that the units of density are grams per mL (g/mL), there is an implied for every statement we can use. For every 1.0 grams of water, there is 1 mL. But for a rate of $0.20\ M*s^{-1}$ that statement is incomplete. For every 0.20 Molarity there is 1 second doesn't make any sense because we fail to address the change component. Rate deals with change and so it is important for students to look at actual graphs that highlight the changes taking place. Rate is the change in molarity relative to the time interval. For every 1 second, there is a change in concentration of 0.20 M.

We want students to have some concrete examples to help them make connections and find patterns for rates. But the reactions we use in kinetics are often excessively complicated. Consider a clock reaction. The clock reaction is a common choice because it has a very sudden change indicating a particular concentration change in the reactants. This allows students to determine a rate via concentration change and time. But that reaction is actually three reactions taking place. There is the production of iodine in a complicated redox reaction (Rxn 1). There is the consumption of the iodine by thiosulfate (Rxn 2). And finally, there is the complex formation between the iodine and the starch (Rxn 3) when the thiosulfate is completely consumed (Rxn 2). A clock reaction is a horrible concrete example in the sense that it requires excessive processing to be able to understand what's happening at the particle and symbolic levels.

On the other hand, a clock reaction is an excellent concrete example at the macroscopic level. It provides a highly visual color change that indicates a specific change in concentration of a reactant. The distinct color change makes it easy to compare rates at different concentrations and temperatures.

Chapter 13 – Kinetics

Rxn 1: $6I^- (aq) + BrO_3^- (aq) + 6H^+ (aq) \rightarrow 3I_2 (aq) + Br^- (aq) + 3H_2O (l)$
Rxn 2: $I_2 (aq) + 2S_2O_3^{2-} (aq) \rightarrow 2I^- (aq) + S_4O_6^{2-} (aq)$
Rxn 3: $I_2 (aq) + starch \rightarrow blue\ complex$

Relative rates are mentioned early in the chapter, and then often set aside. The key concept with relative rates is that stoichiometry still is in play. If $2A + B \rightarrow 3C$ is a reaction, then the relative rate of disappearance of A will be twice as fast as that of B. C will appear 3 times as fast as B disappears. We want to be explicit that usually we will only study 1 chemical species for rate even though all the chemicals are changing.

In the clock reaction above, we only look for the formation of the blue complex. This complex indicates that the thiosulfate has gone from its initial concentration to zero. The rate of change for the thiosulfate can be used to determine the relative rates of changes for all the chemical species in all three of the reactions. Be prepared for students to struggle with questions comparing relative rates after they've spent time working with rate law determinations.

Why do we determine rates as the reaction begins instead of in the middle? Initial rates are higher than after the concentrations have decreased. The ratios between rates should remain constant, but the signal to noise ratio drops as concentrations decrease. Instantaneous rates are preferred over average rates since average rates are over a time interval where concentration is not constant. Typically, we look at a single chemical that changes color, or a gas produced because these are easiest to measure changes in concentration.

In order to let students wrestle with the subtle distinctions of these concepts, you want a simple phenomenon they can discuss. Food coloring mixed with bleach is a good phenomenon. Students will be able to see the color fade to colorless as the reaction proceeds. Data can be collected with a spectrophotometer to produce a graph (Fig. 13-1). From here you can ask students questions the help them make connections and distinctions.

- What would a concentration vs. time graph look like for bleach if 2 bleach particles react with 1 food coloring particle?
- What if 2 food coloring particles react with 1 bleach particle? Draw what you think the graph would look like if the temperature were increased. Draw another for what would change if the volume doubled but concentration remained constant.
- What part of the graph has the fastest rate? Why?
- How much faster is the rate there than at 40 seconds?
- What might the units of rate be?
- If we ran a second experiment with twice as much food coloring what might change or stay the same in the new graph?

Some introduce kinetics with four factors that influence rate. Concentration, catalysts, temperature, and surface area all impact reaction rate. The four factors introduction can be utilized to enhance the particle collision models and provide many concrete macroscopic connections for students. Students can mix different concentrations of bleach and food coloring to see how rate is impacted. They can try a clock reaction at various temperatures. The teacher can perform the elephant toothpaste demonstration to show how a catalyst impacts the decomposition of

Chapter 13 – Kinetics

hydrogen peroxide. The teacher can also show how steel wool burns in a glorious fashion while a steel bar does not due to limited surface area for the oxygen to collide with.

Observations of these demonstrations give students the opportunity to process the information, make comparisons, and strengthen their model of rate. After students have had opportunities to think critically about rate (and the units of rate!), we are ready to transition to rate laws.

Rate Laws

Kinetics typically begins with some basics of rate before moving into rate laws. Many students do not have a sufficient mental model to draw from. This leads to rate laws being strictly abstract to students. They use algorithms and patterns to make sense of rate laws instead of visualizing the macroscopic or particle levels. You can see evidence of this superficial processing when you provide a question where the concentration quadruples and the rate doubles. You will have students who get 2nd order instead of the correct order of ½. This indicates that students are just working with the relationship between the changes in numbers but are not identifying what variables those numbers go with.

Experiment	[A]	[B]	Initial rate of appearance of C ($M*s^{-1}$)
1	0.01	0.04	0.0008
2	0.02	0.04	0.0016
3	0.02	0.08	0.0064

Table 13-1: Initial concentrations and rate for the reaction $2A + B \rightarrow 3C$

What we often do as teachers is draw students' attention to the changes between experiments (Table 13-1). We highlight that the concentration of A doubles from experiments 1 to 2. Since our rate also doubles, we know that A is first order. But before we start making these comparisons, take a moment and highlight all the information that a student might not understand the relevance of in Table 13-1.

- Do they know why C is chosen for the rate?
- Do they know what the units $M*s^{-1}$ represent?
- Do they know that each experiment represents a single reaction being run?
- Do they know why the words initial, or appearance are used?
- Do they have a particle level image of what is changing between the different experiments and within a single experiment?
- Do they know that volume is constant even though concentration varies?

The zero-order rate law is confusing. Let's assume the reaction $2A + B \rightarrow C + D$ has a rate law that is rate = $k[A]^2$. This means that changing the concentration of B has no impact on the rate. *How can changing the amount of B not cause any change in the rate?* Shouldn't there be more collisions? The answer to this question comes later with elementary steps, but the question can give students

Chapter 13 – Kinetics

an opportunity to think critically. Some students might even figure it out ahead of time.

Before you jump into the abstract manipulations, try asking students to write how they would conduct the experiments in Table 13-1. Have them develop a complete procedure and see what they include. For most teachers, this will be enlightening of how little comprehension they have. It would be even more illuminating to have them make particle representations at t = 0 and t = 10 s.

One problem with students not knowing a couple of items from the chart is that this prevents students from feeling confident in pieces that they might understand. Thus, students stick to the safety of comparing the numbers between experiments and they follow the abstract algorithm. By not providing time for conceptual development, we run the risk of pushing a fixed mindset.

If you don't have the time to develop the students' connections, you can still help them understand the algorithm better by comparing the rate law for different experiments. Our generic rate law for the reaction before we know anything is:

Rate = k $[A]^x[B]^y$

In this rate law, rate is how fast the reaction occurs and has units of $M*s^{-1}$ (per Table 13-1). The k represents the rate constant, which will have units that reflect the orders of the two chemicals. [A] and [B] are the concentrations of the reactants and the x and y are the exponents (usually 0, 1, or 2). For experiments 1 and 2, we have some information that we can put into the generic rate law, but not enough to solve for anything new.

Experiment 1: 0.0008 $M*s^{-1}$ = k $[0.01]^x * [0.04]^y$
Experiment 2: 0.0016 $M*s^{-1}$ = k $[0.02]^x * [0.04]^y$

We can divide the rate law with data from experiment 2 by the rate law with data from experiment 1 to find some new information.

$$\frac{Experiment\ 2}{Experiment\ 1} = \frac{0.0016\ M*s^{-1}}{0.0008\ M*s^{-1}} = \frac{k\ [0.02]^x * [0.04]^y}{k\ [0.01]^x * [0.04]^y}$$

After some simplification we end up with 2 = 2^x. This means that the exponent x is 1 or that the order for A is first order. This process can be repeated with experiments 3 and 2 to find that the order of B is second order.

Benefits of using this method include providing a tool for students to solve for rate laws where the concentrations are not held constant, students might do better with finding the units of the rate constant k, and students might differentiate the components of the rate law (rate, concentration, rate constant) with more precision.

What is the physical meaning of the rate constant? Ask students what factors influence k before you reach the Arrhenius equation. They'll respond better if you ask how a reaction with a large rate constant would differ with a reaction with a small rate constant in real life.

One thing you may notice during this is that students are not clear on what the difference between the rate and the rate constant is. The best way to help students connect what the rate constant represents is to use some specific situations. If you have a very large rate constant what does that mean? Even if concentrations are somewhat small, a large rate constant would mean that the rate is fast.

A large rate constant has bonding implications. If the bonds between the reactants are weak this could lead to a faster rate. We can skip bonding by using the term activation energy. Students should have some concept of having seen that

Chapter 13 – Kinetics

combustion reactions generally require an initial impetus to start the reaction. They will likely not have a complete understanding of activation energy, but a simple model will suffice for the moment.

The rate constant can also be large because the temperature is high. The particles are moving so quickly that they frequently collide with sufficient speed to cause bonds to break. The key here is establishing that different chemical reactions will occur at different rates. Some of those differences are via the concentration of the reactants, and the other factors are combined into a single constant, k.

When teaching students how to solve for k, I find it helpful to have them commit to memory that the 1st order reactions have k with units of s^{-1}. The 1st order kinetics analysis is common due to nuclear decay being first order, and having a baseline known unit makes it easier to distinguish units for 0 ($M*s^{-1}$) and 2nd order reactions ($M^{-1}*s^{-1}$).

Many teachers have such a tight timeline for teaching that they have to remain in abstract mode to save time for other subjects. But if you have the time, have the students do an experiment. The iodine clock reaction is popular, and I recommend doing the experiment by mixing 3-4 variations at once. If there are three reactants, do a control along with three other mixtures where one reactant has its concentration doubled (Fig. 13-3).

Have students draw particle representations of 0, 1st and 2nd order reactions at three different times (0, 10s, and 20s). These diagrams should show evidence of how stoichiometry matters and how the 0-10 s time interval has a faster average rate than the 10-20 s interval. How do the concentrations at times 0, 10 s, and 20 s compare? Does doubling [A] cause rate to increase for A and B, or just for A?

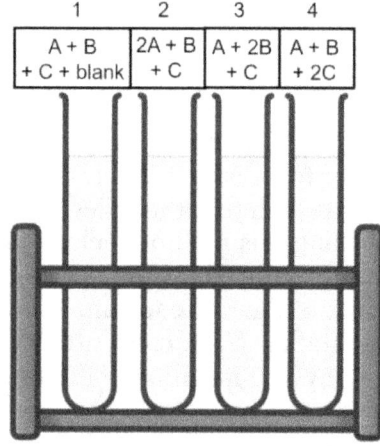

Figure 13-3: Potential set up for a rate law demonstration

Kinetics offers a wealth of opportunity for students to utilize their math knowledge. Students can use calculus knowledge to relate rate laws and integrated rate laws. Students struggle to maintain why there are different equations based on the reaction order. This is a great opportunity for you the teacher to talk about how a zero-order reaction has the same rate no matter what the amount is. A first-order reaction gets slower as the concentration decreases. A second order reaction gets slower at an even faster pace as the concentration changes. It takes them time to translate this into a graph or even two concentrations at different times.

Chapter 13 – Kinetics

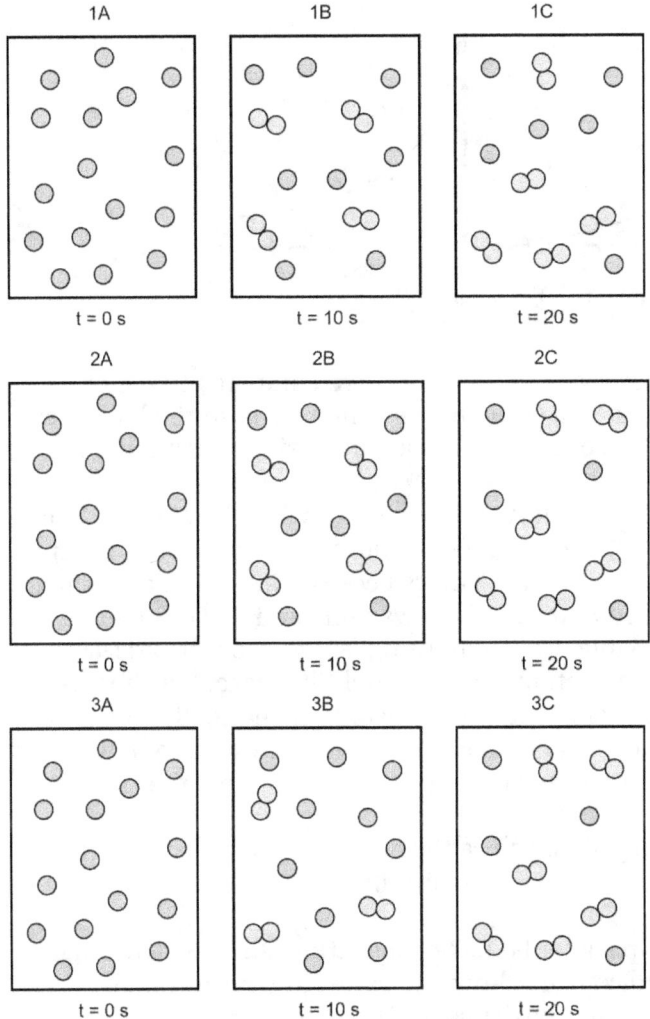

Figure 13-4: Which (1-3) shows a 0-order, 1st-order, and 2nd-order reaction?

One method to get students to make observations about graphs is to have them write down the points of the graph. How does concentration vary on a 0, 1st, and 2nd order graph? What differences do we notice? What patterns are there?

How can you tell just by looking at a concentration vs. time graph if the reaction is 1st or 2nd order? A 0-order reaction has a linear slope, but how does the slope change for a 1st-order or 2nd-order reaction? If the students know that the time intervals are consistent for 1st-order half-lives, ask them to do a Claim, Evidence, Reasoning (CER) for how the curve for a 2nd-order reaction will compare. Then provide the following options (Fig. 13-5).

Chapter 13 – Kinetics

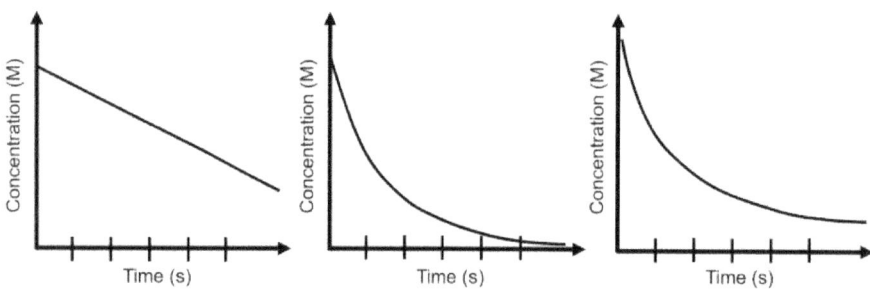

Figure 13-5: Graphs for a 0, 1st, and 2nd order reaction.

Notice how for a second order reaction how the curve compares to the first order reaction. The first order reaction can easily be identified because the half-lives are at consistent time intervals. The second order has an increasing spacing between half-lives.

Have students track the mass of a burning candle throughout the hour. When they plot the mass, they will find that the candle mostly burns at a constant rate. The change in mass is constant with respect to time. This is because the size of the candle has no bearing on how well the oxygen and vaporized wax mix and burn. Students will observe the relatively constant flame. Candle burning is a zero-order process. Now if we increased the oxygen concentration the rate would increase, but there is so much oxygen in the room that the initial concentration will remain mostly constant.

Another great concrete example students can explore is the volume of water left in a draining buret. This will not be zero-order. As the buret drains, the rate at which water escapes becomes slower. The rate of water leaving is proportional relative to the height of the water (or volume since the area of the surface is constant). The higher the water is in the buret, the more pressure there is on the water coming out the tip of buret.

These concrete examples help students make connections to how rates might vary as time progresses. They should consider how rate varies as concentration changes, how reaction order can and can't be determined, and how graphs of rate vs. time and rate vs. concentration would differ for each order. Students might also be tasked with explaining how buret size or candle type would impact the rate constant.

Lastly introduce the integrated rate law equations. These use integration to go from rate vs. concentration to concentration vs. time. This is a big deal because concentration is much easier to measure. If students are provided these equations before they've had some opportunity to engage in critical thinking, they are likely going to be stuck in a superficial understanding of them (which might be acceptable).

When assessing rate laws, you will want to utilize spaced practice. Give students a table of data every other day throughout the unit to determine the rate law, rate constant, and units of the rate constant. Also add in a variety of questions that test the conceptual understanding. Some questions should be simple, but they should include the macroscopic and particle viewpoints. You will also want to stress the differences between concentration and rate while emphasizing how they impact each other.

Chapter 13 – Kinetics

Elementary Steps

Now it's time to reconnect with where we started the unit, collisions. In order for most reactions to happen, there are multiple collisions that must happen. Each of those collisions is called an elementary step. Elementary steps provide concrete images to tie all the abstract reasoning in rate laws together at the particle level.

To start elementary steps, begin by having students put together some particle representations of a reaction. For example, $2NO\ (g) + O_2\ (g) \rightarrow 2NO_2\ (g)$ is a reaction that has three reactant particles. It is extremely unlikely that a reaction will ever happen based on three particles colliding simultaneously. If your students are skeptical, give them three tennis balls and see if three students can throw them so that they all hit at the same time. Even when we are working with quintillions of molecules, a three-particle simultaneous collision is unlikely to be the basis for a successful chemical reaction.

What are some possible steps that could happen with only two particles colliding at a time where we end up with the final products? Have students draw some out. They might need some assurance that the chemicals formed in the middle do not need to conform to the rules they learned earlier in chemistry. This is in the middle of a violent collision. The rules aren't normal.

Students might have the 2NO particles stick to form an N_2O_2 intermediate. This intermediate would need to get hit by the O_2 with the proper orientation to form the NO_2 molecules.

They might instead propose a mechanism where the O_2 hits one of the NO molecules to form an NO_2 and an O radical. In a second collision the O radical combines with a second NO. There are more possibilities depending on the creativity of the students.

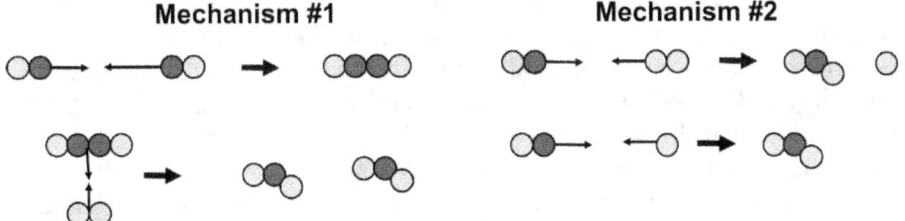

Figure 13-6: Two potential mechanisms for the reaction $2NO + O_2 \rightarrow 2NO_2$

Having students generate possible mechanisms like this helps them understand how the elementary steps represent each collision, they help reinforce the particle level view of what happens during the reaction, and they set you up for the rate-determining step.

Overall Reaction
    $2A + 2B \rightarrow 2C + D$

Elementary Steps
Collision #1 $A + A \rightarrow A_2$
Collision #2 $A_2 + B \rightarrow AB + C$
Collision #3 $AB + B \rightarrow C + D$

Chapter 13 – Kinetics

In the fake reaction above there are three collisions that must occur for the overall reaction to happen successfully. Note that in the midst of collisions, unstable chemicals form that might not fit into a student's construct of chemicals. The reason they can form is because they exist for a brief time after a large collision. Just like you probably won't see totaled cars driving around on the road, but you might see one after a large crash.

Some of those collisions are more likely to occur successfully than others. It turns out that when we first start a reaction, the slowest step is the one that determines how fast the overall reaction that will happen. This slowest step is called the rate-determining step (RDS). An analogy that can help is doing a large number of loads of laundry. Assume that you have a washer that runs in 40 minutes, but the dryer takes 70 minutes. As you continue to run loads of laundry, the dryer is going to determine how quickly you get done. Even if the washer speeds up to 30 minutes, the overall laundry time won't be changed much because we still have to wait 70 minutes for the dryer. But if the dryer were sped up, our laundry would get done faster.

This analogy connects to the rate-determining step because the concentration of the particles in the slowest step will affect the overall rate. Assume Collision #1 is the slowest (A + A → A_2). If I add more B, it's like getting a faster washing machine. It doesn't really change anything about the overall time because I'm still waiting for that first collision. But if I change the amount of A, then I speed up the slowest part (like speeding up the dryer).

It's important to clarify, by slow I don't mean that the collision is occurring slowly. Slow refers to the likelihood that the collisions will result in changes. It is the change in amount that is slow, and this is a result of high activation energy, not collision frequency.

The reason we propose elementary steps is because the rate law can be used to disprove a mechanism. If we do some experiments and find the rate law to be rate = k $[A]^2$, then the mechanism proposed above might be correct. But if we find the rate law to be rate = k [A][B], then we know the proposed mechanism is definitely not correct because [B] wouldn't influence the rate if A + A → A_2 was the rate-determining step. We can't watch a reaction happen live, so the connection between rate laws and mechanisms reveal the curtain for something that is partially invisible to us.

If you've taught organic chemistry, this is a great opportunity to come back to S_N1 and S_N2 reactions. The S_N1 is unimolecular meaning that only one molecule is involved in the rate-determining step (step 1a). This is slightly misleading since molecules don't just split, but the collision with anything in the solvent might cause the molecule to break apart into the carbocation intermediate and leaving group.

$CH_3CHBrCH_3$ → $CH_3CHCH_3^+$ + Br^- Step 1a

Then the nucleophile collides with the carbocation.
$CH_3CHCH_3^+$ + Cl^- → $CH_3CHClCH_3$ Step 2a

If the mechanism is S_N2 the nucleophile collides with the molecule first.
$CH_3CHBrCH_3$ + Cl^- → $CH_3CHClCH_3$ + Br^- Step 1b

Chapter 13 – Kinetics

This means we can determine the rate law to figure out which mechanism occurs. If the rate law is rate = k[CH$_3$CHBrCH$_3$][Cl$^-$] then the S$_{N}$2 mechanism is what happened. If the rate law is rate = k[CH$_3$CHBrCH$_3$] then the S$_{N}$1 mechanism is what happened.

This is a big deal. We can use the macroscopic data we obtain via experiment to determine what is happening at the particle level. Our determinations of the rate law don't just provide insight into the relationships between concentration and rate. The rate law is also instrumental into understanding the chemical mechanism.

Activation Energy

Motion, forces, and position should always be considered in conjunction with energy. What does having a large activation energy inform us about forces, positions, and motion of a set of reactants? A large activation energy means that the bonds between reactants are strong. In order for those bonds to break, a large collision must occur meaning that the particles must be moving fast relative to each other so that the bonding forces are insufficient to cause the needed acceleration for the particles to remain in contact. If a slow collision happens, the particles will move apart a small amount, but the bonding force will be sufficient to keep them together. In a set of particles, there is a wide range of speeds so even at low temperatures it is plausible that some collisions will occur at much higher speeds than typical collisions.

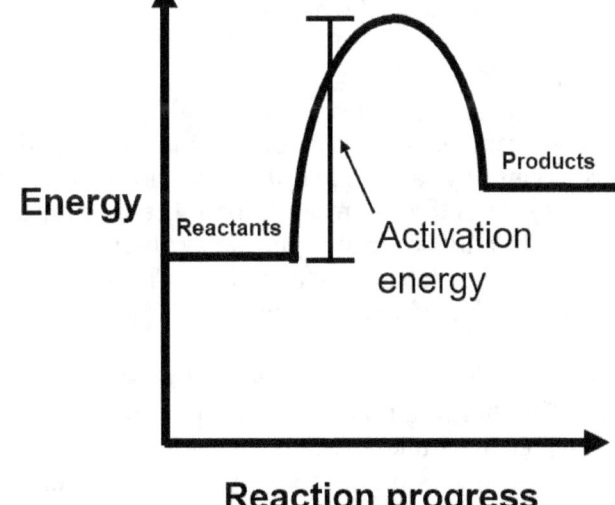

Figure 13-7: Reaction energy diagram

When we tell students that activation energy is the amount of energy needed to initiate a reaction, we give them little information of what is going on. Even if we connect that to a phenomenon they've witnessed (e.g., striking a match), we want them to get a chance to connect this to multiple concrete details. The worst-case scenario is that the student pictures a ball rolling up a hill. Instead, we want to focus the students' attention on particle collisions. Ideally, students would view a simulation with a distribution of particle speeds that change during collisions.

We can also compare enthalpy changes, activation energy, and bond strength. If we alter the bond strength of reactants and products, how will the enthalpy changes and activation energy change? What observations would students take from this chart? Could they translate each scenario into a reaction energy diagram?

Chapter 13 – Kinetics

Reactant bond strength	Product bond strength	Activation energy	Enthalpy change
Strong	Weak	High	Large and +
Strong	Strong	High	Small
Weak	Strong	Low	Large and -
Weak	Weak	Low	Small

Table 13-2: How bond strength impacts activation energy and enthalpy

If the bonds for the reactants are strong, the activation energy will be large. If the bonds for the reactants are weak, the activation energy will be small. This can be connected back to earlier when we talked about the physical meaning of the rate constant, k. The rate constant can be calculated using the Arrhenius equation:

$k = Ae^{\frac{-E_a}{RT}}$.

As the activation energy (E_a) gets smaller, the rate constant k will be larger.

Note that the Arrhenius equation also includes temperature. As the temperature increases the particles move faster and thus have larger collisions. The bond strength and hence the activation energy are not changed as temperature changes, but the number of collisions happening with larger energies does increase. In the Arrhenius equation as T gets larger the exponent becomes smaller and so k becomes larger due to the negative exponent.

$2NI_3$ (s) → N_2 (g) + $3I_2$ (g)

Nitrogen triiodide is a challenging demonstration. When the compound is wet, it will remain stable. As soon as the compound dries completely it can be set off by just about anything. A feather being dropped, or sometimes even a door closing on the other side of the room can initiate the decomposition. The long bonds between the nitrogen and iodine are quite weak; hence there is a very low activation energy required for the reaction to take place. The strong triple bond of the nitrogen makes the enthalpy change exothermic.

Fe_2O_3 (s) + Al (s) → Al_2O_3 (s) + Fe (l)

The thermite reaction is another challenging enterprise that some brave (or reckless) teachers take on. The strong bonding due to the 3+ charge of the iron (III) and aluminum ions produce a very high activation energy. The thermite reaction is difficult to initiate (I find a sparkler works best for me, but also terrifies me to be so close). If you can successfully pull it off, the products are so hot that the iron produced is in the liquid state. This reaction used to be the means by which railroad tracks were joined in regions where it was difficult to heat iron to its melting point. This reaction can be done on a smaller scale by covering one of two rusty steel balls (spheres if students are listening) with aluminum foil and striking them with a glancing blow. When sufficient activation energy is supplied to break some bonds, sparks will fly indicating the formation of product via electron transfer. This is a vivid example for students to see energy going into the system (the glancing strike), and the energy leaving the chemical system (sparks/heat/sound).

Chapter 13 – Kinetics

One issue with the reaction energy diagrams is that it tends to lead us to discussing a single set of collisions at a time. If we go back to the nitrogen triiodide reaction, the solid compound has more than trillions of compounds. Yet most representations in kinetics will just show two molecules. Kinetics is a great opportunity to build student connections between the symbolic levels and macroscopic levels. But to do that well students must understand how particle motion is distributed in a typical sample of chemicals.

In order to really capture the concept of activation energy, you'll want to focus on two pieces of prior knowledge. Students need to have a strong concept of bonds. Here you're assisted when students not only have a picture of a chemical bond, but when they have the flexibility to include vibration. What happens when the two atoms get closer? What happens when they get further? What does a broken bond look like at the particle level?

This ties in with the second component which is temperature. We want students to understand that as temperature increases that motion increases. But we also want students to have a visual concept that speeds are a distribution and that as molecules collide those speeds change based on the relative speed and angle during the collision.

Not all chemicals within a system move at the same speed. Particles are constantly colliding and changing speeds. The overall number of particles with different speeds will remain relatively constant within that sample, but each individual particle is constantly changing from fast to slow, or slow to medium, etc.

The Maxwell-Boltzmann distribution is the primary representation used to track these motions and changes in motion.

Figure 13-8: Maxwell Boltzmann distribution of a sample at two different temperatures

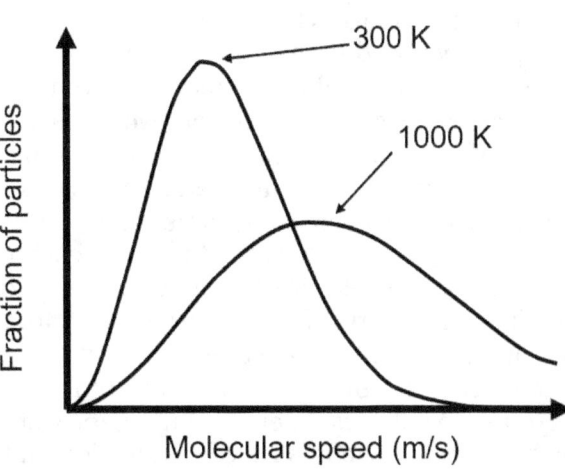

Students struggle to make sense of these distributions. Point out specific speeds on the x-axis and fraction of particles on the y-axis. You will want them to spend some time identifying which particles are the fast ones, which are the slower ones. Students should take a moment to explain why the higher temperature curve has a smaller and broader peak. It can help to shade the portions of the curve that do not overlap.

You can break this down ever further for students by using a constructivist approach. Choose three different speeds and have students estimate how many

Chapter 13 – Kinetics

particles there would be for each curve at those three speeds. After the students make their observations about the numbers, the conclusions are straightforward. The lower temperature peak has more molecules at low speeds. The higher temperature curve has more molecules at higher speeds.

There are a variety of Maxwell-Boltzmann distributions. They can describe a set of molecules at two different temperatures (Fig. 13-8). There can be distributions that look identical but describe two different chemicals at the same temperature. In that case, the mass of the particles leads to different speeds at the same temperature.

When showing a simulation be sure to ask students to watch for a very large collision between two particles that could potentially lead to a reaction. Then do the same thing at a higher temperature. How frequently do the collisions happen? Set a timer for 30 seconds and collect some data to emphasize the point.

A good bridge for students to connect with is the evaporation of water. At a low temperature the distribution of water molecule speeds is shifted towards lower speeds. But a few molecules will have sufficient speed to leave the system and become steam in the surroundings. This is something you can show students by having a puddle of warm and hot water evaporate from a lab table and asking students to write about or draw particle representations of why the hot water evaporates faster.

You can expand their thinking even more by evaporating a different liquid with a lower boiling point. Ethanol has a boiling point of 78 °C, but you should be cautious to not ignite the flammable vapors produced. Do not use methanol. Methanol is a frequent cause of harm to students because of its low flash point and unpredictable expansion of vapors. Methanol vapors can be ignited without a spark or flame.

It's worth noting that the water molecules that evaporate must be at or at least very near the surface. Otherwise, they will collide with other molecules and probably lose speed before they leave the system. This can be a great discussion point to connect with chemical reactions where some collisions will have enough relative speed for the reaction to occur but won't have the proper orientation.

When we put temperature and bonding together the concept of activation energy emerges. As students picture collisions occurring within a small sample of molecules, they will envision some collisions being smaller and a few being quite a bit larger than typical. These tend to be the reactions that lead to product.

Two molecules are racing toward each other, and they collide with a fury. Now what happens? Parts of the molecules start to move apart from one another. The bonding forces pull these pieces back, but they just can't stop them from flying apart. As the particles move apart from each other, they slow down slightly. Energy is converted to potential energy from kinetic as motion changes to separation. This high potential energy chaos is called a transition state or an activated complex. Bonds have been broken, pieces have been separated, and the potential energy is at a peak.

Chapter 13 – Kinetics

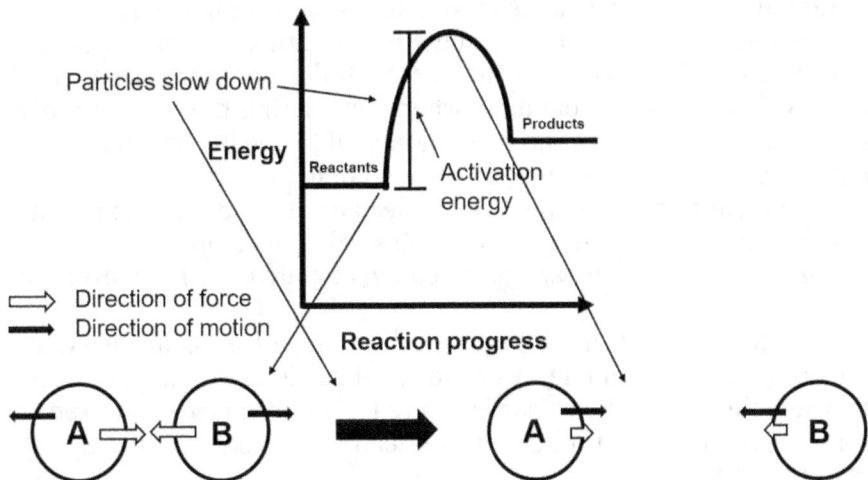

Figure 13-9: Help students connect a particle representation with motion and forces to the reaction energy diagram by asking them to identify what happens with forces, speed, and direction at multiple points.

For some activated complexes, they can form into a slightly more stable molecule that later collides and changes either back into the reactant, or into a product. These molecules with moderate stability are called intermediates. These intermediates are what we see in elementary steps.

Students tend to be able to construct a visual image of intermediates and activated complexes at the particle level. We want to take advantage of that and next add in symbolic representations. We do, however, want to display reaction energy diagrams paired with a set of elementary steps.

Overall reaction $A + 2B \rightarrow AB_2$
Step 1: $A + A \rightarrow A_2$ Slow
Step 2: $A_2 + B \rightarrow AB + A$ Fast
Step 3: $AB + B \rightarrow AB_2$ Fast

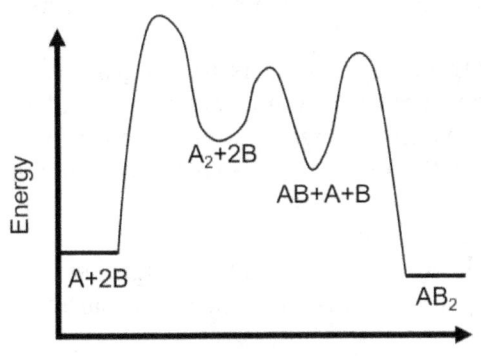

Figure 13-10: Reaction energy diagram for $A + 2B \rightarrow AB_2$

What are the intermediates and activated complexes in this reaction? The A_2 is an intermediate. The AB is an intermediate. You can see those at the local minima (Fig. 13-10). But what would the activated complex be? That would be when A_2 collides with B and the bonds of A_2 are breaking.

Chapter 13 – Kinetics

The exact nature of a transition state is muddy to a student. They may envision a chaotic explosion where the molecule splits into multiple fragments each moving in a different direction. This is difficult to align with the smooth curve on a reaction energy diagram. This is also problematic when considering the two reactants were in contact during the collision and bond formation could have begun then.

The macroscopic levels are challenging because students will think of these collisions in a linear fashion and for only one set of molecules at a time. Getting them to make that leap involves many tools (temperature, Maxwell-Boltzmann distributions, transition states, etc.) that can easily lead to cognitive overload even for simple reactions.

The reaction energy diagram is not intended to be a map of the collisions that happen during a reaction. It is a plot of energy as the set of particles rearrange from reactant to product where the two colliding particles are locked into place at a fixed separation. This is not the reality for all collisions that lead to a reaction. But it does provide a useful set of information.

When students have issues, we want to give them some simple understandings that we can easily agree upon to help them. We can agree that particles separate from each other after a collision occurs. Do they split into individual atoms separated to infinity? No. But as they do separate, we know that the particles slow down as energy is converted from kinetic to electrical potential energy. As the fragments come back together from the activated complex to the intermediate (or product), the fragments speed up. The reaction energy diagram provides a simplistic model that does give students an approachable visual.

*****

The kinetics unit is going to be filled with students doing practice problems. One objective as the teacher is to monitor their understanding of the concepts embedded within those problems. Don't assume if they can find the rate law correctly that they understand what they have done. Ask them to make connections between the concepts highlighted earlier in this chapter whenever possible. Ask them to explain why a unit, or term was included in the question. Why was the word instantaneous put next to rate? Why was the reaction described as 1st order and how does that impact how you might find the concentration at 20 s? What does the units of k being min^{-1} imply about the rate law?

Have students interpret final answers. If the rate constant is 100,000 s^{-1}, what does that mean about the activation energy? Are the reactant bonds strong or weak? Does this change when temperature changes? Does this change when concentration changes? If we were to increase concentration of HCl and IO_3^- by ten each, how much would the rate change?

When you complete your unit on kinetics, take some time to review how frequently you represented reactions with a single set of molecules in a single set of collisions? How frequently did you illustrate large collections of particles that are in various stages of the elementary steps?

Let's assume that a student has an incorrect model where a fast reaction means that a fast collision happens, and the particles quickly turn into products. To them a slow reaction rate is when a slow collision happens, and the particles take a long time to turn into products. In other words, it is not the probability of a successful collision that determines rate, it is the timeframe for the collision itself that causes fast or slow rates.

Chapter 13 – Kinetics

That student sees a reaction energy diagram with a large activation energy and sees a large hill that would take a long time to climb. That student sees a set of elementary steps that have fast written next to them, and they interpret that to mean that the collisions occur quickly. It is easy for a misconception to connect to the algorithms we use in kinetics.

Making an effort to utilize particle diagrams, symbolic representations, and macroscopic experimental evidence is just as important in kinetics as in other units. Don't be fooled by the students' ability to complete a problem correctly as evidence that they have a thorough understanding of those three levels.

Kinetics is an opportunity to link to other chemistry content. The reactions we use can limit this ability but try to find reactions that have multiple applications. The decomposition of NI_3 connects to bonding. Why are rates so fast for acid-base reactions? When $S_2O_3^{2-}$ turns into $S_4O_6^{2-}$ is the sulfur oxidized or reduced? What happens to the electron density of the sulfur atoms?

We are capable of dealing with kinetics separately from thermodynamics. But that doesn't mean they don't share commonalities. Both subjects rely on bonding and temperature at their core. Being able to link the two helps students have two major models to build new ideas about equilibrium from.

Mechanism #1

Mechanism #2

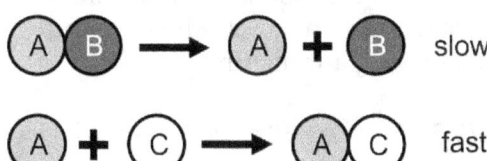

Figure 13-11: Two proposed mechanisms for the reaction AB + C → AC + B

If the rate law for the reaction AB + C → AC + B is found to be rate = k[AB], what does that mean about the two proposed mechanisms (Fig. 13-11)? If the AB bond is very strong, how will that impact E_a, k, and rate? If AC is a very strong bond, how will that impact E_a, k, and rate? If the reaction is run at a higher temperature, how will that impact the Maxwell-Boltzmann distribution, E_a, k, rate, and the Arrhenius equation?

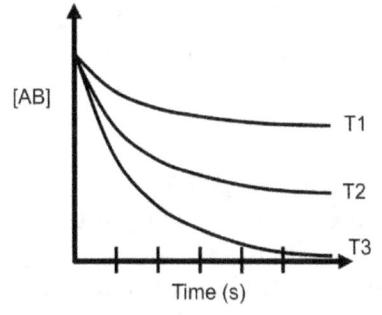 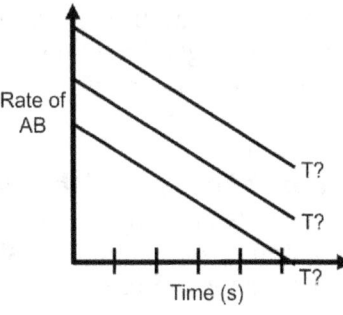

Figure 13-12: Reaction AB + C → AC + B run at three different temperatures

Chapter 13 – Kinetics

Which temperature is the highest of T1, T2, and T3 (Fig. 13-12)? Which temperature is each one on the Rate of AB vs. Time graph? Why are the Rate of AB vs. Time graphs all linear? When would they not be linear?

Student Struggles

1. "I have no idea what I'm doing!" Many students will be able to find patterns from data that they can use to align with algorithms to find the rate law. Many of those students do not have a full picture of what the data is that they are using. But they can tell that one component doubled and the other quadrupled, so they put together that a 2nd order relationship makes sense. Even if they haven't yet noticed that concentration and rates are the two components.

To help students unpack without limiting them to the algorithms, have them draw or set up a set of experiments. The more detail they include the better. Students will understand even better if they can include a follow up drawing that shows how fast the reaction takes place at the particle level.

2. "I don't know how to get the units for the rate constant." Issues with units start with the units for rate. Rate is going to be measured by a change in amount (usually molarity) relative to a time interval. The variety of ways rate and units for rate are presented lead to a lack of clarity for students. Students also do not have a strong sense of what the rate constant (k) is. If we jump in and show them how to find the units, they struggle to place the algorithm due to the cognitive load of the other components. This can be alleviated temporarily by showing different rate constants (with units) for different rate laws. Show students that the units will match for rate = k [A]² and rate = k [A][B]. Show that a first-order reaction always has the rate constant units of inverse time (s^{-1}) or that zero-order reactions always have the rate constant units match the units of rate. Help them notice patterns first.

3. "How do the units for activation energy work?" Activation energy can be found using the equation $k = Ae^{\frac{-E_a}{RT}}$ or by multiplying the slope of the graph ln k vs. 1/T by -RT. Either way, there is a connection between the units of activation energy (E_a) and the units of RT. Those who have learned statistical mechanics will recognize RT as being related to kT where k is the Boltzmann constant. Recall that temperature can be defined as a measure of the average kinetic energy of a substance. The Boltzmann constant allows to go from a measure of energy to the actual amount of energy in Joules. If we multiple the Boltzmann constant by 6.02×10^{23}, we get the gas constant 8.314 with units of $J \ast mol^{-1} \ast K^{-1}$. Notice how the Boltzmann constant gets us the energy per particle, and the ideal gas constant gives us the energy per mole of particles. Either way that we find activation energy, the units will be reported initially as J/mol. This tells us how much energy is required to add to change 1 mole of reactants to the transition state. We can convert this to the commonly reported kJ/mol.

4. "How do we propose a mechanism with elementary steps?" By the time we reach elementary steps, students are often carrying some confusion. They often don't get what an elementary step is. An elementary step is a single collision that happens during the course of a reaction. Reactions do not (typically) occur in a single

Chapter 13 – Kinetics

step. As the chemicals collide, changes occur to charge distribution and particle arrangements. Students struggle to propose these because they don't understand what they are. Students should start by evaluating what the reactants are, what products need to form, and what series of collisions might lead to that transformation. They need to be comfortable forming intermediates that are not stable except during a high-speed collision. All the particles present at the start of the reaction need to be present at each step (although not all must be involved in each step). And finally, they need to make sure the rate-determining step has coefficients that match the rate law.

5. "Is the reaction slow (if the activation energy is high) because the hill is higher to climb?" Sometimes the hill is such an obvious metaphor for a reaction that students take the meaning too literally. It takes time to go up a hill. It takes longer to go up a steeper hill. But that is not what is happening in a chemical reaction. The frequency and speed of collisions is not what alters the reaction rate for high activation energies. Rather, the frequency of collisions that cause a successful reaction is what determines the speed of a reaction.

6. "What does a fast rate look like at the particle level?" If we start with blue reactants and end with red products, a fast reaction is one where blue changes to red in a short time interval. But we must take that one step further. What would cause blue to change to red quickly? Faster-moving particles, and high concentrations would both help. But when we think about the chemical nature of the reactants, the key is that collisions between blue particles must frequently cause a change to red ones. If 4% of collisions leads to a reaction in one vessel, and 1% of collisions leads to a reaction in a second vessel, how will the two rates compare? When students follow this logic well, it helps to add one final question about how bond strength would vary in each of these examples. The reactant bonds would be weaker for the vessel where 4% of collisions lead to a reaction.

7. "I don't know which equation to use to figure out concentration." When students work with the integrated rate laws, they miss the fact that the integrated rate laws depend on the initial rate law. A zero-order rate law gives a different relationship between concentration and time than a first-order rate law. When students miss this detail, they end up with a variety of similar equations and become overwhelmed with trying to choose the appropriate one. Typically, students select an equation based on the given information. But multiple equations will match the givens in these instances so students must know to start by seeking out the rate law.

8. "Why does the rate law match the coefficients in the rate-determining step, but not the overall reaction?" The rate law does not (usually) match the coefficients in the balanced overall reaction because some of the reactants are not involved in the least likely collision needed for the reaction to progress. Think for a minute about how strange it is that some of the reactants in reactions are zero-order. Changing their concentration has no impact on the overall reaction rate. Once we accept that the reaction occurs in a series of steps, we can emphasize that one of those

Chapter 13 – Kinetics

steps will be the least likely to have a successful collision. Anything we can do to impact that step will have a proportional impact on the initial reaction rate.

9. *"How do I find the rate law if neither reactant is held constant?"* If the rate law experiments hold one reactant concentration constant, you can determine the order of the other reactant. From there, add in a new data point that would hold the other reactant constant.

[A]	[B]	Rate (M*s^{-1})
1 M	0.1 M	44
2 M	0.1 M	88
4 M	0.2 M	352

If we know A is 1st order (expt. 1 and expt. 2), and A is never held constant (1M, 2M, 4M). We can add a 4th experiment where A is 2M and B is 0.2 M. The rate will be 176 M*s^{-1} since [B] doubles from expt. 2. Now we can see that B is also first-order.

In the event that no variable is held constant, you have to look for information. If one reactant decreases by 2, and the other increases by 2 you can tell that both are the same order if the rate remains unchanged. If the rate increases, you know that the reactant that increased is a higher order.

You can always assume one of the orders to be 1. Work out a solution and see if the rate constant remains the same throughout. If it doesn't work, start over and assume the order is 2. Repeat this until you find a combination that works. The majority of reactants are 0, 1st, or 2nd-order and it's usually easy to rule out 0-order.

10. *"How can I tell if the reaction is 1st or 2nd-order using only the concentration vs. time graph?"* A first order reaction will have a constant half-life. The concentration vs. time graph should show constant time-intervals where the amount halves. A 2nd-order reaction will have the half-lives space out at larger and larger intervals as the concentration decreases. It is an excellent exercise to ask students to predict how a 2nd-order reaction will compare for the concentration vs. time graph to a 1st-order graph that is provided.

Phenomena

1. Bleach food coloring rainbow - Take six beakers of water and add red, yellow, and blue food coloring to create a rainbow. Mix the red with yellow to get orange, blue with yellow to get green, and blue with red to get purple. Add bleach to all six beakers. The dyes will fade at different rates so you will see the rainbow change from 6 to 3 colors, then to 1, and finally colorless.

2. Nitrogen triiodide - Nitrogen triiodide can be made by mixing ammonia and iodine. While the solid is wet, it is stable. Once the solid dries, a small shift can set off the compound into violently exploding. This compound helps connect bonding with activation energy. The large iodine atoms create long and weak bonds. These are easy to break, so the activation energy is quite small. Thermite works well as the antithesis of this. Thermite has strong bonds and thus has an incredibly high activation energy.

Chapter 13 – Kinetics

3. Draining a buret - A buret drains faster when the water level is higher. The more weight of water pushing down at the bottom leads to higher pressure. As the buret drains the rate can be measured by plotting the remaining volume vs. time. Students can compare plots of volume vs. time, ln V vs. time, and 1/V vs. time to determine if the draining is 0, 1st, or 2nd-order.

4. Burning a candle - A candle will burn at a mostly consistent rate. As the candle burns the mass will decreases as gases leave to the surroundings and are no longer measured. Students can compare plots of mass vs. time, ln m vs. time, and 1/m vs. time to determine if the reaction is 0, 1st, or 2nd-order. The candle wax needs to be prevented from dripping off the balance. The oxygen concentration would impact the reaction, but in most rooms (hopefully!) the oxygen levels will maintain.

5. Elephant toothpaste - Hydrogen peroxide is an interesting chemical. Some students will have had hydrogen peroxide added to a cut (this is not ideal as the peroxide will damage the skin and prevent healing while disinfecting) or use peroxide as a mouthwash. Elephant toothpaste is created by putting concentrated hydrogen peroxide into a graduated cylinder (or other container with a narrow neck) along with some dish soap (food coloring makes it look cooler). When a catalyst (KI or other) is added the rapid generation of oxygen gas creates a foam with the soap solution. This foam can grow quite large.

6. Iodine clock reaction - This reaction is actually three reactions occurring in the same solution. But this is a highly visual representation of the end of the reaction where a sharp change from colorless to blue/black occurs. I recommend doing this with 4 beakers. The first beaker will have a set concentration of iodide, bromate, and acid. Then each subsequent beaker will have one of those components doubled in concentration. This shows students how rate is impacted by concentrations differently depending on the order of each component.

7. Glow stick - A glow stick is another reaction that is highly visual since light is produced. A glow stick can be cut open and emptied into a test tube. There are some chemical and physical (broken glass) hazards. Once the mixture is in a test tube, the mixture can be heated or cooled. The change in light emission mirrors the changes in reaction rate. When the mixture is heated, the peroxide also runs out faster reminding us that stoichiometry is still applicable during kinetics.

8. Lighting a match - The matchstick head has ground up glass, phosphorus, and oxidizing agents mixed up. You can ignite a matchstick even in an environment without oxygen because the matchstick has its own source of oxygen within it. The glass causes the phosphorus to change forms from red phosphorus to white phosphorus (an allotrope). The white phosphorus burns spontaneously in the presence of oxygen. Ask students to draw a reaction energy diagram, then ask them to highlight how the diagram does or does not change due to the glass, oxidizing agent, and phosphorus.

9. Steel wool burning - Surface area matters in reactions. A large piece of iron rusts slowly. But steel wool has a substantial amount of contact between the iron and oxygen. This can be increased further by burning the steel wool in an oxygen rich environment. You can also find videos with extravagant sparks emitted when the steel wool is spun in a circle.

10. Three tennis balls - Give three students each a tennis ball. Instruct them to throw the tennis balls (not in the lab!) so that all three collide simultaneously mid-air. This highlights how difficult and unlikely this is to occur. When we teach elementary

Chapter 13 – Kinetics

steps, this demonstration can be used to justify why elementary steps occur between two particles only. If three particles must react, often two will collide and stick before the third collides with the intermediate complex.

Flashcards

rate = k [A]^2

What are the units of k?

$k = Ae^{-\left(\frac{E_a}{RT}\right)}$

As Ea ↑, what happens to k?

AB + C → AC + B

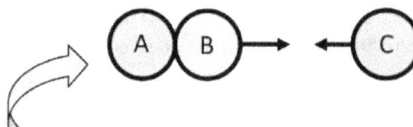

This collision will not lead to a reaction because this collision….

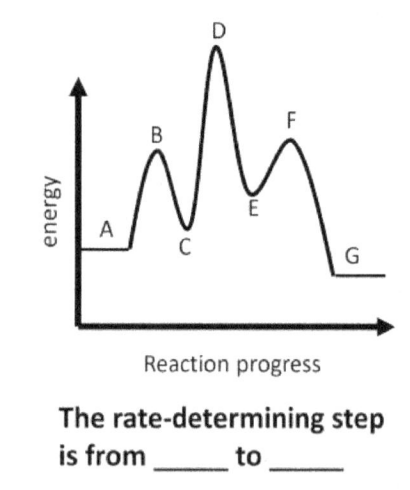

The rate-determining step is from _____ to _____

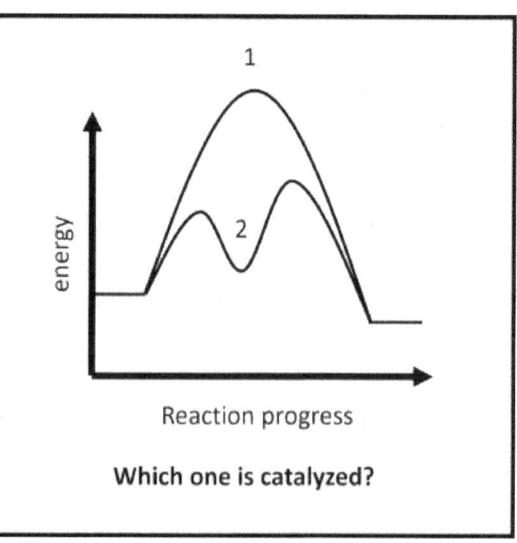

Which one is catalyzed?

Chapter 13 – Kinetics

X + Y → C + E   slow
E + X → C + B   fast

Overall reaction:
2X + Y → 2C + ???

---

rate = k [A]^x

If [A] ↑x3, rate ↑x3

What is x?

Experiment	[A]	Rate (M*s^{-1})
1	0.50	0.040
2	0.50	0.040
3	0.80	0.040

rate = k [A]^x

x = ???

Chapter 13 – Kinetics

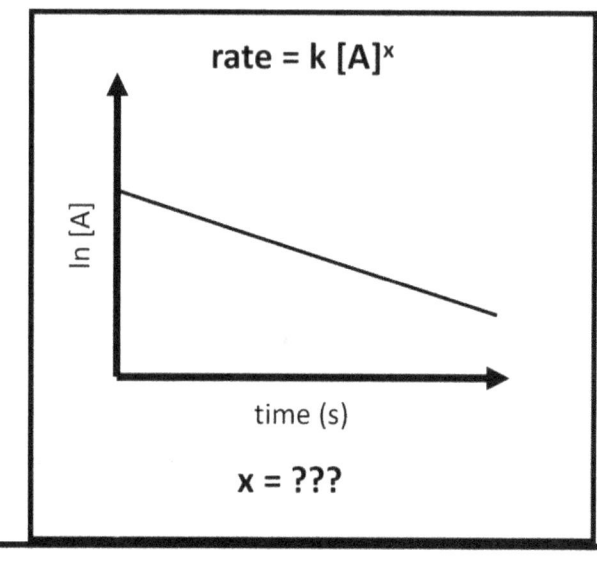

rate = k [A]^x

x = ???

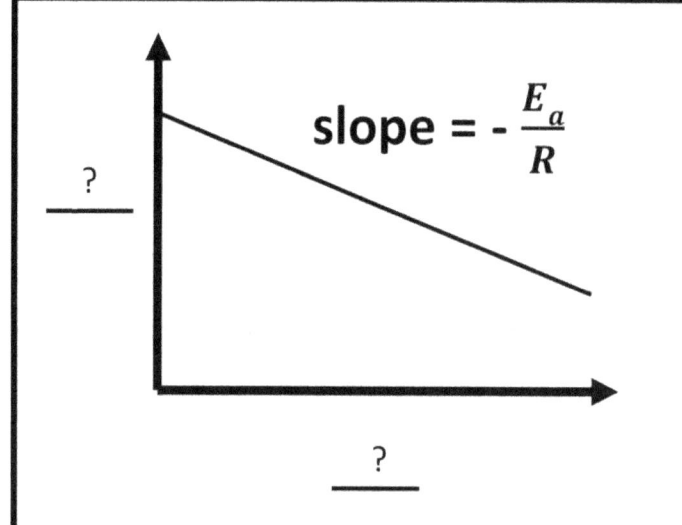

slope = $-\dfrac{E_a}{R}$

Experiment	[A]	Rate (M*s^{-1})
1	0.10	0.002
2	0.20	0.008
3	0.30	0.018

rate = k [A]^x

x = ???

Chapter
Equilibrium

History

Textbooks start equilibrium with the constant K and end with connections to thermodynamics. Historically these two approaches developed intertwined. Georges Lemoine studied hydrogen and iodine combining to form hydroiodic acid $H_2 + I_2 \leftrightarrows 2HI$ (1877). What he found was that the same proportions of chemicals were present after time had passed no matter what the initial starting reagents were. If he started with all HI, or all H_2/I_2, the resulting mixture was the same. He could reach that equilibrium much faster at higher temperatures. Pressure was not found to influence the equilibrium position for the reaction (all species are gases at temperatures used).[1]

That same year Van't Hoff had stated that equilibrium is when the forward and reverse rates become equivalent. He went on to also state that increasing temperature for an endothermic reaction caused a shift in equilibrium position that favored products (1884). Le Chatelier expanded this to include other stresses besides just temperature (1884).

The current equilibrium expressions that we use are similar to what Cato Maximilian Guldberg and Peter Waage researched from 1864-1879. They determined that the concentrations of chemicals were relevant instead of just the types of chemicals being used. They also proposed that each chemical could be raised to an integer value in the equilibrium expression. Their work built on the work mentioned in kinetics by Harcourt and Esson.[2]

While these equilibrium experiments were being conducted, you also have Willard Josiah Gibbs (1875-1878) working through thermodynamics. Much of his free energy model was applied toward equilibrium analysis.[3] Walther Nernst applied Gibbs's work toward high temperature gas equilibrium. Nernst had begun his career by determining how concentration influenced voltage in a battery (1888) before working on entropy and equilibrium. Nernst stipulated the third law of thermodynamics (1906) that allowed chemists to determine entropy values for chemicals experimentally.[4] This helped in predicting equilibrium values. By 1920 he was only able to make a rough predictor of equilibrium constants for different industrial reactions. It was useful, but it's interesting to note that research on equilibrium was not a priority.

The reason why equilibrium should be a priority is that industrial operations would be massively improved if they could create greater yield. Ammonia is a product of nitrogen and hydrogen that has a reactant favored equilibrium. Fritz Haber is famous for finding a set of reaction conditions that allow reasonable amounts of ammonia to form. In 1907 he had a contentious meeting with Nernst that caused him to dedicate a substantial effort to invalidating Nernst's prediction that very limited ammonia would ever be produced from nitrogen and hydrogen.[5]

Haber ended up using a catalyst (Os), high temperatures (but not too high), and extremely high pressures to produce an equilibrium mixture that was about 8% ammonia. The impact of this ammonia production is still felt today. The ability to

Chapter 14 – Equilibrium

take nitrogen from the air and convert it to a chemical form that can be used for fertilizers allowed for greater farming yields that sustained communities in ways that allowed further specialization. Haber also went on to work for the Nazi regime where he tested chemical weapons that used chlorine gas to suffocate opposing soldiers. Multiple staff were killed in testing and Haber tended to appear unaffected by their deaths. His wife Clara shot herself after a heated disagreement and Fritz skipped her funeral.[6]

Rate and Concentration

A bucket filled to the brim has a hole near the bottom. Water leaks out quickly. Later only a small amount of water remains. The water leaks out slowly. The height of the water in the bucket influences how quickly the water drains.

A differential equation is when the function is a part of the derivative. When calculating the rate that water comes out of a hole in a bucket, the height of the water influences the rate. If we represent the rate of change of the height of the water, we will need to include the height into the equation describing the rate of change.

Rate and concentration present the same pattern. The rate of a reaction depends on the concentration. But concentration changes proportionally to the rate. Our first goal for students in equilibrium is for them to verbalize the relationships between rate and concentration. Concentration can be apparent macroscopically while rate is usually more difficult to observe directly (large productions of gases for explosions are an unlikely exception in a classroom). In order to determine rate, we must look at the concentration of particles at two different times. But by determining the rate at two different times, the rate will now differ due to the changes in concentration.

To help students puzzle through the relationship between rate and concentration start with an equilibrium activity. The activity should involve two students doing something that involves opposing processes. In my class we do the following three reversible experiments:

1. Student A puts together 2 hydrogens from a model kit to make an H_2. Student B takes the H_2 model apart to form 2 separate hydrogens. $2H \leftrightarrows H_2$
2. Student A ties together 2 rubber bands. Student B takes the tied rubber bands apart. $2R \leftrightarrows R_2$
3. Student A links 2 paper clips. Student B takes the 2 paper clips apart. $2P \leftrightarrows P_2$

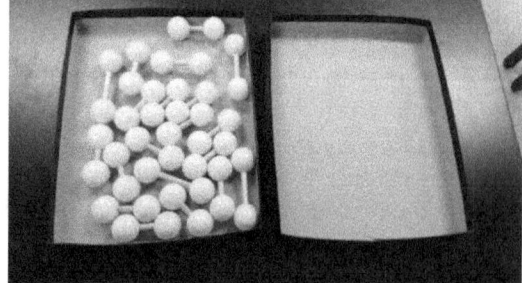

Figure 14-1: Model kits representing $2H \leftrightarrows H_2$

The students track the rates of change as well as the concentrations for six continuous 1-minute time intervals. After students track how many times per minute

Chapter 14 – Equilibrium

a model kit (or rubber band, or paper clip) is constructed or undone, they plot the data to see how rates impact concentration and vice versa.

The 1st trial of H/H_2 begins with 40 H particles and 0 H_2. The 2nd trial starts with 0 H and 20 H_2 (Fig. 14-1). In spite of the different starting amounts, students reach an equilibrium with mostly all H_2 within the first minute for both trials. It is much easier to take apart the model kits than to construct them. For trial 1 the student constructing H_2 can only construct about 5 per minute. The second student is limited to waiting for the model kits to be formed so they can only deconstruct about 5 per minute.

In the second trial, the second student can take apart the 20 H_2 at a rapid rate. Or at least they can do so initially until the amount of H_2 runs low. Student 1 cannot match the rate of the reverse reaction and so we accumulate more and more H until the reverse rate slows, and an equilibrium is established.

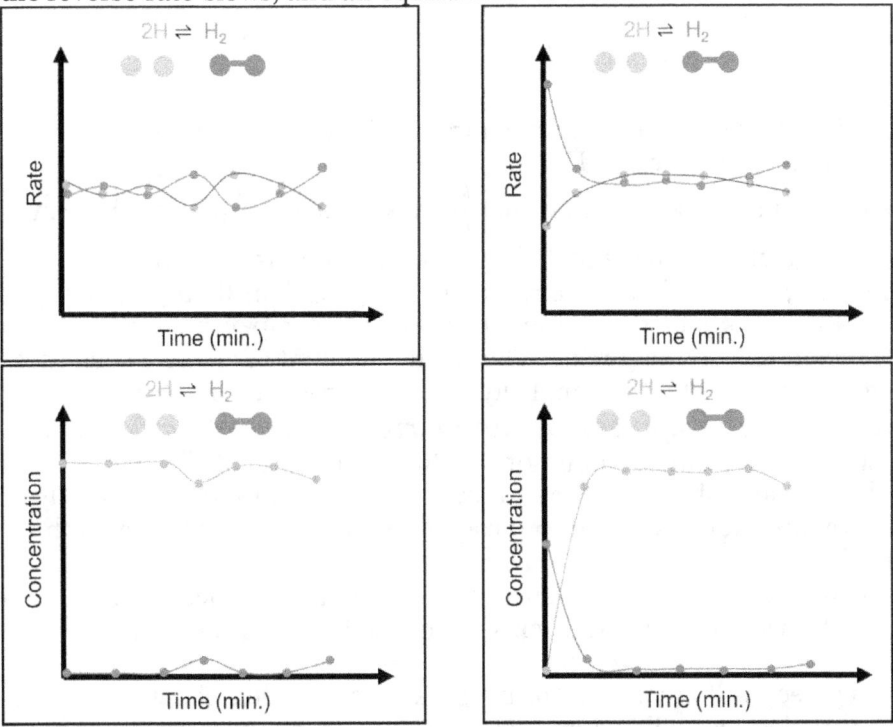

Figure 14-2: Rate vs. Time and Concentration vs. Time for Trial 1 and Trial 2 of model kit experiments

We want students to articulate why the reverse rate starts high in trial 2 but not in trial 1. We want them to explain why that reverse rate slowed down in trial 2. We also want the concepts of equilibrium to emerge. At equilibrium the forward and reverse rates are equal, but the reaction is still ongoing. We call this dynamic equilibrium (as opposed to static equilibrium).

Kinetics helps keep these ideas grounded in concrete examples. Students learned about rate laws. Let's construct rate laws for the forward and reverse reactions. We will assume that the rate laws match the stoichiometry coefficients from the balanced reaction.

For the model kits we would have two rate laws:

Chapter 14 – Equilibrium

$Rate_f = k_f [H]^2$

$Rate_r = k_r [H_2]$

When the concentration of H is large, the forward rate is large. When the concentration of H_2 is large, the reverse rate is large. But large is relative. The forward rate is limited by the difficulty of putting together the model kits. That means that at equal concentrations for products and reactants that the forward rate should be smaller. This implies that the rate constants provide some indication of the reaction difficulty. In this instance we would expect that $k_r > k_f$.

Rate laws give students more explaining power in connecting rate and concentration. But they also lead us forward into some new representations for equilibrium. By definition, the forward and reverse rates are equal at equilibrium. If $Rate_f = Rate_r$ then we also can state that $k_f [H]^2 = k_r [H_2]$. This can be rearranged to give us:

$$\frac{k_f}{k_r} = \frac{[H_2]}{[H]^2}$$

The discussion is much simpler if we ignore relative rates for now (i.e., 2H forming is treated as the same rate as 1 H_2 forming)

Since we know that $k_r > k_f$, we know that $\frac{k_f}{k_r}$ must be smaller than 1. This can be summarized conceptually as saying that if a forward reaction is difficult and a reverse reaction is easy, there will be more reactants at equilibrium than products. The larger amount of reactants increases the forward rate to compensate for the difficulty of the forward reaction occurring. The smaller amount of product decreases the reverse rate to compensate for the simplicity of the reverse reaction occurring.

The preceding paragraph can be summarized by a single letter, K. The equilibrium constant K has multiple definitions including the ratio of the rate constants. It is also the ratio of the products to reactants at equilibrium. Specifically, it is the ratio of the concentrations raised to the appropriate exponents to match the stoichiometric coefficients.

We choose to represent K with products over reactants for concentrations. That also means that K is the forward rate constant divided by the reverse rate constant.

When we dig deeply, there are inconsistencies for comparing people taking apart model kits vs. molecules colliding and reacting. But setting those aside, we have a visual concrete example that we can connect to the abstract symbols. When K < 1 we have a reactant favored reaction (ignoring coefficients/exponents). This constant is related to the relative rate constants of the forward and reverse reaction. We can even take this a step further by looking at the strength of bonds being broken and formed. A reaction with a difficult forward reaction has strong bonds for its reactants. A reaction with a simple reverse reaction has weak bonds for the products.

In all these cases, it does not matter what we start with. We will reach the same ratios of products and reactants when we get to equilibrium. It doesn't matter if we begin with all reactants, all products, or a mixture of both.

After discussing and establishing how rate and concentration impact each other on the path to equilibrium, we now transition to the 2nd experiment using rubber bands. A rubber band is easy to tie together (for some people), and difficult to

Chapter 14 – Equilibrium

take apart. This is the opposite of the model kits. Now the forward reaction is favored, and the reverse is going to be slower.
- What will we have at equilibrium?
- Will it matter if we start with all R or all R_2?
- How will the rate constants k_r and k_f compare?
- Will the equilibrium constant be greater than or less than 1?

Now let's look into when the equilibrium constant (K) is greater than one. This happens in trial 1 with 40 R and 0 R_2 as well as trial 2 with 0 R and 20 R_2. We end up with mostly R_2 or product when we reach equilibrium. Even though this situation is similar to experiment 1 (H/H_2), it is possible that students will still struggle to explain the impact of rate on concentration and vice versa.

The 3rd experiment uses paper clips that are intended to be a balanced set of rates. This does not always happen and will likely vary from group to group. You may have to work to get students competing that have mostly equal abilities with paper clips. You can also end up with varying data which can be its own lesson on the limitations of the activity.

Obviously, the logistics of the activity are flexible, and the teacher is aiming to come up with three different combinations that have a product favored, reactant favored, and a somewhat balanced process. If you ask students which reaction achieves equilibrium, they will be initially drawn to the paper clips because the amounts could be nearly equal at equilibrium. In spite of this, all reactions achieve an equilibrium. It is important to point out that the "equal" of equilibrium does not refer to amounts, but to the rates. Those rates can be equal even when concentrations are not.

This lesson sets up future concepts in equilibrium that you will connect back to later. The landing points for the initial discussion are to introduce the equilibrium constant and promote student thinking about the relationships between concentration and rate. Those two landing points do intersect. What information does the equilibrium constant tell us about rates and concentrations at equilibrium as well as when a system is not at equilibrium? A fast rate causes a large change in concentration. The large change in concentration causes a large change in rate. For any reversible reaction there will always be a point reached where the forward and reverse rates balance. Even if the reverse is much faster, at some point the amount of products will become so small that the forward rate catches up.

Each reaction allows the teacher to press the students about what the equilibrium constant would be, and the physical meaning of a large or small K values. Students also observe that the equilibrium position does not change based on initial quantities, the reaction is still occurring at equilibrium, and that stoichiometry is relevant (20 H_2 still turns into 40 H). These observations are similar to how equilibrium was developed historically.

There is a future learning objective that the teacher should be aware of, but perhaps not articulate yet. If we were to put a stress on an equilibrium system, we would change those rates causing a shift to a new equilibrium being established. If there were an equilibrium with rubber bands, but a rogue student dumped 20 more rubber bands into the pile that equilibrium would be gone. Now the forward rate would increase until many of those rubber bands became tied and the rates became equal again.

Chapter 14 – Equilibrium

Reaction Quotient Q

This raises an interesting question. If we could only look at a snapshot of a chemical system, could we tell if the system was at equilibrium? If we knew the equilibrium constant and the concentrations, then the answer is yes. We could plug those concentration values into the expression, and if they were equal to K we would be at equilibrium. If the results were bigger or smaller than K, we can make useful predictions from that information.

The reaction quotient (Q) is the equilibrium expression being evaluated when we are uncertain if the chemical system is at equilibrium or not. If Q = K, we now know that the system is at equilibrium. But what do we know if Q > K? What does that tell us about the forward and reverse rates? What does that tell us about how concentrations will change in order to establish equilibrium?

Let's assume that in our initial example for model kits we find that the equilibrium constant is 0.0015 for a given set of students. This means that at equilibrium we have 2 completed H_2 products, and we have 36 disconnected H reactants.

$$K = \frac{[H_2]}{[H]^2} = \frac{2}{36^2} = 0.0015$$

Another group has a larger number of model kits for some reason. They have a current arrangement of $4H_2$ and 84 H. Are they at equilibrium (assuming identical rates to the first group)?

$$Q = \frac{[H_2]}{[H]^2} = \frac{4}{84^2} = 0.00057 \qquad Q < K$$

The second group is not at equilibrium and is currently at a point where H_2 are being assembled faster than taken apart to form H. As equilibrium is approached the amount of H will decrease and the amount of H_2 will increase. The second group has a different number of model kit pieces, but should hit an equilibrium with either 8 or 9 H_2 models constructed (leaving 76 or 74 H pieces).

$$Q = \frac{9}{74^2} = 0.0016 \approx K$$

The reaction quotient, Q, should be considered a test that evaluates whether a reaction is at equilibrium. Q also allows us to predict how concentrations will have to change in order to reach equilibrium. Students should be able to explain how forward and reverse rates compare for different combinations of Q and K. If Q > K, the reverse rate is faster than the forward rate. If Q < K, the forward rate is faster than the reverse rate. If students struggle, have them evaluate a system with no products or reactants. If there are no products, Q is 0 and the forward rate is obviously faster since the reverse rate must be 0 as well. If there are no reactants, Q is undefined, and the reverse rate is faster since the forward rate is 0.

Equilibrium is a challenging unit for teachers and students. One method of teaching is to show students samples and definitions for the students to mimic. But truly the students can do a lot of the rationalization on their own prior to quantitative algorithms. If we use an opening activity as the framework, they can develop the concept of Q, Le Chatelier's Principle, or K without providing them definitions or calculations. That process is going to lead to a much more permanent, concrete, and developed set of mental models.

Chapter 14 – Equilibrium

ICE Charts

Once Q has been introduced it is now time for students to start using ICE (Initial, Change, Equilibrium) or RICE (Reaction, Initial, Change, Equilibrium) charts. If students have experienced BCA tables, this is a straightforward transition (in particular if you used the equilibrium description while graphing in stoichiometry).

I prefer the first example of an equilibrium calculation to be one where there are both products and reactants present. We also want to avoid getting bogged down with a quadratic equation calculation that distracts from the chemistry concepts. Assume there is a mixture of 2.40 mol H_2, 2.40 mol Cl_2 and 12.4 mol of HCl in a 2.00 L container. If the equilibrium constant, K, is 53.4, is the mixture at equilibrium? What will the equilibrium concentrations be?

$$H_2\,(g) + Cl_2\,(g) \leftrightharpoons 2HCl\,(g)$$

$$K = \frac{[HCl]^2}{[H_2][Cl_2]} = 53.4$$

$$Q = \frac{[HCl]^2}{[H_2][Cl_2]} = \frac{6.20^2}{1.20^2} = 26.7$$

Q is less than K. This means that the forward rate is faster than the reverse rate. This means that in order to reach equilibrium there will need to be more products and less reactants. Our ICE chart must have increases for the changes in products and decreases for the change in reactants.

$$H_2\,(g) + Cl_2\,(g) \leftrightharpoons 2HCl\,(g)$$

	H_2	Cl_2	2HCl
I	1.20	1.20	6.20
C	-x	-x	+2x
E	(1.20-x)	(1.20-x)	(6.20+2x)

Plugging these values into the equilibrium expression gives the following result:

$$53.4 = \frac{(6.20+2x)^2}{(1.20-x)^2}$$

We can take the square root of both sides to simplify to:

$$7.31 = \frac{(6.20+2x)}{(1.20-x)}$$

A bit more manipulation results in x = 0.276.

Plugging x into the equilibrium values tells us that at equilibrium $[H_2]$ = 0.924, $[Cl_2]$ = 0.924, and [HCl] = 6.76. It is good to double check that the values make sense. All our concentrations are positive, they have changed as Q predicted (reactants smaller, products larger), and the equilibrium values produce a K value of 53.5. The slight difference is due to rounding and uncertainty.

After an initial ICE problem, we want to debrief.
- What numbers were unclear?
- Why was it -2x for HCl?
- How would this change if Q was greater than K or Q was 0?

Chapter 14 – Equilibrium

This initial debrief is not the time to communicate about solving the quadratic or other polynomials. Instead, you're trying to get students to digest the information in the solution. This will work best if students can do some or all the initial calculation, but for many classes that may not be possible.

The second ICE calculation that students would do could then be a problem where there are no products. The reason for this order is that now students will identify that Q is 0, rather than having an initial algorithmic step of assigning the initial products as 0. We want the norm to be that the first step in evaluating an ICE chart is determining the direction of shift. If we start with a calculation that is too obvious, students will fail to make that connection. This will leave them in a worse spot for connecting ICE charts to Le Chatelier's Principle.

ICE charts help display a lot of information. Give students time and questioning to prompt them to make observations. You'll want to direct their attention to the fact the changes are always proportional to the balanced reaction coefficients. Stoichiometry does not become optional just because a reaction is reversible.

You may also wish to give students some common assumptions. If products are not mentioned, we frequently assume they have zero initial concentration. If the equilibrium constant is very large (>>>>1) or very small (<<<1) then we assume that the changes (+x, -x, -2x, etc.) can be negligible relative to the initial amount. If we start a reaction with a concentration of 0.100 M, and the change of -x is -0.00000047 M, we can safely ignore the change.

As the teacher you should make the decision ahead of time for what your objectives are as far as mathematical analysis is concerned. Do you want students to have to solve quadratic formulas? If so, can they use a program to do so or do they need to use the quadratic formula? Will students be expected to solve even more complicated polynomials? If not, then it is important to make sure the questions you use for assessments work out in advance. Starting with no reactants or no products generally simplifies the mathematics. So does starting with equal initial concentrations when there are multiple reactants or products.

ICE charts present challenging algorithms for students. Do not let students lose sight of the particle level ideas that we developed earlier. Ask students to identify which rate is faster initially when only the initial concentrations are known. Ask students how those rates change between initial and equilibrium. Have students draw particle representations of initial and equilibrium systems from ICE chart data.

One of the best ways to connect ICE charts with the particle level models is by using K_{sp} equilibrium analysis. The reason why this is the simplest approach for students is because the transitions occur at the surface boundary between the solution and precipitate. This negates the need for students to have a complete understanding of what happens within the crystal structure or within the solution.

K_{sp} functions very similarly to our initial experiments with model kits, rubber bands, and paper clips. But now we are working with a chemical system so we can begin developing a stronger overlap with the symbolic representations.

Chapter 14 – Equilibrium

$$BaCO_3\ (s) \rightleftharpoons Ba^{2+}\ (aq) + CO_3^{2-}\ (aq)$$

	Ba^{2+}	CO_3^{2-}
I	0	0
C	+x	+x
E	x	x

$$K_{sp} = [Ba^{2+}][CO_3^{2-}] = x^2 = 2.58 \times 10^{-9}$$
$$x = 5.08 \times 10^{-5}\ M$$

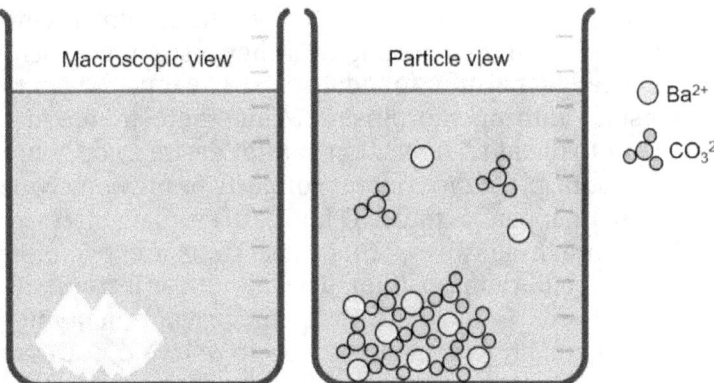

Figure 14-3: A saturated solution of barium carbonate with excess solid in the beaker

We can easily show students a macroscopic view of a saturated solution with some excess solid that precipitates and settles to the bottom of the container. The student could have some misconceptions about the charges of the different species, but otherwise the student will be able to draw a reasonably productive particle level representation of the barium and carbonate ions leaving the surface of the crystal to go into solution, and the ions colliding with the crystal to reform the solid.

The questioning that can be done is powerful.
- What changes at the particle level, macroscopic level, and in the ICE chart if we increase the temperature?
- What does K_{sp} tell us about the solubility and what does it not tell us?
- Why do K_{sp} values differ from $BaCO_3$ to $SrCO_3$ to $CaCO_3$?
- Why do K_{sp} values differ from $BaCO_3$ to $BaSO_4$ to BaC_2O_4?
- Why does the shape of the salt crystal at the bottom of the beaker change over time?
- Why is it difficult to compare solubility using K_{sp} values for $BaCO_3$ and BaF_2?
- What similarities and differences are there to the model kit simulation?
- What similarities and differences are there to a gas phase reaction (e.g., $N_2 + 3H_2 \rightleftharpoons 2NH_3$).

These questions allow for a great assessment of student understanding which is why it is common to use K_{sp} as the final objective when teaching equilibrium. The teacher can use K_{sp} to make connections to acid-base chemistry by asking about the pH of a saturated $Mg(OH)_2$ solution ($K_{sp} = 2 \times 10^{-11}$). Teachers can also introduce the common ion effect where one of the ions is present initially in the solution. How would the solubility of $BaCO_3$ change if it were dissolved in a 0.10 M solution of Na_2CO_3?

Chapter 14 – Equilibrium

Without common ions, there are some patterns that instructors might wish to be aware of. K_{sp} problems can be solved without ICE charts by using the charges of the ions. In the barium carbonate example, the final result of the ICE chart was that $K_{sp} = x^2$. This will be the same result for any salt that has a 1:1 ratio of cation to anion. If the ratio of ions is 1:2 or 2:1 the result will be $K_{sp} = 4x^3$. If the ratio of ions is 2:3 or 3:2 the result will be $K_{sp} = 108x^5$. Students may discern these patterns as they work problems.

Le Chatelier's Principle

Le Chatelier's Principle is a dangerous topic. Students can state the right answer without knowing much about why. The accompanying explanations do not always clarify if the student gets it, or if they just know which phrase to select. That convenience can be harmful for education. The topic is also riddled with language that can inspire anthropomorphism. "When there are too many reactants the reaction wants to shift to the left." A reaction cannot desire to do something. But there can be value to this principle if the content connects to previous concepts. The best way to make sure that happens is to use Q and K to evaluate stresses.

There are three stresses that cause shifts and two more stresses that do not influence the equilibrium position. The first stress introduced is often concentration changes. This is simple to connect with the reaction quotient Q. What you don't want students to do initially is to evaluate how to relieve the stress.

$$2NO_2\ (g) \rightleftharpoons N_2O_4\ (g) \qquad K_c = 242$$

A system of NO_2 and N_2O_4 is at equilibrium. There are 0.400 moles of NO_2 and 38.7 moles of N_2O_4 in a 1.00 L container. If 1.000 mol of NO_2 are added what will happen?

It's best to avoid the terms stress and shift as long as possible. These terms are helpful for communication once we are proficient. But initially we want students to connect these situations with what they learned previously in kinetics and ICE chart calculations. Le Chatelier's principle should not be a standalone concept.

When the NO_2 is added, what happens to the forward rate, reverse rate, concentration of NO_2, the concentration of N_2O_4, the reaction quotient Q, and the equilibrium constant K? These are all the factors to consider without even considering the enthalpy change and temperature. We don't want to oversimplify this by reducing it to a stress of too much reactant, so the equilibrium position shifts to the left.

When the NO_2 is added there will be an increase in the forward rate due to more frequent collisions of NO_2 molecules. The reverse rate remains the same for the moment. Due to the increased forward rate, there will be more N_2O_4 formed and some of the NO_2 will get used up. So, the NO_2 starts with an equilibrium concentration, the concentration is increased by adding NO_2, then the concentration decreases until it is between those two values to re-establish a new equilibrium concentration. The reaction quotient is equal to K prior to the additional NO_2 being added. After the NO_2 is added Q becomes smaller due to an increase in reactants. As the reactant is converted to product Q becomes equal to K once again. Sometimes as we describe these shifts students perceive that we imply that only the forward

Chapter 14 – Equilibrium

reaction happens. Throughout these changes the reverse rate is still non-zero and changes as the shift occurs.

There are a lot of changes occurring that students are not going to consider unless we give them a tool to be able to make connections. Using a graph of concentration vs. time is the best means to do so. The student should then identify how Q varies throughout the graph and how the forward and reverse rates change. At time 0 (Fig. 14-3), what is the value of Q? Which rate is faster (forward or reverse)? How does the forward rate change as time increases? Why? How does the reverse rate change as time increases?

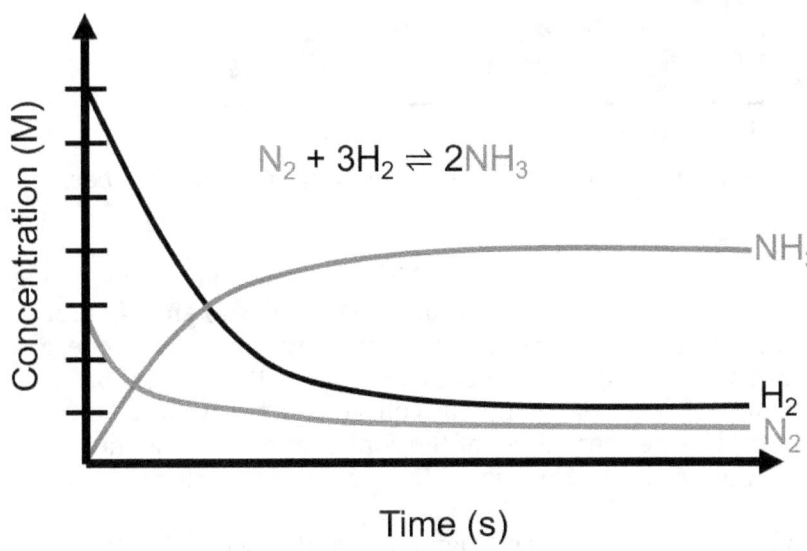

Figure 14-4: Graph of nitrogen, hydrogen, and ammonia concentration

N_2 (g) + $3H_2$ (g) $\rightleftharpoons$ $2NH_3$ (g)

A system of nitrogen, hydrogen, and ammonia is at equilibrium. What happens when some nitrogen is removed from the system?

The forward rate decreases due to fewer collisions between hydrogen and nitrogen. The reverse rate (immediately after the change) remains the same. The reverse rate is now larger than the forward rate. At the instant the nitrogen is removed the nitrogen concentration decreases while hydrogen and ammonia are constant. As the shift proceeds the amount of nitrogen and hydrogen increase while the amount of ammonia decreases. We describe all these changes by saying that the equilibrium position shifts to the left or towards the reactants. Notice that in this example the net change from equilibrium before to equilibrium after is that the nitrogen decreases, the hydrogen increases, and the ammonia decreases.

Chapter 14 – Equilibrium

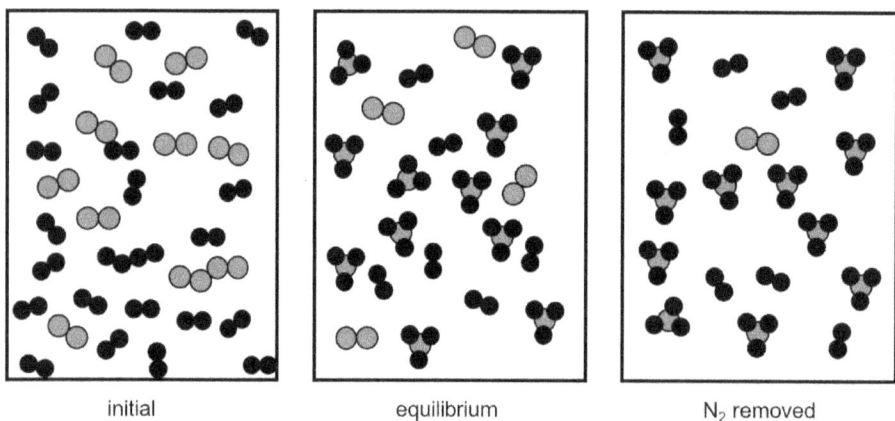

figure 14-5: particle diagrams at time 0, equilibrium, and after nitrogen has been partially removed

What happens when we change the pressure? This depends on how we change the pressure. What happens if our earlier equilibrium mixture of NO_2 and N_2O_4 has neon gas added? Nothing happens. The Ne gas would collide with the NO_2 some, and with the N_2O_4 as well. But neither of those results in a change in the forward or reverse rate. The additional Ne does not change the collisions between the NO_2 and N_2O_4 molecules. But if we decrease the volume of the container to increase the pressure, a shift occurs. Those collision frequencies change, and the effects will not be equal for both rates.

The best way to look at the effects of changing pressure by changing the volume is to go back to Q. When we change the volume of the container to 0.500 L instead of 1.000 L the concentrations of both gases change. Now the $[NO_2]$ = 0.800 M and the $[N_2O_4]$ = 77.4 M. Even though both concentrations doubled, the reaction quotient has the concentration of NO_2 squared. Our Q value changed from 242 to 121 by halving the volume. This means that the forward rate has increased by more than our reverse rate has increased. The reaction will shift producing more N_2O_4 from the NO_2 until Q = K again.

Note that for a reaction with the same number of gas molecules on both sides that the rates would have changed by equivalent amounts. Q would remain equal to K, and no shift would occur. H_2 (g) + Cl_2 (g) ⇌ 2HCl (g) does not experience an equilibrium shift when volume is altered.

After all of this is understood, we can now start to use shortcuts. If a change in volume results in a lower pressure, the reaction mixture will shift to the side with more gas particles. If a change in volume results in a higher pressure, the reaction mixture will shift to the side with fewer gas molecules. Equilibrium systems with equal numbers of gas molecules for products and reactants will not shift when volume changes. The rates still change, but they change the same for forward and reverse.

You may think that stresses are a hindrance to student learning. These provide students with a means to skip over the thinking and quantitative rationalization of Q vs. K. But consider that when we initially investigated equilibrium in history, we had not yet constructed many of the tools we use now. Moles were not a thing. To observe a shift in equilibrium position back then is quite similar to the stress and shift response that the students are utilizing today.

The third stress is to change the temperature. Note that here we are not adding nor removing any chemicals from the system. This means that if a shift occurs the value of K must change. Temperature is the only way to change the value of K. K can be altered by changing the coefficients of the reaction, but this does not change the value since the equilibrium concentrations are identical regardless of how the reaction is described.

When we raise the temperature the particles move faster, collide more frequently, and have more collisions with sufficient activation energy. This occurs for both reactants and products. The forward and reverse rates both change, but they do not change equally. The mathematics behind how they change comes from a Maxwell-Boltzmann distribution.

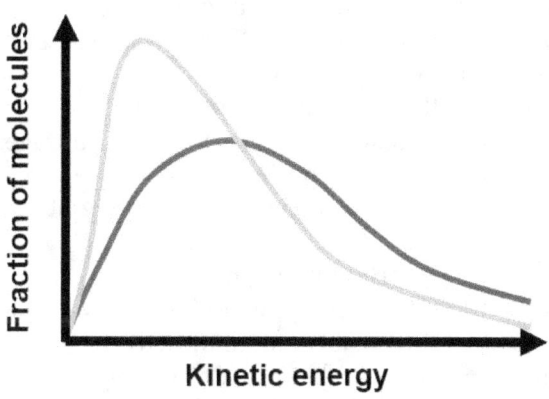

Figure 14-6: Maxwell-Boltzmann distribution for a sample at two different temperatures. As the temperature rises more particles have sufficient activation energy

Which activation energy is higher? Is it the reverse reaction or the forward reaction? It turns out that depends on whether the reaction is endothermic or exothermic. An exothermic reaction has a higher activation energy for the reverse reaction while endothermic reactions have a higher activation energy for the forward reaction. In equilibrium analysis, we can ignore intermediates.

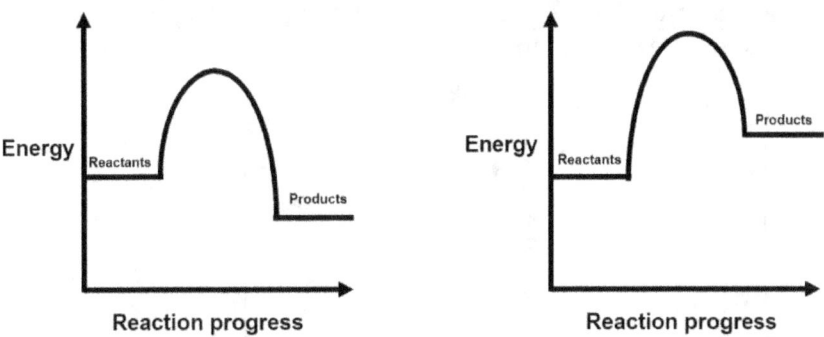

Figure 14-7: Reaction energy diagrams of endothermic and exothermic reactions

We can combine Fig. 14-6 and Fig. 14-7 to get more insight into how equilibrium positions will change when temperature is changed. In an endothermic reaction, raising the temperature causes an increase in the number of reactants and products that have sufficient activation energy to react. But the reactants increase by a greater proportion causing the forward rate to speed up more than the reverse rate. The equilibrium position shifts to the right. The stress of added heat is resolved by shifting to the right.

When we have an exothermic reaction, and we increase the temperature the reactants and products both have a larger proportion of molecules with sufficient activation energy to react. Both rates increase, but the fraction of products that have sufficient activation energy has increased more than the fraction of reactants. Thus, the reverse reaction increases by a larger amount than the forward reaction. The reaction shifts to the left causing K to decrease. As energy is added, the stress of the energy causes the equilibrium to shift to the left which cools the reaction back down.

When we cool the reaction, the opposite happens. An endothermic reaction shifts to the left and an exothermic reaction shifts to the right. If students write the energy term into the reaction as a thermochemical equation, they could treat the energy term as a reactant (endothermic) or product (exothermic) and just pretend that heating or cooling is the addition of or removal of that term. This will help them correctly predict the direction but will not serve as an explanation.

The application of the temperature shift is that we can maximize the amount of product or reactant that we desire by controlling the temperature to our advantage. This is often discussed about ammonia production which Fritz Haber maximized by running the reaction at high pressure and low temperature (relatively speaking at least, you've got to break that triple bond).

$$N_2 (g) + 3H_2 (g) \rightleftharpoons 2NH_3 (g) \qquad \Delta H = -92 \text{ kJ/mol}$$

Adding a catalyst will cause both forward and reverse rates to increase due to lower activation energy. By lowering the energy of a transition state or activated complex, the catalyst increases the number of collisions with sufficient energy to overcome the activation energy. But the rates increase equally for both forward and reverse reactions. Adding a catalyst can lead to equilibrium being achieved faster, but a catalyst will not affect the position of equilibrium.

How is that the case for a catalyst but not for temperature changes? The Maxwell-Boltzmann distribution (Fig. 14-3) can help us provide some insight. When temperature changes, the shape of the curve changes while the activation energies do not. When a catalyst is added, the shape of the curve remains the same, but both activation energies change by the same amount.

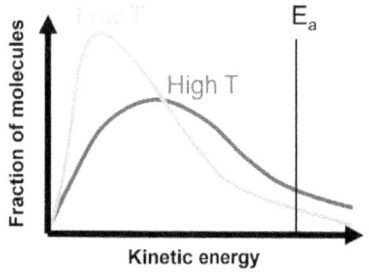
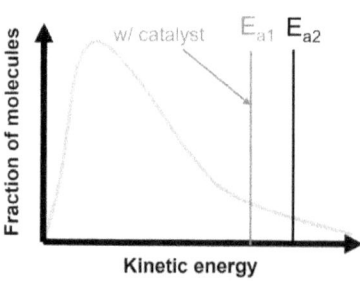

Figure 14-8: Higher temperatures result in a greater proportion of molecules having energy equal to or greater than the activation energy (E_a). Adding a catalyst (E_{a1}) reduces the activation energy barrier (E_{a2}).

Chapter 14 – Equilibrium

The alternative to starting an equilibrium unit with kinetics is to start with energy and thermodynamics. The kinetics is far more concrete. For an introductory student that concrete nature is critical as it allows the student to connect the particle level with the symbolic and macroscopic. A purely mathematical treatment is less effective as the initial method for equilibrium. If you begin the unit by describing the equilibrium constant K and the equilibrium expression you will have students struggling to identify the components of your descriptions. This can lead to misconceptions developing, and students may develop less confidence in their ability to rationalize the concepts. Some students will appear to develop strong understanding using strictly mathematical frameworks, but this can limit their transition to the particle and macroscopic levels.

If you do start equilibrium without kinetics, you can highlight select regions of graphs of concentration vs. time to help assess student understanding.

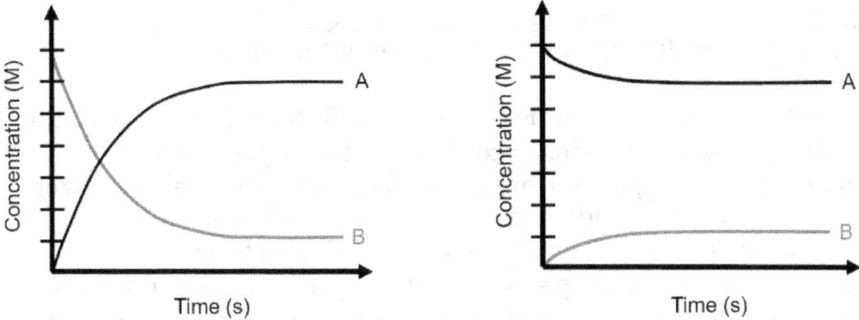

Figure 14-9: graph of concentrations of A and B vs. time

Ask the students to determine which rate is faster when A and B cross on the first graph. Ask them which rate is faster at 100 s in the first graph. Your goal with that question is to elicit that the reaction is still ongoing and that the rates are not zero. In the second graph does the reverse rate increase or decrease after time zero? Why do the concentrations in graph 2 end up the same as graph 1 even though they started with the opposite amounts?

A proper challenge for students with strong understanding would be to have them produce a rate vs. time graph from the concentration vs. time graphs. By doing a series of multiple-choice options you can open this activity up to a wider range of initial abilities.

At some point students will likely ask why solids and liquids are left out of the equilibrium expression. There are two ways to address this question. The first is to highlight that the concentration of a liquid or solid is mostly constant even when amount varies. It's true that the values in equilibrium expressions are technically activities and not concentrations. Prior to this point concentration had always been a comparison of two chemicals. There are 1.0 moles of copper (II) sulfate per 1.0 L of solution. The concentration of a solid on the other hand is 1.0 moles of copper (II) sulfate per the space that the 1.0 mol occupies. This is more ambiguous. We can help a little by talking about how doubling the amount of a solid or liquid would also double the space that it occupies.

Chapter 14 – Equilibrium

But I prefer to view this question through an additional lens using physical equilibrium of vapor pressures. When a liquid is in a sealed container, some of that liquid will vaporize. Some of that vapor will condense back to the liquid. These opposing processes will establish an equilibrium where the rates of condensation and the rates of evaporation are equal.

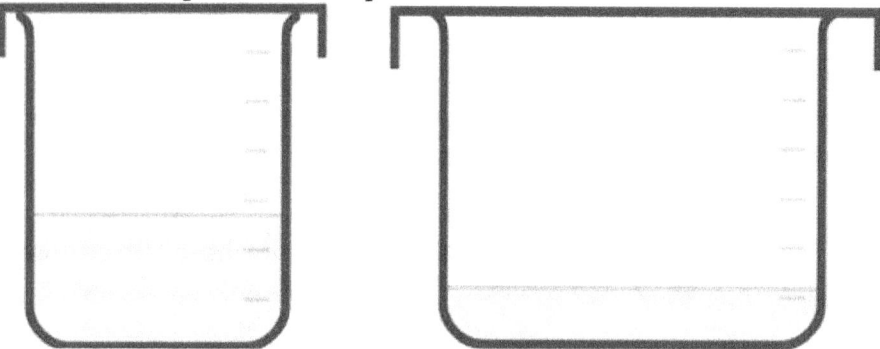

Figure 14-10: Containers A and B both have liquid water at the bottom

Container B is longer and so the water spreads out further. The condensation and evaporation both occur at the surface. In container B the larger surface area means that more molecules will vaporize from the surface, but there is also a larger surface for the vapor to collide with and condense. Container A has a smaller surface so there will be less evaporation at the surface, but there will also be less condensation. The end result is that the pressure of the vapor in the container does not change regardless of the shape or amount of the water. Adding more or taking away water does not change the vapor pressure as long as some water remains. If the water in both A and B is halved or doubled, the rates would remain identical and thus the equilibrium achieved would be the same.

If we extend container B to be higher, then the vapor would travel further up and there would be more vapor. But the equilibrium pressure would not change even though there is more vapor (n) because the additional vapor is contained in extra space (V). It would take longer to reach the equilibrium, but when the rates of evaporation and condensation are equal, the pressure would be the same ($\frac{nRT}{V}$ is constant). Students could discuss how changing the size of the container, shape of the container, amount of liquid, etc. all still result in the same equilibrium vapor pressure. After we can go back to the reaction H_2O (l) ⇋ H_2O (g) to highlight that the amount of water does not alter the equilibrium position, nor does it add any value to our insights of equilibrium. This gives some great opportunities for students to make concrete drawings of particle diagrams.

In kinetics, surface area matters for rate. But once equilibrium is established, the surface area arrangement is irrelevant. Thus, aqueous solutions and gases influence relative rates at equilibrium. But the amount of a solid or liquid is irrelevant so long as there is enough to not run out prior to equilibrium being achieved.

Calcium carbonate decomposes to produce carbon dioxide.

$CaCO_3$ (s) ⇋ CaO (s) + CO_2 (g)

If we put a chunk of limestone into a container the limestone will decompose. As the amount of carbon dioxide increases more collisions of the carbon dioxide will occur with the calcium oxide to reform the limestone reactant. If we break the chunk

Chapter 14 – Equilibrium

of limestone into tiny pieces, we will increase the surface area. This will increase both the rate of the forward and the reverse reaction equally. If we add more limestone, we will see equivalent changes in both the forward and reverse rates.

Again, we see a more concrete analysis through rates. One plausible explanation for why solids and liquids are omitted from equilibrium expressions is that their impact on equilibrium amounts does not modify as surface area changes. Surface area changes will affect the rates of forward and reverse reactions equally and so we do not need to consider solids and liquids in the equilibrium expression. Note that using rates and kinetics also highlights the fact that some solid or liquid is of course necessary to achieve equilibrium. If there is an insufficient amount of solid present, then the rates will not be able to become equal.

Minor Details

Equilibrium has a lot of quick components that you will instruct on at some point during the unit. The equilibrium expression is based on how the reaction is written. What happens when we write the reaction backwards or with fractional coefficients?

$$H_2 (g) + I_2 (g) \rightleftharpoons 2HI (g) \quad K = 0.0120$$

What will the equilibrium constant for the reaction written below?

$$HI (g) \rightleftharpoons \tfrac{1}{2}H_2 (g) + \tfrac{1}{2}I_2 (g) \quad K = ?$$

A set of equilibrium concentrations that results with the equilibrium constant K = 0.0120 in the first reaction would be $[H_2] = 2.00$, $[I_2] = 4.00$, and $[HI] = 0.310$.

$$K = \frac{[HI]^2}{[H_2][I_2]} = \frac{0.310^2}{(2.00*4.00)} = 0.0120$$

If we change our coefficients and reverse the reaction our new equilibrium expression would be:

$$K = \frac{[H_2]^{\tfrac{1}{2}}*[I_2]^{\tfrac{1}{2}}}{[HI]} = \frac{(2)^{\tfrac{1}{2}}*(4)^{\tfrac{1}{2}}}{(0.310)} = 9.12$$

Note that if we take the original K and raise it to the exponent -0.5 we get:

$$0.0120^{-0.5} = 9.12$$

Reversing the reaction causes us to invert the equilibrium constant. Doubling the coefficients leads us to square the equilibrium constant. Halving the coefficients leads us to take the square root of the equilibrium constant.

After this problem is concluded direct students to notice that how they describe the reaction has no bearing on what happens at the macroscopic level. When they mix hydrogen and iodine vapor, the equilibrium values will be established whether we write coefficients of 1,1,2 or 1, ½, ½ (reversed and scaled). The different results are due to us describing the system differently.

Mathematical manipulations of the equilibrium constant can be used early in conjunction with establishing what K is. You can also use it as a review where you get an opportunity for spaced practice. Either way, it is helpful for students to plug actual numbers into both expressions to see how K changes and that the equilibrium rates and concentrations are not altered.

Chapter 14 – Equilibrium

Bonding impacts most aspects of chemistry and equilibrium is no exception. Stronger bonds will be more difficult to break. The triple bond of N_2 will cause slower forward reaction rates for the synthesis of ammonia. Thus, we find the equilibrium constant of this reaction to be less than one except at unusual conditions.

$$N_2 (g) + 3H_2 (g) \leftrightharpoons 2NH_3 (g)$$

Bonding relates to both the kinetic model of equilibrium, and to the thermodynamic model of equilibrium. Whether we use rates or energy to explain equilibrium, the bonds between the atoms and ions give us the rationale. This easily relates to the opening activity with model kits and rubber bands.

When discussing rates, we are comparing the bond strength of reactants and products to the activated complex. We are comparing the activation energy of the forward and reverse chemical processes.

But when using thermodynamics to explain equilibrium we are comparing the bond strengths of reactants directly to the bonding of the products. Both of these can be used because the activated complex is the same for both. Comparing the energy of the reactants, the activation energy, and the products comes out similarly to just comparing the energy of the reactants and products. Both reach the same transition state in between.

K_p can be related to K_c using the ideal gas law. PV = nRT can be rearranged to form $\frac{n}{V} = \frac{P}{RT}$ where $\frac{n}{V}$ is the concentration of the gas (using the volume of the container in place of a solvent). Sometimes it is highly convenient to measure equilibrium amounts of gases using pressures and we want to be able to convert between K_c and K_p. I personally find it simpler for students to just convert pressures into concentrations to calculate K_c rather than convert.

Teaching equilibrium should focus on ICE chart calculations and Le Chatelier's Principle. But rates should be incorporated frequently throughout the unit. Be wary of students getting answers correct without an underlying conceptual understanding. Make sure to think about what you consider understanding to mean for you and your students. For me, being able to identify what rates do at a variety of points in the reaction is a big deal.

There are a lot of connections between equilibrium and thermodynamics. Whichever unit gets taught last should be where those connections go. I prefer to teach thermodynamics last so that I can connect equilibrium, thermodynamics, and redox calculations all at once with bonding remaining a key theme.

Student Struggles

1. "I have no idea what I'm doing!" There are a lot of numbers and manipulations that students can get lost in. Sometimes the issues are from an earlier concept such as stoichiometry or kinetics. The best place for students to start struggling in equilibrium are with the relationships between concentration and rate. You want students to explain shifts by articulating how rate and concentration change, and how change for each affects the other. The concentration and rates both start high. What happens to the concentration of a reactant that has a high rate of conversion to product? What happens to the rate in the forward direction as the concentration decreases? Why? This approach minimizes mathematical processing but focuses the student on mathematical reasoning by exploring the cause-and-effect

Chapter 14 – Equilibrium

relationship between concentration and rate. This sets the student up with some knowledge to attach the mathematical processing to.

2. "I can predict the equilibrium position shift, but I can't explain why." Le Chatelier's Principle can be dangerous when students only know how to predict what shift occurs and now why that shift occurs. When students are stuck here, we want to direct them to evaluate systems using Q and K. If a reactant is added, what happens to Q? If Q < K, how do the forward and reverse rates compare? If the forward rate is faster, how will the quantities of reactants change?

This is easiest to do for stresses that alter the amounts of reactants and products. Changing the volume of a gas (and consequently the pressure) should always be considered through the lens of the reaction quotient.

3. "I can set up an ICE chart, but I can't solve for x." The point of equilibrium instruction is to understand chemistry. It's unlikely to produce a meaningful level of understanding when a student struggles through a polynomial or even a quadratic expression. Students should use an online equation solver or problems with simpler mathematical manipulations should be used.

4. "When does K change, and when does it not?" The value of the equilibrium position only changes when the temperature changes or when the chemical reaction changes. Adding a catalyst, changing the initial concentrations, or changing the volume will all produce the same equilibrium result. Changing the way that the reaction is written will impact what the value of K is that we use (typically). But the actual mixture of chemicals at equilibrium will remain the same. The equilibrium expression has changed and therefore the equilibrium value changes by the same proportion to maintain the same value for the equilibrium mixture.

5. "Why did the rates in kinetics depend on the rate-determining step, but in equilibrium we can use the balanced reaction?" The big difference between kinetics and equilibrium is how much of each chemical you have. At the start of a reaction when Q is 0, the rate-determining step is the crux of how fast the forward rate is. When we've built up amounts of the reactants, products, and intermediates (Q is approaching K), the individual steps have less impact on the overall rate. When we reach equilibrium, the reactants and products are sufficient to determine the forward and reverse rates.

6. "How do I know if it should be +x or -x?" If this question comes up, share the question with the class. What happened here is that the student has not realized that starting with no product is the same thing as starting with a Q of 0. Without that knowledge, the student has missed that the first thing to do in starting an ICE chart is determine which direction the reaction is going to shift in order to achieve equilibrium. Any reaction without initial products has to shift right to make Q = K. Because many equilibrium questions start with Q = 0, students will think they can start an ICE chart without determining Q until they get to a problem with both reactants and products.

Chapter 14 – Equilibrium

7. *"Do I have to do an ICE chart if I can get the answer without it?"* There are some patterns that students may find that allow them to skip the ICE chart. A K_{sp} calculation for a salt that has one of each ion (e.g., AgBr) is going to end up as $K_{sp} = x^2$. A K_{sp} calculation for a salt that has 1 cation and 2 anions (e.g., $Cu(OH)_2$) is going to end up as $K_{sp} = 4x^3$. Except that only happens when we start with no product. If instead we're dealing with a common ion, that will no longer be true. If you feel that the student will be able to differentiate situations such as having a common ion, by all means allow them to skip over the ICE chart. But early on it's important that students demonstrate the ability to organize the chart before skipping ahead to shortcuts.

8. *"What do I do once I find x?"* This student has a good understanding of the manipulations and algorithms, but they don't really know what it is they are doing. This student would benefit from drawing macroscopic representations of the initial and equilibrium lines of the ICE chart. These students tend to also struggle to link the value of x with the solubility in K_{sp} calculations. For AgCl, solving for x tells you the solubility of silver chloride as well as the ion concentrations at equilibrium for Ag^+ and Cl^-. But for $Cu_3(PO_4)_2$, the value of x tells you the solubility of copper (II) phosphate only. The x must be multiplied by 3 for the equilibrium concentration of copper (II) and x must be multiplied by 2 for the equilibrium concentration of phosphate. When you are working out solutions, make sure to be explicit about linking the solution for x back to the ICE chart and reaction.

9. *"Does changing the temperature impact the forward or reverse rate more?"* The forward and reverse rates differ at equilibrium based on how big of an energy gap there is from the initial set of chemicals to the transition state (activation energy). The activation energy differs in the forward and reverse directions. When the particles speed up or slow down, the number of collisions that result in a reaction change. Those changes are not felt evenly for both directions. Adding more energy is going to have a bigger impact on the rate for the direction with the larger activation energy. This is a result of some calculus on the Maxwell-Boltzmann distribution as well as the Arrhenius equation. For a high school student, it should be sufficient to state the shift, and that raising/lowering the temperature causes the forward/reverse rate to increase more for an endothermic reaction. Raising/lowering the temperature causes the reverse/forward rate to increase more for an exothermic reaction.

10. *"Why doesn't water impact the equilibrium constant?"* Solids and liquids are omitted from equilibrium expressions. There are two common rationales for this. The first is that the concentration of a solid or liquid does not vary with amount. A big puddle of water has the same concentration of water as a small puddle. A small pile of copper shot has the same concentration of a sheet of copper metal. The other explanation is that the equilibrium expression technically does not use concentrations. Instead, activities are used. Activities are similar to concentrations, but they are unitless and incorporate the impacts of interactions between dissolved ions. A 1.0 M solution of NaCl would be expected to form 1.0 M Na^+ and 1.0 M Cl^-, but in reality, some of the ions remain in contact causing both values to deviate from the theoretical amounts. The activity describes the real value.

Chapter 14 – Equilibrium

Phenomena

1. **Cobalt chloride complex ions** - Cobalt chloride complexes are blue, but when water ligands replace the chloride the color changes to pink. A mixture of the two will look purple. A purple solution will shift in color upon addition of chloride, water, or removal of chloride (Ag$^+$) or water (acetone). A temperature change will also produce a color change. Have students predict the color change that will occur for each change before demonstrating each. Disclaimer - the addition of water shifts the other concentrations but does not influence the rates since it is a liquid.

2. **Amusement Park rides** - A popular roller coaster will have a rate of how many riders go on each coaster, how many coasters there are, and how long the ride takes for each coaster. When more riders ride the coaster than enter the queue, the length of the line will get shorter. When riders enter the queue at a faster rate than the riders exiting, the length of the line will get longer. When the two opposing rates are equal, the line length will remain constant. This is a vivid example of dynamic equilibrium that students experience (in some form) on a consistent basis.

3. **Sulfite content of wine** - Sulfite ions are dissolved in wine as a preservative. But the sulfite undergoes an equilibrium where sulfite is converted to sulfur dioxide gas ($2H^+ + SO_3^{2-} \rightleftharpoons H_2SO_3 \rightleftharpoons H_2O + SO_2$ (g)). The more basic the wine is, the more the equilibrium position shifts in favor of the sulfite. The more acidic the wine is, the less sulfite exists in solution at equilibrium. The sulfite in basic wine can be precipitated using strontium ions (if the solution is too basic, strontium hydroxide can be formed as well). The strontium sulfite can be filtered and measured to determine the relative amounts of sulfite at various pH levels in wine.

4. **Iodine sublimation and deposition** - Take a large test tube and put some solid iodine in it. Then put a small test tube partially into the large test tube and secure with lots of tape. The tape should seal the iodine in between the two test tubes. When the large test tube is heated gently with a water bath, the amount of iodine vapor increases. This is visible since the iodine gas is purple. If cool water is added to the small test tube, solid iodine will crystallize on the small test tube. The rates of sublimation and deposition can be used to discuss the equilibrium amount of vapor depending on the temperature of the two surfaces.

5. **Nitrogen dioxide** - There used to be thick glass tubes that were filled with nitrogen dioxide. The nitrogen dioxide sets up an equilibrium with dinitrogen tetroxide ($2NO_2$ (g) $\rightleftharpoons N_2O_4$ (g)) where the nitrogen dioxide is a brown gas, and the dinitrogen tetroxide is colorless. The color of the two gases will clearly shift in color upon heating or cooling.

6. **Copper (II) complex ions** - Copper (II) ions will turn green with an excess of chloride ligands and a deep blue color when ammonia ligands are dominant. The addition of both HCl and NH$_3$ will also cause white smoke to develop as the hydrochloric acid and ammonia form ammonium chloride salt. If the ammonia is dilute, copper (II) hydroxide can also form.

7. **Silver precipitates** - The K$_{sp}$ for AgBr is 7.7x10^{-13}. The K$_{sp}$ for AgI is 8.3x10^{-17}. Ask students to predict what color will form if a small amount of silver ions are added to a mixture of sodium bromide and sodium iodide. Silver iodide is a bit more yellow and silver bromide is a bit more cream colored.

Chapter 14 – Equilibrium

8. Beaker transfer - Fill one of two large containers with water (add food dye to make this more visible). Scoop water from the first container using a medium sized beaker (500 mL ish). Once the water is poured into the second container, take a scoop from the second container using a smaller beaker (200 mL ish). At some point, the medium beaker will no longer fill all the way and the two opposing rates will become equal. At this point you can pour the medium scoop into the smaller beaker and the beaker should be full.

9. Precipitation filtration - Prepare a partially soluble compound such as silver sulfate (Ag_2SO_4, $K_{sp} = 1.4 \times 10^{-5}$) by mixing 0.1 M solutions of $AgNO_3$ and Na_2SO_4. Filter the precipitate. The solution that passes through the filter paper should have sufficient sulfate dissolved to form a precipitate with 0.1 M $Ba(NO_3)_2$ solution. The $BaSO_4$ K_{sp} value is 1.1×10^{-10}.

10. Acid-base equilibria - General equilibrium examples are fewer than acid-base demonstrations. Feel free to find a couple acid-base examples that would be applicable to your equilibrium unit to share some of the wealth.

Flashcards

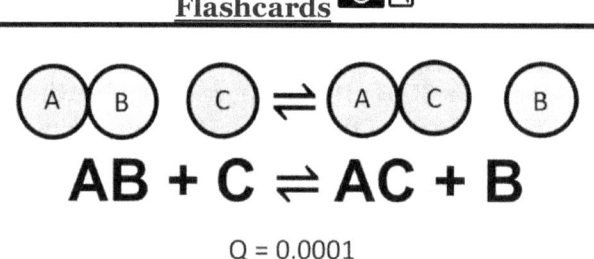

$$AB + C \rightleftharpoons AC + B$$

Q = 0.0001
K = 40,000

The bond between A-B is stronger than A-C
The bond between A-C is stronger than A-B
The bond strengths are about equal

$$3A + B \rightleftharpoons 2C + 8kJ$$

If an equilibrium mixture of A, B, and C is heated, the equilibrium constant K will:

a. Increase
b. Decrease
c. Remain the same

Chapter 14 – Equilibrium

$CaCO_3\ (s) \rightleftharpoons CaO\ (s) + CO_2\ (g)$

What is the equilibrium expression for this reaction?

$A \rightleftharpoons B \quad K_c = 2.2$

	A	B
I	4.0	2.0
C		
E		

Which chemical will have a –x for change?

How does Q compare to K at this time?

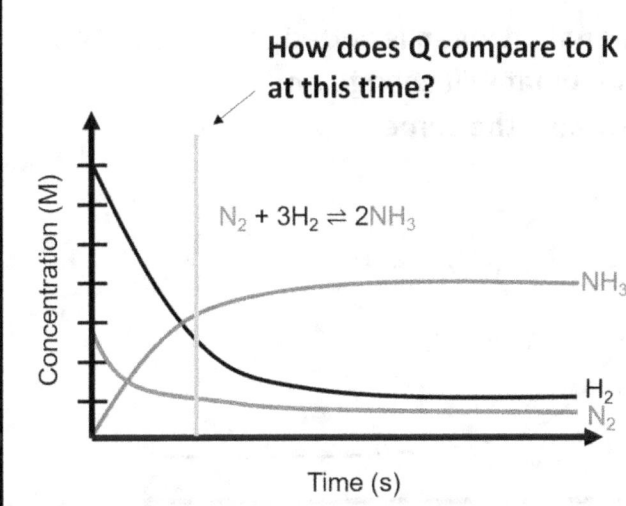

$N_2 + 3H_2 \rightleftharpoons 2NH_3$

Which reaction will shift toward more product if pressure is reduced by an increase in volume?

a. $H_2(g) + I_2(g) \rightleftharpoons 2HI\ (g)$
b. $N_2\ (g) + 3H_2\ (g) \rightleftharpoons 2NH_3\ (g)$
c. $MgO\ (s) + SO_2\ (g) \rightleftharpoons MgSO_3\ (s)$
d. $C\ (s) + O_2\ (g) \rightleftharpoons 2CO\ (g)$

Chapter 14 – Equilibrium

A ⇌ B

Q = 6.0
K_c = 2.2

Which direction will the reaction mixture shift?

	$Al(OH)_3$ ⇌ Al^{3+} + $3OH^-$	
I	0	0
C	+x	+3x
E	x	3x

K_{sp} = ???

If additional nitrogen is added at this time, what will change and what remains the same?

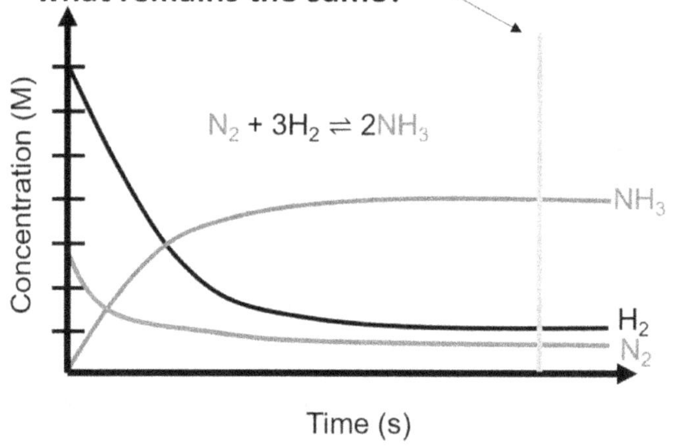

[Graph: Concentration (M) vs Time (s) for $N_2 + 3H_2 ⇌ 2NH_3$, showing NH_3 increasing, H_2 and N_2 decreasing to equilibrium]

Is the forward or reverse rate faster here?

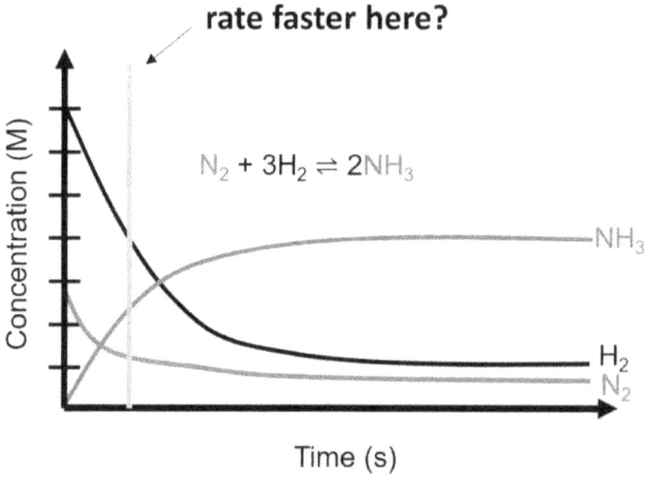

[Graph: Concentration (M) vs Time (s) for $N_2 + 3H_2 ⇌ 2NH_3$, with vertical line indicating a time before equilibrium is reached]

Chapter 15 P
Acid-Base Chemistry

History

We knew about acids for a considerable portion of chemical history. The properties made it easy to categorize even when it was challenging to explain what made something an acid. Acids had a unique taste (sour), reacted with bases, changed the color of an indicator, and caused bubbling when added to a carbonate.

Antoine Lavoisier used nomenclature that we still use today (sulfuric acid for H_2SO_4). Lavoisier wouldn't have known the exact formula, but they were able to ascertain that sulfuric acid had more oxygen than sulfurous acid. Lavoisier named oxygen which means acid generator and he also named hydrogen which means water generator.[1] Lavoisier was early in believing that water was not an element, rather it was a combination of the two elements. This was confirmed by Cavendish (1784). The theory that all acids contained oxygen was disproved by Humphry Davy (1811) when he verified that HCl did not have oxygen. Claude Berthollet had likewise shown that HCN had no oxygen (1787).[2]

Jöns Jacob Berzelius completed (1814) the dualism theory[3] developed by Lavoisier and Davy. Lavoisier proposed that an acid was a combination of a positive radical with negative oxygen.[4] Davy added that a base was a positive metal with negative oxygen. Berzelius combined these ideas by having a salt be a combination of the acid and the base. It's critical to note here that Lavoisier is defining an acid to be the component gas that would be added to water. SO_3 would be the acid, not H_2SO_4. The base would be Na_2O, not NaOH. The salt would thus be Na_2SO_4. This theory was supported by the evidence that when aqueous salts had electricity run through them that acids would form at the positive terminal and bases at the negative terminal.

The dualism theory held for some time in spite of the flaws since it provided a means to explain why some chemicals attracted to each other. Our first big improvement to understanding acids and bases comes from Svante Arrhenius (1884) in his dissertation. He studied conductivity of dissolved substances and proposed that an acid be defined as a substance that increases the concentration of H^+ ions when dissolved in water. A base would be similarly defined as a substance that increases the concentration of OH^- ions when dissolved in water. His dissertation received substantial criticism and he nearly failed. But later as he worked to clarify his work, he ended up receiving the Nobel prize for it (1903).[5]

Søren Sørensen proposed the logarithmic pH scale to track concentrations of hydrogen ions (1909).[6] The Bronsted-Lowry definition (1923) and Lewis definition (1923) for acids and bases were both released in the same year. The three definitions of acids and bases show how we progressed from a definition based on experimental evidence (Arrhenius), to a definition focusing on chemical rearrangements (Bronsted-Lowry), to a definition focusing on electronic changes (Lewis).

To this day the field of acids and bases continues to influence a variety of chemistry fields. The ability to change the charge of a substance makes acid-base chemistry applicable to redox chemistry, catalysts, and they are even used in mass spectrometry. The taste of acids and bases influences food chemistry. And the color

Chapter 15 – Acid-Base Chemistry

changes of acid-base indicators such as anthocyanins highlight a captivating artistic side of chemistry.

Bonding Model

Acid-base chemistry is a loaded unit. There are a lot of challenging equilibrium problems coupled with a variety of concepts. I recommend starting with at least some of the conceptual components first. It is also important to know where your students will start. Most of them associate acids and bases with pH. The pH scale is introduced to many students at a very young age. They are not ready to develop a conceptual understanding of acids and bases at that time because the abstractions are not developmentally appropriate.

An interesting tidbit that can be used to initiate doubt is that the pH of pure water is not always 7. At higher temperatures, neutral water has a pH below 7 and at cold temperatures the pH is above 7. Many students are surprised to learn that the pH range is not 0-14 or 1-14. There are acids called superacids, such as fluoroantimonic acid, that can have pH values below -20. If you have concentrated sulfuric acid or saturated sodium hydroxide solution the pH values lie outside the range of 0-14 as well.

Students have a lot of ideas that have formed solely based on the sequencing of the words acids and bases. Because we list acids first traditionally, many students have an idea that acids are dangerous and reactive, but that bases are protectors that neutralize acids. This is reinforced if they learn about acid rain prior to chemistry. Part of undoing this misconception is having students develop a better understanding of what a base is.

I have had the best success with acids and bases by using a bonding model. Many teachers begin the unit by focusing on the definitions. I recommend starting with H_2 and HF.

Figure 15-1: Two protons and two electrons, an H_2 molecule

What makes a hydrogen molecule (H_2, Fig. 15-1) stick together? The protons repel each other. The electrons repel each other. But the protons and electrons attract. When we add everything up, we find that as the two atoms approach the attractive forces overpower the repulsive forces until the nuclei get too close. I want students to focus on how all three of those combine to make a molecule stable.

If the nuclei get too close, the repulsion dominates. If they are too far from each other, the attractive forces diminish. The nuclei oscillate about a point in

Chapter 15 – Acid-Base Chemistry

between those two extremes. The reason why the two nuclei are able to stay close is because the electrons can be positioned between the two nuclei.

Students should consider the fundamental question of why atoms can stick in the first place. Bonds happen due to large attractive forces between electrons and nuclei. This isn't about plugging into Coulomb's Law. Rather the attractive forces are larger when the positions of the electrons and protons are positioned so that particles that attract are closer than particles that repel.

What happens though, when the two bonded atoms are not the same? How does HF differ from H_2? When we replace one of the hydrogen atoms with a fluorine atom, there are more electrons present. In Fig. 15-2 we have omitted all electrons except for one valence electron for each atom. The other valence electrons result in some repulsion. But the biggest effect is the addition of 8 more protons to the nucleus. Now there is a substantially larger force on the bonded pair of electrons. Those electrons draw closer to the fluorine than the hydrogen. This "polarized" bond differs from the one in the H_2 molecule. When we look from the perspective of the H in HF, it is further away from those electrons. The attractive forces between the hydrogen ion and the bonded electrons are weakened.

If something else came along that had a negative charge, the H^+ ion (proton) could detach from the fluoride. This is what we use as our initial model and definition of an acid. An acid is a chemical that has a weak enough bond to an H^+ that something else can remove the H^+. There are many perks to using this definition. I can easily expand this to a Lewis acid definition later because we've focused heavily on the interactions between electrons. The acid is introduced in a way that easily sets up what a base is (a substance with negative charge that can attract a proton). I have set up acids to be defined more as a verb than as a noun. We are setting up strength trends. Not only can students differentiate between a strong acid and weak acid, but they'll also understand why and notice trends.

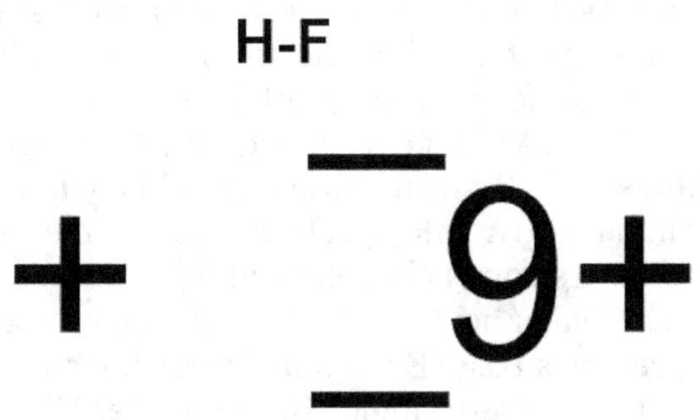

H-F

Figure 15-2: An HF molecule where the only the bonding electrons are shown

$+ \quad \bar{}9+$

This model also allows students to differentiate charges better. Many students struggle immensely with charges and subscripts. Providing a reaction such as HF → H^+ + F^- often leaves students perplexed on where the charges came from and how to predict them in the future. The bonding model gives them a better opportunity to have an image that will allow them to progress forward.

Chapter 15 – Acid-Base Chemistry

This sets up students to link strong and weak directly with the actions of acids and bases. We want students to think of a strong acid as something that does the action of an acid well. A weak acid is not very good at being an acid. A strong base is good at being a base. A weak acid has a stronger bond with the proton and a strong acid has a weaker bond with the proton. When we make this explicit students can formulate a visual model of what happens during acid-base reactions.

We can summarize these ideas by connecting them to a visual model with victims and thieves. Have the students construct an object or set of objects that a group of people might covet and try to steal from one another. Given the timing of this book, I will use a briefcase of vaccines for covid-19 as my set of objects. Let's assume there are 10 doses, but that there are 20 people who want the vaccines.

The vaccine represents an H^+ ion or proton. When a person has a dose of vaccine, they are a potential acid. Someone who does not have a dose is a potential base. The bases take a dose from an acid. Once the transfer is complete, the base person now can act as an acid. We call the relationship between the person before and after the exchange a conjugate acid/base pair. Steve with the vaccine is the conjugate acid of Steve without the vaccine (base).

For the people, there are a variety of factors that might influence their ability to take a dose from someone else, or to hold onto a dose once they have one. The person might be desperate, or they might be physically fit so they can overpower someone else and escape quickly. There might be a chemistry teacher with an exceptionally strong intellect that helps her convince others that she should have priority.

Person B is good at taking the doses. They are a stronger base. Once they have the dose, it will be difficult for them to relinquish the dose to someone else. Person B would be a weaker acid (or conjugate acid). Person A is not good at taking the doses (relative to the others) and is a weaker base. If they do end up with a dose, they will likely have the dose taken away. They are a stronger acid (or conjugate acid).

Moving back to chemistry, a strong acid is something that forms a weaker interaction with a proton (H^+). The stronger acid will relinquish the proton to a base. A stronger base will form a strong interaction with a proton (H^+). We can make this symbolic by using HA as a generic acid and A^- as a generic base. A strong acid (HA) is likely to change to A^-. A strong base (A^-) is likely to change to HA. A weak acid (HA) will often remain as HA unless a strong base comes along. A weak base (A^-) will often remain A^- unless a strong acid comes along.

HF is a weaker acid and HCl is a stronger acid. HCl is more likely to have its proton removed by a base. Even water (which is not a good base) will remove the protons from nearly every single HCl molecule. HF on the other hand will sometimes lose a proton to water but retains most without a stronger base being added. If we were to compare F^- and Cl^- we would find that the fluoride ion (F^-) exhibits a much larger capability to remove a proton from an acid.

Reaction 1: $HA + A^- \rightleftharpoons A^- + HA$

Reaction 2: $HA \rightleftharpoons A^- + H^+$

It is important to include the base in reactions. Reaction 2 is a common symbolic representation in classes, but it supports a faulty idea that an acid exists without a base. Acids do not split apart without something to take the proton away. In our earlier analogy, no one just drops a vaccine dose and runs away (actually this could happen, HA spontaneously splitting into ions could not). They are always

Chapter 15 – Acid-Base Chemistry

taken. An acid cannot exist without a base. Protons do not magically separate from a conjugate base. A collision with a base has to occur.

We've now set students up to be familiar with the concepts of conjugate acids and bases, strength, strength trends, and the three definitions (Arrhenius, Bronsted-Lowry, Lewis). Students are also developing a connection between the symbolic level and the particle level where they can visualize an H^+ as moving back and forth between the acid and a base.

*****

When I took chemistry in high school (1998-2000), acids and bases were introduced via solubility. A strong acid completely dissociated in water. A strong base completely dissociated in water. A weak acid partially dissociated in water. There are two big flaws with using water solubility. The first is that a base dissolving in water is very hard to link to a base reacting. The second is that most students struggle to learn from symbols. To a teacher, the difference between ammonium and ammonia is obvious. The NH_4^+ has one more proton than the NH_3. We know the Lewis structures of both and have seen the reactions where they change into each other. But students do not have the wealth of knowledge to easily track those changes.

Some teachers teach dissolving as a physical process. But when sodium hydroxide (NaOH) dissolves, the hydroxide reacts with water. We don't discuss this much because the product is the same as the reactants.

$$H_2O + OH^- \rightleftharpoons OH^- + H_2O$$

In the above reaction, the reactant hydroxide is taking an H^+ from the reactant water molecule. The water turns into a hydroxide ion, and the hydroxide ion turns into a water molecule. But is this what are referring to when we say a strong base is something that completely dissociates in water? Magnesium hydroxide has the same hydroxide, but the hydroxide is less soluble because of a greater attraction to the magnesium ions. Is the magnesium hydroxide not a strong base because of its limited solubility? Or is it a strong base because of its ability to attract protons? Does the hydroxide in $Mg(OH)_2$ react with the water and the product hydroxide replace the original in the structure? I don't know some of these answers, but I also don't care to know because clearly this is not the path to maximizing student understanding.

A secondary hazard to defining acids and bases via solubility is that students (and often teachers too) will omit the water from their mental construct.

1. $HF \rightarrow H^+ + F^-$
2. $HF + H_2O \rightarrow F^- + H_3O^+$
3. $NaOH \rightarrow Na^+ + OH^-$
4. $NaOH + H_2O \rightarrow Na^+ + H_2O + OH^-$

There is a big difference between 1 and 3 vs. 2 and 4. In 1 and 3 there is no water pictured. This is problematic because a HF does not dissociate on its own and the implication that it could does not mesh well with the collision model for kinetics. This means that in order for a student to use this reaction, they are going to struggle to connect the particle level representations with other content areas. This pushes them to remain in abstract thought instead of utilizing the concrete thinking available to them otherwise. Because students are often particularly weak in symbolism, a learning opportunity is lost.

We should be wary of imbalances between acids and bases. If you write out strong acids and weak acids reacting with water, you shouldn't only show weak bases

Chapter 15 – Acid-Base Chemistry

while omitting strong bases. This will further entrench the idea that acids are dangerous, and bases make the dangerous acids safe by neutralizing them.

NH_3 (g) + H_2O (l) → NH_4^+ (aq) + OH^- (aq) is common.
Why not Na^+ (aq) + OH^- (aq) + H_2O (l) → Na^+ (aq) + H_2O (l) + OH^- (aq)?

Neutralization is a great chance to test these models that we've developed. We can prepare students to see how the anions and cations in salts stem from an acid and base neutralization reaction. We can also begin the mathematical components of the unit by connecting with prior knowledge about stoichiometry.

We can think of all acid-base reactions as being of the generic form acid + base. We shall symbolize acids as A and bases as B. We have the following possibilities for acid-base reactions.
A + B →
A + H_2O →
B + H_2O →
H_2O + H_2O →

A neutralization reaction is the first combination. *The acids and bases can be weak or strong, neutralizations go to completion, and the products are nearly always a salt and water.* The cation and anion of the salt can be acidic or basic depending on the strength of the original acids and bases being neutralized. K_3PO_4 is a basic salt since the phosphate comes from a weak acid (H_3PO_4). NH_4Cl is an acidic salt because the ammonium comes from a weak base (NH_3).

There is a shortcut to balancing neutralizations that can give students a quick boost of confidence. When balancing a neutralization reaction, if you balance the number of H^+, OH^- and H_2O particles, everything else will automatically be balanced.

_____H_3PO_4 + _____$Mg(OH)_2$ → $Mg_3(PO_4)_2$ + _____ H_2O

In this example, there are 3 H^+, 2 OH^-, and 1 H_2O in the skeleton equation. To balance, you would use coefficients to make 6 of each.

$2H_3PO_4 + 3Mg(OH)_2$ → $Mg_3(PO_4)_2 + 6 H_2O$

When I first taught International Baccalaureate (IB) Chemistry, I was surprised at first to see that they wanted students to know that neutralization between a strong acid and a strong base were exothermic reactions. But if you've followed along with how we've defined acids and bases, this makes for a great connection. A strong acid is something that has a very weak bond to a proton. A strong base has a very strong attractive force on a proton. When we combine them, the protons experience a large net force pulling the protons from the strong acid to the strong base. Because of the strong net force, the particles speed up. As the particles move faster, they can now collide with particles in the surroundings (the water in the solution) and the temperature of the surroundings increases. This is a quality connection between bonding, acid-base chemistry, and thermochemistry.

Acid-Base Calculations

When introducing pH, you'll want to use a pH square (Fig. 15-3) to help students organize the various relationships. All these equations stem from the

Chapter 15 – Acid-Base Chemistry

equilibrium of the autoionization of water. The term autoionization is unhelpful to students learning about pH and so it is recommended to identify the reaction as "water reacts with itself."

$$H_2O + H_2O \leftrightharpoons H_3O^+ + OH^-$$

You are better off writing both water formulas (instead of $2H_2O \leftrightharpoons H^+ + OH^-$), so that you can highlight that a collision between them is what leads to the transfer of a proton. It is helpful to take things one step further and show how the Lewis structures need to be arranged in order for the proton to transfer from one molecule to another.

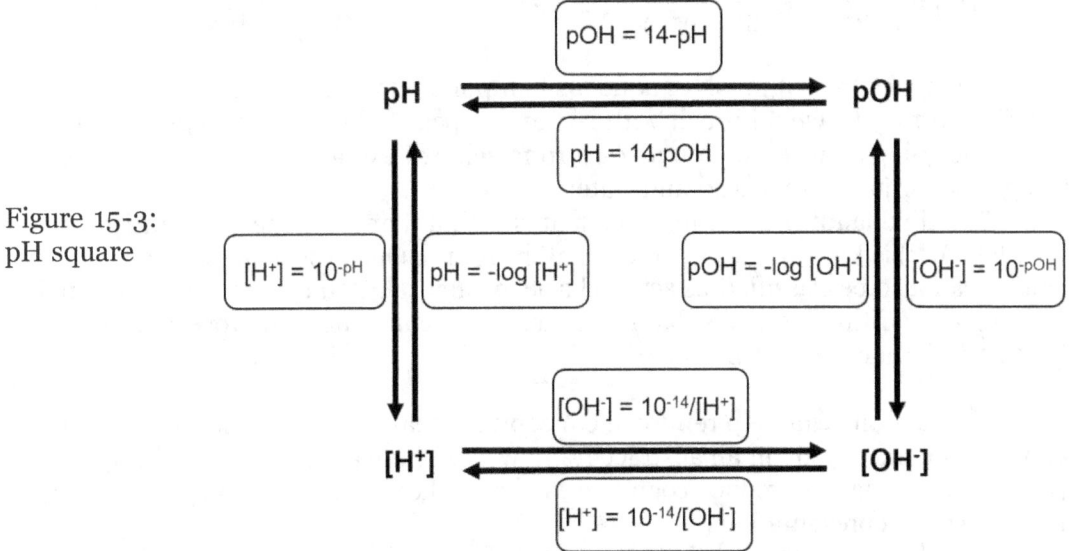

Figure 15-3: pH square

The equilibrium expression for the reaction of water with itself is $K_w = [H_3O^+][OH^-]$ and has a value of 1.0×10^{-14} at room temperature. As temperature increases the number of collisions resulting in proton transfers increases faster than the reactions to transfer the proton back to a hydroxide. Thus, the K_w value increases as temperature rises. This means that for hot water, the pH and pOH will both be less than 7 for neutral.

Neutral should not be defined by having a pH of 7, but instead when there are equal amounts of hydroxide and hydronium present. At standard temperature the pH and pOH are both 7 and so sometimes people use pH of 7 as a simple means of communicating neutrality. But in chemistry we want to start getting students prepared to dive into a deeper level of precision.

At 100 °C, K_w is now 5.1×10^{-13} or 51 times larger than what K_w is at room temperature. This means that at 100 °C the concentration of H^+ and OH^- is over 7 times higher than at room temperature (7.1×10^{-7}). The pH and pOH are both 6.1.

For pH, pOH, pK_a, pK_b, and pK_w we find a trend that the letter p represents the negative logarithm of whatever follows. This means that pH is the -log[H^+] and pK_a is the -log K_a. Students are quick to pick up on this and even if they don't truly understand how a logarithmic scale works, they can plug into a calculator sufficiently.

The larger struggles are moving from pH or pOH to the concentrations. To undo a logarithm, they must raise to the power of ten (in this case the logarithm is

Chapter 15 – Acid-Base Chemistry

base ten). One frustrating result is that students will not calculate completely when the pH or pOH has a decimal. Students will write [OH⁻] = $10^{-3.6}$ instead of 2.51×10^{-4}.

Students might also struggle with the calculations from K_w = [H⁺][OH⁻]. They struggle to do the algebra since there are so many exponents. It can be helpful for them to solve for the variables first so they can see that the unknown concentration is found by dividing K_w by the known quantity.

If the hydroxide concentration of an ammonia solution is 8.33×10^{-4} M, find the concentration of H⁺.

K_w = [H⁺][OH⁻]

$$[H^+] = \frac{K_w}{[OH^-]} = \frac{10^{-14}}{(8.33 \times 10^{-4})} = 1.2 \times 10^{-11}$$

Any reaction that involves aqueous components will have this reaction occurring. It may be what is dealt with last and appear to be in the background, but this will be going on whether you mix a strong acid with weak base, strong acid with strong base, or just a weak acid and water.

The final landing point for pH squares is that if any 1 of the 4 quantities (pH, pOH, H⁺, OH⁻) is known, the other 3 can all be found. Once those are found, the solution can also be classified as acidic, basic, or neutral. You can also stress the point of having a logarithmic scale is that it is much more convenient to process numbers like 5.1 and 8 than 4.3×10^{-9} and 1×10^{-7}.

*****

Calculations can help reinforce conceptual ideas or they can supplant them by keeping students strictly in an abstract framework. There are a few helpful ways to set up calculations that will reduce cognitive load and allow students to stay connected to more concrete representations.

For all equilibrium calculations, use an ICE (or RICE) chart. This will transition well if your students used BCA tables in stoichiometry and/or ICE charts during equilibrium. ICE stands for Initial, Change, and Equilibrium. The (balanced) Reaction can be included at the top if you wish to use the acronym RICE instead.

Avoid the Henderson-Hasselbach equation initially. You can calculate the pH of buffers easily using an ICE chart. The Henderson-Hasselbach is a nifty derivation, and it should be shown later so students will know that the ratio of [HA] to [A⁻] must change by a factor of 10 for the pH to change by 1 unit. This will help with titration curves. But you can get more understanding with students if they only need two systems to calculate.

Don't have students use the Henderson-Hasselbach equation. Instead have them use BCA tables for neutralization reactions (reactions that go to completion), and ICE charts for reactions that establish an equilibrium with significant quantities of products and reactants.

Neutralization reactions occur whenever there is a strong acid or a strong base present. Equilibrium will occur whenever there is a weak acid with water or a weak base with water. The presence of a conjugate acid or base is also going to be equilibrium. Calculations organize nicely when you use BCA and ICE tables.

Chapter 15 – Acid-Base Chemistry

Neutralization Reactions (use BCA)
SA + SB (Strong Acid + Strong Base)
SA + WB (Strong Acid + Weak Base)
SB + WA (Strong Base + Weak Acid)
SA + H_2O
SB + H_2O

Equilibrium Reactions (use ICE)	Equilibrium Constant Used
WA + H_2O	K_a
WB + H_2O	K_b
WA + H_2O + CB (Conjugate Base)	K_a (or K_b)
WB + H_2O + CA (Conjugate Acid)	K_b (or K_a)
H_2O + H_2O	K_w

Some neutralizations end with products that set up an equilibrium reaction (all of them do if we include K_w). If the strong acid (SA) is the limiting reagent, the weak base (WB) will be in excess. But the product of the strong acid and weak base that react will be the conjugate acid. In order to calculate we would do a stoichiometry calculation for the neutralization. Then the products and excess would require an equilibrium calculation.

Another item to be wary of is that this will be one of the few times that students are given a reaction that starts with both reactants and products. When asked to find the pH of a mixture of HF and F^-, some students will treat both as reactants. They will set up the following incorrect reaction HF + F^- ⇌ HF + F^- instead of HF + H_2O ⇌ H_3O^+ + F^-.

Sample questions
1. What is the pH of 100.0 mL of 0.10 M $HC_2H_3O_2$ (K_a = 1.8x10^{-5})?
2. What is the pH of 50.0 mL of 0.10 M KOH?
3. What is the pH of 20.0 mL of 0.10 M KOH mixed with 50.0 mL of 0.10 M $HC_2H_3O_2$?
4. What is the pH of 50.0 mL of 0.10 M KOH mixed with 20.0 mL of 0.10 M $HC_2H_3O_2$?

Before we look at the solutions to these questions, start by assigning each of these to a category discussed above (neutralization or equilibrium). You'll want students to begin problems in a similar fashion where they start by assigning acid/base, strong/weak, and equilibrium/stoichiometry. I am often surprised by how much more students struggle with this than the calculations themselves.

1. What is the pH of 100.0 mL of 0.10 M $HC_2H_3O_2$ (K_a = 1.8x10^{-5})?

	$HC_2H_3O_2$	+ H_2O ⇌	$C_2H_3O_2^-$	+ H_3O^+
I	0.10		0	0
C	-x		+x	+x
E	0.10-x		x	x

Chapter 15 – Acid-Base Chemistry

$$K_a = 1.8 \times 10^{-5} = \frac{[C_2H_3O_2^-][H_3O^+]}{[HC_2H_3O_2]} = \frac{[x][x]}{[0.10-x]} = \frac{x^2}{[0.10-x]} \approx \frac{x^2}{0.10}$$

$$x = 0.0013 \qquad -\log x = pH = 2.87$$

If the quadratic formula is used instead of ignoring the -x in the denominator, the value of x goes from 0.00134 to 0.00133

This entire calculation can be simplified into two equations:

$$K_a = \frac{x^2}{[HA]} \text{ and } -\log x = pH$$

Before a student uses those (or becomes aware of their existence) it is important they learn the key details within the structure of the full solution. Students will have questions about why water is omitted and why the -x is ignored. The teacher should always point out to students that the x value is equal to the $[H_3O^+]$ at equilibrium which is why the pH is calculated using -log x. The teacher must highlight that values in an ICE chart are in molarity units while BCA tables use moles. Stoichiometry is about relative amounts of chemicals, while equilibrium is based on concentrations since the density of particles is most important for frequency of collisions and reaction rates.

Students might also wonder how to do significant figures with logarithms, how the problem changes if a weak base is used, how would changing temperature change things, and where do K_a values come from? Several students will struggle with some uncertainty about the final products being an acid and a base and not being clear on why the conjugate base ($C_2H_3O_2^-$) does not neutralize the hydronium. You'll see evidence where students exaggerate this by having a weak acid react with its conjugate base (e.g., HF + F- ⇌ F- + HF). This misconception will be reduced when the focus on defining acids is based on transfer or protons instead of by properties.

The teacher may wish to highlight the difference between the initial and equilibrium states in the context of the previous units on equilibrium and kinetics. Acid-base reactions achieve equilibrium quickly. Students may struggle with the lack of time for change from initial to equilibrium when observing experiments.

There is a lot of content within that question. Take some time for students to process the results and highlight key components before students reduce it to an algorithm.

2. What is the pH of 50.0 mL of 0.10 M KOH?
   50.0 mL of 0.10 M KOH
   There are a couple of valid ways to view this question. One is that 0.005 moles of KOH were dissolved in water.

	KOH (s)	→	K⁺ (aq)	+	OH⁻ (aq)
B	0.005		0		0
C	-0.005		+0.005		+0.005
A	0		0.005		0.005

Chapter 15 – Acid-Base Chemistry

We don't know that the solid was dissolved technically. There are some alternatives where we could have run electricity through KF solution, added potassium metal to water, or the KOH could be a product of a double replacement reaction.

Either way, the end result is that we have 0.10 M K$^+$ and 0.10 M OH$^-$.

	H$_2$O	+	H$_2$O	⇌	H$_3$O$^+$	+	OH$^-$
I					10^{-7}		0.10
C					-9.99999x10^{-8}		-9.99999x10^{-8}
E					10^{-13}		0.10

The end result is that the large addition of hydroxide results in an equilibrium concentration of the amount added. We can therefore plug 0.10 M OH$^-$ into the K$_w$ expression.

K$_w$ = 1x10^{-14} = [H$_3$O$^+$][OH$^-$] = x * 0.10   [H$_3$O$^+$] = 10^{-13} M
-log (10^{-13}) = pH = 13

I used to oversimplify this calculation and immediately take the -log of 10^{-13}. Students weren't given time to connect the conceptual components to the algorithm. Students would later struggle to start these questions after they have been working with weak acid and weak base equilibria to my surprise.

A helpful consideration for teachers is to think about what the pH of a solution that has 10^{-8} moles of HCl added to 1 L of distilled water. The pH can't be 8 or adding an acid would have made the solution basic. Instead, we have 10^{-7} moles of H$^+$ to start, and add another 10^{-8} moles for a total of 1.1x10^{-7} M H$^+$ with a pH of 6.96.

3. What pH is 20.0 mL of 0.10 M KOH mixed with 50.0 mL of 0.10 M HC$_2$H$_3$O$_2$?

By now students should be able to identify the starting reagents. Here we mix a strong base with a weak acid. The weak acid is in excess. We will end with some weak acid remaining and some of the conjugate base as product. An equilibrium will be established between those two species and water.

But first we need to neutralize the strong base and a BCA table is best for that. All calculations in the BCA table need to be done using moles and to convert back to concentrations we will be using 70.0 mL or 0.0700 L.

	KOH	+	HC$_2$H$_3$O$_2$	→	H$_2$O	+	KC$_2$H$_3$O$_2$
B	0.0020		0.0050				0
C	-0.0020		-0.0020				+0.0020
A	0		0.0030				0.0020

The 0.0030 moles of excess acetic acid and 0.0020 moles of potassium acetate become 0.0429 M and 0.0286 M (we'll round more later) when the 0.0700 L are factored in. These are used to set up a second calculation using an ICE chart.

Chapter 15 – Acid-Base Chemistry

	$HC_2H_3O_2$	+	H_2O	⇌	H_3O^+	+	$C_2H_3O_2^-$
I	0.0286				0		0.0429
C	-x				+x		+x
E	0.0286 - x				x		0.0429 + x

The equilibrium values for the acetic acid and acetate can ignore the change by x since x is much smaller than the initial values.

$$K_a = 1.8 \times 10^{-5} = \frac{[C_2H_3O_2^-][H_3O^+]}{[HC_2H_3O_2]} = \frac{[0.0429][x]}{[0.0286]}$$

$$x = 1.2 \times 10^{-5}$$

$[H_3O^+] = x \qquad -\log x = pH = 4.92$

For those that are tracking along with the Henderson-Hasselbach (HH) equation ($pH = pK_a + \log \frac{[A^-]}{[HA]}$) you'll note that the pH is slightly above the pK_a (4.74) which makes sense since there is more conjugate base than weak acid.

Doing calculations this way instead of using the HH equation allows students to use the same process as they would for a weak acid or a weak base. This also gives a lot of guidance towards the transition between neutralization and the subsequent equilibrium. It also postpones the students from having to identify buffers until later during titration curves. This is easier for students and more effective for learning. Avoid the Henderson-Hasselbach (HH) until later.

Students might immediately notice that the concentrations used in the ICE table could have been kept in moles. The conversion from moles to molarity involves dividing by the volume and both the weak acid and conjugate base have the same volume. If they can keep things organized, it's teacher discretion on how you wish to handle this incorrect shortcut that produces the correct answer.

Some students will struggle with the potassium spectator ion. You could start the neutralization reaction with OH- instead of KOH, but some of the students will struggle with how to identify which ion should be omitted. The cations of strong bases (K^+, Na^+, Ba^{2+}, etc.) and anions for strong acids (Cl-, Br-, SO_4^{2-}, NO_3^-, etc.) would be a good list for students to start with.

Students could alternatively use acetate as the reactant instead of acetic acid.

$C_2H_3O_2^- + H_2O \rightleftharpoons OH^- + HC_2H_3O_2$

This work, but the acetate will lead to hydroxide as a product and will need K_b ($pK_b = 9.24$) to be used instead of K_a. The hydroxide product means that students will calculate pOH instead of pH. The end result is still pH = 4.92 (pOH = 9.08).

A helpful reminder to students that struggle with these options is that they need an equilibrium constant in order to do the calculations. There are only three types of equilibrium constants that they are familiar with in acid-base chemistry: K_a, K_b, and K_w.

The only equilibrium reactions possible are:

$HA + H_2O \rightleftharpoons H_3O^+ + A^- \qquad K_a$
$B + H_2O \rightleftharpoons OH^- + BH^+ \qquad K_b$
$H_2O + H_2O \rightleftharpoons OH^- + H_3O^+ \qquad K_w$

Chapter 15 – Acid-Base Chemistry

When students become overwhelmed, this limited selection of options can help them initiate a solution to a problem. Stress this point to them during instruction.

4. What is the pH of 50.0 mL of 0.10 M KOH mixed with 20.0 mL of 0.10 M $HC_2H_3O_2$?

	KOH	+	$HC_2H_3O_2$	→	H_2O	+	$KC_2H_3O_2$
B	0.0050		0.0020				0
C	-0.0020		-0.0020				+0.0020
A	0.0030		0				0.0020

The 0.0030 moles of excess hydroxide and 0.0020 moles of acetate become 0.0429 M and 0.0286 M (we'll round more later) when the 0.0700 L are factored in. These are used to set up a second calculation using an ICE chart.

	$C_2H_3O_2^-$	+	H_2O	⇌	OH^-	+	$HC_2H_3O_2$
I	0.0286				0.0429		0
C	-x				+x		+x
E	0.0286 - x				0.0429 + x		x

Our treatment here has the hydroxide excess as an initial amount. The +x is negligible, so our final concentration of hydroxide is still 0.0429. This step is often skipped by teachers when there is an excess of strong acid or strong base. But now students can see clearly why the weak base product is irrelevant. Our final pH would be 13.63.

If we look back over these four sample calculations, we find that there is a substantial amount of structure provided. The students must learn to identify acids and bases as strong or weak, but from there a general format is followed. The organization gives the teacher opportunities to connect the mathematics to content and allows students to focus. This focused reflection leads to more productive practice and more efficient improvement.

Buffers

Be prepared to discuss buffers multiple times. Students struggle with the concept and the symbols used. I find it works well to use calculations to teach the concept. A buffer is a mixture of two chemicals. One is a weak acid, and the other is the conjugate base. Alternatively, we could say that one is a weak base, and the other is the conjugate acid. There are two common ways to prepare a buffer. The first is to mix the two chemicals. If I mix 0.10 M $HC_2H_3O_2$ with 0.10 M $NaC_2H_3O_2$ I will end up with a buffer. The second preparation method is to mix an excess of weak acid with strong base. If I mix 100.0 mL of 0.10 M $HC_2H_3O_2$ with 50.0 mL of 0.10 M NaOH I will end up with a buffer.

The next question is why do we care? What does a buffer do? A buffer mixture can keep pH changes to a minimum when a small amount of strong acid or strong base is added. That property is important in chemistry, but it's really important in

Chapter 15 – Acid-Base Chemistry

biology. Your blood has a bicarbonate/carbonic acid buffer that maintains a slightly different pH in your lungs and tissues. This causes hemoglobin to change shape so that the heme groups release O_2 in your tissues and pick up CO_2. The pH is around 7.6 in your lungs where the heme group releases CO_2 and picks up O_2. Your ability to breathe in O_2 and breathe out CO_2 is contingent upon a buffer.

The first buffer example you do or that students do should use an ICE table. The Henderson-Hasselbach equation is great for showing students how buffers maintain pH, but it is not the initial representation you want students to connect with mathematically.

50.0 mL of 0.100 M HF has 30.0 mL of 0.100 M KOH mixed in. Determine the pH of the resulting buffer solution.

0.0500 L * 0.100 M HF = 0.00500 mol HF
0.0300 L * 0.100 M KOH = 0.00300 mol KOH

This is a neutralization reaction (WA + SB) so we will begin with a BCA table.

	HF (aq)	+ KOH (aq)	→	KF (aq)	+ H_2O
B	0.00500	0.00300		0	a lot
C	-0.00300	-0.00300		+0.00300	+0.00300
A	0.00200	0		0.00300	a lot

The KF product is in solution so there will be 0.00300 mol of both K^+(aq) and F^-(aq).

The F^- product and the excess HF form a buffer solution. We can describe this using either as the reactant. Students will be confused as they typically do not write a mixture as one reactant and one product. Obviously, the reaction is not $F^- + HF \rightleftharpoons HF + F^-$. Instead, we choose from the following 2 reactions.

Rxn 1: F^- (aq) + H_2O (l) $\rightleftharpoons$ HF (aq) + OH^- (aq)
Rxn 2:   HF (aq) + H_2O (l) $\rightleftharpoons$ F^- (aq) + H_3O^+ (aq)

Rxn 1 uses the K_b equilibrium constant and Rxn 2 uses K_a. The product of K_a*K_b gives us K_w (10^{-14} at 298 K) so if we know one of the constants, we can easily calculate the other. For this example, we will use Rxn 1 to differentiate from an earlier sample problem.

	F^- (aq)	+ H_2O (l) $\rightleftharpoons$	HF (aq)	+ OH^- (aq)
I	0.0375		0.025	0
C	-x		+x	+x
E	0.0375		0.025	x

$$K_b = \frac{10^{-14}}{K_a} = \frac{10^{-14}}{(6.6 \times 10^{-4})} = 1.5 \times 10^{-10}$$

Chapter 15 – Acid-Base Chemistry

$$K_b = \frac{[HF][OH^-]}{[F^-]} = \frac{0.025\,x}{0.0375} = 1.5 \times 10^{-10} \qquad x = 2.25 \times 10^{-10}$$

$-\log x = \text{pOH} = 9.65 \qquad \text{pH} = 14 - \text{pOH} = 4.35$

Again, students may struggle with why the hydroxide and HF do not neutralize each other. They do react some (reverse rate ≠ 0), but there is a very small amount of hydroxide present, and the fluoride is reacting with the water setting up an equilibrium between the two reactions.

A great way for students to practice with buffers is to start by asking them to identify whether a mixture of chemicals will be a buffer or not. If not, explain why.

1. 10 mL of 0.10 M HCl + 10 mL of 0.10 M NaOH
2. 15 mL of 0.10 M HCl + 20 mL of 0.10 M NH_3
3. 10 mL of 0.10 M HF + 20 mL of 0.10 M NaOH
4. 10 mL of 0.10 M HF + 10 mL of 0.10 M NaOH
5. 10 mL of 0.10 M $HC_2H_3O_2$ + 10 mL of 0.10 M $NaC_2H_3O_2$

To help organize student ideas, we want them to verbalize that a buffer can neutralize a small amount of strong acid or strong base without a substantial change in pH. Mixture 1 does not have anything to neutralize either with. Mixture 2 is a buffer because the NH_3 can neutralize strong acid while the NH_4^+ product can neutralize strong base. Mixture 3 has an excess of hydroxide so there is nothing to neutralize strong base added. Mixture 4 has only the conjugate base so nothing to neutralize strong base. And mixture 5 would have acetic acid to neutralize strong base and acetate to neutralize strong acid. Only 2 and 5 are buffers.

Titration Curves

Titrations are designed for stoichiometry involving solutions. We can determine an unknown concentration of one chemical by adding a standard solution until the reaction is complete. In acid-base chemistry the unusual shape of the pH curve and the vivid color changes often steal the show.

Titration curves have multiple details, complicated calculations, have multiple reactions going on at once, and are a struggle to connect to the particle level. I prefer to start with the most obvious visual feature for instruction. Why does the titration curve for strong base added to a strong acid have an inflection point?

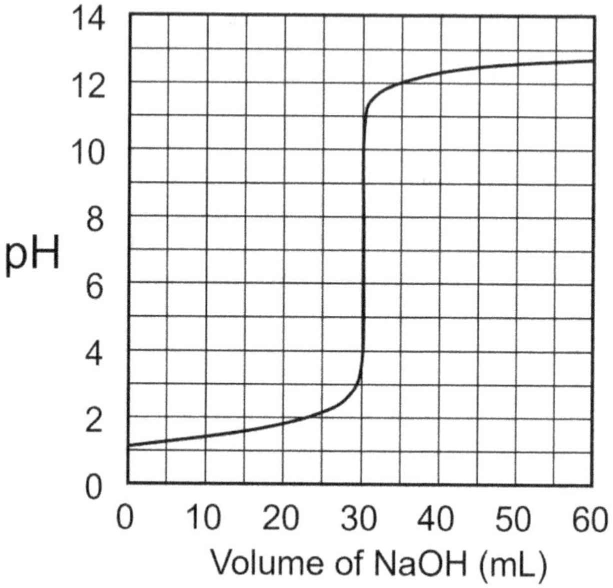

Chapter 15 – Acid-Base Chemistry

Figure 15-4: Strong acid + strong base titration curve

The shape of the titration curve results from the logarithmic nature of pH. It is important for the teacher to share some understanding of what is happening.

When the 0.10 M NaOH is added to the 50.0 mL of 0.10 M HCl the HCl begins in excess. As more NaOH is added two changes result. The first is that the amount of acid decreases. The second is that the total volume of solution increases which dilutes the concentration further.

The curve starts with a pH of 1. Before any base is added we have 0.10 M HCl and -log (0.1) = 1. In order for the pH to become 2 we would need the H^+ concentration to decrease to 0.01 M. If we ignore the volume changes this would require 0.09 moles of the HCl to be neutralized which means we would need to add 45 mL of NaOH (including volume changes gets us to a pH of 2.28). We don't see the pH change by one unit until almost 90% of the reaction has completed.

Volume of base added (mL)	pH
0	1
41	2
49	3
49.9	4
49.99	5
49.999	6
50	7
50.001	8

Table 15-1: The calculated pH for different volumes of 0.10 M NaOH mixed with 50.0 mL of 0.10 M HCl

Chapter 15 – Acid-Base Chemistry

A drop of solution is usually about 0.05 mL (assuming 20 drops gives 1 mL). That means that the pH changes from below 5 to above 9 with the addition of 1 single drop. And this brings us to where we can loop back in with students. When a student sees a titration curve they are drawn to the curve. But we need to guide them to look at the x-axis. If we draw a line down from the curve at pH 5 and pH 9 the students can see that the massive change in pH occurs with a very tiny volume of base added. This large change in pH can be highlighted with a colorful acid-base indicator.

The tiny change in volume at the equivalence point is not obvious to students. You might assume they notice now and later expect them to apply this in a weak acid + strong base titration. But most won't notice it unless prompted. Students tend to view the entire graph at once without focusing in on individual points or axes. Help them sort out the details.

We call the colorful change of the indicator an endpoint. The endpoint is usually precise because of the large shift in pH that happens when equivalent amounts of reactants are present. An indicator is merely a weak acid that has a different color than its conjugate base.

$$HIn + B \rightleftharpoons BH^+ + In^-$$

HIn is one color, In^- is a different one. For phenolphthalein, HIn is colorless and In^- is pink (phenolphthalein changes by 2 H^+ from colorless to pink). For bromothymol blue HIn is yellow and In^- is blue. These indicators have K_a values, and the color change occurs within 1 pH unit around the pK_a. When the pH is 1 more than the pK_a value, there are 10 In^- for every 1 HIn. When the pH is 1 less than the pK_a value, there are 10 HIn for every 1 In^-.

The goal with indicators is to have the pKa close to what the pH will be in the middle of the inflection. The term used to describe this point is the equivalence point when equivalent amounts of acid and base have been mixed. The indicator you use should have a pK_a close to 7 so the end point and equivalence points are equal. You have some wiggle room since the pH can change from under 5 to over 9 with the addition of a single drop.

When we replace the strong acid with a weak acid, things get interesting. When we use a strong acid, we have the excess strong acid until we get to the equivalence point and then after we have excess strong base. But with a weak acid we get a couple more interesting regions to the titration curve.

Chapter 15 – Acid-Base Chemistry

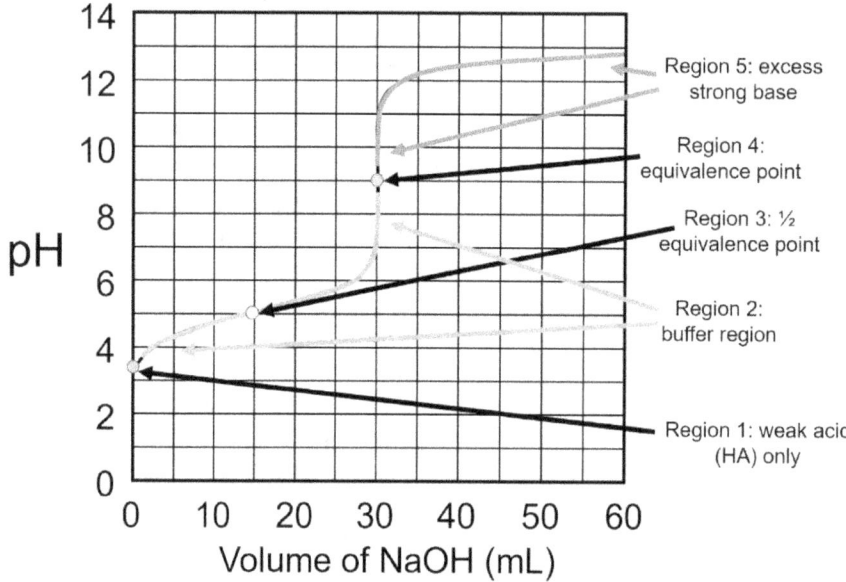

Figure 15-5: Five regions of a weak acid-strong base titration curve[7]

Region 1: At the beginning we only have the weak acid in solution. The weak acid reacts with water and the equilibrium is described by the K_a expression.

Region 2: After some strong base is added but before we reach the equivalence point, we have a buffer region. The strong base is neutralized, and we are left with excess weak acid and its conjugate base. The pH is steady during this region, especially when large amounts of weak acid and conjugate base are present. This region has the weak acid, conjugate base, and water present along with hydroxide and hydronium. The equilibrium can be represented by the K_a or K_b expression. Typically, K_a is used since this expression uses H^+ or H_3O^+ instead of OH^-.

Region 3: Halfway to the equivalence point is a specific point called the (wait for it...) half-equivalence point. Students will find it helpful if you point out that halfway refers to the volume of base added (15 mL vs. 30 mL, Fig. 15-5). They might initially view halfway as halfway along the curve, or halfway between the initial and equivalence point pH values. Students don't automatically draw the points on the curve back to the x-axis and it really helps them if you guide them to do so during instruction.

The half-equivalence point is part of region 2. It is in the buffer region, but it stands out for a couple of reasons that make it worthy of its own label. The pH at the half-equivalence point is equal to the pK_a of the weak acid. Some explain this using the Henderson-Hasselbach (HH) equation (pH = pK_a + log $\frac{[A^-]}{[HA]}$) The concentrations of A^- and HA are equal, so the ratio is 1, and the log of 1 is 0. Hence the pH is equal to the pK_a.

This can also easily be rationalized without the HH equation by just using the K_a expression.

$$K_a = \frac{[H^+][A^-]}{[HA]}$$

Chapter 15 – Acid-Base Chemistry

Since [A⁻] = [HA] we can simplify to K_a = [H⁺] and taking the negative logarithm is both sides gives us pK_a = pH.

The result is highly useful. We can now measure the K_a for an acid just by collecting pH data during a titration and identify the pH at the half-equivalent point.

The half-equivalence point is also where the maximum buffering capacity exists. As the amounts of HA or A⁻ decrease, the pH shifts begin to accelerate. Buffering capacity provides us with a relative idea of how much protection exists for the current mixture. The half-equivalence point is the maximum buffering capacity because it has the largest quantities of both weak acid and conjugate base. Of course, the buffering capacity will also depend on the concentrations of the particular mixture. A buffer with 0.1 M HA and 0.1 M A⁻ will have a lower buffering capacity than a mixture of 2.0 M HA and 2.0 M A⁻.

Region 4: The equivalence point is when there have been equivalent amounts of strong base and weak acid. At this point there is now the conjugate base and water present. The K_b equilibrium expression is used to determine the pH. Students often mix up properties of region 3 and region 4. The fact that the conjugate base and weak acid are equivalent at the half-equivalence point trips up their system 1 processing.

There are a couple of potential mistakes that students will make when calculating the pH at equivalence. They may forget to account for dilution as the total volume has changed since the beginning. They need to know how to find the K_b value using K_a and K_w. And they need to make sure that when they solve for x and take the -log x that they remember that value is for the pOH and not the pH.

If 50.0 mL of 0.10 M $HC_2H_3O_2$ is being titrated with 0.10 M NaOH the equivalence point will happen when 50.0 mL of hydroxide has been added. But at this point, the concentration of the acetate conjugate base is not 0.10 M. There are 0.0050 moles of acetate, but now there are 100.0 mL of total solution so the concentration would be 0.050 M instead. Giving students a problem where the concentrations are not in a simple ratio might expose some student struggles with calculating the final concentrations. This may encourage them to use the structures such as BCA tables.

It is also noteworthy that the pH at equivalence is no longer 7. The presence of the conjugate base means that the pH will be higher than 7 and this will play a role in choosing an appropriate indicator. The pK_a of the indicator should be higher than 7 and the reverse would be true for a titration between a weak base and a strong acid.

Region 5: After the equivalence point, we have excess strong base. The moles of excess strong base must be calculated using a BCA table. Then the concentration of the hydroxide should be calculated using the total combined volume of the initial weak acid solution and the added strong base. Then the equation pOH = -log[OH⁻] can be used and then subtracted from 14 to find pH.

Before students start practicing calculations to find pH, they should start just by identifying what region of the curve they will be in. Give them a set of scenarios and just have them identify the region.

Samples:
1. What is the pH of a 150 mL solution of 0.023 M HCN (K_a = 6.2x10⁻¹⁰)
2. What is the pH when 24.0 mL of 0.20 M $HC_2H_3O_2$ (K_a = 1.8x10⁻⁵) has 12.0 mL of 0.10 M NaOH added?

Chapter 15 – Acid-Base Chemistry

3. What is the pH when 100.0 mL of 0.50 M methylamine (K_b = 4.4x10^{-4}) has 100.0 mL of 0.25 M HCl added?
4. What is the pH when 100.0 mL of 0.50 M methylamine (K_b = 4.4x10^{-4}) has 50.0 mL of 1.0 M HCl added?
5. What is the pH when 24.0 mL of 0.10 M $HC_2H_3O_2$ (K_a = 1.8x10^{-5}) has 12.0 mL of 0.20 M NaOH added?

Let's assume students correctly choose which region these belong to. It is helpful as an extension for them to then evaluate what chemical species are present in large amounts, which reactions are occurring, and what equilibrium expressions are needed to solve for pH.

*****

"Oops I used the wrong indicator" is a demonstration that can help students connect ideas previously learned with the titration curve. If you run a titration of a weak acid and strong base the pH at the equivalence point will be greater than 7. If you use methyl red as an indicator (pK_a = 4.9) the color will change from red to yellow well before the equivalence point is reached.

Likewise, if you run a titration with ammonium hydroxide and hydrochloric acid using the indicator thymolphthalein (pK_a = 9.9) you will again see the color change from blue to colorless before the equivalence point. This allows you as the teacher to emphasize the distinction between endpoint and equivalence point while highlighting the fact that we want them to be as similar as possible. This is another chance to draw students' attention to the x-axis of the titration curve to show them how the volume is the key information being used for the titration purposes. This is especially true if the pH is not being measured since the endpoint is then the only indicator of reaction progress.

Many teachers ask students to choose the appropriate indicator for a given titration. But showing why an inappropriate indicator would fail is even more illuminating.

Every time you teach acids and bases you should do a personal reflection after titration curves. What prerequisite information was helpful for students and what should be changed for the next time? Did they struggle writing formulas? Could they easily write the conjugate forms of acids and bases? Were they able to use the products of a neutralization reaction to find pH? Could they have done more of the initial thinking on their own? If you do these reflections, you'll improve the entire unit every year you teach.

Strength Trends

The strength of an acid correlates with the K_a or pK_a value. For this reason, it is ideal to first introduce the concept of strength prior to these values. Otherwise, students will be quick to connect the two and will not need to struggle through the concept. But those values do make the discussion faster, so it is reasonable to circle back to the trends of acid strength after some mathematical calculations involving the equilibrium constants has taken place.

There are four trends and a couple of them overlap. Two principles guide these trends. The more polar the bond is, the stronger the acid. The longer the bond is, the stronger the acid.

Chapter 15 – Acid-Base Chemistry

HF is a stronger acid than H_2O which is a stronger acid than NH_3 which is a stronger acid than CH_4. The primary distinction is the polarity of the bond. Since fluorine has the largest electronegativity difference with hydrogen, the H-F bond is the most polar and so HF is the strongest acid of the group.

But HF is the weakest of the hydrogen halides. HF is weaker than HCl which is weaker than HBr which is weaker than HI. HCl, HBr, and HI are all strong acids (relative to hydronium). If you had a tug of war on an H^+ ion, the Cl^- would win against Br^- or I^-. This goes against the trend we saw for polarity so something else must be at work here. The issue here is that iodide is the largest of these ions and this creates a long bond with the proton that reduces the attraction between the iodide and hydrogen ions.

The wrench with many of these trends is that for many acids, the proton being removed is attached to an oxygen. For these compounds the bond length a non-issue since the bond length is unlikely to vary much from compound to compound.

Two trends deal with oxyacids. The first is that the more electronegative the central atom, the stronger the acid ($HClO_4 > H_2SO_4 > H_3PO_4$). The more electronegative central atom will cause a shift in electron density that makes the O-H bond more polar (trend 1). The second trend is the more oxygen atoms surrounding the central atom, the stronger the acid ($HNO_3 > HNO_2$, $H_2SO_4 > H_2SO_3$).

The second trend is more difficult for teachers to explain and many resort to using the conjugate base stability due to resonance. Sulfate (SO_4^{2-}) has more resonance stabilization than sulfite (SO_3^{2-}).

$$H_2SO_4 + H_2O \rightarrow H_3O^+ + HSO_4^-$$
$$H_2SO_3 + H_2O \rightarrow H_3O^+ + HSO_3^-$$

The bisulfate (HSO_4^-) and bisulfite (HSO_3^-) products both have a negative charge, but the bisulfate has three oxygen atoms that the negative charge is delocalized over while the bisulfite only has two oxygens for the negative charge. In order for students to use the reactants to explain the trend we would have to look at much more unusual resonance structures and so we tend to use the final product instead.

This can be done in the earlier trends as well. Fluoride is less stable than iodide because the fluoride has a negative charge contained in a much smaller volume of space. Thus, the fluoride conjugate base is less stable than iodide.

But we can explain just by using the Lewis structure of the acids. We know that a stronger acid means the bond between the hydrogen and oxygen is weaker. This must be caused by the resonance from the additional oxygen, but the structures are a little more challenging prior to the reaction.

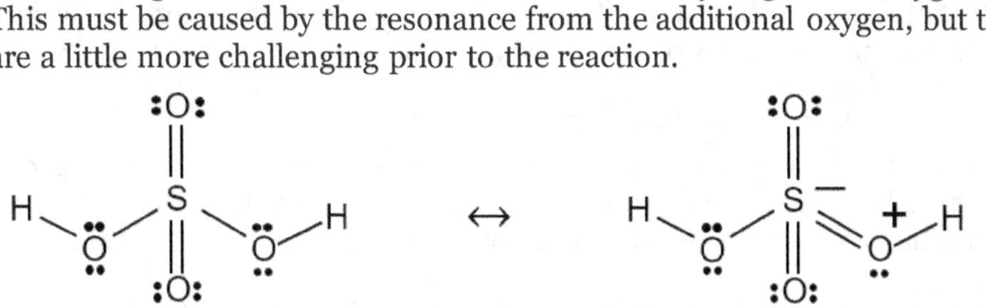

Figure 15-6: Sulfuric acid (H_2SO_4) and a resonance structure showing why the O-H bond weakens. The negative charge is delocalized over the sulfur and two oxygens as well.

Chapter 15 – Acid-Base Chemistry

Note how the formation of the double bond between the oxygen and sulfur creates a shift of electron density away from the hydrogen (Fig. 15-6). This weakens the bond more, and since sulfuric acid has an additional oxygen to distribute the negative charge, the electron density experiences a greater shift for sulfuric acid than sulfurous acid.

When you are working with an acid such as $HClO_4$ it is worthwhile to make sure students know where the hydrogen is attached. Most students won't even consider it unless you give them a prompt that requires them to do so. Drawing a Lewis structure would suffice, but if you go through trends without this some will be missing a key link.

When we look at trends we might also add in the trend for oxides as we move across the periodic table. Metallic oxides (Na_2O) are basic while nonmetal oxides (SO_2) become acidic when added to water. The name oxygen means "begets acid" since Antoine Lavoisier thought that all acids had oxygen.[8] As we move from left to right the trend gradually shifts from basic to acidic. In the middle of the periodic table are some oxides that are what we call amphoteric. These act as acids and bases.

It is easiest for students to deal with the two components separately. Aluminum oxide is one example of an amphoteric substance. The oxide is a base. When we mix oxide ions and water, we have the conjugate base (O^{2-}) of a strong base (OH^-). Thus, the following reaction takes place:

$H_2O + O^{2-} \rightarrow OH^- + OH^-$

The water loses a proton to the oxide to form two hydroxide ions. This is why the aluminum oxide and other metallic oxides are bases.

But what does the aluminum do exactly? It doesn't have any protons available. So why does adding Al^{3+} to water make the water more acidic?

The short answer is that the aluminum reacts with water. But the process is new for students. The aluminum ion is small and has a large amount of charge. This means that negative charge will experience strong attraction to the Al^{3+}. The oxygen atoms in water molecules show a large affinity for the aluminum ions. The water molecules form what is called a coordinate covalent bond where both electrons of the bond come from the water (instead of sharing one electron each).

Six water molecules will surround each aluminum ion and bond to it in an octahedral arrangement. This clump of 6 water molecules and aluminum ion is called a complex ion and it would be represented as $[Al(H_2O)_6]^{3+}$. The 3+ charge applies to the entire complex ion.

The bond between the oxygen and the aluminum weakens the bonds between the oxygen and hydrogen atoms. The complex will produce H^+ ions as these hydrogen split off from the complex.

$[Al(H_2O)_6]^{3+} + H_2O \rightarrow [Al(H_2O)_5OH]^{2+} + H_3O^+$

Cations that are small with a lot of charge have similar reactions. These tend to appear in the middle of the periodic table in the transition from metals to metalloids. Aluminum, iron, and zinc all form acidic solutions (depending on the anion).

You can do an excellent demonstration with zinc hydroxide ($Zn(OH)_2$). If you form zinc hydroxide precipitate, the zinc ion (Zn^{2+}) can react as an acid and the hydroxide (OH^-) reacts as a base. If you mix HCl with a suspension of zinc hydroxide precipitate the precipitate will dissolve. If you mix an excess of hydroxide with the

Chapter 15 – Acid-Base Chemistry

suspension the precipitate again dissolves. $Zn(OH)_2$ is amphoteric in that it reacts as an acid and as a base.

$Zn(OH)_2$ (s) + 2HCl (aq) → $ZnCl_2$ (aq) + $2H_2O$ (l)
$Zn(OH)_2$ (s) + $2OH^-$ (aq) → $[Zn(OH)_4]^{2-}$ (aq)

The reaction with HCl falls into both the Bronsted-Lowry and Lewis acid definition. The second reaction only works for a Lewis acid definition. The electron pairs from the hydroxide are forming a bond to the zinc ion. There is no proton transfer anywhere. Yet the similarities with the first reaction are undeniable when the demonstration is performed.

Note how trends naturally leads into the acid-base definitions. It is at this point where we can establish the Arrhenius, Bronsted-Lowry, and Lewis definitions. We can also compare them to our original bonding model (Fig. 15-1, Fig. 15-2. The Arrhenius is similar but limited to examples in water. The Bronsted-Lowry is extremely similar, and we could even do some spaced practice with examples of acid-base reactions now. We could review conjugate acids and bases, how formulas change with the addition or removal of a proton and look at acid-base strength of the resulting anions and cations.

The Lewis definition is an impressive definition that includes the most reactions. But it is not the best for aligning with much of the calculations in the unit. The Bronsted-Lowry is much easier to connect to the particle level. These connections make this definition more concrete for instructional purposes. It's common to limit the use of the Lewis definition to cases where only that definition works. It is helpful for students to see examples of complex ions as well as some synthesis reactions such as $BH_3 + NF_3 \rightarrow BH_3NF_3$.

Properties and Loose Ends

Starting the acid-base unit is difficult because there are so many ideas to develop that it becomes challenging to choose a starting point. One starting place is a bonding model to define what acids and bases are. Many teachers instead start with properties of acids and bases. Acids taste sour, change the color of an indicator, react with metals to make hydrogen gas, and causes bubbles to form when added to carbonates. Bases taste bitter, change the color of an indicator, and feel slippery when aqueous solutions are on your skin. These can be connected to students' experiences with different foods, and these can easily be demonstrated as introductory phenomena.

The logic for starting with properties is that they can be a simple beginning with lots of observation and evidence. They also highlight the need for two characteristic groups of chemicals. These acids and bases then get defined using the historical definitions through Arrhenius, Bronsted-Lowry, and Lewis models.

When starting the unit with properties we want to constantly highlight the observation that acids and bases differ while gradually working towards a refined definition of what these two groups are. Along that path we want to give students frequent opportunities to practice working with the symbolic representations of acids and bases.

Students might memorize a list of strong acids and strong bases. Students might practice writing the conjugate bases of acids and conjugate acids of bases.

Chapter 15 – Acid-Base Chemistry

Students should practice writing formulas of acids and bases, as well as the formulas for the resulting ions after a proton has transferred.

$HCl + H_2O \rightarrow Cl^- + H_3O^+$

$HCl + KOH \rightarrow KCl + H_2O$

$HF + H_2O \leftrightharpoons F^- + H_3O^+$

$NH_3 + H_2O \leftrightharpoons OH^- + NH_4^+$

Avoid representing acids breaking apart without a base present ($HCl \rightarrow Cl^- + H^+$) as this conflicts with what happens at the particle level. It can also be helpful later to spend some time introducing a generic acid (HA) and a generic base (B). Student understanding can be enhanced by showing how the A⁻ and B can be equivalent as well as HA and BH⁺.

Remember that students have built a framework about charge. Acids and bases will test it. Until now students have usually experienced compounds as neutral overall. But many of the symbolic representations include fragments that carry a net charge. We want to address this by pointing out that an HSO_4^- or a CO_3^{2-} does not exist without something to support that charge. The ion is either in a solvent or with another ion of opposite charge.

If we consider gaseous ammonia reacting with gaseous hydrochloric acid, we might see how we can improve student understanding.

1: $NH_3 (g) + HCl (g) \rightarrow NH_4^+ + Cl^-$

2: $NH_3 (g) + HCl (g) \rightarrow NH_4Cl (s)$

The first representation is better for highlighting the acid-base reaction. We see the transfer of the proton and can clearly identify the conjugate acid and base. But the second representation better matches the macroscopic observation of white smoke forming. The second representation fits better with previous concepts. And this tradeoff is littered throughout the acid-base unit. Make this explicit with your students. Talk about multiple representations and the pros and cons of each.

How many chemicals are there in pure water? There is H_2O along with tiny quantities of OH^- and H_3O^+. But is there anything else? It depends on how we make pure water, but even after distillation there are likely a few other things present. Gases dissolve in water, although often in tiny amounts. Distilled water will have oxygen and carbon dioxide gases. The carbon dioxide will combine with water to form carbonic acid.

Because of carbon dioxide, rain tends to be acidic. Rain commonly has a pH near 5.6. But there are other gases that can combine with water in rain to make even more acidic water. These gases are commonly referred to as SO_x and NO_x gases.

One way these gases get produced is through the burning of coal. Many chemicals adsorb to the surface of coal. In fact, you can even buy activated charcoal to deodorize funky smelling air (teachers take note!). When we burn coal, there tends to be quite a bit of stuff stuck to the coal that burns with the coal. Sulfur burns to make SO_2 and SO_3 which then combines with water in the air to form sulfurous and sulfuric acid (H_2SO_3, H_2SO_4).

When teaching acid rain, the reactions are much simpler for carbon dioxide and SO_x gases than they are for NO_x gases.

$H_2O + CO_2 \rightarrow H_2CO_3$

$H_2O + SO_2 \rightarrow H_2SO_3$

$H_2O + SO_3 \rightarrow H_2SO_4$

Chapter 15 – Acid-Base Chemistry

These reactions are all balanced without needing coefficients and only have the one product. It is the nitrogen oxides that are a bit more challenging, and you don't gain much by using them.

$H_2O + 2NO_2 \rightarrow HNO_3 + HNO_2$

NO is a free radical and can react with oxygen to form NO_2. If you've ever added concentrated nitric acid to copper the brown gas that is produced is the NO_2. NO forms if the nitric acid is dilute.

You can make an acid rain simulation by mixing dilute sulfuric acid with a sulfite or nitrite solution in a petri dish. If you mix a couple drops of each solution in the center, the NO_x or SO_x gases will spread through the dish. You can put tiny drops of indicator in concentric circles to watch the color change. I use bromocresol green which changes from blue, to green, to yellow as the acidic gases mix with the indicator solution (Fig. 15-7). NH_3 can be used to change the color back to blue.

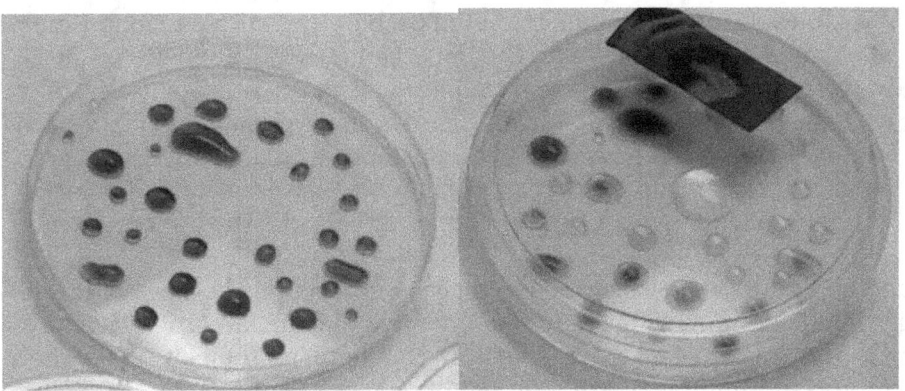

Figure 15-7: Acid rain petri dish experiment

Where I live in Michigan the lakes have limestone beds. These form a protective buffer from acidic rain. This can also be simulated by putting a small pile of $CaCO_3$ in the petri dish to show how the indicator adjacent to the $CaCO_3$ is unaffected.

There is a lot of content and difficulty in the acid-base unit. At the end we do want to connect some ideas to previous content. What was the point of all of this and why do we care about acids and bases? Now is a good time to do an overview of chemical reactions. Originally, we defined chemical reactions by changes in chemical formulas or by inclusion in a set of five or so categories (synthesis, decomposition, single replacement, double replacement, combustion).

But now we can do a more general categorization. A chemical reaction is either a redox reaction, an acid-base reaction, or a precipitation reaction. Each of these three categories involves changes in charge allocation. In redox reactions, electrons move from one species to another. In precipitation reactions, cations and anions move relative to each other. And in acid-base reactions we see that electrons are shifting to form a new coordinate covalent bond. This is the Lewis definition but remember that the Lewis definition incorporates all three definitions as subset groups.

Chapter 15 – Acid-Base Chemistry

Chemical reactions are about a shift in charged particles (electrons or ions). That shift must create a new substance, but our fundamental understanding of what a chemical reaction is should include a view of what type of charged particle is changing position. If we go back to organic chemistry or the original reaction types, all these reactions can be classified as either precipitation, redox, or acid-base.

A combustion reaction is a redox reaction where electrons are shifting from carbon and hydrogen to oxygen. Addition of a halogen to an alkene is both a redox reaction since electrons are shifting to the halogen from carbon, and an acid-base reaction since the carbon of the alkenyl group is forming a coordinate covalent bond with the halogen. An S_N2 substitution reaction is an acid-base reaction where the nucleophile is the Lewis base and the carbon attached to the leaving group is the Lewis acid. A single replacement reaction between Cu and $AgNO_3$ is a redox reaction where electrons move from Cu to Ag^+. A double replacement reaction between barium nitrate and sodium sulfate is a precipitation reaction where barium ions form a crystal lattice structure with sulfate ions.

We want to draw attention to the fact that in each of these three reaction types we are looking at charged particles moving. The impetus for the reactions to occur should lie in how strongly the reactants are at pulling on these charged particles. A precipitation reaction compares how well the barium cations can pull on sulfate ions relative to the waters of hydration. A redox reaction compares how strongly copper (II) ions can pull on electrons relative to silver ions. An acid-base reaction compares how the attraction of a nucleophile to a carbon compares with the leaving group and the carbon.

Acids and bases are frequently used as catalysts in reactions because of their ability to modify charge and then later return to the original state. Sulfuric acid can donate a proton to make a chemical more positively charged, and then later take the proton back to reform the sulfuric acid.

A wonderful example to tie these ideas together is with laundry. When companies develop new stain removers, they look to find chemicals that will add charge to a stain, but not to the dye on the fabric. Once the charge of the stain is altered, the stain will likely dissolve in the water. Something like bleach will do this to the stain, but also to many of the dyes used.

A clog in a pipe might have lye or NaOH added (this isn't recommended as it can damage the pipe). The base reacts with grease slowly and as charge is added more and more of the clog becomes soluble in water until the pipe eventually clears. The ability of acids and bases to add charge is taken advantage of frequently in all sorts of routine chemical engineering.

Student Struggles

1. *"I have no idea what I'm doing!"* The toughest part of acids and bases is usually figuring out pH from a variety of combinations of acids and bases. When we are mid-titration, the student must identify strong vs. weak, acid vs. base, limiting vs. excess, and use appropriate equations with the chemical reactions. There is a lot for the student to process. It is recommended that teachers spend some time having the students just identify what there will be in a solution before doing all of the calculations. If I mix 15.0 mL of 0.10 M HF with 22.0 mL of 0.050 M NaOH, what will I have?

Chapter 15 – Acid-Base Chemistry

2. *"How do I tell how many hydrogens there should be?"* When teaching about conjugate acids and bases, some students will struggle to change both the formula and charge simultaneously. Going from H_3PO_4 to $H_2PO_4^-$ is tricky for them. Connecting $H_2PO_4^-$ with KH_2PO_4 can also cause issues for those who are still not keen on the spectator ion concept. Adding particle representations and using Lewis structures can both help add visual components to this that can give these students a starting point to draw from.

3. *"Are all Arrhenius acids also Bronsted-Lowry acids?"* Students may wonder why an Arrhenius definition is even necessary in the first place. Why do we need this definition when the Bronsted-Lowry is even better and includes all the Arrhenius acids and bases? The Arrhenius definition helps show how our model of what an acid or base is has improved. As history progresses, we move from a list of substances that are acids and bases (Lavoisier, 1780s), to the Arrhenius model based on ions in solution (1887), and the Bronsted-Lowry and Lewis definitions both came out in the same year (1923).

4. *"Can I just use the Henderson-Hasselbach equation instead of an ICE chart?"* When students learn about buffers it is valuable for them to work through the calculations using ICE charts. This helps them connect buffer calculations with weak acid or weak base calculations that they've already done. This also helps connect the calculations with the reactions taking place. The Henderson-Hasselbach is helpful for students to understand what has to happen for pH to change by 1 unit, but otherwise the equation obscures much of the chemistry.

5. *"Would a pH of 10^{-8} M HCl be 8 or something else?"* The -log [HCl] would be 8, but that's not what the pH is. This question arises when we forget that water does not start with 0 hydronium concentration. This is a natural discrepancy because we assume 10^{-7} is negligible. We make the initial concentration to be 0 in many calculations by assigning all products to be 0. The actual hydronium concentration would be 1.1×10^{-7} with a pH of 6.96.

6. *"If acids are dangerous, does that mean bases are safe since they neutralize dangerous acids?"* This is common for students to think this. Bases are dangerous in a similar fashion to acids. Both cause changes in charge by moving protons. This increase in charge can damage the structure of an object, and the change in charge can change solubility in a way that causes deterioration. Bases are used to clear clogged drains in plumbing. They work by removing protons which causes grease, fat, and other insoluble substances to become water soluble. Acids and bases should be considered as two sides of the same coin and not as a dangerous substance and something that neutralizes the danger.

7. *"How do I tell if an acid is weak?"* Using absolute terms such as strong or weak is not accurate. We should say stronger, or weaker instead. But there are some acids that are consistently stronger than so many other acids (especially hydronium) that they get the label of strong. You likely provide students a list of 5 or 7 strong acids to memorize. Students should also then use that list for weak acids. Any acid that is not on that list should be assumed to be weak. H_3PO_4 isn't on the list. We can

Chapter 15 – Acid-Base Chemistry

assume that it's a weaker acid. H_2CO_3 is absent from the list. We can assume that it's a weaker acid. For strong acids that are not on the list provided to students, we must be explicit that they are strong in assessment questions or prompts.

8. "If the pH can be below 0 or above 14, why do most pH scales say they go from 0-14?" The number 14 is relevant because the pK_w value is 14. This means that at room temperature, the sum of pH and pOH will be 14. But that does not restrict the scale of pH to a maximum of 14 nor a minimum of 0. Concentrated acids that exceed 1.0 M will have a negative pH. Concentrated bases will have a pH above 14. Superacids have pH values that are absurdly low. Nonetheless students will come to your class thinking that the pH scale goes from 0-14 (I've also seen a few 1-14). Make it clear to them why this was poorly presented to them while helping them expand the range.

9. "Can water be completely pure?" No. Water will always have tiny amounts of gases dissolve in it. Water will also react with itself to form hydroxide and hydronium. Unless we allow an absurdly small number of particles to be a system, water on a macroscopic scale will never be completely pure. The gases that dissolve have an impact as well. Carbon dioxide causes distilled water to be slightly acidic. Other gases can lower pH even more. Dissolved oxygen has an impact on organisms within the water. Dissolved solids impact the physical properties of the water.

10. "When does the pK_a and pK_b combine to make 14?" This is only true when the pK_a is of an acid, and the pK_b is of the conjugate base. The reason this combination works is that the combination of the weak acid + water reaction with the weak base + water reaction results in the autoionization of water ($2H_2O \rightarrow H_3O^+ + OH^-$). Since the pK_w is 14, the pK_a and pK_b must add up to 14 ($K_a * K_b = K_w = 10^{-14}$).

Phenomena

1. Making paneer or cheese - Heat milk and add acid to it (citric acid or vinegar both work). The acid combines with phosphate groups in casein. The casein is mostly hydrophobic, but the phosphate groups allow the casein to interact with water. When the phosphate groups become protonated, the nonpolar portion of the casein attracts to other casein particles more than water and clumping occurs as the casein separates from the rest of the milk.

2. Goldenrod paper - Some goldenrod paper functions as an acid-base indicator. I've read that you can fashion your own using turmeric, but I've only had success with purchasing the paper. Adding a slightly basic solution to the paper causes the color to change from goldenrod to red. If you use ammonia, the color will slowly fade back as the ammonia vapors leave the paper.

3. Milk of Magnesia - Milk of magnesia is mostly magnesium hydroxide. $Mg(OH)_2$ is insoluble in water so you must stir to suspend the particles in water. Add universal indicator and the tiny amount dissolved will still cause the indicator to be purple. When acid is added, the indicator flashes to red since the acid is water soluble. However, the suspended particles react with the acid slowly. The result is a flash of red followed by a gradual shift through the colors of the rainbow. The transitions to blue and purple take longer.

Chapter 15 – Acid-Base Chemistry

4. Acid-rain petri dish - Have students arrange concentric circles with dots of indicator leaving a space in the middle of the dish (Fig. 15-7). By mixing dilute solutions of potassium nitrite and sulfuric acid, a cloud of NO_x gases will cause the indicator drops to change color to the acidic form. Adding a drop of ammonia reverses the color change.

5. Strong vs. concentrated, weak vs. dilute - Take 4 graduated cylinders. Put concentrated (1-3 M) HCl into the first, concentrated $HC_2H_3O_2$ into the second, dilute (0.01-0.1 M) HCl in the third, and dilute $HC_2H_3O_2$ into the 4th. The first container is strong and concentrated. The second is weak and concentrated. The third is strong and dilute. The fourth is weak and dilute. Add indicator to each. Measure the pH of each or conductivity of each. Then finally add sodium bicarbonate to each. Both of the concentrated solutions will react rapidly with the baking soda and generate lots of bubbling and carbon dioxide. The dilute solutions will react with much less effervescence.

6. Distilled vs. tap water - Take 2 graduated cylinders and fill one with distilled water and the other with tap water. Add indicator and a stir bar to both. The distilled water may start out as slightly acidic due to dissolved carbon dioxide. The tap water may start out as slightly basic due to dissolved substances remaining after the treatment process. When a tiny amount of strong acid is added to teach, the indicator in the distilled water changes to its acidic form. The tap water on the other hand is likely to stay neutral until a much larger quantity of acid has been added. Many tap water sources are naturally buffered with bicarbonate from a limestone base to the water source. Treatment facilities may also buffer the water as part of the treatment process.

7. Amphoteric zinc hydroxide - Place stir bars in two small beakers. Mix equivalent amounts of zinc chloride solution (0.1 M) and sodium hydroxide (0.2 M) solution so that zinc hydroxide suspension is formed. To one $Zn(OH)_2$ suspension add strong acid (0.5 M HCl). The suspension dissolves. To the other $Zn(OH)_2$ suspension add strong base (0.5 M NaOH). The suspension also dissolves. This shows that $Zn(OH)_2$ is amphoteric in that it can act as an acid and a base. The zinc ions can form complex ions with hydroxide ligands. The hydroxide ions are capable of neutralizing acid.

8. Pop can in hydroxide - Stick a pencil through the tab of an empty aluminum pop can. Use the pencil to suspend most of the pop can in a beaker of concentrated sodium hydroxide. The aluminum will react with and dissolve in the hydroxide slowly. After a day goes by, the plastic liner will be all that remains of the section submerged in the solution. A redox reaction must take place to turn the aluminum from neutral to charged. After that the aluminum ions are complexed by hydroxide ligands.

9. Green cabbage - Acid-base indicators are everywhere. Flowers, vegetables, fruits, and many other objects have chemicals that change color depending on whether they are protonated or not. The generic reaction for an indicator is $HIn \rightleftharpoons In^- + H^+$ where HIn is one color and In^- is a different color. A simple indicator that is easy to prepare is red cabbage juice. Purchase some red cabbage and extract solution from it by boiling it in water. The cabbage juice will have a variety of colors at different pH values due to the anthocyanins in the cabbage. The cabbage indicator does start to smell shortly after preparation.

Chapter 15 – Acid-Base Chemistry

10. pH curve - The pH curve during a titration is not intuitive. Students need to spend time understanding why the curve changes more quickly near the equivalence point. They need to process how the endpoint and equivalence point differ in the context of what each is as well as by the difference of volume of solution added. Students need time to determine what chemicals are present at various points and what reactions are occurring with those chemicals. Throughout these analyses, students must focus on points on the graph, identify changes that occur, and identify what is present at different points along the curve. For this to happen students would be best served to collect or at least see the generation of a titration curve in the lab. They benefit even more from collecting or seeing a variety of titration curves to see how changes in sequence, concentration, and strength impact the curve.

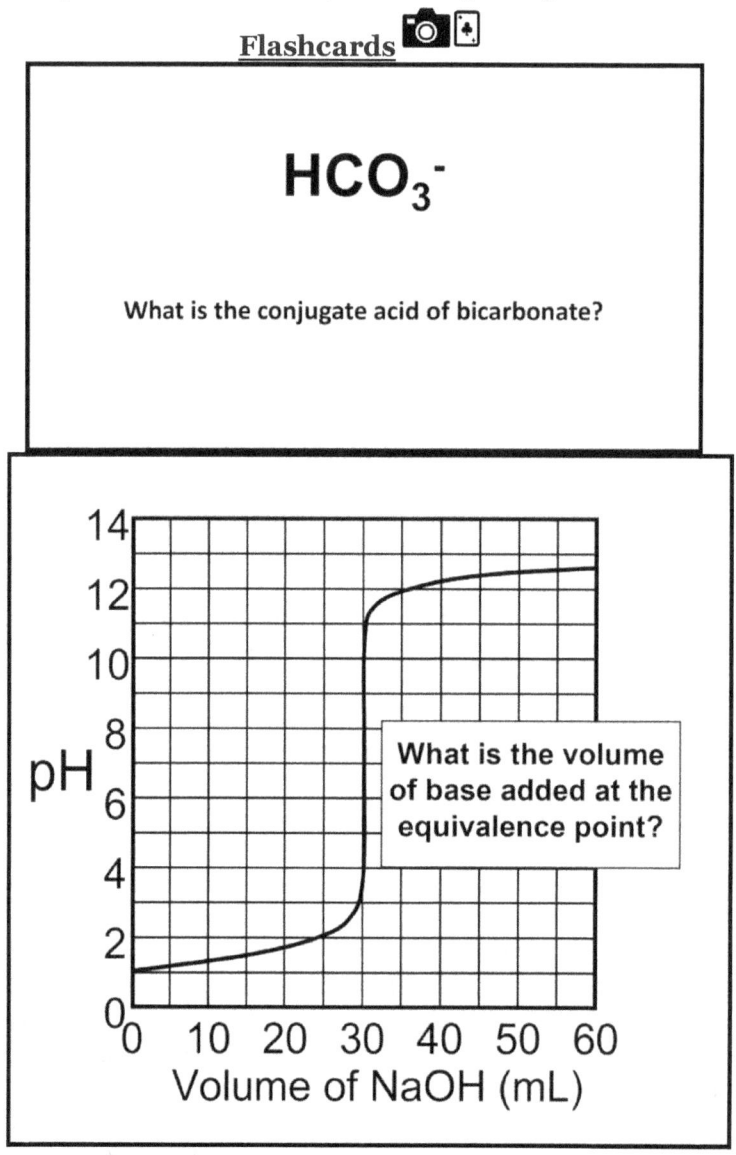

Chapter 15 – Acid-Base Chemistry

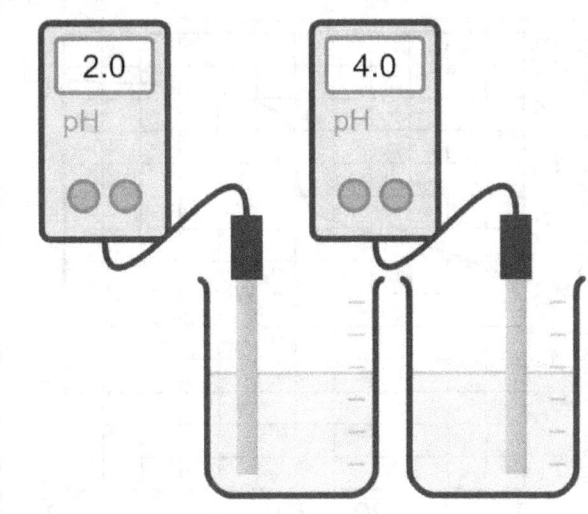

The first beaker has _____ times as many H_3O^+ particles

$3KOH + H_3PO_4 \rightarrow K_3PO_4 + 3H_2O$

KOH is a _____ base
H_3PO_4 is a _____ acid

Is K_3PO_4 acidic, basic, or neutral?

Which is stronger?

Hypochlorous acid
HClO

Perchloric acid
$HClO_4$

Chapter 15 – Acid-Base Chemistry

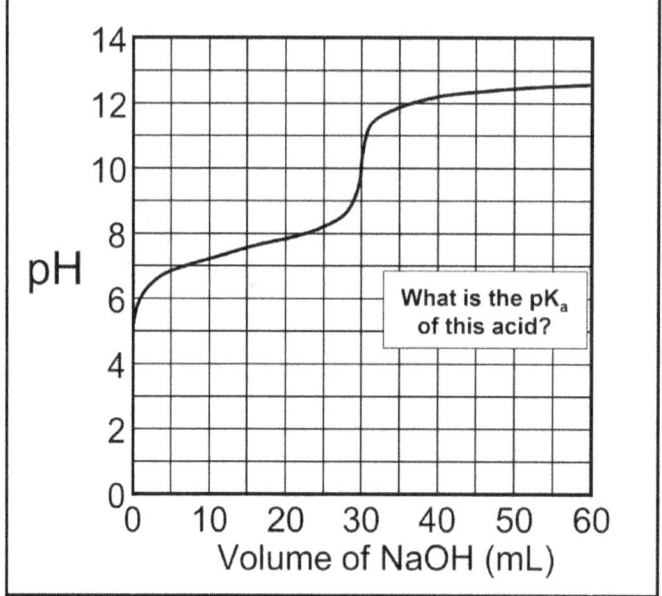

What is the pKa of this acid?

$CO_2 + H_2O \rightarrow$???

$Na_2O + H_2O \rightarrow$???

$ZnO + H_2O \rightarrow$???

_____ is a buffer mixture

a. $NH_3/NaOH$
b. $HCl/NaOH$
c. $HC_2H_3O_2/C_2H_3O_2^-$
d. $H_2PO_4^-/Cl^-$
e. H_2O/OH^-

Chapter 15 – Acid-Base Chemistry

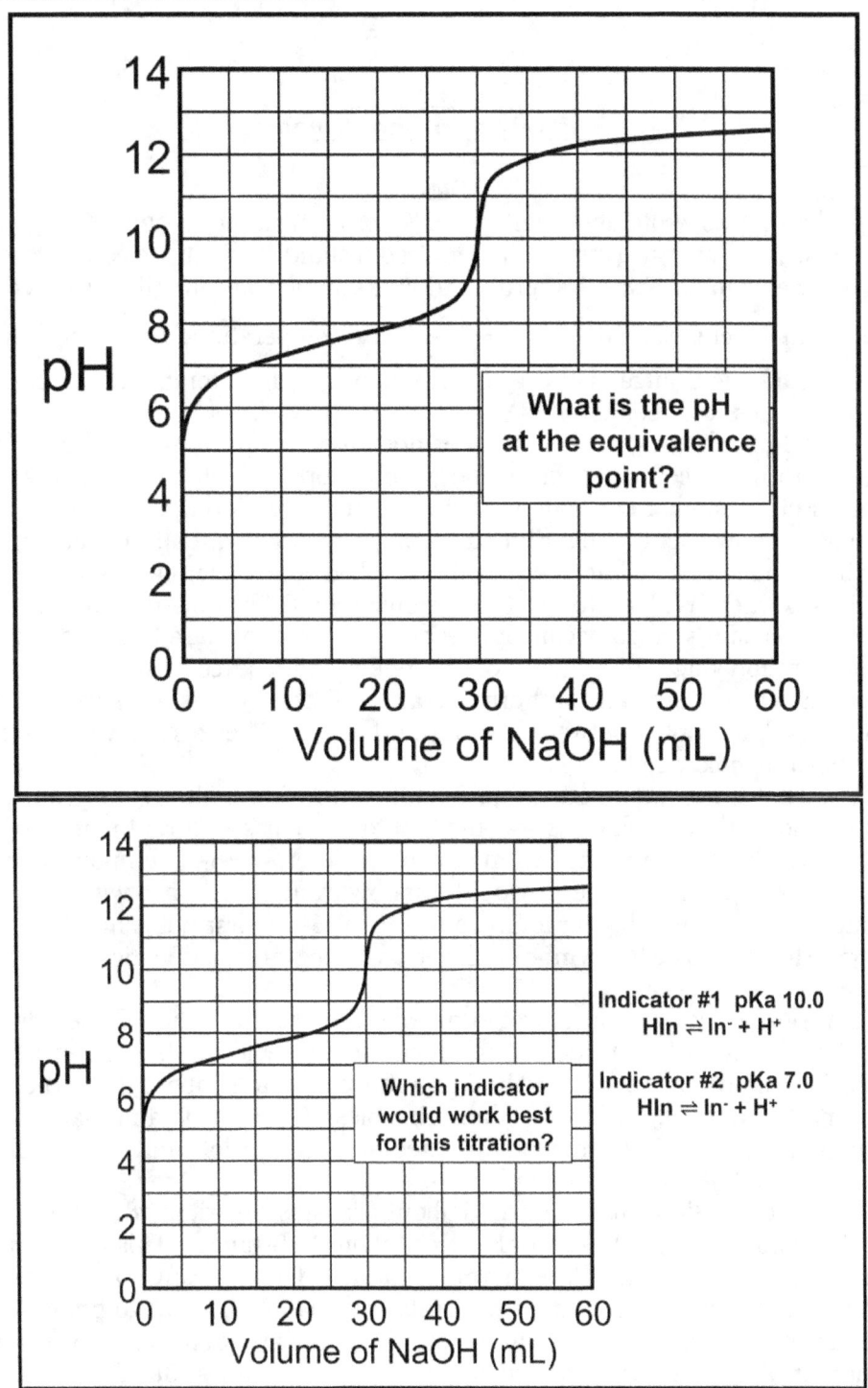

Chapter 16 S
Entropy and Spontaneity

History

Gilbert Lewis wrote about entropy in Thermodynamics (1923). He described how the concept of energy distribution required a standard for us to compare a system to. That standard was first proposed by Rudolph Clausius (like Boyle, the youngest of many children (18)), $dS = \frac{dQ}{T}$. At a low temperature, an input of energy (dQ) leads to a large change in energy distribution. At a high temperature, the same energy input is less consequential.[1]

Unfortunately for Clausius, no one understood his proposal. He had selected the term entropy as a combination of energy and "trope" which means to transform or change. But the similarity to energy led scientists to think that he was referring to energy and not how energy is distributed.[2] Clausius worked out all the details, but he was unable to build an audience because they lacked conceptual understanding.

This is where Willard Josiah Gibbs stepped in. Gibbs understood the relevance of Clausius's entropy concept. And he was able to share that understanding with others by applying it to a variety of concrete scenarios. Ice melting, a precipitation reaction, and an exothermic reaction that produces a gas were all able to help chemists differentiate between entropy and energy. They also saw how entropy could be used to make predictions.

Prior to Clausius, Sadi Carnot and Emile Clapeyron both used engine cycles to study heat and work. But Clausius was the first to move past caloric theory so that heat and work could be converted from one to the other. In 1854 he published work that included the claim that energy would not flow from a substance with lower temperature to one with a higher temperature. In 1865 he first published the details of entropy. He chose S as the symbol without providing any justification or rationale for the decision.

Clausius struggled to interpret entropy at the particle level. He used the term "disgregation" to describe "the degree in which the molecules of the system are dispersed." In the 1860s we still had limited information to support the particle model of matter. Energy, work, and other principles of thermodynamics are emerging, but with frequent error and disagreement. Reversible and irreversible processes made all of this more complicated.

Nonetheless Gibbs first published about Clausius's work in 1873. From 1875-1878 Gibbs published Equilibrium of Heterogeneous Substances. This paper was more of a book, and it analyzed the thermodynamics of a wide range of processes. The analysis assigned chemical potentials to substances that Gibbs used to predict what would and would not happen for a variety of mixtures. Parts of his analysis were the foundation of what would grow to become statistical mechanics. But Gibbs was also able to combine the entropy, absolute temperature, and internal energy into a single new term called free energy.[3] When this free energy change was negative for a process, the process would proceed without work being required.

Chapter 16 – Entropy and Spontaneity

Entropy

I know there are teachers hoping that I'll provide a definition of entropy that magically unites the oversimplified "disorder" with the overly complex mathematical descriptions. One that circles between messy bedrooms with the spontaneity of vinegar reacting with baking soda. I do not believe that such a phrase exists. My reductive description of entropy is the combination of different ways a system can use energy coupled with the amount of energy the system has to use.

A larger molecule has more entropy than a smaller one because it has more ways to use energy. More particles and more space for particles lead to greater entropies because there are more ways to use energy. Higher temperatures correlate with larger amounts of energy and thus show higher entropy values.

Entropy is a contrived tool. It does not exist in real life, but it can be used to predict what will and won't happen. Entropy can be defined mathematically in two different manners. The original definition by Rudolph Clausius was that $dS = \frac{dQ}{T}$. That definition is still functional, but a more general approach is to use $S = k_B * \ln W$ where W is the number of possible ways for something to be arranged without any macroscopic changes.

Part of the problem with teaching entropy is that the definition just isn't very important. It is the purpose of entropy that is more critical for students to understand. There is some value to be derived from exploring those two equations with specific examples. But for a novice student little is clarified by providing those two equations. So, let's travel back in time to their origin.

Clausius came up with the idea of entropy in the 1850s. The name was meant to be energy combined with trope which means to change. When he first published his work on the idea nobody understood it. It wasn't until Willard Josiah Gibbs came along later and used entropy with concrete examples that people began to understand how entropy differs from energy.

Gibbs published On the Equilibrium of Heterogeneous Substances around 1876. In this work he detailed specific chemical changes and how entropy played a role in them. It was this work that clarified the entropy concept that Clausius had developed but failed to communicate to others in a way that they could understand. No wonder we struggle with entropy in chemical education!

The reason why Gibbs work helped so many understand entropy was because he applied entropy towards concrete examples in chemistry. We should emulate this in our classrooms. It is less important for students to define entropy and more important for them to know that entropy increases when a gas expands or when a solution is heated.

Entropy is a highly technical mathematical description of matter and energy. Its purpose is to be able to predict what changes will happen. It is not a direct cause of anything. Rather it is a means to describe changes that are taking place. Systems do not have entropy, rather systems can be assigned an entropy value that we can use to make predictions about what will happen next.

A ball is rolling along a straight path with constant velocity. Sometime later it reaches another ball, and they collide. The second ball continues rolling and the first ball comes to a stop. What made the second ball move and the first ball stop?

Chapter 16 – Entropy and Spontaneity

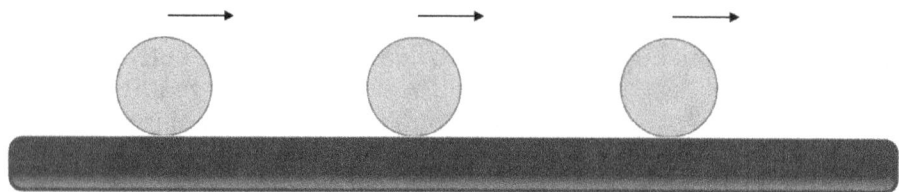

Figure 16-1: A ball rolling at constant velocity. What makes it roll in a straight line?

You probably wouldn't say that entropy caused that to happen. But when we replace the balls with billions of particles, we have a harder time assigning forces, motions, and collisions. We must seek another path to explain what we observe even though we've just scaled up the number of motions.

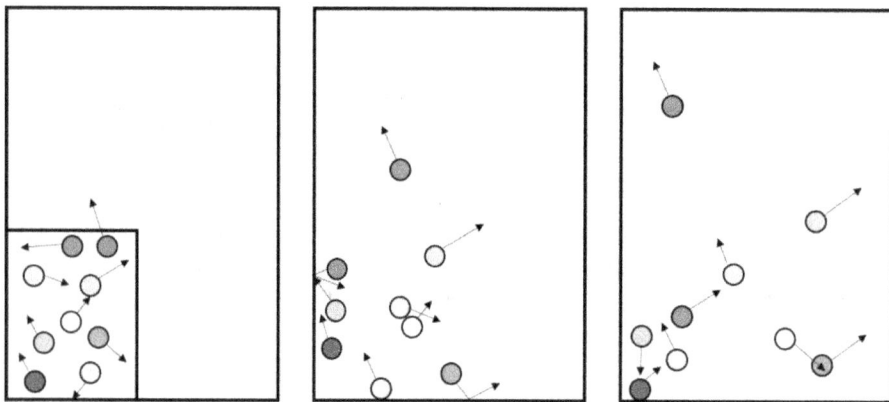

Figure 16-2: Gas particles starting in a confined space and spreading out upon release

Let's take a set of gas particles in a confined space (Fig. 16-2). When the confined space is opened, the particles spread out to fill the new larger space. Why do the particles expand to fill that empty space? Is it because of entropy? Did the vacuum pull them? No. Neither of these are accurate. The particles just moved there because that was how they were moving in the first place. Much like the ball that hits the second ball because it was moving towards it. The initial velocities and positions cause the particles to end up where they end up.

We can look a little bit deeper. The initially empty space does not have any particles. When particles move towards that space there is nothing to collide with them. But if particles move towards the space with lots of particles, they are more likely to bump into something and redirect. Or at least initially that will be the case. If a particle is moving left, there must be a particle to its left for it to collide and change direction.

When I took organic lab in undergrad, I remember staring at the sink attachment we used for vacuum filtration. It puzzled me! How did this thing create a vacuum? Everything else I had seen used some form of motor. But all this had was an attachment to running water. Finally, it clicked. The water could push air particles down and away. But no air particles would come back. That's how all vacuums work, they push particles away while preventing their return.

Chapter 16 – Entropy and Spontaneity

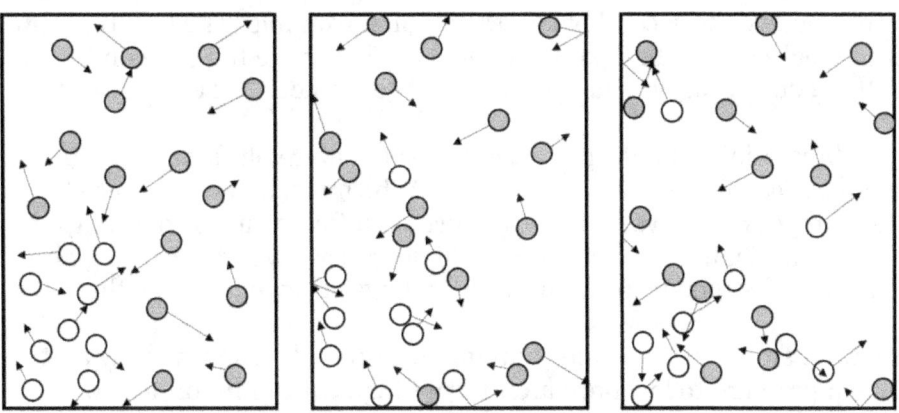

Figure 16-3: Mixing of two gases. Why do they mix?

There are two ways for us to make predictions about a system. The more difficult yet more accurate way is to know every position and motion of each particle. We can then predict where everything will go. But as the number of particles becomes larger, this becomes implausible. Instead, we assign probabilities. If there are fewer particles in one section, there is a greater probability that new particles will enter that region. We can design equations that predict how particles will move based on probability and we can easily test those predictions to see if they need revision. Because the number of atoms in a system are typically overwhelmingly large, we see very little deviation from these predictions. Entropy functions as a mathematical language of these predictions.

One of the best thought experiments to teach students about entropy and particle motion is to work through what happens when you breath. Breathing is an experience we all have (hopefully!), breathing is loaded with misconceptions, and yet we rarely use it in class. Take in a deep breath right now. It feels like you are pulling in the air. But you are not exerting any forces on the air particles rushing through your mouth into your lungs. You can't pull on gas particles. There's even a famous phrase that many of us chemistry teachers use. "Science never sucks!" There is no such thing as suction. But most teachers reach a point where it becomes such a struggle to explain something that their intuition creeps in thinking that suction works in some special scenario.

So how does breathing work? While you cannot control air particles, you can expand and contract muscles. Your diaphragm moving causes your lungs to get bigger. Now there is more space in your lungs. Molecules move into that new space since nothing is there to prevent them and they were moving that way in the first place. The lung expansion leads to a reduced pressure, or alternatively it leads to fewer molecules leaving your mouth. But we tend to overfocus on the pressure in the lungs and not on the fact that those air particles were on the way towards your mouth whether you expand your lungs or not.

If you just stood there with your mouth open would air particles move in and out of your mouth? Of course they do. There are air particles moving towards your face right now that will end up going into your mouth or nose. When we "breath in" we don't pull those in, we prevent them from being pushed away by reducing the number of collisions that counter their motions.

Chapter 16 – Entropy and Spontaneity

Let's say we remove the second ball in the original example. The first ball now continues on its original straight-line path (Fig. 16-1). What made the ball continue in that motion? Nothing did. It was already going, and it continued. Did entropy cause it to happen? No.

When you "breath in" did entropy cause the air to enter your lungs? No. But the entropy value that describes the air did increase. Which means we can predict that when you expand your lungs you will experience an influx of air to your lungs. When you contract your diaphragm, your lungs will decrease in volume causing air to leave as there is no less obstruction from the surroundings relative to inside the lungs.

There are ways students will avoid thinking this through. They may say that air travels from high pressure to low pressure or from high concentration to low concentration. These are observations, but not explanations. The phrasing is similar to that used in chapter 2 where students talk about air pressure equalizing. Push students past these to where they can understand why air shifts from high pressure to low pressure.

There are two ways we can use entropy. We can develop a complex mathematical treatment using statistics. Or we could just share some of the common increases in entropy and make qualitative predictions. A gas expanding into a larger volume can be described with an increase in entropy. Salt dissolving in water (usually) can be described with an increase in entropy. Unless these changes are accompanied by the surroundings experiencing a decrease in entropy, they will likely happen.

Some themes quickly emerge. The more gas particles you have, the higher the entropy is. The larger a molecule is, the larger its entropy tends to be. The more space something occupies, the higher the entropy tends to be. The larger the total energy is, the higher the entropy is. Do we need the term disorder? I think it can be helpful, but it's probably not necessary depending on the style of the teacher.

I don't particularly care for the messy room analogy. The perks are that rooms tend to get messier without intervention, but the placement of large objects as a microstate seems like an unnecessary stretch. I think this analogy improves as the system becomes smaller with a limit on the number of possibilities.

When I was in college a professor was teaching about entropy and made a claim that is common and worth exploring. He claimed that in the desk in front of him that all the particles were moving randomly with varying amounts of energy. But it is possible that all the thermal energy could be concentrated into one of the particles. But this doesn't happen because it is probabilistically prohibited. They use this to claim that all distributions of energy are equal in probability but that so many different distributions make this one scenario highly unlikely. Something along the lines of, if the energy of every particle in this desk were all in one spot that spot could fly through the air at near light speed, but this won't happen because the probability is so low. I believe he even calculated the odds.

The professor's claim highlights difficulties with teaching entropy. If the motion of every particle converged until just one particle had all the energy, we would have to trace back what existed prior to that. In order for all the particles to be moving and then many to be stationary while one moves, those particles would have to have a sequenced and aligned set of collisions. This will never happen and not

Chapter 16 – Entropy and Spontaneity

because of probability. The initial conditions that would lead to this state don't exist and aren't possible to construct.

If you consider how the particles would have to move to all hit one particle at the surface, there would have to be a coherent pulse that converges in that one direction balanced from all sides. Prior to the 1 particle moving up, the 5 particles surrounding the 1 would have to be converging at high speeds. Prior to that the 20-25 particles surrounding those would have to be converging on those 5. Prior to that the 100-200 particles surrounding those would have to be converging on those. Quickly it becomes clear that unless we start with that extremely odd configuration of motion it will never become that final state where 1 particle has all the energy. The fact that particles are already moving in all directions prohibits this.

2nd Law of Thermodynamics

What would be required to make a set of particles all move in a single direction? The answer is that it would take even more particles colliding with them. If you want particles to have a more coherent set of motions, you obtain it by having a larger number of particles collide with them. Afterwards those surrounding molecules will have less coherent motions, and the tradeoff tends to result in an overall increase in entropy.

The second law is a very interesting digression at the particle level. If you are unfamiliar with the Maxwell's demon scenario, a demon has a trap door that allows only slow particles to go left and fast particles to go right. This would violate the second law because it would have cold become colder and hot become hotter. Eventually Maxwell conceded this thought experiment because there is no such demon. But a complete concession is also wrong.

If a fast particle hits a slow particle, usually the fast particle slows down and the slow particle speeds up (assuming ideal behavior and elastic collisions). But not always. If you've ever calculated 2-dimensional collisions with conservation of momentum you know that it depends on the component speeds. A fast particle might be moving mostly in the y-direction. The fast particle glances off the slow particle causing the slow particle to transfer more motion in the x-direction. The fast particle would speed up and the slow particle would slow down. While this is less common than the fast particle slowing down, this does happen. There are an unbelievably large number of collisions between particles. Entropy on a very localized level can decrease with no accompanying increase elsewhere. This is why the 2nd Law should always include "tends" for the entropy of the universe to tend toward a maximum.

Much of entropy stems from the fact that particles move in all different directions (I prefer different to random). What if there were a rectangular shaped container that had only a tiny number of particles in it. The container is connected to a second empty container that has a closed stopcock between them. But in this theoretical experiment the particles in container A are moving in an unusual manner. They are spaced apart from each other and are moving straight up and down. When they hit the top of the container they collide and reverse direction towards the bottom. They have no left and right motion (Fig. 16-4).

Chapter 16 – Entropy and Spontaneity

Figure 16-4: Container A has 7 particles that are moving directly up and down. When the stopcock is opened, will the particles move into container B?

When I open the stopcock are the gas particles going to move from one container to the other? No. They will continue moving up and down forever. But entropy predicts they will move. Entropy is wrong here because it relies on a distribution of directions. That distribution would normally come about from movement in a variety of directions and collisions. Here we have neither of those.

Not only does this thought experiment help us with understanding the limitations of entropy, but this can also be further expanded into understanding the 2nd law of thermodynamics. If we wanted to make a small set of particles that did move in such a fashion, how would we go about it? We would need to collide particles with them until they moved with no motion in the xz plane. By doing this we would have to have surrounding particles increasing in entropy so that the overall entropy would likely go up. Basically, to organize the motion of a set of particles we cause surrounding particles to decrease systematic organization of motion.

When we state the second law, we should always emphasize the word "tends". The entropy of the universe tends towards a maximum. This allows for the minor reductions in entropy that are inevitable from the rare collisions where speeds diverge to the fluctuations of a system at a dynamic equilibrium.

One of the primary understandings of the 2nd law of thermodynamics stems from the original function Clausius discovered $dS = \frac{dQ}{T}$. What this relationship implies is that adding energy to different systems results in different entropy changes. Adding energy to a system with an already high entropy results in a smaller change. Think of honking a car horn in busy traffic. You are adding in a disturbance to the traffic, but overall, it is a minor addition. But if you were to honk a car horn inside a library the result will be much more disruptive (and hilarious!).

Likewise, adding energy to a system with high entropy has a smaller change than adding energy to a system with low entropy. If you have a very cold ice cube, adding 10 J of energy makes a big difference to the motion of the particles. But if you have warm water and add 10 J nothing big changes. If you have hot steam and add 10 J there is even less that changes. As we approach 0K the impact of energy increases rapidly (with respect to entropy).

If we mix 100 g of water at 80 °C with 100 g of water at 20 °C the water will become 50 °C if no energy is lost to the surroundings. The hot water cooled down, so its entropy decreased. The cold water heated up, so its entropy increased. The total entropy change was not zero though.

The cold water increased entropy by about 40.7 J/K while the hot water decreased by about 37.1 J/K (calculated using differentials and average temperatures). The overall change is approximately +3.6 J/K for the system.

Just like with honking the car horn, the energy going into the cold water made a more substantial change than the energy leaving the hot water. Does this mean that

Chapter 16 – Entropy and Spontaneity

we should use entropy to explain what happened? No. There is a better explanation available. The hot water particles were moving faster and therefore their collisions tended to impart more motion to the slower moving water particles. This continued until a thermal equilibrium was established.

Let's transition from particles moving in all directions to work. Work and energy can easily fall into a cycle like the cello is a big violin and a violin is like a little cello. They are often circularly defined, and this keeps them abstract. To combat this, we want students (and teachers!) to explain what they think is happening with work and energy using concrete examples.

There are two definitions of work. Work is the product of force and displacement, and work is the amount of energy transferred except for heat. There are some technical additions that can be made to those definitions, but they aren't needed here.

When an object has a net force acting on it, the object can either change its speed and kinetic energy, or the object can change its position in a field (electric or gravitational) and its potential energy. If I push on a baseball the ball might move at a higher speed as I push more and more. But if I push the baseball up and away from the Earth, the baseball might continue moving at the same speed but gain in gravitational potential energy.

The 1st law of thermodynamics states that energy is conserved. The written form of the law is $dU = dW + dQ$ where U is a measure of internal energy (the total energy of a system), dW is the work done to the system, and dQ is the heat added to the system.

All these definitions for work that we use in physics do not easily translate into chemistry. The best way to build a visual representation of work in chemistry is to focus on the particle level.

The difference between heating and work at the particle level is based on the directionality of the particles. If the particles all move faster in a single direction, that is energy gained from work.[4] If the particles move in all different directions faster, that is energy gained from heating. Often both changes happen because in order to push particles in a single direction you would need a set of particles to push in only that direction. We would need to start with a highly unusual set of particles in the surroundings like in Fig. 16-4 where direction is aligned.

One of the ways we do work is by having a combustion reaction occur in a piston. What happens in this event is that the combustion reaction occurs which produces fast moving product particles. These particles push the piston against the pressure of the surroundings. But the products of the combustion reaction do not have highly organized motion, they have faster motion than the surroundings. This means that some of the collisions of particles with the piston result in heating instead of work.

There are ways to manipulate how much work and heat result. A smaller temperature/pressure difference between the system and the surroundings will lead to more work and less heat.

Under what conditions would all heat be converted into work? If we think at the particle level, this becomes quite clear. In order for thermal motion to change into work we need to organize the motion. Thermal energy is when a set of particles moves in a variety of directions. Work is when a set of particles moves in a single direction. An excess of thermal energy at a boundary can lead to gas particles pushing more in

Chapter 16 – Entropy and Spontaneity

one direction and thus we can obtain work. For thermal energy to convert 100% into work, the thermal energy would have to involve all the particles moving in the same direction. Even if this was possible, it would no longer be defined as heat, the particles all moving in the same direction would be work.

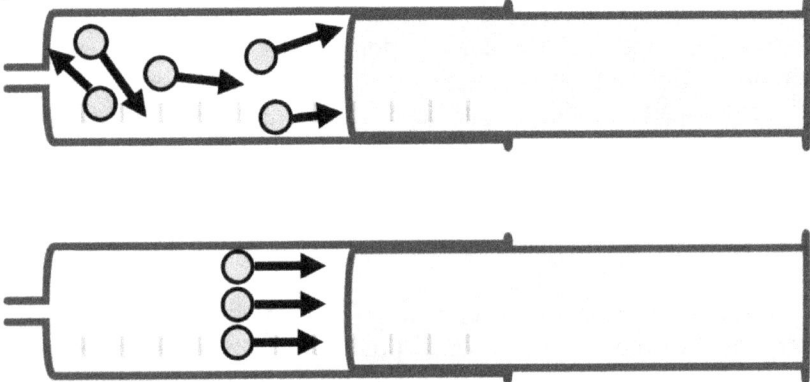

Figure 16-5: The top syringe has particles that transfer energy as work and heat. The bottom syringe would exclusively be work. Transferring heat only would require the gas volume to not expand.

Let's assume a system transfers energy to the surroundings exclusively as work. At the particle level, the particles would have to have a perfect alignment of their motion as shown in the bottom syringe (Fig. 16-5). This is never the case. The surroundings would have to have similar alignment as well or the collisions would create motion in a variety of directions. But this does help us understand why we always "waste" energy as heat when we try to produce mechanical work. This also primes us to understand how we can limit heat and maximize work.

What would the opposite of this look like where only heat is transferred, and no work is done? The particle collisions would not result in any coordinated motion. Heat without work is common as we would expect based on our observations and experiences.

Gibbs Free Energy and Spontaneity

At this point perhaps you question the purpose of entropy. Entropy's value primarily comes from making predictions about systems with large numbers of particles. To maximize the value of entropy we combine entropy, temperature, and enthalpy into a single variable called Gibbs free energy. Gibbs free energy gives us insight into how a set of particles will change moving forward.

Gibbs free energy should be considered as a mathematical description of a system of particles just as entropy was. We should not use Gibbs free energy to explain why something happened if we want to maximize understanding. But we can safely assume that changes in Gibbs free energy will correlate with what will happen (ignoring kinetics). If Gibbs free energy decreases, that change should proceed without anything else needing to happen. We call such changes spontaneous or thermodynamically favorable.

We can think of Gibbs free energy as a combination of entropy changes of the system combined with the entropy changes of the surroundings.

$dG = dH - TdS$

Chapter 16 – Entropy and Spontaneity

Remember that $dS = \frac{dQ}{T}$ so if we consider dH to be dQ (assuming constant pressure) we get:

dG = -TdS$_{surr}$ - TdS$_{sys}$

The first entropy term describes the change of the surroundings, and the second term describes the entropy change of the system. Since T is in Kelvins and must be positive, this means that if the sum of the entropy changes is positive for the universe, the Gibbs free energy will be negative.

The reason why the first term has a negative sign is because the original dH term was in reference to the chemical system, but we are now looking at the entropy changes of the surroundings. If dH was negative for the chemical system, this would mean that the exothermic process would add energy to the surroundings. If dH was positive, we would have an endothermic process where entropy of the surroundings would decrease.

We now arrive to how we can talk to students about Gibbs free energy. There is no productive definition that we can provide them that's going to make all of this clear. What we want to stress is not what Gibbs free energy is, but how we can use it. We can assign a Gibbs free energy change to a process, and if the value is positive or negative that gives us useful information to predict to what extent that change will occur.

If ΔG is negative, the change will occur spontaneously. By spontaneously we mean that it will happen and that no work is required to make it happen. If you just mix the chemicals, the reaction will happen. If ΔG is positive, we would need to input work to get the reaction to occur.

ΔG = ΔH - TΔS

If ΔH is negative, energy will be released to the surroundings and the entropy of the surroundings will go up. This favors spontaneity. If ΔS is positive, the entropy of the system increases, and spontaneity is favored. If both of those are true, the reaction will occur spontaneously. If neither of those are true, the change will not occur without work being input. If one is true, then what happens depends on the temperature.

ΔH	ΔS	Spontaneous
-	+	always
+	-	never
+	+	only at higher temperatures
-	-	only at lower temperatures

Table 16-1: Combinations of ΔH and ΔS with impact on spontaneity

Remember that adding entropy to a system with a lot of energy results in a smaller impact than adding entropy to a system with less energy. When both ΔH and ΔS are +, the energy is leaving the surroundings. ΔH being + means that the surroundings lose energy, but this results in a smaller impact on entropy if the temperature is large.

Chapter 16 – Entropy and Spontaneity

When we say "spontaneous only at lower temperatures" what we mean is that there is a cutoff temperature and below that temperature the change is spontaneous. Each reaction or change has a different point where the spontaneity changes.

When both ΔH and ΔS are -, the energy is moving from the system to the surroundings. If the surroundings are at a low temperature, the resulting change in entropy becomes larger and enough to offset the entropy losses of the system.

1) $2NO_2$ (g) → N_2O_4 (g)   ΔH° = -   ΔS° = -
2) Na_2CO_3 (s) → Na_2O (s) + CO_2 (g)   ΔH° = +   ΔS° = +
3) C_2H_5OH (g) + $3O_2$ (g) → $3H_2O$ (g) + $2CO_2$ (g)   ΔH° = -   ΔS° = +
4) H_2O (l) → H_2O (g)   ΔH° = +   ΔS° = +

Recall that when Clausius originally wrote about entropy that the confusion caused people to miss the point of what he was describing. We run the same risk with our students. We want to take an approach like Willard Gibbs did where we use lots of concrete examples so that students can develop these abstract ideas.

In reaction 1, a bond is formed between two radicals. Bond formation is exothermic as discussed previously in chapter 7. The entropy of the system decreases since the number of gas particles decreases. This reaction will be spontaneous at some temperature and any other temperature that is smaller than that cutoff temperature. Above that temperature the reaction will not be spontaneous. But the reverse process would be.

In reaction 2, several bonding changes are taking place. But overall, the reaction is endothermic. The entropy of the system increases as the reaction proceeds because the number of gas particles increase. At high temperatures we expect this change to be spontaneous. The entropy of the system increasing is favorable. But the energy leaving the surroundings results in a decrease of the entropy of the universe unless the temperature of the surroundings is high enough for the surrounding's entropy decrease to be offset by the increase of the entropy of the chemical system.

Reaction 3 is interesting. Here we have an exothermic burning process where the entropy is increasing as the gas particles increase from 4 to 5 during the reaction. The entropy of the system increases, and energy goes to the surroundings to the entropy of the surroundings increases. This process is spontaneous always. But we know that ethanol does not always just start burning so why doesn't the reaction automatically take place?

The reaction does not require work to occur, but the activation energy is too high for the reaction to start without a supply of initial energy. Once the mixture is sparked, the reaction will proceed without any further interference because the energy of the first products to form will supply the activation energy for the next sets of reactants to overcome activation energy. This is one of two warnings about the limitations about using free energy and entropy.

Vaporization of water (change #4) is an endothermic process where the system increases in entropy. This is a great example to show the usefulness of entropy and free energy quantitatively. There is some temperature where above this happens spontaneously, and below the change is not spontaneous. We will transition from the one to the other when ΔG changes from - to + or when ΔG = 0.

ΔG = ΔH - TΔS

If ΔG is 0 then

Chapter 16 – Entropy and Spontaneity

$\Delta H = T\Delta S$

We can look up standard enthalpy and entropy of formation values to find how they change from liquid to gas.

$\Delta H° = +44 \frac{kJ}{mol}$ and $\Delta S° = 118.8 \frac{J}{mol*K}$ (or $0.1188 \frac{kJ}{mol*K}$)

$+44 \frac{kJ}{mol} = T \, 0.1188 \frac{kJ}{mol*K}$

T = 370 K

The transition for ΔG being zero would be 370 K which is 97 °C. This means we would expect at 97 °C that we would have an equilibrium of water in the liquid and gaseous states. The value differs from the experimental value of 100 °C since the standard enthalpy and entropy values assume we are at standard temperature.

But Gibbs free energy is not just a predictor of whether something will occur or not. It also has a numerical value. Does $\Delta G = -240 \frac{kJ}{mol}$ mean that something will happen twice as much as $\Delta G = -120 \frac{kJ}{mol}$? No. Both of those things will happen spontaneously, but the first change can also produce twice as much work as the first change. If battery A has a larger Gibbs free energy than battery B, battery A can do more work than battery B. It is unlikely this will matter in an introductory class and most of our efforts can be spent on spontaneity instead.

The free energy is the maximum amount of work that can be derived from the process. If one battery had a free energy change of -240 $\frac{kJ}{mol}$ this means that if the electricity generated is used just right that 240 kJ can of work can be extracted (per mole of reaction as written). But as the resistance to the electricity varies the amount of work decreases and the thermal energy increases.

The exception to spontaneity is that sometimes there is such a large obstacle between the initial and final states and the process does not occur. Diamond should turn into graphite, but the in between is too difficult to accomplish and so the rate of conversion is effectively zero unless massive pressures and/or temperatures are utilized. What this means is that graphite is more thermodynamically stable than diamond. But the strong bonds in diamond are not going to be able to break in a manner where the structure can rearrange unless there are extreme temperature and/or pressure conditions. You can call this a kinetics limitation because that bond breaking is the activation energy. This is a particularly good example because even if a few bonds broken, they are much more likely to reform the diamond structure rather than change to coal because the other bonds are still in place and have a set structure. Under standard conditions the conversion isn't going to happen regardless of how patiently you wait.

Q and K

What happens if ΔG is zero? In order to give students a proper chance at understanding the answer to this question we need to start by clearly distinguishing ΔG° from ΔG.

ΔG° describes a system at standard conditions. Standard conditions (1 atm, 298 K, 1 M) give us a way to compare the relative stability of the chemicals before and after the change. You can think of this as before a football game starts, how the two teams compare on paper. ΔG on the other hand compares any situation. This could be

Chapter 16 – Entropy and Spontaneity

during the middle of a game where it is raining and the worse team has a surprising lead, or it could be before the game starts.

$\Delta G°$ compares the thermodynamics of the reactants and products on equal footing. ΔG describes a specific pile of chemicals. There are two common reports we draw from the two values that conflict for students.

$\Delta G°$	K	ΔG	Shift to equilibrium position
+	<1	+	Left, more reactants form
0	About 1	0	None, at equilibrium
-	>1	-	Right, more products form

Table 16-2: $\Delta G°$ tells us about equilibrium values and ΔG about position

$\Delta G°$ tells us what will happen when we get to equilibrium. ΔG tells us whether we are there yet or not. In our analogy with the football teams, $\Delta G°$ would compare the likelihood of each team winning and ΔG would tell us what is happening during a specific game. Perhaps team A has a 70% chance of winning a random game, but during this game team B has a 2-goal lead and now the odds are that team B has a 60% chance of winning.

There are some flaws in this analogy because there is no winning for a chemical reaction. But the focus we're trying to draw to students is that $\Delta G°$ describes a general comparison under specific conditions. ΔG on the other hand describes what is happening at a given moment.

$\Delta G°$ tells us what we'll have when equilibrium is achieved. ΔG tells us whether the chemical mixture is there or not (Table 16-2). These two ideas are very similar and abstract so we should expect students to be confused.

When students have had a chance to explore these ideas, we can then help by adding in an old concept. If we go back to equilibrium we had the equilibrium constant K, and we had the reaction quotient Q. Both had the same calculation, but K was for a system specifically at equilibrium, while Q described a specific mixture of chemicals that might not necessarily be at equilibrium. By comparing Q and K we could make predictions about what would happen.

If K is...	then....	If Q is....	then...
<1	At equilibrium we have mostly reactants	<K	The reaction shifts right
About 1	We have lots of reactants and products at equilibrium	=K	The reaction is at equilibrium
>1	At equilibrium we have mostly products	>K	The reaction shifts left

Table 16-3: What Q and K tell us about a chemical reaction

Note the similarities between Table 16-3 and Table 16-2. There are some obvious correlations between K and Q with $\Delta G°$ and ΔG.

Chapter 16 – Entropy and Spontaneity

We can further connect those four variables with each other by comparing two equations.

Eq. 16-1 $\Delta G° = \Delta G - RT \ln Q$
Eq. 16-2 $\Delta G° = -RT \ln K$

$\Delta G°$ gives us information about what we will have at equilibrium. Will we have more products or more reactants? ΔG on the other hand describes where we are in that progress towards equilibrium. If we take Eq. 16-1 and set the conditions that the reaction is at equilibrium, ΔG becomes 0 and Q becomes K. This results in Eq. 16-2.

If we are not at equilibrium, then we can predict how the shift will occur using ΔG. If ΔG is +, then the RT ln Q term must be larger than RT ln K. If Q > K, then the reaction will shift towards the reactant side.

If ΔG is -, then the RT ln Q term must be smaller than RT ln K. If Q < K, then the reaction will shift towards the product side.

This language is confusing to a student because prior we had envisioned reactions as either happening spontaneously or not. Now we're discussing shifts which do not mesh with the vision students had at the particle level. A nonspontaneous reaction has particles collide and no reactions take place in their heads. But now we're saying that if ΔG is + that the reverse reaction is taking place at a faster rate than the forward. This implies there are collisions where the reaction is taking place, but the frequency of the reverse reaction is faster.

This is a legitimate limitation of the entropy and Gibbs free energy models. Remember that these models can be used to predict, but the reality of what happens is based on the combination of particle positions, relative speeds, and electrostatic forces.

Notice how $\Delta G°$ and ΔG start to deviate from our original interpretation as we link to equilibrium. Originally our model was that ΔG predicted whether a reaction would happen spontaneously. Spontaneously meant that we needed to not input work for the reaction to happen. But now we're looking at processes that occur without work in both directions and we're focusing on shifting.

$\Delta G°$ fits better with the original model since $\Delta G°$ predicts our relative quantities once equilibrium is achieved, but even there we have a dynamic equilibrium where the reverse reaction should require some form of work according to our initial model. Perhaps it might be helpful for us to consider a large collision between two particles as a potential source of work in this case that can cause a nonspontaneous reaction to occur.

E and E°

Lo and behold there is another topic that relates to $\Delta G°$, redox chemistry. The voltage of a redox reaction as electrons shift from one chemical to another can be related to $\Delta G°$ and ΔG. This makes sense to connect these topics since voltage is the amount of energy electrons gain as they move from one position to another. If the voltage is positive, the electrons are moving in a favorable direction, and we would expect the reaction to be spontaneous. If voltage is negative the electrons would have to move against a net force. This would only happen if an external source supplied work to move the electrons.

Chapter 16 – Entropy and Spontaneity

Electrochemical cells that have a +E and a -ΔG are labeled galvanic cells. These proceed spontaneously and can be used to produce useful electrical work. Cells with a -E and a +ΔG are electrolytic cells. These must have an external source provide work for the reaction to happen. Electrolytic cells are common because of electroplating metals.

$E°_{cell}$ tells us how the two oxidized species compare at pulling on electrons under a standard set of conditions (298 K, 1 atm, 1 M). When we look at a copper and zinc cell, the standard +1.1 volts gives us a numerical representation of how much more the Cu^{2+} ions pull on electrons relative to the Zn^{2+} ions. As electrons travel from the Zn atoms to the Cu^{2+} ions the electrons gain +1.1 J of energy per coulomb of charge.

$E°_{cell}$ can be related to $\Delta G°$ and the equilibrium constant K. This is confusing because we tell students that K describes a system at equilibrium while the other two describe systems obviously not at equilibrium.

The resolution to this is that all three are comparative measurements of products and reactants via energy. And they do not just describe what we have at equilibrium, they also are used to predict what will happen to get to equilibrium. K doesn't just describe a system at equilibrium, the constant also gives us comparison of the relative energies of products and reactants. We can consider K as the combination of equilibrium concentrations, but we can also consider K as a prediction as to how concentrations must change in order to achieve equilibrium. If we know that Q = 10, we do not know how the chemical mixture will change. But if we know that Q = 10, and K = 0.023, now we know that more reactants will form and more product will be consumed to achieve equilibrium.

$E°_{cell}$ and $\Delta G°$ also give us information about what is going to happen at equilibrium. If $E°_{cell}$ is + then the reaction is going to favor products when equilibrium is reached. If $\Delta G°$ is -, then the reaction is going to favor products when equilibrium is reached.

What we're seeing in this analysis is that students need to do more than just plug in numbers to really digest and understand the subtle distinctions between these constants. They need to compare how these values change with varying conditions. If $E°_{cell}$ is +1.10 V for a reaction, and our current E_{cell} is +0.84 V, what do we have and what will happen?

E_{cell} gives us a numerical representation comparing two specific half-cells under a specific set of conditions. Neither $E°_{cell}$ nor E_{cell} is used to exclusively describe equilibrium systems, but both give us information about what we will have at equilibrium ($E°_{cell}$) and current progress towards equilibrium (E_{cell}).

E_{cell} is a fascinating topic that gets undermined by Le Chatelier's Principle. If you have a battery and you alter one of the concentrations the voltage of the battery changes. This should be explored at the particle level. Why would changing the concentration of Cu^{2+} ions in a Cu/Zn cell alter the voltage?

This question should start with looking at why we use concentration in the first place. Why does concentration matter instead of total quantity of chemical? Concentration being the critical variable means that the contact between the dissolved ions and the electrode are what dictate the voltage.

If we use a larger volume of solution, but leave the concentration identical, the number of collisions between cations and the electrode will happen at the same frequency for a given area. It doesn't matter if the contact area increases, this will

Chapter 16 – Entropy and Spontaneity

change the total frequency of collisions, but not the frequency for each given amount of surface area of the electrode. In order to alter the voltage, we would need to increase or decrease the frequency of collisions of cations with a given area of the electrode itself.

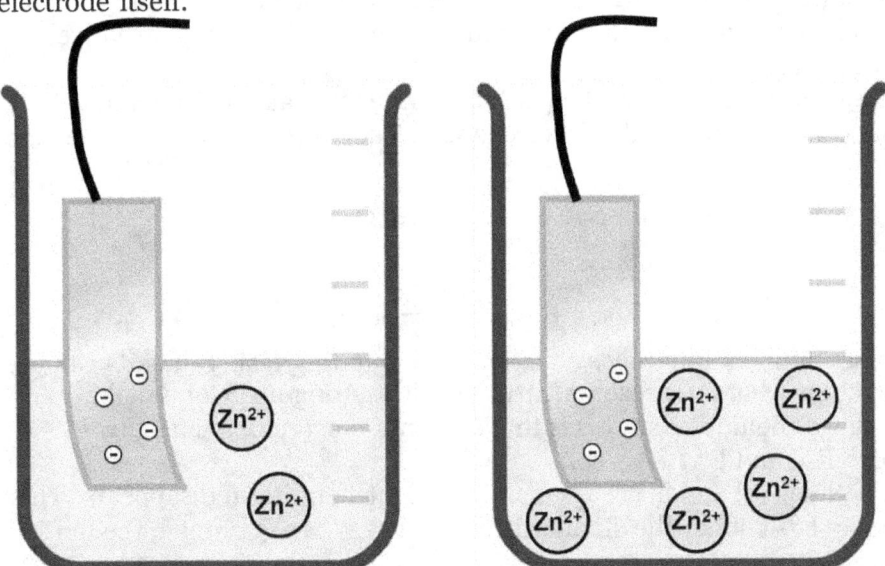

Figure 16-6: As the concentration of cations changes, the pull on electrons in the metal changes as well. This results in changes to the potential difference when a circuit is completed.

When we decrease the frequency of collisions of cations with the surface of the electrode, we decrease the pull on electrons from that side. But the electrons could be moving away or towards that side. If we decrease the concentration of Cu^{2+} ions, now we have less pull on electrons from the cathode. This means that the electrons moving towards the cathode experience less force from the Cu^{2+} ions, but still experience the same force from the Zn^{2+} ions that are colliding with the anode.

If we decrease the concentration of Zn^{2+} ions instead, now the electrons experience less pull from the anode that they are leaving (Fig. 16-6). The electrons experience a greater net force as they move towards the cathode. Because of this greater force, the electrons gain more energy and hence the voltage increases.

Take a note of how concrete these particle representations are. They are very simple to understand. But contrast that with what is usually given as an explanation. Instead of explaining voltage changes in terms of positive charges attracting electrons, we use the reaction quotient (Q) and Le Chatelier's Principle.

If the concentration of Cu^{2+} ions decreases to 0.10 M, the Q value goes from 1 to 10. This causes the shift towards more reactants to get back to standard conditions and thus the voltage decreases.

It pains me to type that last explanation. It is a cheap explanation that avoids understanding. We're using Le Chatelier's Principle without ever working toward equilibrium.[5] We're comparing a non-equilibrium state for the final conditions of 1M solutions with a non-equilibrium state for the initial conditions of 0.10 M Cu^{2+} ions. Usually this is presented without a particle representation which makes the leap from

Chapter 16 – Entropy and Spontaneity

changing forward and reverse rates to a change in voltage very difficult for students to formulate.

The entire basis for the explanation is that stresses have thus far correlated with equilibrium shifts. That's a dangerous assumption to base understanding off of. It will work to produce an answer, but we should at minimum have an alternative working model as the teacher.

We do however need the Nernst equation in order to make mathematical predictions about concentration and voltage changes. There are two forms to the Nernst equation.

$$\text{Eq. 16-3 } E_{cell} = E°_{cell} - \frac{RT}{nF} \ln Q$$

$$\text{Eq. 16-4 } E_{cell} = E°_{cell} - \frac{0.0592}{n} \log Q$$

Just to plug numbers into these equations is a daunting task. Students can use the wrong R, not know what n represents, struggle with setting up Q, and make mistakes with signage. Behind all of this is that students must understand the difference between E_{cell} and $E°_{cell}$.

If Q is 1, both log (1) and ln (1) are 0 and the right-hand term disappears. This leaves us with $E_{cell} = E°_{cell}$ under the standard conditions. When Q is not 1, the fun begins.

The n component is important as it balances out the possible variations of Q based on the communication of the reaction. If we look at a reduction ½ reaction for $Cu^{2+} + 2e^- \rightarrow Cu$ we can define Q as $1/[Cu^{2+}]$, but we could also write the reaction as $2Cu^{2+} + 4e^- \rightarrow 2Cu$ and our Q value will change to $1/[Cu^{2+}]^2$. The n value also changes also though, keeping the values constant. If we assume the $[Cu^{2+}] = 0.0030$ M, we can plug in to find the voltage to be:

$$E_{cell} = +0.34 \text{ V} - \frac{0.0592}{2} * \log\left(\frac{1}{0.0030}\right) = 0.265 \text{ V} = 0.27 \text{ V}$$

If we write the reaction as $2Cu^{2+} + 4e^- \rightarrow 2Cu$ the plugged-in equation would be:

$$E_{cell} = +0.34 \text{ V} - \frac{0.0592}{4} * \log\frac{1}{0.0030^2} = 0.265 \text{ V} = 0.27 \text{ V}$$

The Nernst equation can also be used for the entire redox reaction, or the two half-cells can be calculated separately.

$Cu^{2+} + Zn \rightarrow Cu + Zn^{2+}$	$Q = \frac{[Zn^{2+}]}{[Cu^{2+}]}$	$E°_{cell} = +1.10$ V
$Cu^{2+} + 2e^- \rightarrow Cu$	$Q = \frac{1}{[Cu^{2+}]}$	$E°_{cell} = +0.34$ V
$Zn \rightarrow Zn^{2+} + 2e^-$	$Q = [Zn^{2+}]$	$E°_{cell} = +0.76$ V

If the $[Zn^{2+}]$ is changed from 1 M to 0.044 M and the $[Cu^{2+}]$ is changed from 1 M to 2.32 M the voltage will increase.

$$E_{cell} = +1.10 \text{ V} - \frac{0.0592}{2} * \log\left(\frac{0.044}{5.32}\right) = +1.161 \text{ V} = +1.16 \text{ V}$$

Chapter 16 – Entropy and Spontaneity

We can also calculate each half-cell individually:

$$E_{anode} = +0.76 \text{ V} - \frac{0.0592}{2} * \log(0.044) = 0.800 \text{ V}$$

$$E_{cathode} = +0.34 \text{ V} - \frac{0.0592}{2} * \log\left(\frac{1}{5.32}\right) = 0.361 \text{ V}$$

What is happening is that the [Cu^{2+}] increase creates a larger pull towards the cathode and the [Zn^{2+}] decrease leads to a weaker pull as the electrons move away from the anode. Both contribute to the increase in positive voltage.

We can conclude that there are relationships between $E°_{cell}$, K, and $\Delta G°$. These terms can be related mathematically or conceptually as a means to predict how a system will change in order to achieve equilibrium. These three terms come with three similar variables (E_{cell}, Q, and ΔG) that are used to describe a more specific set of conditions for a chemical system. The three terms provide a compelling case to teach thermodynamics last to connect all three ideas.

Applying these in a non-mathematical setting can highlight the challenges that students who engage critically with these 6 terms can expect to experience. Here we will look at one of each of the three general reaction types with regards to each term.

Example #1 Redox
  $Cu^{2+} + Sn \rightarrow Cu + Sn^{2+}$
  $E°_{cell} = +0.48$ V
  $K = 1.7 \times 10^{16}$
  $\Delta G° = -92.6$ kJ/mol

The $E°_{cell}$ being positive tells us that the Cu^{2+} ions pull on electrons more than the Sn^{2+} ions. The numerical value of +0.48 V gives us some idea of how much more the Cu^{2+} ions pull. The equilibrium constant is extremely large which informs us that at equilibrium there will be nearly no reactants. The Gibbs free energy being negative informs us that the reaction will proceed spontaneously without any input of work. The numerical value of -92.6 kJ/mol tells us how much work could be extracted in theory from the chemical system under standard conditions.

If we were working with a chemical system that was not under standard conditions, we could use E_{cell}, Q, and ΔG to evaluate what needs to happen in order to reach equilibrium. If ΔG is negative, the reaction will continue in the forward direction to reach equilibrium. If Q is smaller than K, the reaction will continue in the forward direction to reach equilibrium. If E_{cell} is positive, the reaction will continue in the forward direction to reach equilibrium. The numerical values give us a sense of progress towards equilibrium.

The standard values give us a comparison of the products and reactants. The non-standard values give us a roadmap to what will proceed within a specific chemical mixture.

If we were to reverse the reaction to $Cu + Sn^{2+} \rightarrow Cu^{2+} + Sn$ the values would change to:
  $E°_{cell} = -0.48$ V
  $K = 5.9 \times 10^{-17}$
  $\Delta G° = +92.6$ kJ/mol

Chapter 16 – Entropy and Spontaneity

If we start with copper metal and tin (II) ions the reaction would not proceed without the addition of external work. The most likely source of that work would be a voltage source that could push the electrons away from the copper and towards the tin (II) ions. If that were done, the electrons would return to the copper if they remain connected after the external voltage source were removed.

Example #2 Acid-base
$$HC_2H_3O_2 + H_2O \rightleftharpoons C_2H_3O_2^- + H_3O^+$$
$K_a = 1.8 \times 10^{-5}$
$\Delta G° = +27\ kJ/mol$

In this example we cannot calculate the voltage because we do not have a transfer of electrons taking place. However, there is a charged particle moving from one chemical species to another. We can qualitatively describe this transfer of a proton in similar terms to how we would describe a redox reaction. The proton has a greater attraction to the acetate ion than the water molecule.

The voltage for the reaction would be negative, but we cannot separate this reaction into half-reactions like we can for redox reactions. This is because the charged particle is moving between two species that are mixed. In a battery, we can separate the anode and cathode by a wire that the electron can move within. But in this reaction, there would not be a way to put the acetic acid in one location, the water in another, and allow the proton to move between the two so that a voltage could be measured.

If we assume (just for fun) that the charge of the proton can be substituted in as the value n = 1, we get $E°_{cell} = -0.28$ V.

The Gibbs free energy tells us that in order for the reaction to happen work must be put into the system. From a student's perspective this is a little confusing since we know the reaction goes from all reactants to equilibrium very quickly. Remember there is what we observe to happen (the reaction proceeds to a dynamic equilibrium) and our attempts to describe what is happening (the Gibbs free energy change is positive). Our model has some limitations that make this an imperfect but not useless match.

The equilibrium constant tells us that when we reach equilibrium, we expect to have more reactants than products.

Note that if we were to write this reaction backwards
$C_2H_3O_2^- + H_3O^+ \rightleftharpoons HC_2H_3O_2 + H_2O$ our values would change even though nothing about the chemical reaction changes. Our equilibrium constant would be the reciprocal (K = 5.56×10^4), and $\Delta G°$ becomes -27 kJ/mol. If we were to mix acetate with strong acid, the reaction would proceed.

Example #3 Precipitation
$$BaSO_4\ (s) \rightleftharpoons Ba^{2+}\ (aq) + SO_4^{2-}\ (aq)$$
$K_{sp} = 1.10 \times 10^{-10}$
$\Delta G° = +56.8\ kJ/mol$

Chapter 16 – Entropy and Spontaneity

Again, we find a very small equilibrium constant. This means at equilibrium we will find very little product. The reactants in this case do not factor into the equilibrium amounts since there is only solids that have a nearly constant activity.

The Gibbs free energy tells us that work is required in order to get the reaction to proceed. This can be confusing with the reversible reaction that is reactant favored. The voltage becomes even more confusing and hidden. Now we have two charged particles moving away from each other in a solid crystal to become surrounded by water molecules in solution. But we can see that the driving force behind these changes is still electrostatic attractions.

If the $\Delta G° = -nFE°$ equation were to function and we arbitrarily choose n = 1 we would get $E°_{cell}$ = -0.589 V. There is a greater electrostatic force of attraction between the barium and sulfate ions in the solid crystal than there is with those ions dissolved in water. Work is required to produce the change and that work can happen infrequently through collisions with the water.

When we reverse the reaction to be Ba^{2+} (aq) + SO_4^{2-} (aq) ⇌ $BaSO_4$ (s) we are now looking from the perspective where the dissolved ions are coming together to form the solid crystal. Now the product is more favorable, the equilibrium constant is large, and the $\Delta G°$ is negative.

Many teachers operate under the premise that kinetics and thermodynamics function separately from each other. The reality is that thermodynamics, bonding, and kinetics are all intertwined heavily, but they can be analyzed separately from each other.

A mixture of hydrogen and oxygen gas does not burn unless a spark is added. The bonds of the oxygen and hydrogen molecules are too strong to be broken on a consistent basis at room temperature. The bond strength is not only relevant to kinetics, but also has obvious implications for the thermodynamics.

The conversion of diamond into graphite is spontaneous. The bond strength for both species is enormous. So even though the process is slightly favorable, it's not going to happen unless you have extreme conditions. Even if you have a single bond within a diamond structure break from during a collision, the carbon is unlikely to move far enough away that it won't just settle back into place with the rest of the structure.

You could use that as a challenge to students. Create a particle representation that shows why it is difficult to change diamond into graphite. Then explain the difficulties in terms of $\Delta H°$, $\Delta S°$, $\Delta G°$, activation energy, and bond strength. What conditions would it take to establish an equilibrium between the two allotropes?

If you do manage to teach thermodynamics last, you can review a lot of earlier topics with it. This is a great opportunity for some spaced practice, retrieval practice, and tying up earlier topics that didn't go well. If I burn methane, what do we know about $\Delta H°$, $\Delta S°$ and $\Delta G°$? Why do I need to ignite the mixture? How is burning a redox reaction and how does this differ from a battery? Could I make a battery using methane?

How do $\Delta H°$, $\Delta S°$, and $\Delta G°$ factor into S_N1 vs. S_N2 mechanisms for a secondary leaving group? If S_N1 and S_N2 mechanisms have the same initial and final chemicals, what factors play a role in determining which will happen? What would a reaction energy diagram look like?

Chapter 16 – Entropy and Spontaneity

If dilute hydrochloric acid is poured on an impure sample of limestone, some bubbles start to form, and the mixture increases in temperature. What does this tell us about ΔH°, ΔS°, and ΔG°? What would you need to measure to determine how much limestone is in the sample?

In this book I have split thermodynamics into three different chapters. The three big ideas can be thought of as prerequisite information, enthalpy, and entropy/spontaneity. Often when students fall apart in one of the latter two sections teachers assume it is a mathematics issue. This is almost always incorrect. The issues are that students don't have a strong sense of what the purpose and limitations are for enthalpy, entropy, and Gibbs free energy. The other issue is that students are likely to carry weak understandings of the prerequisite information.

Heat, temperature, specific heat capacity, systems and surroundings are all rife with misconceptions. Enthalpy can be used to expose those, but often some will slip through the cracks. But when a student struggles with the more complicated terms, it helps to have them identify what they see and think of first so that the teacher can understand their perspective.

Thermodynamics has many abstract ideas. When students struggle, the best path to success is using lots of concrete details. At the conclusion of the unit have students put together arguments and evidence that they would use if they were to somehow travel back in time 200 years and had to convince scientists of concepts such as entropy or Gibbs free energy. What examples or observations could you show them? What counterexamples might they use to disagree? Which chemical reactions would be challenging to fit into their schema?

Student Struggles

1. "I have no idea what I'm doing!" The concepts in thermodynamics are abstract. Often students will feel unconfident in the material even when they can successfully make predictions and explain why. In this unit the ability to connect to the particle level is more challenging than many others. In lieu of particle diagrams, we want to show lots of concrete examples of reactions to help students see commonality behind these ideas. Don't have students stress over technical definitions, but instead look at the usefulness of things like entropy and enthalpy. The major purpose of entropy and Gibbs free energy is to make predictions about what will and what will not happen. Can the student apply ΔS and ΔG to specific phenomena?

2. "I just don't understand what entropy is." When we get to abstract ideas in science, we often construct an abstract phrase to match the term. These phrases add little to understanding. What we want to emphasize with entropy is how entropy varies as changes occur. How does entropy change when something melts, or temperature decreases, or volume of a gas increases? What types of molecules have large standard entropies relative to others in a table of thermodynamic values? Recall that Gibbs normalized using entropy by providing many concrete examples. Clausius on the other hand remained obscure by using mathematical proofs as his means of explanation. Students who try to understand entropy without the context of a concrete example find themselves stuck. Guide them to always using an example.

Chapter 16 – Entropy and Spontaneity

3. *"Is free energy a type of energy?"* There are two uses for free energy. By tracking how free energy changes we can make predictions about what we expect a chemical system to do. If free energy were to decrease, we can predict that the reaction is thermodynamically favorable and will happen if it is not limited by reaction rate. The value of free energy tells us the maximum amount of energy we can obtain from the change that we can use for work. We don't often use this second piece in introductory chemistry. Free energy is energy. The units are kJ/mol where the per mole part obviously scales up how much energy can be extracted for work as we do more of the chemical reaction.

4. *"Why does the temperature matter for being thermodynamically favorable?"* In thermodynamics we talk a lot about energy. The two big energetic changes that we're tracking are how much kinetic energy the particles have and how much potential energy changes as bonds break and form. Temperature changes how fast the particles move and collide, which we should expect to implicate whether a reaction will occur. The best examples of this are in phase changes. When we increase the temperature until we reach a phase change, we then see the particle motion becoming so fast (or slow) that intermolecular attractions are no longer able to accelerate the particles sufficiently for them to remain in their positions. The surroundings also play a role here. Recall that adding +Q to surroundings at 200 K has differing impact than 373 K.

5. *"I'm struggling with units when changing between K, $\Delta G°$, and $E°$."* The RT component has the units of J/mol. The equilibrium constant has no units. $\Delta G°$ is reported in kJ/mol. Changing units from K to $\Delta G°$ or vice versa is relatively straightforward. We just need to watch for conversions between kJ and J. The changing between $\Delta G°$ and $E°$ is a little more difficult. Here the units of the Faraday constant are Coulombs (C) per 1 mol of electrons. But a C is also 1 J/V. When we multiply $E°$ by F, we end up with J/mol. This means that all the equations easily translate units as long as $\Delta G°$ is reported in J/mol.

6. *"How do they come up with the standard entropy values for the data tables?"* The standard entropy values can be calculated once we assume that the entropy at absolute zero is 0 (this isn't always the case, but most values are very close to 0). From this starting point the heat capacity is used. The integral of C/T vs. T results in the standard entropy value (if T is evaluated from 0 to 298 K). This shows why larger molecules tend to have larger entropy values.

7. *"I heard that there's a tiny probability that my clothes will fold themselves in the dryer. Is that true?"* No. No it's not. There is a tiny probability if you assume all states are equivalent, but this assumes we are ignorant of our initial conditions. The way that a dryer rotates clothes does not lead to folding and it's not going to happen. You could cheat by setting the initial conditions to have folded clothes that are glued together or something similar. But the ignoring of the mechanism proves faulty without intervention.

Chapter 16 – Entropy and Spontaneity

8. "I can plug into the equations, but I'm not sure what the bigger point of this chapter is." Physical chemistry is predicated upon searching for useful relationships between variables. Here we have defined two new technical terms (Gibbs free energy and entropy) and we have found these two terms to be useful in predicting what will happen for chemical reactions. Because the two terms are abstract, it feels to students like they aren't taking anything useful away from the analysis. Momentum is a similar concept. Momentum is a relationship between mass and velocity that allows us to make predictions of what objects will do after a collision has taken place. It even allows us to do so without knowing all the details within the collision itself. This type of processing allows us to work on defining systems and surroundings, being explicit with our assumptions, and recognizing the limitations of our model. In our analysis here, kinetics adds a component that we cannot address easily using free energy and entropy.

9. "Why don't the voltages add up for the following reactions?"
1. $Cr^{3+} + e^- \rightarrow Cr^{2+}$ ($E° = -0.42$ V)
2. $Cr^{2+} + 2e^- \rightarrow Cr$ ($E° = -0.90$ V)
3. $Cr^{3+} + 3e^- \rightarrow Cr$ ($E° = -0.74$ V)

Gibbs free energy changes are additive. The $\Delta G°$ values for reactions 1 and 2 add up to reaction 3. When we look at the equation to change $\Delta G°$ to $E°$ we see that each value must be adjusted by the number of electrons in the half-reaction (and the Faraday constant). Since each reaction has a different number of electrons involved in the reduction, the additive nature of $\Delta G°$ will not carry over to $E°$.

10. "Will a catalyst impact thermodynamics?" What happens in the middle of the reaction does not have any bearing on thermodynamics because we define thermodynamic quantities to compare the initial and final states only. Adding a catalyst does not impact the reactants nor the products. In a bonding frame, thermodynamics describes the bonds at the beginning and the end. A catalyst changes how those bonds rearrange, but not the initial or final arrangement.

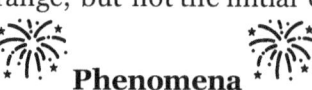
Phenomena

1. Rubber ducky - When you squeeze a rubber ducky it fills quickly with air, but slowly with water. What's the difference? Water molecules move at higher speeds than nitrogen and oxygen molecules at room temperature!

2. Supersaturated sodium acetate - A large quantity of sodium acetate can be dissolved in a small amount of water when heated. If the solution cools slowly, the sodium acetate will stay dissolved below the temperature where it should crystallize. When a seed crystal is added, the sodium acetate will rapidly crystallize. The flask will feel warm as the crystallization occurs. Ask the students what the signs will be for ΔG, ΔH, and ΔS.

3. Cobalt complexes - These were previously suggested for Le Chatelier's principle in the equilibrium unit. Here we want to use these to distinguish between ΔG and $\Delta G°$. $\Delta G°$ compares the stability of the two complexes at standard conditions. If you prepare a solution where all aqueous components start at 1.0 M, $\Delta G°$ will tell you how the reaction will proceed. ΔG on the other hand tells how the reaction will change with other starting quantities. By altering the concentrations of the initial amounts,

Chapter 16 – Entropy and Spontaneity

we are then using ΔG to predict what will happen by comparing ΔG (current conditions) to $\Delta G°$ (equilibrium conditions).

4. Zinc reacting with HCl - As zinc metal reacts with hydrochloric acid hydrogen gas is evolved and the flask will warm considerably. Ask students to determine what the signs will be for ΔG, ΔH, and ΔS.

5. Sodium bicarbonate decomposition - Sodium bicarbonate will decompose to make sodium carbonate, steam, and carbon dioxide gas when the solid is heated strongly. The carbon dioxide can be confirmed using a splint test. Ask students to determine what the signs will be for ΔG, ΔH, and ΔS.

6. Boiling water - Start by asking students to determine what the signs will be for ΔG, ΔH, and ΔS. Then have students use thermodynamic tables to calculate what $\Delta H°$ and $\Delta S°$ are. When an equilibrium between liquid and gaseous water exists, ΔG will be 0. Have them solve for the Kelvin temperature must be to have ΔG be 0 using the values for $\Delta H°$ and $\Delta S°$. They will find that the temperature is 370 K or 97° C. This is very close to the boiling point of water. Why don't they match more precisely?

7. Vinegar and baking soda baggie - Students may have done this in the stoichiometry extension chapter earlier. This reaction was confusing as it was endothermic yet spontaneous. Chemists had observed most spontaneous reactions to be exothermic. Ask students to explain how the entropy of the system and surroundings are changing. Then have them compile the entropy changes to explain how this endothermic reaction can be spontaneous. At what temperature would this reaction not be spontaneous (assuming no phase changes)?

8. Plating gold and silver pennies - Pennies can be heated in a zinc hydroxide or zinc chloride solution to plate a thin layer of zinc on the penny. This will make the penny look silver. The penny can then be heated causing the zinc and copper to mix forming brass that looks like gold. This can also be done using electrolysis. If the penny is connected to the negative terminal of a 9V battery (positive terminal should connect to a piece of Zn metal in the same solution), the zinc will electroplate onto the penny. This is how some currency is made since it can limit the amount of expensive metal used to be on the surface only. Ask the students what the sign of $\Delta G°$ is since the external battery is required. How will altering the concentration impact the thermodynamic values?

9. Mg burning - This demonstration has a wide array of applications. It makes sense to show this multiple times over the course of the year so that students can enhance their knowledge of the reaction as they learn and develop stronger chemistry models. Here the reaction decreases the amount of gas so entropy lowers. The reaction is highly exothermic. Under what conditions will the reaction be spontaneous? What might happen if we heat the product strongly?

10. Zn + CuCl$_2$, Cu + ZnCl$_2$ - Add a strip of zinc to a copper (II) solution. The reaction occurs spontaneously. Add a strip of copper to a zinc ion solution. The reaction does not occur. What are the signs of $\Delta G°$ for both reactions? What does this imply about the impact of reversing a reaction on the thermodynamic signs? How would we describe the values of E° and K for both reactions?

Chapter 16 – Entropy and Spontaneity

Flashcards

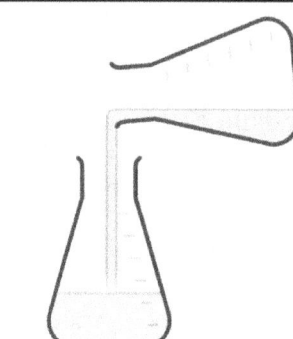

As the solutions mix: a precipitate forms and the temperature rises

What are the signs for
ΔS
ΔH
ΔG

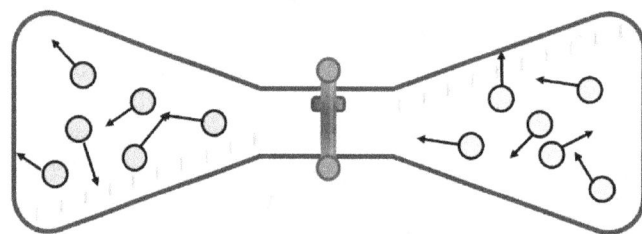

The top container mixes faster, is there a larger change in entropy for the top container?

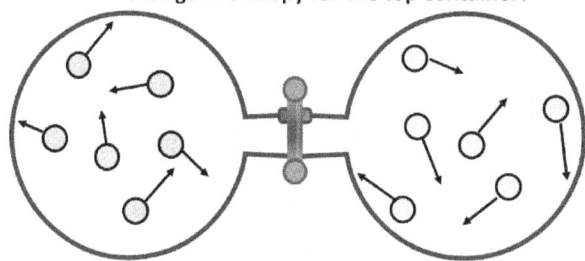

$$\Delta S = +$$
$$\Delta H = +$$

When will ΔG be negative?

A. At all temperatures
B. At no temperatures
C. Only at low temperatures
D. Only at high temperatures

What will ΔS be for the surroundings?

1. Positive
2. Negative
3. Only at low temperatures
4. Only at high temperatures

Chapter 16 – Entropy and Spontaneity

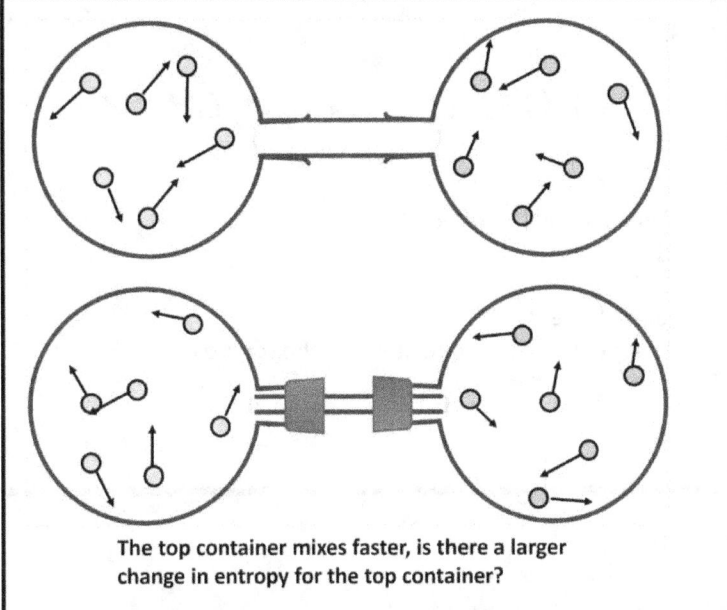

The top container mixes faster, is there a larger change in entropy for the top container?

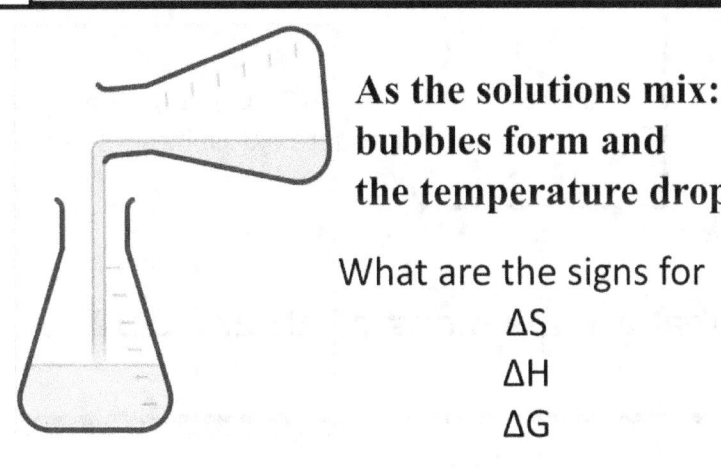

As the solutions mix: bubbles form and the temperature drops

What are the signs for
ΔS
ΔH
ΔG

Which changes will have an increase in entropy?

A. $2NO_2\ (g) \rightarrow N_2O_4\ (g)$
B. $NaCl\ (aq) \rightarrow NaCl\ (s)$
C. $Zn\ (s) + 2HCl\ (aq) \rightarrow ZnCl_2\ (aq) + H_2\ (g)$
D. $H_2O\ (g) \rightarrow H_2O\ (l)$
E. $NaHCO_3\ (s) + HC_2H_3O_2\ (aq) \rightarrow NaC_2H_3O_2\ (aq) + H_2O\ (l) + CO_2\ (g)$

Chapter 16 – Entropy and Spontaneity

AB + C → BC + A   ΔH = +

Which bond is stronger?
1. A-B
2. B-C
3. Both are the same
4. It depends on the relative number of moles

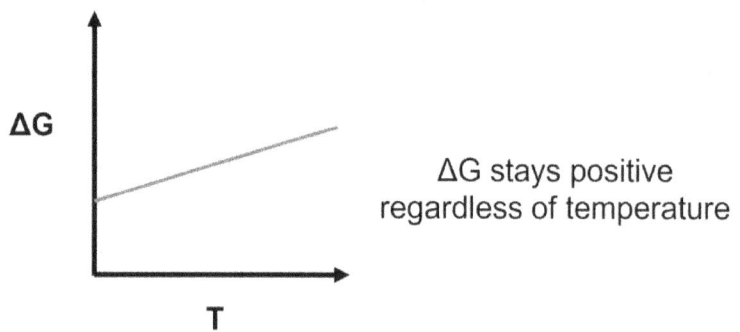

ΔG stays positive regardless of temperature

What are the signs of ΔH and ΔS?

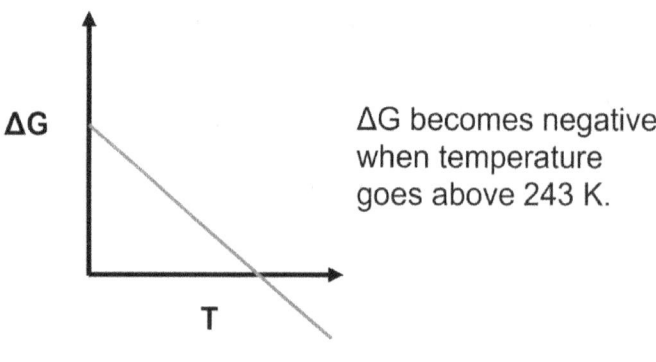

ΔG becomes negative when temperature goes above 243 K.

What are the signs of ΔH and ΔS?

Chapter 16 – Entropy and Spontaneity

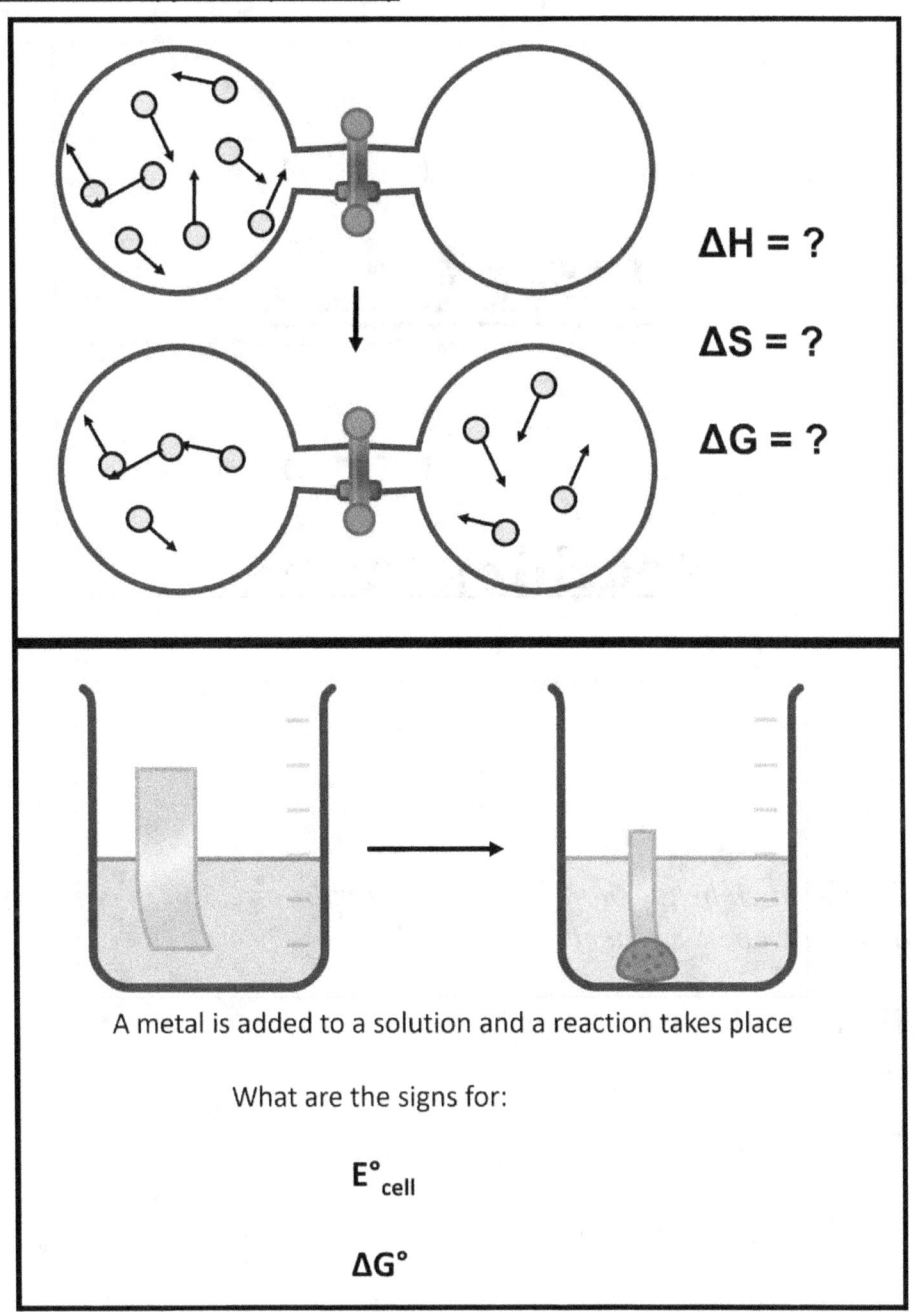

$\Delta H = ?$

$\Delta S = ?$

$\Delta G = ?$

A metal is added to a solution and a reaction takes place

What are the signs for:

$E°_{cell}$

$\Delta G°$

Part III

Teaching Primer

"You do this much better than I was able at your age, but of course you also had a much better teacher than I did"

- Hermann Kopp to August Kekulé

Chapter
Teaching

Teaching might be the most complex profession. Teachers must be proficient in chemistry, apply the cognitive science of memory, and engage attention of adolescents. Teachers must facilitate the introduction of new concepts, support time for learning, review forgotten material, and assess the efficacy of the learning that transpired. On top of these are endless logistical challenges such as absences, late work, range of prerequisite information, classroom culture, administrative mandates, and school culture. I remain wary of anyone who professes to have a secret ingredient that makes all of teaching simple.

Instead, we should expect and embrace the challenge of teaching chemistry. It is always possible to improve upon our work. It is critical to find joy in our current state. Success in teaching chemistry is an incredibly rewarding accolade. In this final section I am going to disaggregate components of quality teaching. I do not do this to be reductive, but the reality is that strong content knowledge and instruction can still fall apart with poor assessment. All the components of teaching must be considered, reflected upon, and improved whenever possible. If you find a topic you would like to see more depth, please make use of the notes and citations.

Grading

Too often teachers receive little time to process grading policies. Grading assessments is a long and arduous task. By the time we complete a set of assessments our brains are ready for a break. Prior to that exhaustion, we should consider the accuracy of our assessments, as well as the logistics of how we assign final grades from those assessments.

This means that we must start with philosophy. What will we value in assigning grades, and how much will we value each component? Homework, classwork, participation, tests, quizzes, lab reports, and projects are all common components of grades. But should they be? A grade in chemistry class should reflect the student's understanding of chemistry.[1] Does completion of homework indicate student understanding of the content itself?

It has become a norm for teachers to use grades as incentives to exert control on student behaviors. We use grades to reward effort, participation, and compliance. This elicits a host of issues. When a student struggles in a class where the grade is a measure of learning, the feedback guides the student to learn more. But in a classroom where grades are incentives, the student receives feedback that guides them toward point grubbing.

Grading students based on effort sells the ability of the student short. They are all capable of learning chemistry. It will take more time and effort for some of them, but there is no concept in this book that any person is incapable of mastering. Grade them based on what they learn and give them some flexibility to take the time to figure it out.

If we consider any certification process, we would assume that competency is the primary focus of assessment. For a driving test, the student must demonstrate proficiency with driving to earn a passing grade. We would not provide additional points for completion of homework to elevate a student who failed the driving

Chapter 17 – Teaching

portion. We would not want a surgeon who participated in class but failed to achieve competency in surgery.

When we assign grades based on learning, we often incite fear among students, teachers, and parents. You should have a plan for this. But the plan should not undercut the ability for students to learn. When a student struggles, they must be able to work back from their struggles. But they shouldn't be able to do so without learning.

Grading is difficult and imperfect. But it is imperative that critical reflection on assessment is part of teaching chemistry. Using the best teaching methods is insufficient. If your assessment quality doesn't match, learning will ultimately suffer. Assessments should be somewhat predictable, but not too predictable. If your students can tell what the test questions will be from the review guide, then learning is limited. If your review the day before the test tells students exactly what the test is, you're excusing them from learning for the two weeks prior to the test. But a test also can't be erratic where a student does not know how to prepare.

We should not be quick to evaluate tests based on student perception. Students who study by reading notes and highlighting will find tests to seem more novel than they were. Recall that rereading causes a false sense of competency due to the familiarity with the notes or text. Students who study with retrieval practice will outperform their perceived results. Therefore, we always want to use data to inform our measure of success with a critical eye to the validity of that data.

Grading should reinforce good teaching. A critical problem in many schools is that the feedback teachers receive from grading is ineffective at helping teachers to improve. When people first learn to drive, they receive quality feedback about how much to steer the wheel to make a turn or how hard to hit the brakes. If you turn the wheel too much or too little, the outcome is readily apparent.

But feedback for teachers is tricky. Student opinions of lessons are not good feedback. Remember that learning is counterintuitive. When students learn a lot, they feel less confident. When they learn little and see information they recognize, they feel overconfident. A teacher's system 1 will naturally pick up on that confidence in their own personal evaluation of their teaching. Be skeptical that the proverbial "light bulb going on" is a valid indicator of quality learning.

Teachers that use grades to coerce students into selected behaviors or for assignment completion will receive feedback about management instead of learning. The best way to use feedback to improve teaching is to have assessments that inform how well the students have learned. Teachers must be wary of assessments that are too similar to review materials. Even when assessment uses retrieval, students can be retrieving abstract information and algorithms that they have minimal concrete understanding of. Sometimes students can be successful on an assessment because they recognize the phrasing from a lesson in spite of not understanding. A vocabulary matching section can be rife with these issues since students might not understand definition nor term but can associate the two easily.

As a teacher you should be able to demonstrate your effectiveness using your assessments. You should be able to use results to show what you taught and how well students learned the standards. Your assessments should be scrutinized not just on average score, but in how well they measure student learning of specific items. A common problem that disrupts this is to have too many learning objectives in a single

Chapter 17 – Teaching

question. If a question contains a conversion, a significant figure rounding, a calculation, and an explanation, the teacher can't easily decipher what went wrong when a student makes an error. This leads to grading that is more punitive than feedback oriented.

Our students frequently provide us with evidence that we are not grading them based on learning. When students seek a minimum requirement of achievements for an A or a B, they are telling us that their expectations for grades are not based on what they learn. Teachers should learn to push back against this expectation even if resistance is high. We should not be grading students on practice, completion, homework, discussion, attendance, nor behavior.

Instead, homework should be assigned and completed when it is necessary for the student to learn. The test should provide students with feedback about how their current studying and learning is progressing, so they know whether they need more practice, more studying, or greater engagement during class time.

To help students connect your grading with feedback for their learning, it is important to give them grades quickly. If you cannot return grades quickly, provide some feedback to the class within two days of the assessment whenever possible. One way to do this is to be efficient with how you give feedback. It is a waste of time to write individual comments for students on their tests or assessments. *They are unlikely to read them, unlikely to learn from them, and this takes an inordinate amount of your time.* Instead find a few key ideas, misconceptions, or great explanations. Share these with all your students. This requires less work so you can give feedback in a more timely manner while students can engage with feedback in a meaningful interaction. Too often, high quality student work exists in a vacuum. Show that work off so that others can see what high quality work and thought truly looks like. Students should learn more from success than from mistakes.[2]

Chemistry is my favorite subject. In physics the student can remain in abstract mode so much, that they can avoid learning the physics and yet still be successful. When I was in high school and college, I did very well in physics in spite of not always understanding physics well. But I could decipher which algorithm to use and find a solution without the concepts. Biology is so concrete. The systems are so complicated that to switch between abstract and concrete involves too many steps. But chemistry is the best subject to be able to link the abstract and the concrete. What does this look like macroscopically, at the particle level, and symbolically? Because chemistry is such a good learning opportunity it is imperative that we give students the ability to use chemistry class to master learning and improve metacognition.

Standards-Based Grading

I began using standards-based grading (SBG) in 2015 and I added reassessments in 2018. This has helped immensely with many of the issues from the previous section. Standards based grading helped my assessments change from punitive to informative. Previously when students did poorly on a test there was little recourse available to them. Students felt that they did poorly because they weren't smart enough. Very few students were able to improve even if they worked diligently.

The first thing that SBG does is that it organizes the feedback between you and your students. You are better able to communicate what they were supposed to learn. They are better able to reflect on what they need to improve.

Chapter 17 – Teaching

When a student made a mistake on a test prior to SBG, the feedback usually was that the question was difficult. It was difficult to ascertain what concepts the students struggled with because many questions had multiple concepts being assessed at once. SBG gives you the opportunity to organize your tests more effectively.

Prior to SBG my test data was mostly what the average score was and maybe a little bit of information about the extremes. Now I get a breakdown by standard. It becomes easier to adjust my tests for the future years. I get better feedback about what concepts need more time during lessons.

When students get their test scores, they have to look to see which standards they did the best and worst on. Instead of getting a 50, or a 70, or a 95, they get a series of scores that help them reflect on what they were best prepared for and where they fell short.

Each standard is assessed on a four-point scale. A 4 indicates quality understanding and communication of that understanding. A 3 indicates partial or good understanding with some minor issues. A 2 indicates emerging understanding. A 1 would indicate limited understanding and a need for intervention. These obviously vary in application depending on the specific concept being assessed.

One perk to the 4-point scale is that teachers are much more consistent. Evaluating understanding on a scale of 0-100 is erratic and hopeless. Because we are unable to differentiate between an 86 and an 87, we become comfortable with a wider range of variability between evaluation of student work.[3]

Allowing students to reassess on tests adds a completely new dimension. Reassessments will be more work (especially the first year), but you will get more student learning. I have found it worthwhile to do reassessments and cut other forms of work to maintain a reasonable workload. I no longer grade labs, homework, participation, or classwork.

When a student can reassess the test anxiety plummets. If they get to a test and are unsure how to do something, they know they will be able to give it another try. This helps with test performance, this helps with attendance, and it really changes grading from punitive to feedback.

Standards-based grading can help with students who only want to know what they need for an A and nothing else. What do I need to do to get an A? You need to learn chemistry and be able to communicate it effectively.

This makes it easy to eliminate bad grading policies. There is no need for extra credit. If you want a higher grade, learn more chemistry. Students may reassess if they require more opportunities to improve. There is no need to grade behavior or completion of assignments. Your grade reflects the understanding of chemistry.

In my experience, participation and completion of assignments is superior when these are not graded. There is no purpose to cheat on a homework assignment, so instead the student can use practice as a means to assess their progress and learning. This can lead to a healthy environment where the student is not constantly seeking out transactions for points.

There are multiple perks to reassessments. The students are now engaging in spaced practice. While they learn about thermochemistry, they are also studying to reassess on gas pressure. Each reassessment is a new opportunity for retrieval practice. Test anxiety can decrease. Reassessments are a policy that would align with a growth mindset.

Chapter 17 – Teaching

The teacher can now reinforce the importance of classroom content. We learned how to name chemicals because future chemistry content will depend on your ability to do so. If you did not learn it the first time, you must go back and try again.

The logistics of doing reassessments are challenging. You need a system to communicate student intent to do a reassessment, an organized system of answer keys, space for reassessments, student motivation to take reassessments, a system to record and update grades, and you need a set of reassessments that are aligned to standards but differ from the original assessment.

That is a lot of work and any way you can make that work is progress towards better grading. No matter what you do, there will still be people upset with you not taking one more additional step. This is why I recommend cutting out other forms of grading if you move to reassessments. It's more effective to grade SBG with reassessments than to grade tests and every worksheet, homework and discussion that students do. Make sure you remove some things from your plate to make room for improved assessment strategies.

If there are multiple teachers, then some of this work can be split up. You can ask administrators for help. A testing site where students can take reassessments would save you time from proctoring.

Teachers may find a disconnect between who they think should be motivated to reassess and who is. I strongly encourage you to look past this. If a student has a B and reassesses to improve to an A that's a good thing. If a student has a 97% and they reassess to get a 99% that is a good thing. This is not just for struggling students. It is about giving students multiple opportunities to learn and to demonstrate that learning. It is also about changing the classroom culture to center learning. That student that bumped up to a 99% may see a big payoff from that additional understanding in a future chemistry class. They might have also been able to push their metacognition when they don't get that opportunity often.

If students are not motivated, you might consider doing a set of reassessments during class to get them that initial success that breeds future motivation. You might also consider access and if your after school reassessments are unavailable. Knowing that a second chance is available can help reduce student anxiety during the initial assessment even if that student does not reassess.

Part of reassessing is communicating that it's more important that you learn, as opposed to having to learn on a constricted timeframe. A practical application that I have found extremely successful is having quizzes aligned to the same standards as the test. This allows me to replace a quiz score with an improved test score automatically. A student receives a quiz score of 2/4 on unit 3 standard 2. When they test, they receive a 3 out of 4 on the same standard. I would now set both scores to be a 3. If they reassess and receive a 4, both scores would update to a 4. It doesn't matter to me that they had a lower understanding earlier, it only matters that they learned it.

In the past I would have students skip class to get more study time for a 4-point quiz. Now I'm incentivizing attendance and minimizing anxiety for students. I'm also sending a healthy message via this grading policy.

Chapter 17 – Teaching

Modeling Instruction

I was trained in modeling instruction in 2015. It has been a career altering professional development for myself and many other teacher colleagues that I have great respect for. I am open about being biased in favor of modeling instruction.

Modeling instruction is a constructivist approach to teaching that has students analyze their mental models of a phenomenon. Through discussion, observation, and refinement these mental models are developed into scientific models by developing a consensus around what was observed and how we communicate the ideas. The scientific models then serve to explain data and observations as well as to make predictions that can be tested.

It is challenging to define what precisely the model is. Models are not limited to physical representations, but rather can be thought of as a complex idea related to scientific content. A model for density might include graphical interpretation, particle representations, mathematical predictions, and proportional reasoning. *None of these individually serve as the model*. Rather it is the sum of them in concert. The particle model of matter can be enhanced by including charges, can indicate motion, can have different elemental particles, differentiate isotopes, and include quantized mechanics for electrons to explain interactions with light. Or the particle model can include just some of those traits depending on the level of complexity required.

A scientific model should be based on evidence. The model should explain phenomena. And the model should allow for predictions that can be tested. Leading students to develop models from evidence helps to undermine their misconceptions. In direct instruction, a student can easily create an exception for a discrepancy between a misconception and the instruction. With modeling, the student is forced to reconcile using evidence. By listening to peers explain their mental models, the student is more likely to find conflict with their misconceptions in prior knowledge.

The sense-making that students do while analyzing their mental models provides students the means to experience cognitive memory development. To store something in memory, the student must sense something (sensory memory), place the experience into short-term memory (STM), then connect the experience with existing prior knowledge in long-term memory (LTM). During discussion the students are frequently grappling with ideas in their STM by connecting to prior knowledge in their LTM. This creates opportunities for students to construct long-lasting memory about conceptual information without resorting to rote memorization where ideas remain abstract and are easily forgotten.

The typical sequence for modeling begins with an experiment or phenomenon that students perform. The learning objectives for the experiment are not clear to students as they do the experiment. They then construct a representation on a whiteboard.

The whiteboards are a source of initial student models that are discussed in a circle. The teacher facilitates students making observations about what is on their whiteboards as well as what they saw in the experiment. The teacher pushes back on students using either elaborative interrogation or talk moves[4] where they question students using techniques designed to increase the length and authenticity of the student response. This helps students identify what they know and what they do not know yet.

Chapter 17 – Teaching

Following an initial discussion, students will be given a task such as a worksheet. They must use the current model they developed in the discussion to find solutions. The solutions are often displayed on the whiteboards and a second circle discussion is used to assess the learning and processing that took place.

This cycle is called modeling because the students are developing, enhancing, and deploying their mental models. The opposite of modeling would be to start with the final explanation or algorithm in mind. This would be presented to students, and they try to acclimate to the teacher's mental model. This can be problematic if the student's initial model is too far removed (which is common). In that case the student will either experience cognitive overload, or they will work towards an understanding that is abstract and separate from their experiences and understanding. A student may assign two groups of truths. Here is real life, and here is what I say in an academic setting. Physics sets this up when we stipulate unusual requirements such as zero friction, zero air resistance, or massless strings with point particles. Too much abstraction breeds misconceptions that can cause frustration and fixed mindsets.

Another issue with starting with a final algorithm or "answer," is that there are assumptions behind the accuracy of the final answer. Teachers tend to limit their presentation of these assumptions and students tend to miss many of them entirely. Oxygen is a diatomic molecule. This assumes somewhat standard temperature conditions, that we're not talking about ozone, and that the oxygen is elemental. These assumptions are obvious to teachers, especially an experienced one. But they are not obvious to a novice learner.

In the modeling cycle, the students explain their assumptions behind their mental models. When a set of assumptions fails to agree with evidence or previously established models, we can revise. This process allows students to better understand the limitations of the content they are learning.

When teachers are able to help students express their ideas, the teacher is no longer limited by the curse of knowledge. They are better able to perceive how a novice learner would see chemistry. This helps them guide the novice learner into the appropriate methodology and thinking. Teachers receive more feedback from students through formative assessment, and this helps teachers to develop elite ability to question student thinking.

Chemistry modeling instruction follows the same sequence of history. It turns out that chemists had similar stumbling blocks to understanding chemistry that high school students have. In our sequence, the particle model of matter is developed by gradually increasing the complexity of the particle. At first the particles are spheres (density). Then motion and energy are added (pressure, basic thermochemistry). Then different types of particles are proposed (elements + compounds, relative mass + moles). Then charge is added via electrons and the plum-pudding model (naming, reaction types, stoichiometry). Historically these topics are explored simultaneously where composition is deduced via reaction stoichiometry, and types of reactions are classified based on composition of reactants and products. Finally, the structure of the atom is explored with the Rutherford model, the Bohr model, and the quantum mechanical model (atomic structure/periodicity).

There are many cognitive science tools we can apply in education. Modeling instruction offers teachers the opportunity to use them all in an effective manner. Teachers can use retrieval practice. They can demonstrate spaced practice by cycling through material and using agreed upon models to test new models. Teachers can

Chapter 17 – Teaching

minimize cognitive load by focusing on a narrow, yet complicated line of questioning. The art of balancing simple (avoiding cognitive overload) with complex (making sure the material reaches long-term memory) is a staple of modeling instruction. We can observe patterns in numerical data while minimizing vocabulary. We can construct particle diagrams while ignoring algorithms and vocabulary. Finally, we can develop vocabulary that chunks these meaningful ideas into a powerful retrieval cue.

In modeling instruction, the teacher is frequently utilizing multiple representations to assess the students' understanding. This provides multiple data points for the teacher to determine what might be going awry while helping students receive feedback and guidance on how to narrow their focus within an area of struggle.

I remain impressed with the level of detailed analysis that I see from students in the modeling approach. The connections they make, the strong memory formation, the analysis of complex systems, and the brilliant questions they produce are beyond anything else I have seen from high school students in my many years of teaching and connecting with other amazing chemistry teachers.

Modeling instruction has a strongly recommended 3-week training. These trainings are what teacher professional development (PD) should look like. They challenge teachers to think critically about their content and their pedagogy choices. Teachers learn about assessment, leading discussions, mental models, and enhance their content knowledge at the same time. A lot of teacher professional development is too quick with too little effort from the teacher. Even when we enjoy a PD session, it is unlikely to lead to a considerable change in teaching. Modeling workshops are a powerful teacher development tool that consistently receive rave reviews from participants.

Discussions

The goal of student discussion is not to get students to state correct answers. The teacher should aim to get students to express their thoughts authentically with sufficient length to get a true sense of what a student believes.

Teacher: "What type of reaction is this?"
Student: "Combustion"
Teacher: "Correct, does anyone have any questions?"

That isn't a discussion. There is no way to know if the student actually understands anything about the chemistry. Let's take a look at some techniques to enhance student discussion.

1. Wait time

It takes time to think. If you ask a question and get an immediate answer, then no one besides to the person who answered needs to think. Instead give students thirty seconds before a student answers. After a student speaks, wait for a moment in case they had more to say. Waiting will get students to speak more, and their longer responses will include more authentic commentary.[5]

Waiting after a student speaks can cause them to think what they said was wrong. In the event that this causes anxiety, assure your students that you are going to wait to make sure that they have finished speaking. You can even justify yourself by explaining that your intent is to get them to speak for longer periods of time and that if you cut them off, they'll respond with brevity.

Chapter 17 – Teaching

2. Elaborative interrogation

Many students come into chemistry class with the impression that answers to science questions are a single word.
Why does this object move the way that it does? Gravity.
Why do my hands get warmer when I rub them together? Friction.
How do plants get bigger without eating? Photosynthesis.

This is an important habit to break. Giving a one-word answer inhibits learning even if the answer is partially correct. Students are allowed to make abstract associations between words and phrases without having the chance to explore what they actually think about these terms.

Elaborative interrogation is a process where the teacher asks students to elicit more information. It can be somewhat similar to the persistent toddler asking "why?"

Teacher: What happened to the salt?
Student: It dissolved.
Teacher: What does that mean that it dissolved?
Student: It's in the water.
*Teacher: *pauses and waits**
Student: The pieces of the salt break up into smaller pieces that go into the water.
Teacher: Wasn't the salt already in the water before it dissolved?
Student: Yes, but the pieces of salt get smaller, so they can fit between the water particles.
Teacher: (to class) go ahead and draw in your notebook what you think dissolving looks like at the particle level based on what you heard "Student" say.

3. Talk Moves

Getting students to talk in class can be a daunting task for teachers. Talk moves are questions that teachers can use to increase participation, listening, length and authenticity of student responses.

"Could somebody repeat what (student's name) just said?"
"Tell me more about that"
"What do you see in your head when you visualize that?"
"If I understand you correctly, you're saying (try to repeat what student said)?"
"How many of you agree with what (student's name) just said? Does anyone disagree?"
"What is your evidence for that claim?"
"Who can add on to what (student's name) just said?"

Notice how many of these questions ask students to respond to what the student just said. You want to push students to interact with other students' ideas, so they engage as listeners. Notice how these questions offer opportunities for students to speak about what they think rather than serve as a compliance tool or punitive nature.

Chapter 17 – Teaching

4. The role of the teacher

In many traditional classrooms the role of the teacher is to disseminate information to the class. The teacher is the expert who tells students information and helps clarify challenging pieces. There is a different role teachers can opt for during discussions.

A teacher who leads a discussion is trying to get students to verbally express what they think. Students may be more comfortable discussing what they think when the teacher is not viewed as someone who has "all of the answers and explanations." Students don't want to explain chemistry to an expert. That can be intimidating. If I act like I don't understand chemistry, even though they know I'm pretending, students feel more comfortable helping me to understand. If I act like "Mr. KnowItAll" then students are going to feel intimidated.

Your role as the teacher becomes to learn what your students think. You'll want to push back against your students to help them clarify their thoughts, make connections, and to use their prior knowledge. This can be helped by being an expert in your content area, but will be different in how your knowledge is being used.

5. Talk to your neighbor

When conversations get stuck, ask students to turn and talk to their neighbor for a minute about the question. This gives them a practice run before they talk to the whole group. Talk to your neighbor engages all the students in your class. Instead of 1 student speaking and 29 listening you have 15 speaking and 15 listening.

Talk to your neighbor is helpful for students with anxiety about speaking in front of 30 peers. If a student appears hesitant when called to speak, let the class talk with their neighbors for a minute. This gives the student a practice run where they can prepare by talking with one or two students nearby first.

As your students complete their discussions, you can warn a couple of students that they will be the ones to speak first. You can also ask students to talk about what they heard their neighbor say.

Chemistry is a difficult subject and students are going to be hesitant to put their thoughts out for the class to judge. Talk to your neighbor is an excellent tool to help create a culture in your classroom where students are comfortable taking risks.

6. Confusing questions

"Confusing questions" is an advanced technique. Once you've hit a point in your class where students are comfortable, you can try asking questions that are confusing. You are not clear what response you are looking for. This can be unsettling for students, but there is an upside.

When your questions are confusing, students have to think through a variety of possible explanations and evidence to try and clarify what you are asking. If you have a group of students willing to be vulnerable, there can be a lot of learning.

Let's assume you just heated some water and have a heating curve.
Question #1 When the temperature is constant, is the energy going into the water increasing thermal energy or phase energy?
Question #2 What happened that the temperature and energy started to shift differently at times?

Chapter 17 – Teaching

The first question is straightforward. Students will know what the teacher is looking for in the response and they'll know what the question refers to. The second question is confusing. Students will know that they should talk about temperature and energy but aren't clear on what they should be looking for about each to talk about. It is not clear what the response should focus on.

This can push students to search a bit deeper for observations that might be helpful in trying to start a discussion. This will also lead to different students making a variety of comments and observations that will give the teacher more to work with. You will potentially learn more about what students actually think about energy and temperature from the second question. You can always clarify the question later if students become frustrated. This is an advanced technique that should be reserved for teachers with experience leading discussions as well as advanced content expertise.

7. Phenomena

Students need something to talk about. If you want students to discuss something, it should be interesting. Some ambiguity gives the students opportunities to be creative.

A phenomenon is a set of concrete observables. This can be a set of data, a demonstration, or an experiment where they are able to take in the information and provide explanations of what they are seeing or sensing. The opposite of starting with a phenomenon would be starting with an abstract concept and trying to connect the abstract idea to concrete experiences. Both can function but starting with the same experience for all students allows them to have a common focal point while still making connections to other experiences.

Discrepant events make wonderful phenomena. When the end result does not match up with expectations students become curious as to why. You would be surprised at what is discrepant to students. Sometimes after years of teaching we lose some of the magic of pushing a syringe in and watching the compressed air push the plunger back towards us when we release. Even something as simple as a candle burning has the potential for a large number of observations and discussion.[6] There is a starter list of phenomena at the end of this chapter and each content chapter if you are looking for ideas.

8. Cold call on students

It can be intimidating to call on shy students. Students might be hesitant to speak. But the teacher needs to be in control of who is speaking. That doesn't mean that it isn't terrifying for students to have to explain chemistry in front of peers.

If you do not call on students, you will get a sample of students that speak far more than other students. This establishes a dynamic where the "smart kids" explain things and everyone else learns from them. That doesn't make for productive discussion and has a negative impact on the students' perceptions of themselves. Instead, you need to shift student expectations to help them be comfortable talking.

The key here is that students expect the people talking to be the ones that know the answer to the question. You want to frame speaking as letting the teacher know what you think. Don't say things like "we learn from mistakes" or similar phrases. Cold calling isn't about producing right and wrong answers. It's about revealing student thinking. You want to know what the picture is in their head. And

Chapter 17 – Teaching

you want to know what assumptions they are making. Neither of those should be evaluated in a binary right/wrong approach.

The idea of a single correct answer undermines model development. It undermines discussion. Instead, we want students to express what they see when they see a phenomenon or express what they see when they are asked a question. It's not about the answer to the question, it is about what pieces do you recognize, what methods could be used to solve the question, what evidence do you know of that might be useful? What conditions and assumptions support your ideas? Which ones conflict with them?

If your students are resistant to cold calling, try having a discussion without it and have 1 or 2 students track who speaks (called a spiderweb map). Have them draw the circle and put lines from one speaker to the next. After the discussion is over show them how frequently some students talk and how many never speak. Students may even wish to track the demographics of who speaks the most.

You want to hear what your students think. Assessing students goes beyond whether they got a correct answer, you want to know what they were thinking and to help develop that thinking further. In order to know what your students think you need to ask them what they think, and you need to ask all of them. Letting students volunteer for speaking is rarely going to deliver that.

NGSS

The Next Generation Science Standards were adopted in 2013. The implementation of them in schools has been a mixed bag. The standards are excellent but are limited to assessment standards. This has caused some implementation policies to leave much to be desired.

In chemistry many teachers reacted to the standards by first noticing some key omissions. Topics such as naming, states of matter, gases, and others were missing. How am I supposed to teach a reaction to students who don't know how to name? NGSS was not intended to be a set of instructional standards. The standards are what will be assessed on state tests. The state will not be assessing naming, but if that is a prerequisite piece of information in the view of the curriculum writers, then it should be included in instruction.

One problem that has resulted is that some administrators are viewing a brief set of standards for chemistry and physics and deciding that these courses can function in a much smaller timeframe. Another is that the curriculum being used are often predicated upon coverage of the Disciplinary Core Ideas (DCI).

What the NGSS does is set up a three-tiered system of standards. The Disciplinary Core Ideas (DCI) are the content standards. The Science and Engineering Practices (SEP) are the methods students are to use to solve new problems. The Cross-Cutting Concepts (CCC) are the ideas that transcend science disciplines.

When we combine all three of these into a single assessment, we use the term 3-dimensional assessment to describe it. 3D assessments are incredibly difficult to write and to prepare students for. But we should welcome these challenges.

I think that implementation would be more efficient if we refocused on the intention behind NGSS. The intent was to increase the amount of deep thinking that students do about science content. The combination of DCI, CCC, and SEP are the

Chapter 17 – Teaching

vehicle to get there, but the biggest objective is to replace ineffective rote memorization with deep conceptual understanding and application.

Some teachers view NGSS as a system that will cause cognitive overload. They think that for students to think critically they must first be given a foundation of basics skills and knowledge that students can operate from. There is some evidence behind this, but that evidence is often realized through overly simplistic assessments. But remember there are two sides to good teaching. Keeping things simple enough for students to engage with but maximizing the difficulty to cause maximum permanent change in brain structure.

Deep thinking does not translate between subjects. I personally can engage at a much deeper level when discussing chemistry or education than I can with history or art. This is partially due to a large amount of prior knowledge I have about chemical education. But I also have a lot of experience with critical thinking about those two subjects. Both must be utilized during instruction.

In order to do this with a wide range of skill sets, NGSS starts units with a phenomenon. This allows all students to have a common experience that they can use as a foundation to think critically about. Phenomena tend to be highly engaging and memorable. The teacher can then use that memory to dual code more abstract information.

If a teacher adds an alkali metal to water with phenolphthalein indicator the student will make two key observations. The solution turns pink indicating the presence of a base, and the metal will react violently. The metal may even burst into flame or explode.

Those two observations line up with the two products of the reaction. If the metal used was sodium, the two products are sodium hydroxide (the base), and hydrogen gas (the source of the fire/explosion). Now the student has a visual memory to attach the abstract symbolic representation of the reaction to.

$$2Na + 2H_2O \rightarrow 2NaOH + H_2$$

If we have students mix baking soda and vinegar in a baggie, they will sense the energy changes that follow the endothermic reaction. The teacher can then develop this phenomenon by having students construct a model to represent the energy changes in an LOLOL diagram. The teacher can guide students to construct a particle model to accompany energy changes. The teacher can have students do proportional reasoning to determine how much energy change there was.

Currently there are a wide range of curriculum that have been developed. These have a wide range of effectiveness. Be skeptical and if you have the autonomy to do so, worry more about the pedagogy and commitment to content than coverage and alignment.

POGIL

Process Oriented Guided Inquiry Learning or POGIL is a highly intentional and structured set of worksheets used for teaching.[7] POGIL worksheets provide students with a set of data, statistics, or other representations that compare two or more features. The student is then able to draw out patterns from the models used. These are a guided constructivist form of teaching and learning. For naming compounds, a traditional approach gives the students the rules and has them try problems. A POGIL approach provides students the names and formulas and has

Chapter 17 – Teaching

them search for patterns. After making observations the student then constructs the rules with guidance.

The advantage to POGIL is that students are forced to think harder initially. They have to pull out the valuable information from the models and this requires them to connect to their prior knowledge. This is a wonderful opportunity for students to increase their metacognition as they are adding information to their short-term memory from the POGIL while connecting it to their long-term memory in order to find solutions.

POGIL worksheets are meant to be completed as introductions to material and not as practice. They make for great lesson plans for when you are absent! The activities are most effective when students complete them in groups with defined roles. If you include roles, have one student be an ambassador that can seek help from another group and return with new information to their original group. Other roles could be to have a manager, a timekeeper, and a recorder.

The worksheets have checkpoints in them where the teacher can check in with groups to see how they are progressing. Ideally the teacher does not use these to do direct instruction, but instead refers students back to the information and models in the worksheet itself. The teacher can help students identify what their conflict is but should not answer the questions for the students. Otherwise, the students may reduce their effort and rely on the teacher to do the work for them.

When students are presented with traditional learning techniques, many of them miss out on important details in the symbols used. But when the student constructs the information, the student must learn to analyze the new information by seeking patterns. This increases the observations made relative to direct instruction.

I used to frequently hear students say something along the lines of, "I understand it when you do it, but when I go to do it, I don't know how." This indicates that the student is not putting enough mental effort into the learning. That's not an indictment of the student, rather the teaching method. You can explain something perfectly while a student listens intently, but no major change in the student's brain occurs. POGIL is a quality tool to increase the change in brain structure during an initial learning activity for a unit.

A POGIL assignment can be constructed by a teacher, but there are a growing number of them available for purchase or for free. Some are better than others, and you'll want to do the activity yourself to see what adjustments to the assignment you'll want to make.

While the students are working the teacher has many options of how to spend their time. They can move from group to group to assess the students' struggles and strengths. They can work with struggling or advanced students. The teacher can also attend to any management issues that arise. Teachers should take advantage of the ability to listen to students' perceptions of the content. It is beneficial for us to remember what learning chemistry during the beginning is like.

It is tempting to reduce the difficulty of the task when students struggle so the student will experience some success. But the best thing for the student is usually to give them guidance on how to complete the difficult task. When you circulate the groups, avoid giving students too much information about a question. Instead redirect them back to the model so they can learn. It is their brain that needs to change.

Chapter 17 – Teaching

The POGIL activities for acid-base chemistry and bonding are some of my personal favorites. It can be a struggle to start those topics because there are so many concepts that it is difficult to choose the first few. POGIL will help structure the beginning of the unit by introducing students to topics without experiencing cognitive overload that ultimately ends in abstract association.

These assignments are one of the better ways for students who are absent to catch back up on subject matter they missed. These also function well for lessons for substitutes when the teacher is out. Normally missing the first day of a new unit is horrifying because the students aren't prepared to do anything without instruction. But the POGIL activities have enough scaffolding for students to learn from them without prerequisite information.

Antiracism

A common misconception teachers have is that you only teach chemistry content. The reality of teaching is that numerous roadblocks will exist that prevent students from learning chemistry and part of your job is to help students navigate those roadblocks even if they are not connected to a direct standard.

Chemistry has had problems with racism from overt bigots such as Moseley. But a more frequent problem that should not be ignored is the subtle biases that teachers and students display to our students. Teachers should aim to understand and minimize their biases. We also must teach students about how bias operationalizes racism.

Too often we operate under the assumption that the status quo is fair. If your classroom shows enrollment disparities, achievement disparities, or both then something is wrong with your class. Much of that might come prior to students showing up to your class, but you should still work to make things right. It is convenient for teachers to shift blame elsewhere. Instead of falling into that trap, make it a priority in your class to make things better.

A relevant example is the Buraku caste in Japan. These students did worse in Japan, yet when immigrants came to the US where the caste differences were not noticed by teachers or peer students these differences vanished.[8] The reality of education is that the expectations of a teacher are manifested in their students through a series of small yet impactful interactions.

Studies of bias will send job applications that are identical except for an indication of race. The results show that not only will people review the identical resumes differently, but they are unaware that they do so.[9] The reviewer will find an attribute and will explain their rationale for not hiring based on that.[10] Think of the implications that these results have for how we teach.

When we are teaching students there are several ways that bias can impact our interactions with students. We might mark a grade differently, we might call more frequently on a student, we might respond to questions differently (even with our facial expressions), or we might interact with students differently in a social manner before or after class. But I may not know that I do this, and it will be easy for me to justify my systems as fair in spite of evidence to the contrary.

I need to make sure that when I respond to student questions, that I maintain high expectations for them.[11] I don't want to answer one student by guiding them towards a conceptual idea while working out part of a problem for a different student to help them get started. I shouldn't socialize with white students by exclusively

Chapter 17 – Teaching

asking about sports and clubs, while socializing with Asian students about academics and future college decisions.

You also must consider that a student may have had some history of negative interactions and you may need to be intentional to develop their trust. A student might think you are singling them out whether you call on them frequently or never call on them. Having an open dialogue might be the key that builds student trust, so they feel comfortable learning from you.

One of the biggest omissions in schools is teaching students how to be a good peer to the other students in the class. As students get into high school, they can experience challenges that are unique to their social identity, but we do not teach students how to support each other when those challenges come up. Guidance from an adult is important so that students in class are not only comfortable with their teacher but their peers.

Stereotype threat is a concern in a chemistry class where many negative stereotypes exist that can impact cognitive load during performance.[12] The best ways to undermine stereotype threat are to promote growth mindset, increase representation, and have student access to places where they can discuss academics in a social setting.

There is a difference between being the only female student in a class and being one of two. Having a teacher that has the same race, or religion, or sexual orientation matters. Some of these we can't control, but there are ways to help. Be a supportive colleague to other teachers. And be receptive toward feedback so that people feel heard and included.

Review

Review should provide some clarity for what students are expected to know for their test without too much specificity. Letting students know that they are expected to be able to calculate mass-mass stoichiometry problems is fine. Giving them a review guide and telling them that the test will be the same questions but with the numbers swapped out is too much guidance.

The problem with giving students too much guidance in review is that it communicates that they don't have to learn anything for two weeks because you will tell them all the test answers the day before the test. You are now assessing a student's ability to replicate the review solutions rather than their understanding of chemistry.

A good way to help focus your review is to have students prepare communication of evidence and models. Not only do students need to know chemistry content, but they also need to be able to communicate that information in a manner that gives me confidence that they have a thorough understanding.

One method for review that has worked out well for me is to have students complete a poster of "The Model Thus Far." These posters should have a central theme that shows the updated features to the particle model of matter. Then the central imagery should connect to the phenomena, mathematical representations, key discussion points, applications, and lingering confusions that students have.

This pushes students to go back through what we did in class, the ideas we developed, the experimental evidence, and the questions they feel uncertain about. This forces students to enhance their metacognition. Did I learn what I was supposed

Chapter 17 – Teaching

to and if not, what do I need to do to improve? Another quality review tool are written reflections.

Written Reflections

Reflections push students to go through what we did in class, the ideas we developed, and the experiences that they had. Students will struggle to write quality reflections without guidance. They expect to write a summary of the main ideas. You want to push them to make their reflections personal by focusing on metacognition. They should emphasize their personal understanding instead of organizing rote facts, algorithms, and terms.

In order to get students to write quality reflections you need to start with a good topic. There should be some overlap between at least two of the three features in Johnstone's triangle. Specific heat capacity, gas pressure demonstrations, intermolecular forces, light and quantum mechanics all make good topics for prompting quality reflections.

A bad reflection about gas pressure would list the standard pressures, list the relationships between PVnT, and show sample calculations. A good reflection about gas pressure could involve a detailed explanation of how collisions vary as PVnT change. This explanation should include multiple representations such as graphs, data, particle representations, and writing. If the explanation centers around a demonstration or lab experience, then it will be even better.

Again, you are not trying to get students to write "correct answers." You are trying to get them to retrieve and elaborate on what they think is happening. This requires a prompt that is open-ended that will produce a variety of information. Specific heat capacity is a fantastic prompt for this because your students are unlikely to have fully mastered heat, temperature, and specific heat capacity by the end of the unit. But they will have lots of emerging ideas and those are what you want to read.

Pictures and color coding both enhance reflections by encouraging Mindful Learning.[13] Another tool is to have students write their reflection without using any notes or text. Everything must be retrieved from their brain. Then after they finish, they can use a different color writing utensil to add in pieces that they missed from their notes. These strategies help students use retrieval practice, they help students make connections between ideas, and pictures are much easier for students to remember and express themselves.

Reflections make for great assessments. Too often in chemistry classes, the assessments focus on being able to solve problems, or identify terminology. A reflection allows students an opportunity to write what they think and to be creative. And that level of communication is much more constrained on a test. Written reflections are a great opportunity to enhance teaching and learning in a classroom.

Timely feedback is critical. It is better to give feedback to the whole class based on comments from a few student reflections than to write individual comments for 150 students that they get back a week later. That holds especially true when you can give feedback prior to a summative assessment or review.

Start and End of Class

What happens at the beginning and end of class is crucial.[14] We all know that 50 minutes (or however long your class periods are) is too long for students to remain focused. Even if students are facing the right direction and taking notes, their

Chapter 17 – Teaching

attention is going to shift during that time. Transitions help students focus and this means we should take advantage of the biggest one, the start of class.

There are a wide variety of ways to start class. There are a wide variety of successful strategies on what to do in the first five minutes. Make sure that you are intentional about this time and the opportunity for student learning.

One method of starting class that I find aligns well with multiple pieces of cognitive science is to start class by having a student recap what they remember from yesterday (or the last couple of lessons). Give students a 3-minute window to talk about what they remember. If that's too much for your students, you can choose the student at the end of the class the day before, but you'll lose some of the benefit by removing the retrieval component.

When I have students start the class a few things happen. The other students listen differently to a peer than they do to me. If I talk, they trust me to say the correct thing. But they do not trust their peers. This skepticism can be quite valuable for learning. They wait impatiently for the mildest of errors so that they can correct them. If a letter is not capitalized, or the term used should have been plural they will notice immediately (with glee).

The student selected to present often has to take some time to come up with what they are trying to say. Other students are impatient during this where you can tell they have ideas and thoughts of their own that they would like the student to articulate. They will mouth phrases to the person at the front of the room, they will hold up notebooks with notes from the previous lesson. Even if the rush of presenting limits what the person up front is learning, the rest of the class is learning.

You get an opportunity to see what students took from your lesson. One year we were doing this during the chemical reaction unit. On day 2 the students were explaining that synthesis was when you combine two things ($A + B \rightarrow AB$) and decomposition was when one thing came apart ($AB \rightarrow A + B$). By day 5 they were still giving me generic forms of reactions with general descriptions. This lets me know that they weren't familiar with any specific reactions, and I needed to address that with more demonstrations, symbolic analysis, and pushing them to know concrete examples.

Students are getting spaced practice. They aren't learning something in one lesson and then moving onto a new topic the next day. You can even adjust this so that after an exam, students do a recap of what they recall from unit 2 a month earlier.

When they get to class early, they might ask a peer what we did yesterday. They might review their notes from the day before. Some will even prepare a brief topic statement in advance.

This also sets up your end of class. If students are responsible for what they learned the next day, they are more likely to summarize what we did in class that day. Students will transition from the end of class to the following day by making sure they are prepared in the event they are selected.

Now if you have students complete an exit ticket, they are much more likely to engage in that process with fidelity. They know that feedback looms the next day.

When students leave your class, they immediately start to forget what they learned. Their brains change during your class, but they start to revert when they leave. The end of class is your opportunity to summarize learning so that these reversions are minimized. Summaries are great to end lessons. These can happen by

Chapter 17 – Teaching

students writing down a 3-minute reflection of what was discussed or observed during the class.

A short low stakes assessment can also help students summarize what they learned and what they have not learned yet. A one question quiz that each student does individually or perhaps by talking to their neighbor guides their next steps.

The end of class is a great time to introduce a new term. If you spend a class period discussing mass and volume, at the end of the hour you might introduce the term density and ask students to connect all the ways density might be defined based on our class discussions. At the end of a class on concentration, you might introduce the unit molarity and ask students to draw a representation of a large and a small molarity graphically and at the particle level.

By introducing a term here at the end, you give students a seed for retrieval. The next day they will all recall density or molarity, and this will allow them to retrieve more information from their brain. Note how we are using abstract terms and ideas to solidify concrete details.

Feedback

I want to talk about three ideas related to feedback. Feedback functions best when the student has to do some work for the feedback to be valuable. If a student gets a question wrong and their only conclusion is that they didn't know what they were doing it was poor feedback. Consider a student that gets 8/10 correct answers on a multiple-choice test. If that student is given the test with the 2 marked wrong, they will look at what they got wrong, what the wrong answer they selected was, and if possible, what the right answer is. But imagine instead that the feedback to the student is that they had 2 answers marked wrong, but they don't know which 2. Now the student has to evaluate which answers they were confident in and where potential mistakes are. This feedback is much more impactful for the student's learning.[15]

Secondly, feedback is most effective when given in close proximity to the initial experience. It is inefficient to give specific feedback to students in writing. Do not write comments to students on their tests. Your time is better spent taking in a few responses and highlighting those for all students to see. This gives the feedback in a more efficient and more effective format. For feedback to be effective it should be specific, and it should be as close as possible to the action. Taking the time to write comments to students looks great, but the research doesn't support it. It is better to get students quick feedback where they must actively invest energy and thought to apply to their own work. If my writing has frequent mistakes with comma usage, I am more likely to learn with general feedback than if an editor goes through and changes my commas for me.

The third feedback consideration is that your system 1 learns naturally. If you start driving, you will get better at driving just by driving. If you start hitting a golf ball, your system 1 can learn how to swing the club better just by practicing. Your brain can figure it out better than you (your system 2) can.[16] When you teach, your system 1 will automatically adjust based on your experiences. Think about what feedback you solicit from your students and what that feedback reinforces.

Feedback that the teacher receives from students matters as well. Your goal is for assessment feedback to improve teaching and learning. When you grade a test, you should be able to identify what went well and what did not. But this is difficult to do well. Teacher feedback frequently focuses on classroom management instead of

Chapter 17 – Teaching

learning. As a teacher one of your goals should be that you as the teacher are receiving feedback about student learning more than feedback about behaviors. If you grade based on completion or compliance, then your system 1 is learning how to get more completion and compliance from students. If you grade based on chemistry knowledge, your system 1 is learning how to get students to learn better. If you can analyze your assessment results, your system 2 gets feedback on how to teach better.

Flashcards

We know that retrieval practice is a big component of learning. Yet we rarely spend as much time with retrieval as we do with other steps of learning. Flashcards are a great way to add retrieval practice for those times when you finish a couple of minutes early or need a break in the middle of a long activity or lecture.

Flashcards can be creative. Take a stack of papers and draw some particle representations on them. Put some proportional reasoning onto them. And ask follow-up questions after students answer correctly.

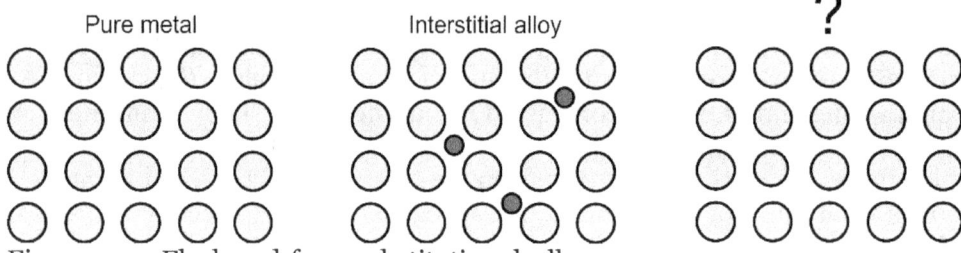

Figure 17-1: Flashcard for a substitutional alloy

You show students the flashcard (Fig. 17-1) and ask what type of alloy (you intentionally avoid asking if it is interstitial or substitutional) is shown on the right. Some students quickly shout out substitutional. You can then follow up by asking for a set of two metals that might make up a substitutional alloy. What about those metals makes them substitutional? If they answer copper and zinc tell them that would make brass. What's an object in the school made of brass (Tuba!). What would be an interstitial alloy? Why do we make alloys?

Something interesting about flashcards is that if you repeat them students will learn the answer without knowing what anything is.

Which of the following is a buffer?
a. 10 mL 0.1 M HCl + 10 mL 0.1 M NaOH
b. 20 mL of 0.1 M HCl + 10 mL 0.1 M NaOH
c. 10 mL of 0.1 M HF + 10 mL of 0.1 M KOH
d. 20 mL of 0.1 M HF + 10 mL of 0.1 M KOH

A confused student hears everyone shout out d. Then later you come back to the card. They remember that the answer was d. This allows them to now look at what features are in d and how they compare to the other options. Answer d has HF instead of HCl, KOH instead of NaOH, and 20 mL of acid instead of 10 mL. Maybe the next time they start to work out why having more acid than base matters. Or why a weak acid works while a strong acid does not. This works particularly well when you

Chapter 17 – Teaching

work in follow-up questions. Follow-up questions also provide time for struggling students to make connections.

But the student now has an abstract piece of information in their brain that they can work from to develop better understanding of the symbols used. They know d is a buffer, so they can now work to understand why d is a buffer.

This is why we see that retrieval practice not only increases rote memorization, but retrieval practice also leads to better student outcomes that involve critical thinking. The students are able to use metacognition of what they know and what they do not to work towards understanding and that process mirrors what we see in critical thinking.

Critical thinking does not easily translate from subject to subject. A student cannot be taught to be a good critical thinker and have that apply to anything the student learns.[17] There must be opportunities for the student to develop critical thinking within chemistry and within science. Many chemistry teachers are frustrated by a lack of mathematical ability when the research shows we should expect mathematical skills to not translate into chemistry without instruction from chemistry teachers.

Flashcards are effective, they fit conveniently into lessons as transitions that increase engagement, and they give teachers opportunities for spaced practice. If you are teaching AP chemistry or IB chemistry, you must prepare students to remember the entire course for a single set of assessments at the very end. Spaced practice is critical so that students are strong in the ideas learned at the beginning of September.

Advice For Struggling Students

Schools frequently provide suggestions for struggling students that are largely ineffective. Chemistry is a difficult subject, yet a lot of advice for struggling students is simple and reductive. Simple solutions are not always effective. At the end of each chapter, you will find content specific struggles that students have.

The most frequent issue that I have seen with students is that they quit way too fast. They say things such as, "I don't know," or "I don't know anything." These are dangerous phrases because the first step to a student learning something new is to figure out what they already know that they can start with. When a student becomes frustrated and shuts down, they eliminate that possibility. When the student becomes frustrated, their amygdala interferes with their learning, and they need to take a break and regroup.[18]

One simple method to re-engage a student who has an overactive amygdala is to ask them a question that they will respond to by choosing yes as the response.[19] When the student chooses to say yes, they are using their prefrontal cortex, and this reduces the fight or flight response. For example, "Would you like me to have someone else answer this question while you think it over for a minute?" or "Would it be helpful for me to ask this question as a multiple-choice question instead?"

Students frequently seek someone else to explain things to them. This is a dangerous precedent. The message that the student receives here is that they are not capable of learning chemistry on their own, they require someone else to have it make sense. They spend their time trying to replicate what someone else said instead of critically examining their own prior knowledge and building from there. It is their brain that needs to change. At the start of each "Student Struggles" section at the end

Chapter 17 – Teaching

of each chapter you will find a common example of students quitting too early in the process

Student learning is influenced heavily by socialization. Students linked their learning of content through the social interactions with their peers.[20] This finding is further confirmed in Whistling Vivaldi where Claude Steele discusses the research of Philip Treisman who found that minority students perform worse in classes in part because they tend to work in isolation.[21]

Think about how many times students say, "When I watch you do it, it makes sense. But when I go to do it, I go blank." They've spent too much time mimicking someone else that they're very unsure of what is in their own head. They lack confidence in learning from their own thoughts. These students often aggressively seek out algorithms to further avoid challenging themselves. They say, "Just show me how to do it and I'll figure it out." They say, "If I could just see what I'm doing wrong on the test, I would be able to learn."

These students will also have a lot of support in avoiding learning. Administrators and parents can push to undercut their learning. You as the teacher might not have the authority to resist that. If you are going to try and push students to do their own thinking, be prepared to persuade and to persuade frequently. Here are ten strategies for struggling students. These are not easy for students to do, but they are effective.

1. The struggling student should write down everything they know. Having a student do this gives them an opportunity for retrieval practice. It helps them articulate what is in their head. It exposes what they do not know. This sets up the learning process of finding what you know, what you don't know, and then trying to use evidence, logic, and observations to link the two. Students should write down what they remember from the previous lesson when they start the next class. Students who are studying should start their session by writing on a blank sheet of paper.
2. The struggling student should try doing problems. Low stakes quizzing is effective at helping students learn and helping students with their metacognition. Doing problems is a method of retrieval practice. Interleaving different problems will help the student learn to focus on identifying information and making a plan for a solution. If the same type of problem is done too frequently, the process will result in the student just learning the algorithm.
3. The struggling student should make predictions. If they are starting a new unit, what do they think is going to come next? What connects to the last unit? What experiences have they had in their lives? What have previous courses taught them about this new unit? What questions do they have before the new unit starts? Making predictions sets up a neural pathway to fill in that information when it does become apparent. That happens even when the prediction is incorrect. It is important for students to get over their fears of having an incorrect thought. Instead, they should focus on developing their thoughts and not assigning them to be correct or incorrect.
4. The struggling student should focus on one thing. Chemistry is difficult. Sometimes a student will come in with an intense focus and determination. They try and learn everything at once, they become frustrated, and they shut down again. Instead, they should focus on one topic. They can connect that

Chapter 17 – Teaching

one topic to other content later, but they should not move on until they understand that topic. Ideally, they would complete an assessment to show their understanding. Success is the best motivator and struggling students need that motivation sometimes. Chemistry class can be grueling when you're struggling, and you only receive negative feedback.

5. The struggling student should use flashcards. To properly use flashcards the student must take the time to read the front and respond before checking the answer. They must retrieve the information. Many students feel that creating flashcards is sufficient, but this is not true. Creating flashcards gives a false sense of security because it is a low-level task that does not require retrieval.[22] Using flashcards is more effective. Flashcards can include things besides vocabulary. Have particle diagrams, symbolic representations, or anything else that mirrors the upcoming assessment.

6. The struggling student should explain what they know to someone else. The other person does not have to understand them for this to be effective. There is a reason why teachers become so good at chemistry. Explaining chemistry to others forces you to think and anticipate. If no one is available to listen, writing could be substituted. The student could also just talk to themselves.

7. The struggling student should use a technique called elaborative interrogation. Elaborative interrogation involves asking questions to try and expose what you understand and what you do not understand yet. The student should keep probing with what they think something is, and what questions they have to help refine their understanding.

8. The struggling student should make representations that are macroscopic, symbolic and at the particle level. Switching between these is a key to understanding chemistry well. Novice students struggle with this, and teachers underestimate their struggles. Free simulations such as PhET simulations are an excellent resource.[23]

9. Some struggling students might be having issues on assessments with stereotype threat. Experiments show that when a student is prompted to think of a negative stereotype prior to an exam, the student will have lower achievement on that exam. The stereotype occupies some of their cognitive load during the exam. Three potential solutions for stereotype threat are for increased representation for the student, forming a peer study group, and reaffirming a growth mindset for the student.

10. The struggling student should learn about how their brain works. Learning is very counterintuitive. Most study methods that feel good don't work well, and the methods that do work well are frustrating. When people learn a lot, they don't feel like they learned. When people learn little, they feel overconfident.

11. Just try something. Too many students are scared to try something because they might be wrong. Try dividing and see what the result is. Does it make sense? Would multiplying give a better result instead? Can you find moles or make a proportion? Usually by trying something the student will begin to see the correct method even if the initial attempt was incorrect.

New Teachers

Everything you teach in chemistry stems from experimental evidence. As you teach chemistry you want to be on the lookout for that evidence. When you tell the

Chapter 17 – Teaching

students something, in the back of your head think about why you make that claim. There are protons. You may be really confident that there are protons, and no one might ever challenge you on that fact. But you should be thinking about how we know there are protons. We didn't always know they existed. Was Rutherford's gold foil experiment the only evidence of protons or did something else happen that was even more specific than just a positive charged dense lump at the center of the atom? Can we actually see protons? What would it mean to see one? Do they have a color? Is a proton in a sodium atom the same as a proton in a hydrogen atom?

For those first few years of teaching, you want to imagine you have an annoying three-year-old that follows you around asking "why?" (also "how do you know?" and "what's your evidence?") in perpetuity. You might even find a student who does this (if you're lucky). Back in the mid-1900s the textbooks were filled with justifications because back then many were skeptical of science and the claims in science. But now textbooks focus on reading levels and simplifying the evidence to improve student understanding. The cost of that is that students forget how to question, how to seek evidence, and how to evaluate evidence. And if you want any shot of being successful at getting them to improve on that you're going to need to take the lead.

Learn from others. So often teachers become isolated in their classrooms. But teacher experience is one of the most valuable resources to a new teacher. Be skeptical and reflective about advice. Read books about teaching. Meet with experienced teachers at conferences and teacher organizations.

A good mentor teacher can mean the world. But remember that your brain is the one that needs to change, not theirs. Don't be so over reliant on their experience and advice that you fail to progress. I would often rely on my mentors to tell me how to respond to student infractions. By not making an initial plan myself, it took me longer to learn how to manage disruptions. Your mentor teacher does not need to be in chemistry or even science. This can allow you to work harder at getting advice by having to translate it into your classroom that will probably differ in many ways.

Classroom management is difficult. Almost no one is immediately great at it. You should expect to need practice and experience. It helps to be direct with students and it might take a while before you become comfortable with telling students what to do. I used to pretend to be an actor that was playing a role of a stern teacher. It helped.

If you feel like you messed up, be honest with the student and try to make amends. Having open communication is helpful, because if you avoid confrontations, you can develop toxic ideas in your head.

An underlying fear I had when I was new was that students would not follow my directions. Over time you learn how to phrase things, use tone, and be respectful with commands so that students follow directives while feeling valued in the process. Another issue I had was that when something happened outside of my expectations, I felt like I had messed up. Students will test boundaries, and you need to be prepared for that to happen without feeling as though your lesson was inadequate. Sometimes your lesson was the problem, but even if you teach "perfectly" you will still experience behavior issues.

Chapter 17 – Teaching

Fascinating Stories

In the book Moonwalking with Einstein, Joshua Foer investigates the International Memory Competition. He finds that the people who have the best memories in the world are not exceptional people. Instead, they are average people who use research on memory to their advantage. Our brains have much more success remembering places than abstract definitions. One of the primary tools used in memory competitions are called memory palaces where the competitors use a location to construct scenes that correlate to the abstract things they must remember. We remember stories even better than places.

There is a fine balance though. During my first year teaching I had a student remark that they could easily pass a final exam if it were only based on the stories I had told in class, but not the chemistry. While it is easier to remember images or events, we have to actually focus the content within that image or event. When teachers use cooking as a history lesson, the students tend to just remember the cooking and not the lesson.[24]

Teachers should have a repertoire of stories to engage students with content. But we must constantly reflect on whether the stories help students connect the content. Here are ten scientists, ten places, and ten interesting phenomena. There are of course many more than these, but if you are new these should help you get started.

Scientists

1. Gilbert Lewis was arguably the smartest and most accomplished scientist to never win the Nobel prize. He was nominated 41 times, but never won. He was hated by many. Lewis and Walther Nernst were known to dislike each other, and it is possible that Nernst's friend Wilhelm Palmaer discredited Lewis's work in thermodynamics to prevent Lewis from winning the prize.[25] Nernst was also intensely jealous of Arrhenius's Nobel prize.[26]

2. Ernest Rutherford not only studied under JJ Thompson, but he also mentored Otto Hahn, Henry Moseley, and worked closely with James Chadwick.[27]

3. James Harris graduated from Cal Tech but was offered a job as a custodian. He told them, "I don't need a job that badly." Later he discovered elements 104 and 105 (rutherfordium and dubnium) at Cal Berkeley.[28]

4. Hennig Brand discovered phosphorus. You might not want to know how. In fact, skip to number 5 before you get too far into this. Brand was an alchemist and was seeking the philosopher's stone. His search begins with large quantities of urine. Large quantities as in multiple bathtubs full. He let that urine sit, evaporate, and form maggots (the smell had to have been brutal). Eventually he took the remaining white material which turned out to have concentrated the phosphorus.[29] Brand was the first modern day discoverer of a new element.

5. George Hevesy worked in a lab in Denmark during WWII. When Nazi occupation of Denmark started it was illegal to remove gold from the country. Hevesy dissolved two Nobel prize medals in aqua regia solution. The orange solution remained until after the war. The gold was precipitated out of the solution and the medals were recast.[30]

6. Antoine Lavoisier did research on different types of air. He produced evidence of a substance in air (we call it oxygen) that was needed for breathing, was consumed during combustion, was produced by plants, and combined with metals to form a more massive substance. Lavoisier was a wealthy tax collector, and he had his head

Chapter 17 – Teaching

removed via guillotine. There is a persistent rumor that he instructed his friend Laplace to count how many times he would blink after his head was detached. There has never been evidence to support this rumor.

7. Marie Curie was born in Poland where she grew up extremely poor. She moved to Paris to study and practice science where she encountered extreme difficulty in obtaining lab space, funding, and respect. She extracted two new elements from a uranium ore called pitchblende. She named these two elements polonium and radium. The discovery was worth the Nobel prize in physics. She later won a second Nobel prize, this time in chemistry. Her daughter Irene also won a Nobel prize for work where she bombarded a chemical with alpha particles to induce radioactivity. Marie Curie remains one of four scientists to win two Noble prizes and she is the only to win both in different scientific fields.

8. Glenn Seaborg and I share Michigan as our home state although he was born in Ishpeming which is a hike to get to from where I live in the lower peninsula. Seaborg worked at Cal Berkley where he discovered or helped discover ten elements. Al Ghiorso was heavily involved with helping Glenn in his work. In a wild story, element 106 was named seaborgium but then later the IUPAC ruled that the element could not be named after a living scientist. Later in a massive compromise they unexpectedly reversed this decision. Seaborg's daughter was driving when the radio announced that the element had been named after her father and so she assumed that he had died. She frantically searched for a phone to make contact before discovering that he was ok.[31]

9. Dmitri Mendeleev was the youngest of 17 children born in Siberia. His mother took him to Moscow after his father passed away, but he was denied admission to the Moscow University. They continued traveling to St. Petersburg where he studied.[32] His work with the periodic law and periodic table became well known not when first published but after one of his predictions of a new element was found to be correct. Element 101 (mendelevium) is named for him.

10. Tsutomu Yamaguchi was a Japanese marine engineer. He is considered to be incredibly lucky and unlucky simultaneously. He lived in Nagasaki but happened to be at work in Hiroshima when the United States dropped a nuclear bomb in 1945. Yamaguchi suffered injuries from the bomb dropped on Hiroshima as well as the bomb dropped on Nagasaki later. He managed to live until 2010 when he finally passed at the age of 93. He is assumed to have had freakishly good DNA repair.[33]

11. Yuri Oganessian has element 118 named for him (oganesson) and for good reason. He discovered, helped discover, or had his method used for every element 104-118. Yuri was born in Russia in the same town as George Flerov who hired Yuri and the two were a dynamic pair that discovered many elements together and both have elements named for them.

12. Thomas Midgely made two major discoveries that both ended up having destructive impacts on the environment. Midgley discovered that tetraethyl lead reduced engine knock. The reason we now sell unleaded gasoline today is because Midgley pushed us into using lead in gasoline before we realized the dangers. His other big discovery are CFCs or chlorofluorocarbons that are used in refrigeration and led to ozone in the upper atmosphere being diminished. That harm is now recovering due to an international effort to reduce the use of CFCs. Once Midgely inhaled a bunch of CFCs to show how safe they were. They are safe for inhalation, but not for the ozone layer.[34]

Chapter 17 – Teaching

13. Lise Meitner was fortunate to study under Ludwig Boltzmann. She was the second female to obtain a Ph.D. in Austria. Due to political persecution, she left Austria to work in Germany where she met Otto Hahn. In 1938 Nazis invaded Austria and this put her life in danger. She fled Germany to Holland and then found work in Sweden. Later Otto Hahn was awarded the Nobel prize for her conclusion about fission. Meitner was nominated 48 times for a Nobel prize in physics or chemistry but received none.[35] In 1997 element 109 was officially named Meitnerium in her honor.

Places

1. Oklo, Gabon - A naturally occurring nuclear reactor was discovered here. The uranium composition here differed from all the other uranium in the world. The reason is that a source of neutrons was causing the uranium-235 to decay faster than normal.[36]
2. Ytterby, Sweden - Four elements are named after this small village located just outside of Stockholm (Ytterbium, Yttrium, Erbium, Terbium). Another three elements were discovered here. Gadolinium is named after Gadolin who discovered the first and predicted the future discovery of more elements mixed in with the ore. Holmium is named for Stockholm, and Thulium is named for Scandinavia.[37]
3. Los Alamos Laboratory in New Mexico was where the Manhattan Project took place. Not only were the first nuclear bombs developed here, but they were also tested nearby. You can purchase trinitite online. Trinitite is glass that came from sand melting during the nuclear bomb tests in New Mexico.[38]
4. Stagg Field was home to one of the premier football teams in the early 1900s. The University of Chicago Maroons were a force to be reckoned with that slowly faded into obscurity until the program was abandoned in 1939. In 1942, a secret lab was built underneath the field for Enrico Fermi to create a working nuclear reactor. Fermi had earlier discovered that slowing neutrons down increased the likelihood of a uranium-235 fission. Using graphite to slow neutrons and cadmium to absorb neutrons he put on a show where he gradually increased the rate of fission of uranium until he achieved a self-sustaining reactor at that site in front of 42 witnesses.[39]
5. In Golden, Colorado there is a Ball manufacturing plant. They make 6 million cans every day. The total number of aluminum cans made each year could stack up to the moon 56 times if the cans could support their own weight.[40] And the number of failures of these cans is almost non-existent. The scenarios of when cans fail can be quite disgusting though. In 2008 Ronald Ball bought a can of Mountain Dew that had somehow managed to have a dead mouse in it. The lab results showed that the mouse was not in the can long enough to have been there at the factory which had sealed the can 74 days prior. Mountain Dew is far too corrosive for the mouse to still exist after sitting in Mountain Dew for 74 days.[41] Which really goes to show impressive the coatings inside of the aluminum cans are. You can dissolve the aluminum away to leave only the coating using an excess of hydroxide.
6. Bellevue Hospital in New York City is where Alexander Gettler and Charles Norris founded the field of forensic science. The pair worked in a lab in the Pathology building where they developed methods to test for a variety of poisons. At the time, lawyers had become proficient at exposing the lack of science in the courtroom. This resulted in many avoiding consequences for fatal poisonings. But Gettler and Norris actively worked to develop reliable scientific tests that would withstand the scrutiny of a jury.[42]

Chapter 17 – Teaching

7. The US Department of Agriculture in Washington, D.C. was home to the poison squad. The poison squad was a group of men recruited to empirically test which food additives would make them sick. Dr. Harvey Wiley orchestrated the tests where varying quantities of borax, formaldehyde, cupric sulfate, and other common preservatives in 1905 were added to the food consumed by some of the poison squad. The results eventually led to the formation of the USDA.[43]
8. William Perkins was a young aspiring chemist when he ran a few extra tests on a sample of coal tar that produced a purple dye that he named mauve. He left his studies to open a factory along the Grand Union Canal in North Greenfield, England. Perkin's work with dyes led to further developments in organic chemistry as dyes were used in medicine, textiles, and as biological stains.[44]
9. United States Radium Corporation in Orange, New Jersey was the first factory of three where women were given radium paint to use. The "Radium Girls" suffered grotesque health issues but managed to sue the company before dying. The management knew of the hazards of contact with radium but persisted and even actively misled the workers to continue.[45]
10. Russell Marker set out to find cheap methods to produce steroids such as cortisone by extracting chemicals from plants and altering the structure through synthesis. His methods provided opportunities for other researchers to find more reaction pathways to produce steroids. Carl Djerassi extended this by producing cortisone, estrone, and estradiol. He then altered the structure of progesterone to make norethindrone. This was the first oral contraceptive. The birth control pill had a revolutionary impact on the world by providing women more control over their bodies. This altered gender dynamics, particularly in employment.[46]
11. The Royal Institute was founded by Henry Cavendish in 1799 in London, United Kingdom. Humphry Davy first isolated multiple elements including sodium and potassium there. Davy was a popular lecturer, and the Royal Society would put on free chemistry lectures for the public. When Davy retired, the next lecturer panicked and ran away 30 minutes prior to his lecture. Michael Faraday stepped in to take his place at the last minute. To this day they continue the lecture series but lock the lecturer in the room for the final 30 minutes so they cannot run away.
12. Sar-e-sang Afghanistan is where many of the few mines for Lapis Lazuli exist.[47] Your students may know about this blue dye from the popular Minecraft game. The blue can be extracted from the ore by soaking in a hydroxide solution. Frequent agitation is needed as well.
13. Your local University has equipment and expertise that you would dream of. And many are more than happy to share both with you. You can also listen to graduate students defend their thesis or dissertation. One of the local Universities near me (Eastern Michigan University) let me take a video of their mass spectrometer, and another (University of Michigan) is allowing me to participate in a polymer lab series they are working to implement in local high schools.

Phenomena

1. Vinegar and baking soda get cold when mixed. If you put them in a sandwich baggie it's easy for students to do this without too much of a mess.
2. A rubber ducky quickly fills with air but takes a longer time to fill with water when submerged.

Chapter 17 – Teaching

3. Burning a small strip of magnesium metal causes a metal (Mg) and a gas (O_2) to combine and form an ionic compound (MgO).
4. A glow stick can be cut open and emptied into a test tube. When the test tube is heated the mixture not only glows brighter, but it also goes out faster.
5. Goldenrod paper changes color to red when exposed to a small amount of base. If you cover parts of the paper with tape, you can reveal invisible messages (great for "promposals" if safety is observed).
6. Flash cotton (gun cotton) will burn rapidly and completely due to a large amount of oxygen stored in nitro groups. Flash cotton can be ignited using superheated steam, which means you can start a fire with water.
7. A strong magnet will deflect electrons in a cathode ray tube. If you have an old tv, set it to the blue screen and the magnet will cause the colors to change between blue, red, and green. Turning the tv off usually resets everything back to blue.
8. A strong magnet will attract $MnSO_4$ but not $ZnSO_4$ showing the paramagnetic effect of unpaired electrons.
9. Many forms of currency from all over the world have images that are revealed under a black light.
10. Steel wool will gain mass when burned. Steel wool can also be burned much more rapidly if put into a container of oxygen gas instead of air.
11. Zinc chloride can be melted at low enough temperatures to run electrolysis or a single replacement reaction without water.
12. 1 liter of distilled water will change pH with a small amount of acid or base added. But 1 liter of tap water will potentially be buffered depending on the origin of the tap water. Where I live the lakes have limestone bases and the carbonate causes the tap water to be buffered much like our blood is.
13. Diet cherry cola can be distilled. The first part of the distillate will be the chemicals that produce the scent of cherry. Diet is recommended because the lack of sugar makes clean up substantially easier.
14. Acetone will evaporate noticeably faster than ethanol. Acetone will cool to a lower temperature, and this can be demonstrated by soaking a small piece of paper towel that is attached to a temperature sensor with rubber bands. Acetone will produce a larger flame when ignited.
15. Zinc hydroxide suspension will become clear when excess HCl or excess NaOH is added. This shows that $Zn(OH)_2$ is amphoteric as the Zn^{2+} can act as an acid reacting with the NaOH, and the OH^- can neutralize the HCl.

Themes

Be creative with your chemistry lessons. We know so much about chemistry that connects with other passions a student or teacher might have. You can use a theme to create more dual coding opportunities during a unit. You can also use a theme as a review of all units at the end of a school year. Have some fun with it. Maybe even see if there's a way to do a field trip or connect with someone who works in the field.

1. Medicinal chemistry
a. *Aspirin is acetylsalicylic acid.* Salicylic acid is the pain reliever, but the acetyl group allows the drug to get to your stomach or intestines before hydrolyzing. The salicylic acid is an irritant. Students can synthesize aspirin, but they can also test old

Chapter 17 – Teaching

aspirin tablets by forming a colorful complex between iron (III) and the salicylic acid to determine how much of the drug has decomposed. If you take old aspirin containers the smell of vinegar will be pungent due to the hydrolysis of the ester form (acetylsalicylic acid) back to the carboxylic acid form (salicylic acid).

b. *Medical imaging has a flood of interesting chemistry connections.* The antimatter in PET (positron-emission tomography) scans, the nuclear magnetic resonance in MRI (magnetic resonance imaging), or even just the X-rays in a CT (computerized tomography) scan. For a fun bonus look at how gadolinium increases the signal in MRI and barium sulfate in an X-ray. Students may have stories to tell although be wary of asking students to share personal medical information.

c. *Blood testing has connections to solubility and separations.* Some tests are done on plasma where the blood is put into a tube with an anticoagulant. Other tests are done using serum where the blood clots and is then separated using a centrifuge. Students can compare the tests done in a breathalyzer test for ethanol with a blood test for ethanol.

d. *Many prescription drugs follow 1st-order kinetics.* This presents an interesting phenomenon where a doctor may instruct you to take a double dose for the initial dose. If the half-life of a drug is 4 hours and I start with a single dose I will have 0.5 of a dose left after 4 hours. If I take the 2nd dose, then I have 1.5 doses (half of the original and the new one). By 8 hours I would have 0.75 doses and would take a new dose to get to 1.75. At 12 hours I would have 1.875 after the new dose. At 16 hours I would have 1.9375 doses. Perhaps you can see the limit we are approaching. But if I took 2 doses initially, I would have 1 left after 4 hours. The 2nd dose would take me back to 2 doses and this would repeat throughout the cycle. If you're going to discuss this with students, please make sure to help them understand that the kinetics of drugs is not uniform, and you are not giving them pharmaceutical advice.

e. *Toxicology is a fascinating topic.* The Paracelsus quote is roughly translated to: "All things are poison and nothing is without poison. It is the dose alone that makes a thing not a poison." Which things are the most poisonous at low doses? How do we know how much poison it takes for someone to die? There are LD_{50} values which is the lethal dose for 50% of people. They are reported in mg/kg where the mg is how much poison, and the kg is how much body weight the person is. Botulinum toxin has the lowest LD_{50} value of 0.000001 mg/kg. So, for me (about 85 kg), if I consumed 0.000085 mg of botulinum toxin, I would have a 50% chance of surviving.
If your students need work with interpreting graphs, this is a great subject to design a unit. No Observed Adverse Effect Levels (NOAEL), Lowest Observed Adverse Effect Levels (LOAEL), lethal dose 50% (LD50), and many more acronyms give your students the opportunity to analyze graphical data in a meaningful manner.

2. Food chemistry

a. *What do we store food in?* Why do so many containers say BPA-free on them? What is BPA and what are the alternatives? Is it true that I shouldn't microwave food while it's in plastic or polystyrene containers? Food storage has a long history. Things are better now than they have ever been, but what issues still exist, and which are marketing hysteria?[48]

b. *Make some fizzy lifting drinks.* Citric acid is the sour component of sour candy. You can purchase pure citric acid (food grade!). If you mix citric acid with baking

Chapter 17 – Teaching

soda you end up with water, sodium citrate (tasteless), and carbon dioxide. Have students determine the correct quantities of citric acid and baking soda. They can mix them in a fruity drink to make fizzy drinks. If they're wrong in either direction they'll find out!

$$H_3C_6H_5O_7 + 3NaHCO_3 \rightarrow Na_3C_6H_5O_7 + 3H_2O + 3CO_2$$

c. *The Maillard reaction occurs when sugars react with protein during cooking.* The nitrogen in the amino acid combines with the carbon of the aldehyde in the open ring of the sugar (only reducing sugars have these). The brown result can be a variety of chemicals and these products have substantial impacts on taste and aroma. You can experiment in a class by heating different amino acids with different sugars to see what aromas result. The color of the crust of bread, coffee, bacon, and many more foods and drinks utilize the Maillard reaction.[49]

d. *What is blue raspberry flavor?* What about green sour apple? How do food scientists discover, isolate, and produce these artificial flavors? Interestingly a sour apple Jolly Rancher has 26 chemicals while an actual green apple has over 2500 chemicals in it.[50] Artificial sweeteners are some of the most researched in the chemicals in the world. People fear them even though they have had consistent evidence of their safety. For the chemistry classroom the discoveries of saccharin, cyclamate, sucralose and aspartame were all discovered by mistake.[51] Sucralose was discovered by Shashikant Phadnis when he misheard his boss say "taste" instead of "test" these compounds. Oops!

3. Weapons and explosives

a. *One German advantage in World War I was adding molybdenum to steel.* This was used for the Big Bertha shells that were so large that the barrel would deform beyond use within a couple of weeks. By adding Mo to the steel, the large Mo atoms would prevent the iron and carbon from slipping past each other even at the high temperatures used for launching the shells.[52]

b. *The world has never been more peaceful than it is right now.* Extreme poverty has diminished for years, life expectancies are rising steadily, and global politics are more stable now than ever.[53] At the same time, the use of rare earth metals has exploded. Some of these rare earth metals have limited supplies. There is an estimate from the American Chemical Society that 44 elements have insufficient quantities for our use just in the next 100 years.[54] If we had to choose between developing new weapons or developing phone technology, how would we make such a decision?

c. *What's the difference between a fire and an explosion?* The rapid expansion of gases is the key difference. When we think of a bomb, we think of gases expanding into a large space in a short time frame. This could incorporate reactions, kinetics, and thermodynamics into a single interesting discussion.[55]

d. *Kevlar is a polyamide polymer.* When Kevlar is embedded in a matrix it can rapidly diffuse molecular motion in all direction. This allows the material to stop bullets. Kevlar used to be used for bulletproof vests. Now it is used for bullet resistant vests. No change has been made to the Kevlar, but the bullets are now better able to penetrate the material.

e. *Fireworks are amazing.* You probably can't set off fireworks in your classroom (seriously, don't do it). But you can show students the various colors by spraying salt

Chapter 17 – Teaching

solutions into flames of Bunsen burners. Never use methanol to produce the colorful rainbow demonstration. Methanol has an extremely low flash point and produces irregular vapor paths that lead to explosions about once a year in chemistry classes. In 2017 a group of preschool children was burned by the methanol demonstration. Just don't use it.

4. Chemistry of art
a. *Why are carrots orange?* The structure of beta-carotene in carrots has alternating single and double bonds. The HOMO-LUMO (Highest Occupied Molecular Orbital, Lowest Unoccupied Molecular Orbital) gap absorbs purple light and hence the carrots appear orange to us.[56]

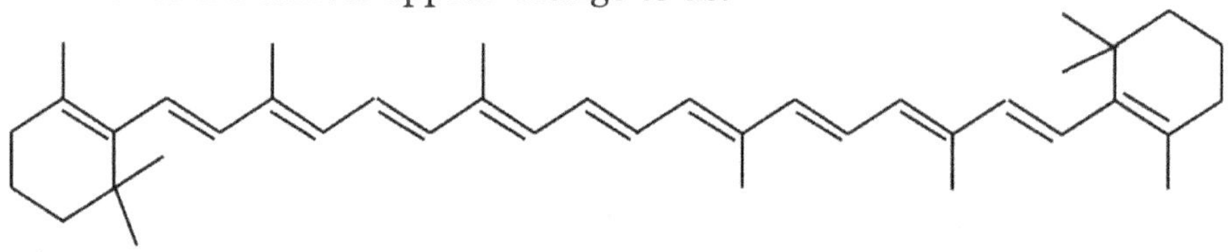

Figure 17-2: Beta-Carotene structure

Pigments and dyes are often made of organic compounds that are designed to absorb visible light. Azobenzenes are another example.
b. *Glass artistry is amazing.* From glassblowing to stained-glass windows, to neon lights, to shaping molten glass there is so much chemistry. This was my favorite field trip to take as a teacher. At one point during a field trip some molten glass dropped onto the concrete floor and burst into flame. I couldn't believe it. How can the glass be on fire? In my head I was thinking that silica is already completely oxidized. Unless the floor has fluorine gas, or there is an additive to the glass it made no sense. The artist looked and said, "Oh yeah, it's actually burning the grease and stuff on the floor. It's not the glass burning."

The use of metals in stained glass window correlates nicely with nanoparticles. To make yellow glass you add silver. You can also make silver nanoparticles by mixing silver nitrate and a reducing agent under the appropriate conditions. The silver nanoparticle colloid is yellow. Gold nanoparticles are red, and this explains why red stained glass was substantially more expensive.
c. *Music has enough physics connections to make an entire course.* But there is some chemistry to be found if you're clever. Violin resin contains abietic acid and you can run a titration on the different types of resin to compare quantities. The resin is not easy to dissolve, and we ended up throwing away some of the glassware rather than clean it.

If you're feeling ambitious you might construct a Rubens tube. A long metal tube with evenly drilled holes that you can fill with methane to create a line of small flames. When a speaker is put next to the tube, the flames change as the pressure changes throughout the sound waves.

5. Environmental chemistry
a. *Environmental racism is a persistent issue in the United States.* You will experience higher levels of pollution in a middle-class Black neighborhood than in a lower-class white neighborhood. The Supreme Court has ruled that unless a company

Chapter 17 – Teaching

explicitly says they are intentionally choosing a site to pollute because of racism, then racist impacts cannot be considered in seeking legal remedy. Students can learn about the lead paint studies in Baltimore done by Johns Hopkins University where they exposed small children to various lead paint removal processes to see how much lead the children would ingest.[57] Or take a look at what computer waste looks like when unregulated.

b. *Generating electricity using wind turbines or solar cells has little drawback once operational.* But what does it take to produce solar cells and wind turbines? The silicon used for solar cells must be 99.9999% pure in order to be functional.[58] This requires mining a specific silica and difficult purification. That isn't to say these aren't superior options for generating electricity, but students should learn all the impacts

c. *In chapter 8 part 3 we discussed a lesson about nuclear waste.* I once had a science instructor ask incredulously, "Would you be ok with them putting nuclear waste in your backyard?" And I would. There is a typical amount of radiation that you are exposed to that we call background radiation. The average amount of background radiation people are exposed to from natural sources is about 360 mrem.[59] The total average amount is a bit over 600 mrem. Living within 50 miles of a nuclear power plant adds less than 0.01 mrem to that total[60] (coal power plants are about 30 times that value).[61] I'd be especially welcoming to nuclear waste in my backyard if it meant I wasn't living near a coal power plant.

d. *Reduce, reuse, recycle.* Those three commands are uneven in environmental benefit. We overvalue recycling and undervalue reducing. It is far better to eat less meat, fly in planes less, and purchase less stuff than it is to recycle a water bottle. Recycling is energy intensive, lots of waste that we send to get recycled does not actually get recycled, and the products of recycling are more expensive. Chemistry is the perfect class to get students to understand some of the limitations of recycling. We can even do experiments where copper or another metal is cycled through a series of reactions to end up the same as it had started.

Plastic recycling in particular has challenges with sorting and contamination. Plastic with dyes added tends not to get recycled. IR (infrared) spectroscopy is now being used in some places to replace resin codes to reduce manual labor and mistakes. There are some experiments where students can do plastic recycling in a lab, but they require some hazardous chemicals and are not simple. The plastic used for pop bottles is called polyethylene terephthalate (PET) if you'd like to search for one. As a much simpler alternative you can take polystyrene foam and by adding acetone, you'll remove the gases trapped within the foam to produce polystyrene.

6. Material Science

a. *Polymers and plastics are versatile subjects.* Previously in this book we've mentioned PET and Kevlar, but polyethylene, polyvinyl chloride, DNA, proteins and other polymers can all be connected to students' lives while they learn some chemistry. Students can learn about how polyisoprene and polybutadiene are used for rubber and tires. They can even do their own cross-linking experiments using sodium borate with glue or polyvinyl alcohol (PVA). A good chemistry teacher life hack is that PVA can be easily dissolved in water by heating it in a microwave. I mix 40 grams of

Chapter 17 – Teaching

PVA in 1 liter of water and microwave for 3 minutes followed by minimal stirring. Usually, the 2nd or 3rd microwave cycle is sufficient.

b. *Superconductivity occurs when the electrical resistance in a substance drops to zero.* There are some ceramics that become type II superconductors below the boiling point of liquid nitrogen. This means you can produce a superconductor in your classroom if you have access to liquid nitrogen (you'll need a Dewar flask for storage of the dangerously cold substance). There are some elements that become type 1 superconductors, but these need to be extremely cold and you likely can't produce the temperatures required in your lab or classroom.

One of the best features of superconductors is how they interact with magnets. Superconductors can be suspended above a magnet so that it levitates. You can find videos of toy trains where a magnetic track locks in a superconductor in a toy train. The train will move around the track until the temperature of the superconductor becomes high enough for the superconducting properties to disappear.

c. *I have a personal bias for nanotechnology as I did my thesis on silver nanoparticles.* A nanoparticle is a size designation between macroscopic and atoms. The nanoparticles I would work with were about 10,000 to 100,000 atoms. The properties of a silver nanoparticle differ from bulk silver and from individual atoms of silver. The surface of the nanoparticles would interact with light differently. The silver nanoparticles I made were yellow, and the gold nanoparticles were red (just like the stained glass). Molecules could be attached to the surface. Adding polyvinyl alcohol tended to cap the nanoparticles and prevent them from aggregating. But there are studies seeing if nanoparticles can have molecules attached that could detect the presence of hormones or other biological molecules. You can make silver nanoparticles in your classroom with very dilute silver nitrate and a reducing agent. You will need to use distilled water and the purer the better.

We used the nanoparticles to enhance the chemiluminescence of luminol. Luminol emits blue light when it is oxidized by peroxide. With the nanoparticles we used ferricyanide instead of hydrogen peroxide.

The point of this is to give you some ideas of creative lessons (or interruptions to lessons!) you can incorporate into your class. There are many others but remember that you represent hundreds of years of scientific progress. We want to share the greatest accomplishments in science and there's no reason to limit what we present of those to ones that are listed in textbooks. The books cited in this previous section are filled with more stories and ideas that can be used. Use them.

Personal History

September of 2006 is when I began teaching chemistry in Lincoln Park High School. Two years later I started an AP chemistry course that lasted for two years before I was laid off due to declining enrollment and state funding. In September 2010 I began teaching Chemistry in the Community and Physics at Plymouth High School. That year I read "Teaching Introductory Physics" by Arnold Arons. It shifted my thinking about physics and chemistry content.[62] After two years I taught AP Chemistry and two years after that I started to teach IB Chemistry HL and Chemistry. The year I began teaching IB Chemistry HL I started to use the modeling curriculum in my Chemistry courses along with shifting my grading to standards-based grading.

Chapter 17 – Teaching

The sequencing of chapters in this book follows my current trajectory that I teach in class. In Chemistry we begin with the simplest models of the atom. We start with whenever a sphere works as the representation for the particulate level. This functions well with gas pressure, elements/compounds/mixtures, simple thermochemistry, and moles. Then we add in the concept of charge (plum-pudding model) before going through naming, chemical reactions, stoichiometry and extensions of stoichiometry (enthalpy, solution chemistry, ideal gas law). Finally, we modify the model from the Thompson model to the quantum mechanical model to learn about atomic structure, periodicity and bonding.

When I teach IB Chemistry HL, we do a near reversal. The unifying theme I use in IB Chemistry HL is +/- charges attract and -/- and +/+ charges repel. We begin with atomic structure and use that to build into periodic trends, bonding, organic chemistry and redox. Those units all have a heavy focus on charges and how they explain properties. From there we shift into kinetics, stoichiometry, equilibrium, acids and bases, and finally we come back to thermochemistry as a means to use energy to link back to the charge relationships.

As a child I loved school and academics. I had wanted to be a math teacher until I took chemistry and AP chemistry in high school. I was frequently told that I would be a bad teacher because I would not be able to control students. The people who said that were partially correct. I was terrible at managing a classroom. I get distracted easily and it was difficult for me to communicate in a direct manner. I try to get better at this, but it's not my strength and it may not ever be. But I managed to do well at teaching despite my shortcomings by reflecting on the chemistry itself. For many years I tried to become so good at explaining things that anyone would be able to understand it. I was frequently puzzled at why this rarely worked. How could the students not get 100% on this? I explained it so well.

Eventually I started to dissect what was happening in the students' heads instead of my explanations. This was slow at first because it is easy to assign blame to the students being lazy or unprepared. Why didn't they listen to me? And why did some students who put forth enormous effort still struggle to learn effectively? But as I learned more about cognitive science, I began to be able to really listen and appreciate what students were thinking. Modeling instruction and standards-based grading gave me frameworks to reflect on this productively.

This book was intended to be a guide to chemistry content for teachers. There are topics that were not included that perhaps should have been. There may be some components that were not critical. But I hope that overall, you found useful advice and frameworks for teaching chemistry. Should you have found an error, or a new way of viewing content that you find helpful, please reach out to provide me feedback.

If you read this entire book, thank you for being a chemistry teacher. Thank you for working to improve your craft. You and your time are both appreciated. Chemistry teacher is the best profession in the world, and I hope this book makes our jobs even better.

Acknowledgements

I want to start by thanking Ariel Serkin, Katy Dornbos, and Michael Farabaugh. Ariel and Katy gave me the encouragement and feedback that I needed. I had no clue how to make images and without Michael's help I wouldn't have finished this book. Thank you to Amy Snyder, Steven Rooney, Kristin Gregory, Dr. Angie Kolonich, Doug Ragan, Dr. Elizabeth Day, Dr. Teresa Bixby, Marc Stephenson, Ben Meachem, and Dr. Timothy Brewer for helping me when I needed assurance and/or editing advice.

Obviously, most of this book is based off my experiences teaching and I struck gold with both mentors and colleagues over the years. I'd like to thank Mandy Straksis, Matt Carey, Eric Calvin, and all the teachers at LPHS for guiding me as a rookie teacher. Mary McMaster had a profound impact on the trajectory of my teaching career as a mentor when I was the only chemistry teacher in the building. Gary Abud trained me in modeling instruction and has been my mentor since then. Thank you both for seeing good in me and helping me to be the best teacher I can be.

I'd like to thank my B pod squad in Plymouth High School especially Becky Kraft, Casey Swanson, and Steven Rooney. You three make coming to work fun even on the worst days. I would never have written a first book let alone this one if it weren't for Dave Fleming. You inspired me to write, helped me learn, and will always be my editor in chief. I could never have read so many books if it hadn't been for the inspiring posters from Amanda Davies. Cheri Steckel has been my principal for over a decade. You believed in me and supported me, and I've only gotten in trouble four times. I am lucky to be surrounded by so many amazing colleagues.

My Twitter account (@IBchemMilam) has surpassed my wildest expectations of connecting with other inspirational teachers (#iteachchem). I learn from the creativity of the Chem Fam constantly. Thank you Kristen, Ariel, Katy, Kristin, Amy, Deanna, Amanda, Doug, Stephanie, Alice, Johanna, Tom, Karen, and everyone else that is brave enough to put your teaching on display.

I want to thank my students who over the years have helped me build a healthy confidence about my work. "To have great poets there must be great audiences." I have certainly had my fair share of great students throughout my career, and I appreciate you. I'd also like to apologize to the students that I did not do a good job of teaching. I think about this frequently and I'm very sorry.

My family is rife with teachers. My mom, Bonnie Milam, just retired from teaching middle school English. Much of what I learned about speaking I learned from listening to my dad, David Milam, preach sermons on Sunday mornings throughout my childhood. My wife, Hillary Milam, is an early childhood special education teacher just as her mom, Tina Tefft, was. Many of my friends at work are teachers. And I have not lost complete hope that one of my children, Emily and David, will become a teacher down the line. Thank you to my family for supporting me as a teacher. Especially to my beautiful wife Hillary who had to watch many terrible tv shows to give me time to write and edit.

Citations and Notes

Introduction
1. BROWN, P. C. (2018). *MAKE IT STICK: The science of successful learning.* BELKNAP HARVARD.
This is the most popular book that teachers begin with. It helps frame how our philosophy of learning should involve the student doing difficult processing without leading you into a trap where things become ineffective.
2. Hammond, Z., & Jackson, Y. (2015). *Culturally responsive teaching and the brain: Promoting authentic engagement and rigor among culturally and linguistically diverse students.* Corwin.
This is similar to Make it Stick and Why Don't Students Like School? as a foundational work. Hammond has a few more specifics about how to apply the basics than the others do. Her definitions of culture are also helpful for teachers to begin integrating their new knowledge into meaningful pedagogical practices.
3. Willingham, D. T. (2010). *Why don't students like school? A cognitive scientist answers questions about how the mind works and what it means for the classroom.* Jossey Bass.
This is the third foundational book teachers can begin with that will help them wrap their minds around how the science of learning can impact decisions for teaching.
4. Kahneman, D. (2015). *Thinking, fast and slow.* New York: Farrar, Straus and Giroux.
The division of the brain into 2 systems was fundamental to setting up how learning is counterintuitive. This is a very technical book with lots of research and situations to engage students or teachers with.
5. Weinstein, Y., Sumeracki, M., & Caviglioli, O. (2019). *Understanding how we learn: A visual guide.* London: Routledge.
This is an excellent resource for students. It has many diagrams, examples, and studies that can guide students into understanding how to engage in productive learning at school. I keep a copy in my classroom for students to peruse.
6. Langer, E. J. (2016). *The Power of Mindful Learning.* Reading, Mass: Addison-Wesley Pub.
This is the best book that I've found for teachers ready to take the next step into elite use of cognitive science in the classroom. Many get stuck in the basics and struggle to move beyond.
7. Sapolsky, R. M. (2018). *Behave: The biology of humans at our best and worst.* New York: Penguin Press.
This is the most thorough of any book on cognitive science. It is very long, and very technical. But it has more research and a very unique application to the research.
8. Lang, J. M. (2021). *Small teaching: Everyday lessons from the science of learning.* San Francisco: Jossey-Bass, an imprint of Wiley & Sons.
This book analyzes small changes that teachers can make to their current systems that produce big results. The author uses a variety of research and methodologies. Ellen Langer's work is implemented nicely in this one.
9. Pink, D. H. (2018). *Drive: The surprising truth about what motivates us.* Edinburgh: Canongate Books.
Many of the books on cognitive science omit motivation and emotional components. There are a few of the first 8 sources that include bits and pieces, but this book helps tie all of the components together.
10. Csikszentmihalyi, M. (2009). *Flow: The psychology of optimal experience.* New York: Harper and Row.
What is the optimal state of productivity and how can said state be accomplished?
11. BROWN, P. C. (2018). *MAKE IT STICK: The science of successful learning.* (p. 172) BELKNAP HARVARD.

"It's one thing to feel confident of your knowledge; it's something else to demonstrate mastery."

12. Weinstein, Y., Sumeracki, M., & Caviglioli, O. (2019). *Understanding how we learn: A visual guide*. (pp. 22-29) London: Routledge.

13. Katz, S., & Dack, L. A. (2013). *Intentional interruption: Breaking down learning barriers to transform professional practice*. Thousand Oaks, CA: Corwin.

14. Agarwal, P. K., & Bain, P. M. (2019). *Powerful teaching: Unleash the science of learning* (pp. 25-91). San Francisco, CA: Jossey-Bass.
This is a thorough collection of classroom practices and research highlighting the effectiveness of retrieval practice.

15. Daniel T. Willingham (2020, August 13). Ask the cognitive scientist: Do students remember what they learn in school? Retrieved May 28, 2022, from https://www.aft.org/ae/fall2015/willingham

16. Willingham, D. T. (2016, January 13). Allocating Student Study Time: "Massed" vs. "Distributed" Practice. Retrieved May 28, 2022, from https://www.aft.org/periodical/american-educator/summer-2002/ask-cognitive-scientist

17. Kahneman, D. (2015). *Thinking, fast and slow*. New York: Farrar, Straus and Giroux.
Veritasium made a video version of system 1 and system 2 that is fantastic for students entitled "The Science of Thinking" https://www.youtube.com/watch?v=UBVV8pch1dM accessed 5/28/22

18. Diemand-Yauman, C., Oppenheimer, D. M., & Vaughan, E. B. (2011). Fortune favors the Bold (and italicized): Effects of disfluency on educational outcomes. *Cognition, 118*(1), 111-115. doi:10.1016/j.cognition.2010.09.012

19. Agarwal, P. K., & Bain, P. M. (2019). *Powerful teaching: Unleash the science of learning* (pp. 106-119). San Francisco, CA: Jossey-Bass.

20. Foer, J. (2011). *Moonwalking with Einstein: The art and science of remembering everything*. Penguin Books.

21. Gabel, D. (1999). Improving Teaching and Learning through Chemistry Education Research: A Look to the Future. *Journal of Chemical Education, 76*(4), 548. doi:10.1021/ed076p548

22. Michaels, S., & O'Connor, C. (2012). Talk Science Primer. *TERC*. https://inquiryproject.terc.edu/shared/pd/TalkScience_Primer.pdf accessed 5/28/22

23. Abud, Gary (2013, March 23). "for every" speak: A cognitive approach to proportional reasoning. Retrieved May 28, 2022 http://abud.me/for-every-speak-a-cognitive-approach-to-proportional-reasoning-in-chemistry/

24. Herron, J. D. (1996). *The chemistry classroom: Formulas for successful teaching*. Washington, D.C.: American Chemical Society.

25. Scerri, E. R. (2016). *A tale of seven scientists and a new philosophy of Science*. Oxford, England: Oxford University Press.
John Nicholson, Anton Van den Broek, Richard Abegg, Charles Bury, John D. Main Smith, Edmund Stoner, and Charles Janet are 7 scientists that made tremendous contributions to the fields of chemistry and physics yet are mostly unheralded. In this book Scerri analyzes their contributions relative to Rutherford, Moseley, Mendeleev, Lewis, Langmuir, Bohr, and Pauli. Scerri proposes a model of science history where individual celebrations are de-emphasized to be replaced by a collective celebration of the forward progress of science.

Chapter 1 - Chemistry Basics
1. A. L. Lavoisier (1965). *Elements of chemistry* New York: Dover.
I'd issue a warning about this book in that the Old English form of "s" is used where an s that isn't terminal looks like an f. Otherwise this book is fascinating as it reveals early development in chemistry when skepticism was high. Illustrations of the experimental apparatus used are included.

2. Gabel, D. (1999). Improving Teaching and Learning through Chemistry Education Research: A Look to the Future. *Journal of Chemical Education, 76*(4), 548. doi:10.1021/ed076p548

This article addresses the challenges of teaching and learning chemistry through a general lens. The development of Johnstone's triangle is highlighted by discussing challenges such as student misconceptions.

3. Frankland, E., & Chaloner, G. (1875). *How to teach chemistry: Hints to science teachers and students being the substance of six lectures delivered at the Royal College of Chemistry in June 1872.* London: J. & A. Churchill.

This book is filled to the brim with experimental evidence. Much of this evidence I would consider to be too hazardous to perform with students today. Nonetheless, this is a fascinating read to compare with current educational practices. In particular, every claim is supported by experimental evidence done by the students and/or teacher.

4. Missing elements - rsc.org. (n.d.). Retrieved April 3, 2022, from https://www.rsc.org/globalassets/22-new-perspectives/talent/racial-and-ethnic-inequalities-in-the-chemical-sciences/missing-elements-report.pdf

5. Partington, J. R. (1989). *A short history of chemistry* (pp. 13-47). New York: Dover Publications.

Partington is the original author dedicated to chemical history. He studied under Walther Nernst. In this opening chapter he analyzes the precursors to chemistry along with how the decision to delineate the start of chemistry is determined. This would be my top recommendation for a first book to learn about chemical history although it might be overwhelming.

6. Kean, S. (2018). *Caesars last breath: Decoding the secrets of the air around us* (pp. 210-222). New York: Little, Brown and Company.

Sam Kean's book is highly applicable for high school chemistry teachers to learn for their own benefit as well as add stories to their repertoire for teaching students. This section contains a detailed analysis of Rayleigh's story that can spice up a dry topic.

7. Pauling, L. (1969). *College chemistry; an introductory textbook of general chemistry* (pp. 249-250). San Francisco: W.H. Freeman.

Nobel prize winning chemist Linus Pauling's text is an excellent contrast to chemistry textbooks today. A few of the models used are outdated, but the analysis and justification used is brilliant. This is a shorter rendition of the tale told by Kean.

8. Arons, A. B. (1997). *Teaching introductory physics*. (pp. 330-331) New York: John Wiley.

Arons has a full development of teaching math in physics along with using ranges of values as a means to conceptually develop significant figures and measurement uncertainty.

9. https://www.youtube.com/watch?v=3g0nNw1xtDo accessed 5/29/22

10. Abud, Gary (2013, March 23). "for every" speak: A cognitive approach to proportional reasoning. Retrieved May 28, 2022 http://abud.me/for-every-speak-a-cognitive-approach-to-proportional-reasoning-in-chemistry/

11. BROWN, P. C. (2018). *MAKE IT STICK: The science of successful learning*. (p. 86) BELKNAP HARVARD.

"When you're asked to struggle with solving a problem before being shown how to solve it, the subsequent solution is better learned and more durably remembered."

12. https://www.youtube.com/watch?v=r2DcfN4EIpg Soquid commercial accessed 5/29/22

13. https://www.youtube.com/watch?v=3ZuMkPsQbjk Steam is always invisible accessed 5/29/22

14. Blum, D. (2014). *The poisoners handbook: Murder and the birth of forensic medicine in Jazz Age New York*. New York: Penguin Press.

15. Blum, D. (2019). *The poison squad: One chemists single-minded crusade for food safety at the turn of the twentieth century*. NY, NY: Penguin Books.

16. KAHNEMAN, D. (2022). *Noise: A flaw in human judgment*. S.I.: LITTLE, BROWN.

17. Gray, T., & Mann, N. (2009). *The elements: A visual exploration of every known atom in the universe.* New York: Black Dog & Leventhal.

Chapter 2 – Gas Pressure

1. Salzberg, H. W. (1995). *From caveman to chemist: Circumstances and achievements* (pp. 155-159). Washington, DC: American Chemical Society.

Salzberg has a heavy focus on the earliest chemists in this book. Lavoisier doesn't start until over the halfway point of the book. Van Helmont gets his proper due being the principal influence on Boyle who is sometimes credited as the first chemist.

2. Partington, J. R. (1989). *A short history of chemistry* (pp. 77-99). New York: Dover Publications.

This is a great read on some chemistry background behind the rivalry between Isaac Newton and Boyle's good friend Robert Hooke. Newton used Boyle's work to showcase his inverse square law and postulated a repulsive force to exist between stationary gas particles. Boyle admitted that the calculations matched experimental evidence, but was non-committal about ruling out particle motion and collisions as the source of pressure.

3. Jackson, J. (2007). *A world on fire: A heretic, an aristocrat, and the race to discover oxygen.* New York: Viking.

This book details the work of Joseph Priestley alongside that of Antoine Lavoisier. Priestley's experimental evidence inspired Lavoisier to replicate many of them. As Lavoisier experimented, he published conclusions that Priestley did not always agree with. Priestley remained in favor of the phlogiston model and often challenged Lavoisier's conclusions in spite of sharing experimental evidence.

4. https://ninja542.github.io/ideal-gas/ accessed 5/29/22

This simulation was made by one of my brilliant students Claudia Chen!

Chapter 3 - Thermochemistry

1. (1965). In 1835567795 1289093861 A. L. Lavoisier (Author), *Elements of chemistry* (p. 175). New York: Dover.

2. Greenberg, A. (2000). *A chemical history tour: Picturing chemistry from Alchemy to modern molecular science* (p. 145). New York: John Wiley & Sons.

This book is a fantastic collection of images from famous chemical texts.

3. Cropper, W. H. (2004). *Great physicists: The life and times of leading physicists from Galileo to Hawking* (pp. 61-65). Oxford: Oxford University Press.

This book has a fantastic section on thermodynamics. I can't recommend this enough. The stories will add so many relevant and engaging details to topics that are a tremendous struggle for teachers and students.

4. Ihde, A. J. (1984). *The development of modern chemistry* (p. 145). New York: Dover.

This is the 2nd of the 2 major chemical history books after Partington's work. Ihde studied under James Bryant Conant and Thomas Kuhn and developed the first course of studies on chemical history at the University of Wisconsin.

5. Eureka! Episode 21 Temperature and Heat
https://www.youtube.com/watch?v=AqDsAVPjgS4

This cartoon does a quality job of illustrating an initial model to distinguish heat from temperature using a similar hypothetical to the phenomenon described.

6. Kuntzleman, T. S., Ford, N., No, J., & Ott, M. E. (2014). A molecular explanation of how the fog is produced when dry ice is placed in water. *Journal of Chemical Education, 92*(4), 643-648.

Chapter 4 - Moles

1. Partington, J. R. (1989). *A short history of chemistry* (pp. 166-174). New York: Dover Publications.

2. Salzberg, H. W. (1995). *From caveman to chemist: Circumstances and achievements* (pp. 219). Washington, DC: American Chemical Society.
3. Marquardt, R. (2019). The Mole and IUPAC: A brief history. *Chemistry International, 41*(3), 50-52. doi:10.1515/ci-2019-0316
4. Ihde, A. J. (1984). *The development of modern chemistry* (p. 287). New York: Dover.
5. Beiser, V. (2019). *The world in a grain: The story of sand and how it transformed civilization.* (p. 109) Riverhead Books.
6. Jackson, J. (2007). *A world on fire: A heretic, an aristocrat, and the race to discover oxygen.* New York: Viking.
7. Johnson, C. E., Yee, G. T., & Eddleton, J. E. (2004). Copper metal from Malachite circa 4000 B.C.E. *Journal of Chemical Education, 81*(12), 1777. doi:10.1021/ed081p1777

Chapter 5 – Naming and Formula Writing
1. Salzberg, H. W. (1995). *From caveman to chemist: Circumstances and achievements* (pp. 36, 150-151). Washington, DC: American Chemical Society. More can be found in Partington pg. 16
2. Ihde, A. J. (1984). *The development of modern chemistry* (p. 112). New York: Dover.
This is the 2nd of the 2 major chemical history books after Partington's work. Ihde studied under James Bryant Conant and Thomas Kuhn and developed the first course of studies on chemical history at the University of Wisconsin.
3. Wothers, P. (2020). *Antimony, gold, and Jupiter's wolf: How the elements were named* (pp. 35-37, 153). Oxford, United Kingdom: Oxford University Press.
This book is fantastic to add interesting trivia pieces about chemical nomenclature.
4. Ihde, A. J. (1984). *The development of modern chemistry* (p. 587). New York: Dover.
5. Wothers, P. (2020). *Antimony, gold, and Jupiter's wolf: How the elements were named* (pp. 50). Oxford, United Kingdom: Oxford University Press.
6. Kean, S. (2010). *The disappearing spoon: And other true tales of madness, love, and the history of the world from the periodic table of the element* (p. 234). New York: Little, Brown and Company.

Chapter 6 – Reaction Types
1. Partington, J. R. (1989). *A short history of chemistry* (pp. 103-108). New York: Dover Publications.
2. Partington, J. R. (1989). *A short history of chemistry* (pp. 88, 139, 149). New York: Dover Publications.
3. Jackson, J. (2007). *A world on fire: A heretic, an aristocrat, and the race to discover oxygen.* New York: Viking.
For a deeper account of Priestley and Lavoisier's experiments and relationship.
4. Ford, L. A., & Grundmeier, E. W. (1993). *Chemical magic* (p. 51). New York: Dover Publications.

Chapter 7 – Stoichiometry + Extensions
1. Levere, T. H. (2001). *Transforming matter: A history of chemistry from alchemy to the buckyball* (p. 127-135). Baltimore: John Hopkins University Press.

Chapter 8 – Atomic Structure
1. Ihde, A. J. (1984). *The development of modern chemistry* (p. 235). New York: Dover.
Here you can read an account of how Kirchoff and Bunsen discovered Cs and Rb using emission spectral lines.
2. Partington, J. R. (1989). *A short history of chemistry* (p. 236). New York: Dover Publications.

3. Pauling, L. (1969). *College chemistry; an introductory textbook of general chemistry* (p. 166). San Francisco: W.H. Freeman.
4. Ihde, A. J. (1984). *The development of modern chemistry* (p. 502). New York: Dover.
5. Brock, W. H. (1993). *The Norton History of Chemistry* (pp. 241-242). New York: W.W. Norton.
6. Rocke, A. J. (2010). *Image and reality: Kekulé, Kopp, and the scientific imagination*. Chicago: The University of Chicago Press.
This book has a very detailed analysis of the mid-1800s using organic chemistry as the vehicle by which many chemical concepts were understood.
You can also find an account in Partington's work on page 240.
7. Abrams, B. (n.d.). Beyond bohr. Retrieved May 29, 2022, from https://www.bu.edu/chemed/resources/beyond-bohr/
8. Pauling, L. (1969). *College chemistry; an introductory textbook of general chemistry* (p. 123). San Francisco: W.H. Freeman.
9. Schwichtenberg, J. (2020). *No-nonsense quantum field theory: A student-friendly introduction*. Karlsruhe, Germany: No-Nonsense Books.
10. https://www.desmos.com/calculator/kapzs3jyni accessed 5/29/22
11. Scerri, E. (2021, September 03). Eric Scerri's lecture to Padova University. Retrieved May 12, 2022, from https://www.youtube.com/watch?v=3Kh1gOjLlUk
12. Brock, W. H. (1993). *The Norton History of Chemistry* (pp. 332-4). New York: W.W. Norton.
13. Ihde, A. J. (1984). *The development of modern chemistry* (p. 491)
14. Johnson, G. (2014). *The ten most beautiful experiments* (pp. 60-74). New York: Vintage Books.
15. Ihde, A. J. (1984). *The development of modern chemistry* (pp. 489-491)
16. Chapman, K. (2019). *Superheavy*. Bloomsbury Publishing USA.
This is an incredibly engaging book about the history of superheavy element discoveries as well as the future of them. There are so many interesting tales.
17. Scerri, E. R. (2013). *A tale of seven elements*. Oxford: Oxford University Press.
18. https://www.gvsu.edu/targetinquiry/ accessed 5/29/22
19. https://web.archive.org/web/20060309210852/http://www.einstein-online.info/en/spotlights/binding_energy/binding_energy/index.txt accessed 5/29/22
20. https://www.desmos.com/calculator/l7v0bbfzhn accessed 5/29/22

Chapter 9 – Periodic Trends
1. BOYLE, R. (2015). *Sceptical Chymist* (pp. 21-28). JEFFERSON PUBLISHING.
This book is an incredibly difficult read. The narrative of elements continues beyond page 28. On page 32 Boyle talks about reducing iron metal from ore using charcoal (and mercurius dulcis and alcalizate salt). Even when the chemical nomenclature is current, the writing style is not the most reader friendly. You can find a good summary of Boyle in Salzberg's From Caveman to Chemist pages 161-171 focusing on elements starting on page 168.
2. Ihde, A. J. (1984). *The development of modern chemistry* (p. 236)
3. Scerri, E. R. (2020). *The periodic table: Its story and its significance*. New York, NY: Oxford University Press.
This is the best periodic table history of any book that I've read. You'll find details on which scientists did what and when along with lots of chemistry that high school students can connect to. The comparisons of Mendeleev and Meyer's periodic tables are from this text.
4. Brock, W. H. (1993). *The Norton History of Chemistry* (p. 324). New York: W.W. Norton.
5. Scerri, E. R. (2013). *A tale of seven elements*. Oxford: Oxford University Press.
This book covers the search and discovery of these 7 missing elements after Moseley's work. When students notice that Tc has no stable isotopes I always reference the stories from this book and the many claims of element 43's discovery.

6. Chapman, K. (2019). *Superheavy*. Bloomsbury Publishing USA.
This is a fantastic book about the discovery of the superheavy elements. The experimental details of how they were produced along with all of the drama in selecting names.
7. Ball, P. (2017, October 9). Immense oganesson projected to have no electron shells. Retrieved April 10, 2022, from https://www.chemistryworld.com/news/immense-oganesson-projected-to-have-no-electron-shells/3008104.article
8. Abraham, D. S. (2017). *The elements of power: Gadgets, guns, and the struggle for a sustainable future in the rare metal age*. New Haven: Yale University Press.
This book has a lot of statistics relevant to rare earth metals. Projections of how much of them we will use and the current challenges of mining and isolating various precious metals.
9. Aldersey-Williams, H. (2011). *Periodic tales (2011): A cultural history of the elements, from arsenic to zinc* (p. 26). New York: Ecco.
10. Cann, P. (2000). Ionization energies, parallel spins, and the stability of half-filled shells. *Journal of Chemical Education, 77*(8), 1056. doi:10.1021/ed077p1056
11. Pauling, L. (1969). *College chemistry; an introductory textbook of general chemistry* (pp. 188-190). San Francisco: W.H. Freeman.

Chapter 10 - Bonding
1. Partington, J. R. (1989). *A short history of chemistry* (pp. 196-204). New York: Dover Publications.
2. Rocke, A. J. (2010). *Image and reality: Kekulé, Kopp, and the scientific imagination*. Chicago: The University of Chicago Press.
3. Brock, W. H. (1993). *The Norton History of Chemistry* (p. 226). New York: W.W. Norton.
This quote from Laurent states: "The validity of a theory is judged by the progress in science that it brings about. Now when we consider the immense advantages which the [dualistic] theory possess for nomenclature, for the learning of chemistry, and now its application to organic chemistry, we would still be constrained to use it, even if it should be demonstrated that it is false and the [unitary] theory true."
4. Partington, J. R. (1989). *A short history of chemistry* (pp. 240, 297). New York: Dover Publications.
5. Rocke, A. J. (2010). *Image and reality: Kekulé, Kopp, and the scientific imagination*. (pp. 150-155) Chicago: The University of Chicago Press.
6. Levere, T. H. (2001). *Transforming matter: A history of chemistry from alchemy to the buckyball* (pp. 142-148). Baltimore: John Hopkins University Press.
7. Ihde, A. J. (1984). *The development of modern chemistry* (pp. 436-439). New York: Dover.
8. Davey, S. (2009) *The legacy of Lewis. Nature Chem* 1, 19 https://doi.org/10.1038/nchem.149 accessed 5/22/2022
9. Pauling, L. (1969). *College chemistry; an introductory textbook of general chemistry* (p. 165). San Francisco: W.H. Freeman.
10. Kean, S. (2010). *The disappearing spoon: And other true tales of madness, love, and the history of the world from the periodic table of the element* (pp. 88-97). New York: Little, Brown and Company.
11. Jean, Y., & Volatron, F. (2010). *An introduction to molecular orbitals*. Oxford: Oxford Univ. Press.
This was a very approachable book on molecular orbital theory that would be very appropriate for a high school chemistry teacher interested in learning more.

Chapter 11 - Organic Chemistry
1. Brock, W. H. (1993). *The Norton History of Chemistry* (p. 212). New York: W.W. Norton.
2. Ihde, A. J. (1984). *The development of modern chemistry* (p. 619). New York: Dover.
3. Brock, W. H. (1993). *The Norton History of Chemistry* (p. 241). New York: W.W. Norton.
4. Brock, W. H. (1993). *The Norton History of Chemistry* (p. 201). New York: W.W. Norton.

5. J. Michael McBride (2009) Lecture 21 starting at 33:37
https://www.youtube.com/watch?v=Xo_IkUx2BPU&list=PL3F629F73640F831D&index=21 accessed 5/22/2022
6. Liebig, J. V. (1851). *Familiar letters on chemistry and its relation to commerce, physiology, and agriculture ... edited by J. Gardner. 16 letters.* (3rd ed., pp. 98-105). London: Taylor, Walton, and Mayberly.
7. Partington, J. R. (1989). *A short history of chemistry* (pp. 304-315). New York: Dover Publications.
8. Kean, S. (2010). *The disappearing spoon: And other true tales of madness, love, and the history of the world from the periodic table of the element* (pp. 173-185). New York: Little, Brown and Company.
9. Brock, W. H. (1993). *The Norton History of Chemistry* (pp. 540-544). New York: W.W. Norton.
10. https://royalsocietypublishing.org/doi/pdf/10.1098/rsbm.1972.0012 accessed 5/22/2022
11. Ihde, A. J. (1984). *The development of modern chemistry* (p. 496). New York: Dover.
12. Salzberg, H. W. (1995). *From caveman to chemist: Circumstances and achievements* (p. 229). Washington, DC: American Chemical Society.
13. Salzberg, H. W. (1995). *From caveman to chemist: Circumstances and achievements* (p. 248). Washington, DC: American Chemical Society.
14. Kean, S. (2010). *The disappearing spoon: And other true tales of madness, love, and the history of the world from the periodic table of the element* (p. 182). New York: Little, Brown and Company.
15. Kean, S. (2018). *Caesars last breath: Decoding the secrets of the air around us* (p. 183). New York: Little, Brown and Company.
16. Brunning, A. (2016). *Why does asparagus make your pee smell?: Fascinating food trivia explained with science.* (pp. 53-54) Berkeley, CA: Ulysses Press.
17. Masland, R. H. (2021). *We know it when we see it: What the Neurobiology of Vision tells us about how we think.* London: Oneworld Publications.
18. https://www.stereoelectronics.org/webSC/SC103.html accessed 5/30/22
19. Flynn, A. B., & Ogilvie, W. W. (2015). Mechanisms before reactions: A mechanistic approach to the organic chemistry curriculum based on patterns of Electron Flow. *Journal of Chemical Education, 92*(5), 803-810.
20. Ihde, A. J. (1984). *The development of modern chemistry* (p. 619). New York: Dover.
21. Infrared Spectroscopy (2022) https://chem.libretexts.org/Bookshelves/Physical_and_Theoretical_Chemistry_Textbook_Maps/Supplemental_Modules_(Physical_and_Theoretical_Chemistry)/Spectroscopy/Vibrational_Spectroscopy/Infrared_Spectroscopy/Infrared_Spectroscopy accessed 5/30/22
22. Data used from https://orgchemboulder.com/Labs/Experiments/SpectroscopyWorksheet.pdf accessed 6/8/22

Chapter 12 – Redox Chemistry
1. Partington, J. R. (1989). *A short history of chemistry* (p. 184). New York: Dover Publications.
2. Aldersey-Williams, H. (2011). *Periodic tales (2011): A cultural history of the elements, from arsenic to zinc* (p. 177). New York: Ecco.
3. Ihde, A. J. (1984). *The development of modern chemistry* (p. 76). New York: Dover.
4. Ihde, A. J. (1984). *The development of modern chemistry* (p. 291). New York: Dover.
5. Ihde, A. J. (1984). *The development of modern chemistry* (p. 451). New York: Dover.
6. Johnson, G. (2014). *The ten most beautiful experiments* (pp. 60-74). New York: Vintage Books.
7. Cullen, D. (2021, April 12). Energizer lab with virtual options. Retrieved June 8, 2022, from https://www.chemedx.org/activity/energizer-lab-virtual-options

8. https://www.youtube.com/watch?v=p_zujnBJk_s accessed 6/8/22

Chapter 13 - Kinetics
1. Levere, T. H. (2001). *Transforming matter: A history of chemistry from alchemy to the buckyball* (p. 163). Baltimore: John Hopkins University Press.
2. Brock, W. H. (1993). *The Norton History of Chemistry* (p. 539-540). New York: W.W. Norton.
3. Ptáček, P., Opravil, T., & Šoukal, F. (n.d.). A brief introduction to the history of Chemical Kinetics. Retrieved May 22, 2022, from https://www.intechopen.com/chapters/62152
4. Levere, T. H. (2001). *Transforming matter: A history of chemistry from alchemy to the buckyball* (pp. 163-4). Baltimore: John Hopkins University Press.
5. Wisniak, J. (2010). The history of catalysis. from the beginning to Nobel prizes. *Educación Química, 21*(1), 60-69. doi:10.1016/s0187-893x(18)30074-0

Chapter 14 - Equilibrium
1. Ihde, A. J. (1984). *The development of modern chemistry* (p. 409). New York: Dover.
2. Brock, W. H. (1993). *The Norton History of Chemistry* (pp. 539-40). New York: W.W. Norton.
3. Cropper, W. H. (2004). *Great physicists: The life and times of leading physicists from Galileo to Hawking* (p. 109). Oxford: Oxford University Press.
4. Cropper, W. H. (2004). *Great physicists: The life and times of leading physicists from Galileo to Hawking* (pp. 126-132). Oxford: Oxford University Press.
5. Kean, S. (2018). *Caesars last breath: Decoding the secrets of the air around us* (pp. 51-58). New York: Little, Brown and Company.
6. Kean, S. (2018). *Caesars last breath: Decoding the secrets of the air around us* (pp. 60-67). New York: Little, Brown and Company.

Chapter 15 – Acid-Base Chemistry
1. Wothers, P. (2020). *Antimony, gold, and Jupiter's wolf: How the elements were named* (p. 115). Oxford, United Kingdom: Oxford University Press.
2. Ihde, A. J. (1984). *The development of modern chemistry* (p. 158). New York: Dover.
3. Partington, J. R. (1989). *A short history of chemistry* (pp. 199-202). New York: Dover Publications.
4. Partington, J. R. (1989). *A short history of chemistry* (p. 136). New York: Dover Publications.
5. Crawford, E. (1996). *Arrhenius: From ionic theory to the greenhouse effect*. Canton, MA: Science History Publications/USA.
6. Ihde, A. J. (1984). *The development of modern chemistry* (p. 547). New York: Dover.
7. This notion was based off the work of Amy Zitzelberger of Hazel Park High School https://slideplayer.com/slide/15292559/ accessed 6/9/22
8. Wothers, P. (2020). *Antimony, gold, and Jupiter's wolf: How the elements were named* (p. 115). Oxford, United Kingdom: Oxford University Press.

Chapter 16 – Entropy and Spontaneity
1. Lewis, G. N., Randall, M., Pitzer, K. S., & Brewer, L. (2020). *Thermodynamics*. Mineola, NY: Dover Publications.
2. Cropper, W. H. (2004). *Great physicists: The life and times of leading physicists from Galileo to Hawking* (pp. 93-105). Oxford: Oxford University Press.
This book has the best thermochemistry history I have seen. I recommend reading the entire sequence.
3. Cropper, W. H. (2004). *Great physicists: The life and times of leading physicists from Galileo to Hawking* (pp. 106-123). Oxford: Oxford University Press.

4. Atkins, P. W. (2007). *Four laws that drive the universe* (p. 32). Oxford: Oxford University Press.
This book is worth reading in its entirety. A masterpiece of taking abstract thermochemistry and clarifying those ideas using concrete representations.
5. Scerri, E. R. (2019). Five ideas in chemical education that must die. *Foundations of Chemistry, 21*(1), 61-69. doi:10.1007/s10698-018-09327-y

Chapter 17 – Teaching
1. Feldman, J. (2019). *Grading for equity: What it is, why it matters, and how it can transform schools and classrooms*. Thousand Oaks, CA: Corwin, a SAGE Company.
This is the best book I've read on grading. I strongly recommend reading the entire thing.
2. Sagarin, R. (2012). *Learning from the octopus: How secrets from nature can help us fight terrorist attacks, natural disasters, and disease*. New York: Basic Books.
3. Feldman, J. (2019). *Grading for equity: What it is, why it matters, and how it can transform schools and classrooms*. (pp.78-84) Thousand Oaks, CA: Corwin, a SAGE Company.
4. Michaels, S., & O'Connor, C. (2012). Talk Science Primer. *TERC*.
https://inquiryproject.terc.edu/shared/pd/TalkScience_Primer.pdf accessed 5/28/22
5. Cartier, J. L., Smith, M. S., Stein, M. K., & Ross, D. K. (2013). *5 practices for orchestrating productive task-based discussions in science*. Reston, Virg.: National Council of Teachers of Mathematics.
This book contains many quality transcripts of discussions that allow a teacher to prepare how they might find their voice in leading a discussion in their own class.
6. Faraday, M. (2005). *The Chemical History of a candle*. New York: Barnes & Noble Books.
7. https://pogil.org/ accessed 6/10/22
8. Yglesias, M. (2010, May 03). Buraku. Retrieved June 10, 2022, from https://www.theatlantic.com/politics/archive/2007/11/buraku/47131/
9. Bertrand, M., & Mullainathan, S. (2003). Are Emily and Greg more employable than Iakisha and Jamal? A field experiment on labor market discrimination. doi:10.3386/w9873
10. Sapolsky, R. M. (2018). *Behave: The biology of humans at our best and worst*. New York: Penguin Press.
For a thorough analysis of exactly how bias functions I cannot recommend this book enough. The detailed analysis is crucial to undermining the many misconceptions about how racism and other biases work.
11. Hammond, Z., & Jackson, Y. (2015). *Culturally responsive teaching and the brain: Promoting authentic engagement and rigor among culturally and linguistically diverse students*. California: Corwin.
Hammond's warm demander profile is particularly fitting.
12. Steele, C. (2011). *Whistling Vivaldi: How stereotypes affect us and what we can do*. New York: W.W. Norton.
13. Langer, E. J. (n.d.). *Out of the question: The Power of Mindful Learning*. Reading, Mass: Addison-Wesley Pub.
14. Whitman, G., & Kelleher, I. (2016). *Neuroteach: Brain science and the future of education*. Lanham: (pp. 58-65) Rowman et Littlefield.
15. Hendrick, C., Macpherson, R., & Caviglioli, O. (2019). *What does this look like in the classroom?: Bridging the gap between research and Practice*. Melton, Woodbridge: John Catt Educational.
16. Gallwey, W. T. (1974). *The inner game of tennis*. New York: Random House.
17. Willingham, D. (2007). Critical thinking - Why is it so hard to teach? AFT
18. Immordino-Yang, M. H. (2016). *Emotions, learning, and the brain: Exploring the educational implications of Affective Neuroscience*. New York: W.W. Norton & Company.
19. Goulston, M. (2015). *Just listen: Discover the secret to getting through to absolutely anyone*. New York: AMACOM.

20. Nuthall, G. (2007). *The hidden lives of learners*. NZCER Press.
21. Steele, C. (2011). *Whistling Vivaldi: How stereotypes affect us and what we can do*. New York: W.W. Norton.
22. Scientists, L. (2019, January 28). Episode 2 - retrieval practice. Retrieved May 28, 2022, from https://www.learningscientists.org/learning-scientists-podcast/2017/9/6/episode-2-retrieval-practice
This is a podcast and from 15:45-17:40 they discuss their study comparing flashcard creation with flashcard use.
23. https://phet.colorado.edu/ accessed 5/28/22
24. Willingham, D. T. (2010). *Why don't students like school? a cognitive scientist anwers questions about how the mind works and what it means for the classroom*. Jossey Bass. "Memory is the residue of thought"
25. Jensen, William B. (2017) The Mystery of G.N. Lewis's missing Nobel Prize https://chemistry.as.miami.edu/_assets/pdf/murthy-group/gnl_jensen-2.pdf accessed 6/11/22
26. Crawford, E. (1996). *Arrhenius: From ionic theory to the greenhouse effect*. Canton, MA: Science History Publications/USA.
27. Andrade, E. N. (1978). *Rutherford and the nature of the atom*. Gloucester, MA: P. Smith.
28. Chapman, K. (2019). *Superheavy*. (pg. 160) Bloomsbury Publishing USA.
29. Aldersey-Williams, H. (2011). *Periodic tales (2011): A cultural history of the elements, from arsenic to zinc* (pp. 113-118). New York: Ecco.
30. Kean, S. (2010). *The disappearing spoon: And other true tales of madness, love, and the history of the world from the periodic table of the element* (pp. 214-215). New York: Little, Brown and Company.
31. Chapman, K. (2019). *Superheavy*. (pg. 204) Bloomsbury Publishing USA.
32. Scerri, E. R. (2020). *The periodic table: Its story and its significance*. New York, NY: Oxford University Press.
33. Kean, S. (2012). *The violinist's thumb and other Lost tales of love, war, and genius, as written by our genetic code* (pp. 54-71). New York: Little, Brown.
34. Couteur, P. L., & Burreson, J. (2014). *Napoleons buttons: 17 molecules that changed history*. (pp. 312-313) New York: Jeremy P. Tarcher/Penguin.
35. Chapman, K. (2019). *Superheavy*. (pg. 122) Bloomsbury Publishing USA.
36. Kean, S. (2010). *The disappearing spoon: And other true tales of madness, love, and the history of the world from the periodic table of the element* (p. 324). New York: Little, Brown and Company.
37. Kean, S. (2010, July 16). Ytterby: The Tiny Swedish Island That Gave the Periodic Table Four Different Elements. http://www.slate.com/articles/health_and_science/elements/features/2010/blogging_the_periodic_table/ytterby_the_tiny_swedish_island_that_gave_the_periodic_table_four_different_elements.html
38. Beiser, V. (2019). *The world in a grain: The story of sand and how it transformed civilization*. Riverhead Books.
39. Mahaffey, J. (2010). *Atomic awakening: A new look at the history and future of nuclear power* (pp. 128-133). New York: Pegasus.
40. Waldman, J. (2016). *Rust: The longest war*. (p. 78) New York: Simon & Schuster.
41. Waldman, J. (2016). *Rust: The longest war*. (p. 75) New York: Simon & Schuster.
42. Blum, D. (2014). *The poisoners handbook: Murder and the birth of forensic medicine in Jazz Age New York*. New York: Penguin Press.
43. Blum, D. (2019). *The poison squad: One chemists single-minded crusade for food safety at the turn of the twentieth century*. NY, NY: Penguin Books.
44. Garfield, S. (2014). *Mauve: How one man invented a color that changed the world*. New York: W.W. Norton &.

45. Moore, K. (2018). *The radium girls: The dark story of America's shining women*. Naperville, IL: Sourcebooks.
46. Couteur, P. L., & Burreson, J. (2014). *Napoleons buttons: 17 molecules that changed history*. (pp. 202-223) New York: Jeremy P. Tarcher/Penguin.
47. Finlay, V. (2004). *Color a natural history of the palette*. New York: Random House.
48. Waldman, J. (2016). *Rust: The longest war*. New York: Simon & Schuster.
49. Hartings, M. (2017). *Chemistry in your kitchen*. Cambridge: The Royal Society of Chemistry.
50. Holmes, B. (2017). *Flavor: The science of our most neglected sense* (pp. 164-165). New York: W.W. Norton & Company.
51. Gray, T. W., & Mann, N. (2018). *Molecules: The elements and the architecture of everything* (pp. 156-173). New York: Black Dog & Leventhal.
52. Kean, S. (2010). *The disappearing spoon: And other true tales of madness, love, and the history of the world from the periodic table of the element* (pp. 88-91). New York: Little, Brown and Company.
53. Rosling, H., Rosling, O., & Rönnlund, A. R. (2020). *Factfulness: Ten reasons we're wrong about the world - and why things are better than you think*. New York: Flatiron Books.
54. Abraham, D. S. (2017). *The elements of power: Gadgets, guns, and the struggle for a sustainable future in the rare metal age*. (p. 12) New Haven: Yale University Press.
55. Kean, S. (2018). *Caesars last breath: Decoding the secrets of the air around us* (p. 176). New York: Little, Brown and Company.
56. Brunning, A. (2016). *Why does asparagus make your pee smell?: Fascinating food trivia explained with science*. (pp. 52-53) Berkeley, CA: Ulysses Press.
57. Washington, H. A. (2020). *A terrible thing to waste: Environmental racism and its assault on the American mind*. New York etc.: Little, Brown Spark.
58. Beiser, V. (2019). *The world in a grain: The story of sand and how it transformed civilization*. (p. 109) Riverhead Books.
59. https://www.atsdr.cdc.gov/ToxProfiles/tp149-c6.pdf accessed 6/12/22
60. https://www.nrc.gov/about-nrc/radiation/related-info/faq.html accessed 6/12/22
61. Hvistendahl, M. (2007, December 13). Coal ash is more radioactive than nuclear waste. Retrieved June 12, 2022, from https://www.scientificamerican.com/article/coal-ash-is-more-radioactive-than-nuclear-waste/
62. Arons, A. B. (1997). *Teaching introductory physics*. New York: John Wiley.
Some images were constructed using chemix.org and ChemDoodle

Summary of history readings:

Partington and Ihde both compiled masterpieces of chemical history. The Norton history guide was written later and can fill in some additional details. However, there are many other valuable books to read. Sam Kean does a wonderful job of storytelling that is particularly helpful for teachers. The thermodynamics section of Great Physicists by Cropper has been valuable for teaching a topic that could use some personalities to spice up the instruction. Scerri's books about the periodic table help illuminate its creation as well as the electronic structure that correlates with the periodic table. Reading directly from some of the famous chemists themselves is also enlightening. The assumptions and conditions behind many of their claims help teachers to understand the development of models in a new manner.

About the Author

Scott Milam just finished his 16th year of teaching (29 to go!). He graduated from the University of Michigan (Go Blue) in 2006 with a BS in Chemistry, minor in physics, and certifications to teach chemistry and physics. He taught chemistry, AP chemistry, and physical science at Lincoln Park High School from 2006-2010. He started teaching at Plymouth High School in September of 2010 where he taught chemistry, physics, chemistry in the community, AP chemistry, and later switched to IB Chemistry HL. In 2010 he completed his MS in Chemistry from Eastern Michigan University and wrote his thesis on silver nanoparticles.

In 2019 Scott published his first book, The No Teacher Left Behind Club. This first book is a comedy about a group of teachers trying to get fired in a school system that can't afford to lose teachers. In 2017 Scott was named the Michigan Science Teacher of the Year. In 2019 and 2021 he was a finalist for the prestigious PAEMST award (Presidential Award for Excellence in Mathematics and Science Teaching).

The Twitter account @IBchemMilam is where you can find several of the projects that Scott works on. He has written and performed numerous chemistry song parodies on his YouTube channel (Scott Milam). He films and designs activities for PIVOT Interactives. He has led numerous professional development opportunities on antiracism, modeling instruction, grading, cognitive science, and chemical history. He has helped co-teach a course on Modeling Instruction under the tutelage of Ariel Serkin and has co-led multiple workshops on chemistry modeling instruction with Gary Abud. Scott enjoys reading books, writing books, learning new things, and teaching science.